114080 X

M/L

Volume I
Diseases of Cereals and Pulses

PLANT DISEASES OF INTERNATIONAL IMPORTANCE

U. S. SINGH

Division of Plant Pathology, International Rice Research Institute, Los Banos, Laguna, Philippines

A. N. MUKHOPADHYAY

Department of Plant Pathology, G. B. Pant University of Agriculture and Technology, Pantnagar, India

J. KUMAR

Department of Plant Pathology, Hill Campus, G. B. Pant University of Agriculture and Technology, Ranichauri, India

H. S. CHAUBE

Department of Plant Pathology, G. B. Pant University of Agriculture and Technology, Pantnager, India

PRENTICE HALL, Englewood Cliffs, New Jersey 07632

Library of Congress Cataloging-in-Publication Data

Plant diseases of international importance / edited by Uma S. Singh
. . . [et al.].

    p.   cm.
   Vol. 3 edited by Jatinder Kumar . . . [et al.]
   Includes bibliographical references and index.
   Contents: v. 1. Diseases of cereals and pulses —   v.
3. Diseases of fruit crops
   ISBN 0–13–678582–4 (v. 1). — ISBN 0–13–678566–2 (v. 3)
   1. Plant diseases. I. Singh, Uma S. II. Kumar, Jatinder.
SB731.P67 1992   91-22598
632'.3—dc20   CIP

Editorial production
and interior design: ***bookworks***
Acquisition editors: *Ken Tennity* and *Betty Sun*
Editorial assistant: *Maureen Diana*
Cover designer:
Copy editor: *Karen Verde*
Marketing Manager: *Alicia Aurichio*
Prepress buyer: *Mary Elizabeth McCartney*
Manufacturing buyer: *Dave Dickey*
Indexer: *WordFinders*

© 1992 by Prentice-Hall, Inc.
A Simon & Schuster Company
Englewood Cliffs, New Jersey 07632

Printed in the United States of America

10 9 8 7 6 5 4 3 2 1

ISBN 0-13-678582-4 ✓   VOL 1

ISBN 0-13-678582-4

90000>

9 780136 785828

Prentice-Hall International (UK) Limited, *London*
Prentice-Hall of Australia Pty. Limited, *Sydney*
Prentice-Hall Canada Inc., *Toronto*
Prentice-Hall Hispanoamericana, S.A., *Mexico*
Prentice-Hall of India Private Limited, *New Delhi*
Prentice-Hall of Japan, Inc., *Tokyo*
Simon & Schuster Asia Pte. Ltd., *Singapore*
Editora Prentice-Hall do Brasil, Ltda., *Rio de Janeiro*

CONTENTS

4 Blast of Rice

Dr. J. C. Bhatt, Plant Pathology Department, Vivekananda Parvatiya
Krishi Anusandham Shala I (AR), Almora 263 601, India

Dr. R. A. Singh, G. B. Pant University of Agriculture and Technology,
Department of Plant Pathology, Pantnagar 263145, India   **80**

5 Brown Spot of Rice

Dr. N. K. Chakrabarti, Deep Water Rice Project (IRRI), c/o Rice
Research Station, Chinsurah 712 102, Hoogly, W.B., India

Dr. Sujata Chaudhary, Department of Botany, Kalyani University,
Kalyani, W.B., India   **116**

6 Sheath Blight of Rice

Dr. M. K. Dasgupta, Reader Plant Pathology Laboratory, Department
of Plant Protection, Palli-Siksha Bhawan (Institute of Agriculture),
Visva-Bharti, Sriniketan 731 236 W.B., India   **130**

7 Bacterial Blight of Rice

Dr. S. Devadath, Principal Scientist, Central Rice Research Institute,
Cuttack, 753 006, Orissa, India   **158**

8 Rice Tungro

Dr. S. Mukhopadhyay, Plant Virus Research Center, Department of
Plant Pathology, Bidhan Chandra Krishi Viswavidyalaya, Kalyani
741235, W.B., India   **186**

9 White Tip Disease of Rice

Ms. E. B. Gregon, Department of Plant Pathology, IRRI, P.O.
Box 933, Manila, Philippines

Dr. J. K. Mishra, Mycological Research Unit, Department of Botany,
Sri J.N. Mahavidyalaya, Lucknow 226 109, India   **201**

10 Downy Mildews of Maize

Dr. S. S. Bains,
Dr. H. S. Dhaliwal, Plant Pathologist, P.A.U.,
Regional Research Station, Gurdaspur 143 521, Punjab, India   **212**

PREFACE

Plant diseases are as old as agriculture and are important to man because they cause damage to plants and plant products. If we go by numbers, plant diseases run into thousands. However, only a handful of them are of much economic and historic significance and consume a major proportion of the funds and scientific power invested on plant diseases. This is because of their widespread occurrence and devastating effects. During the past two decades we have seen dramatic advances in the study of many internationally important diseases. Unfortunately, however, the information generated on these diseases is widely dispersed in diverse scientific journals, in several languages. This book is one of four volumes that attempts to bring this information together, to summarize and evaluate recent developments, to integrate them with significant developments of the past, and to attempt some projections for the future.

Leading plant pathologists from over 20 countries having expertise on individual diseases have contributed chapters for four volumes. All the volumes, therefore, contain highly authoritative and thought-provoking articles by experts in their respective fields. The number of topics presented and the indepth coverage they receive are ample testimony that plant disease research has come of age. The text, besides serving as a reference source for research workers or scientists already in the field, might also interest those who work in allied areas. We have considered the growing numbers of postgraduate students who are interested in plant diseases of international importance.

While dealing with each disease, pertinent facts about the history and distribution (including a disease distribution map), symptoms, etiology, epidemiology and disease cycle are given; control measures used in different countries have been combined into programs which have been made as simple as possible. In presenting their own research findings and those of others, contributors were encouraged, where appropriate, to draw conclusions and propose hypotheses that might stimulate additional research or otherwise further our understanding of the diseases. The chapters have been profusely illustrated and literature pertaining to each chapter has been cited for readers who may need supplemental information on specific matters.

In preparing the four volumes, the editors/contributors drew heavily on numerous sources of information. Many of them are given as references or credited throughout the book, but space limitations preclude mentioning all of them. The editors/contributors wish to acknowledge their indebtedness to these sources. Numerous illustrations are reproduced from the literature and from several other sources. Wherever necessary, the source has been duly acknowledged.

The response of contributors to our request for contributing articles was so overwhelming that we received more articles than normally expected. Editors are greatly thankful to all the eminent scientists for their valuable contributions in making this project a success.

Editorial responsibilities in the preparation of these volumes were shared as follows: A. N. Mukhopadhyay was responsible for preparation of the volume on *Diseases of Sugar, Forest, and Plantation Crops*; H. S. Chaube was responsible for *Diseases of Vegetables and Oil Seed Crops*; U. S. Singh was responsible for *Diseases of Cereals and Pulses*; and J. Kumar was responsible for *Diseases of Fruit Crops*. Finally, all the volumes were reviewed by each editor to bring uniformity to the text.

We are indebted to Kenneth J. Tennity, Senior Editor; Ms. Maureen Diana, Editorial Assistant; and other skillful staff at Prentice-Hall for their excellent cooperation in the creation of this four-volume set. Our sincere thanks to P. K. Mukherjee and T. K. Misra, who spent many hours reading the manuscripts, and to D. B. Parakh for performing many other essential functions. Finally, our special thanks and appreciations go to our wives Namita, Sumitra, Neeta, and Urmila, who were highly cooperative and supportive of our efforts during the last five years.

Of course, it would be unrealistic to suppose that the text is free from errors. Notification of errors of omission or commission will be greatly appreciated.

U. S. Singh
A. N. Mukhopadhyay
J. Kumar
H. S. Chaube

1

KARNAL BUNT OF WHEAT

Elizabeth J. Warham

CIMMYT, Apdo. Postal 6–641, Mexico 06600 D.F.

1-1 INTRODUCTION

Karnal bunt is caused by *Tilletia indica* Mitra (synonym *Neovossia indica* (Mitra) Mundkur), a floral-infecting organism that partially infects seed of bread wheat,[1] durum wheat, and triticale.[2] When *T. indica* was first reported on wheat in India during the early 1930s,[1] it appears to have been limited in its distribution and unimportant. Since then, it has spread extensively in Northern India to Northern Pakistan, Southern Nepal, and parts of Iraq[3] and Mexico.[4]

Indian scientists report that the disease is becoming more prevalent and serious in that country. Karnal bunt has also had serious implications for plant quarantine of germplasm and seed distributed from Mexico to the United States and Canada.[5] In the last few years the more widespread occurrence and effect of the disease on plant quarantine has aroused much concern.

1-2 DISTRIBUTION AND ECONOMIC IMPORTANCE

Karnal bunt was first reported in 1930 on wheat near the North Indian city of Karnal in Haryana State.[1] The disease may have been observed as early as 1909 when Howard and Howard[6] reported the occurrence of a bunt on wheat at Lyallpur (now Faizalabad, Pakistan), where *T. indica* was found during 1930–1934.[1]

In 1934 Karnal bunt was reported in a virulent form at Karnal[7] and in 1941 the disease was found on wheat in Sind Province, Pakistan, and in Uttar Pradesh and Delhi, India.[8] During 1943 the disease was prevalent in the Punjab state of India and the Punjab and North West Frontier Provinces of Pakistan.[9] By 1969 it had spread extensively in the Indian states of Jammu and Kashmir, Punjab, Haryana, Himachal Pradesh, Uttar Pradesh, Delhi, and Rajasthan, and the Northern Areas and Punjab of Pakistan. Although all wheat cultivars grown in the area were susceptible, the large-scale cultivation of Mexican wheats with uniform flowering and high nitrogen applications was believed to be responsible for the increased distribution of the disease.[10] By 1985, the disease was also found in West Bengal, Gujarat, Bihar, and Madhya Pradesh states of India.[11]

The United States Department of Agriculture (USDA) reported *Tilletia indica* on wheat entering United States from India[12] and Afghanistan[13,14] in 1949 and 1955 respectively, and in 1974 the disease was present in Iraq.[3] The disease was supposedly intercepted in India on wheat from Lebanon, Mexico, Sweden, Syria, and Turkey.[15,16] However, although infected wheat grains were imported into India from Lebanon and Mexico,[17] the disease has not been found in Sweden, Syria, or Turkey (Figure 1-1). The USSR was then reported to have taken precautions to prevent the disease from entering that country.[18]

In Mexico, the disease was first found in the Yaqui and Mayo valleys of the northwestern state of Sonora during the early 1970s.[19] Until 1982, only trace amounts were found in farmers' grain,[20] but in 1983 Karnal bunt was found in US Crop Quality Council and Centro Internacional de Mejoramiento de Maiz y Trigo (CIMMYT) research plots in the Yaqui and Mayo valleys. There are no reports of the disease in any wheat-growing regions of the United States.[21] In 1983, the Animal and Plant Health Inspection Service (APHIS) of the United States imposed regulations that required all wheat and triticale seed entering the United States from Mexico to be treated with pentachloronitrobenzene (PCNB) fungicide and accompanied by an APHIS permit.[22] Any CIMMYT germplasm to be grown by scientists in the United States had to be grown in greenhouses under strict

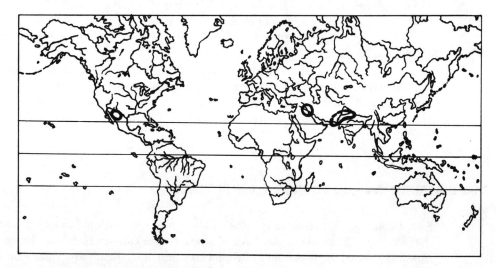

Figure 1-1 Distribution of *Tilletia indica* worldwide.

regulations requiring fumigation of soil and greenhouse. During 1984, these regulations were revised for CIMMYT: if Karnal bunt was found in the CIMMYT international nurseries after testing, APHIS would allow the subsequent nurseries to be grown only in United States greenhouses, as the previous year; but if Karnal bunt was not found, the nurseries could be grown in the field under certain restrictions.[5,23] Today, these restrictions are still imposed but every effort is made by CIMMYT to ensure that the international nurseries are completely free of the disease. Also, the Mexican government imposed legal restrictions to control the development of the disease, including regulations on sowing wheat in infected areas.[20]

The Canadian Quarantine Service also imposed a similar set of restrictions with the belief that if the disease did become established in Canada, export markets would be adversely affected by import regulations of countries like Russia and China.[24] The USDA devised an action plan to be used for guidance in implementing eradication procedures for Karnal bunt should an outbreak of the disease occur in the United States.[25]

In India, the earliest loss estimate was 20% damage in experimental plots at Karnal on a number of wheat cultivars.[7] Later, Munjal[10] quoted an overall incidence of 0.6% in 1969–1970 for an estimated one third of the wheat production area in Northern India, and a loss in grain yield of 0.2%, equivalent to 40,000 metric tons of grain per year. During recent years, surveys in the Punjab have found the following percentages of samples collected infected with Karnal bunt: 1975 (27%), 1976 (34%), 1978 (42%), 1979 (71%), 1980 (27%), 1981 (66%), 1982 (68%), 1983 (54%), 1985 (13%), and 1986 (53%). In the epidemic year of 1978–79, the maximum disease incidence was 58% on HD2009.[26]

In surveys conducted within the Yaqui Valley, Sonora, Mexico during 1982, 1983, 1984, and 1985, 8%, 68%, 1%, and 72% of the samples collected were infected, respectively. The losses for the farmer are minimal, given that in the Yaqui Valley only 10% of the farmers have more than 1% of their grain infected with the disease in a year with severe infection, and the majority of these have less than 3% infection. In the Yaqui Valley the highest incidence recorded is 20%, but this is a very infrequent occurrence.[27]

1-3 SYMPTOMS

Not all of the spikes on a plant are infected by the disease, and within a spike only a few spikelets are bunted[7,28–30] (Figure 1-2). Infected plants generally show a reduction in the length of spikes and in the number of spikelets produced.[31] In severely infected spikelets, the glumes may spread apart near maturity, exposing the bunted grains, but this is not a common symptom.[30,32,33] Infected grains are irregularly distributed in the spike; some are completely infected but most are partially infected[34] (Figure 1-3). Completely bunted grains may be observed in the wheat spike at hard dough stage by the appearance of shiny silvery black infected grains. Partially infected grains cannot be detected until the spikes are threshed. Freshly collected infected grains smell like rotten fish due to the production of trimethylamine.[32]

As the amount of infection increases, the grain weight correspondingly decreases,[35–37] with the difference in weight between heavily and slightly infected grains as much as 50%.[38] An increase in disease severity decreases the 1,000 grain weight, and at low disease severity the losses are greatest for cultivars with the smallest grains.[39]

Figure 1-2   A wheat spike infected with *Tilletia indica*.

Figure 1-3   Wheat grains infected with *Tilletia indica*.

Some studies investigating the relationship between Karnal bunt infection and seed germination found that as infection increases, seed germination decreases. Seeds with moderate to severe infections tend to produce a greater percentage of abnormal seedlings.[39–42] A more recent study has shown that the influence of infection on seed germination (germination under favorable/optimum conditions) appears to depend on the wheat cultivar and on the age of the seed.[43] In general, however, there is only a slight reduction in seed germination even with completely bunted seeds. In contrast, there is a significant effect of infection on seed vigor (germination under unfavorable conditions) with a more pronounced effect in older seed.[43]

At a certain level of infection *T. indica* can affect the quality of bread, chapaties, and so on. One researcher believes 1% infection is sufficient to reduce the quality of chapaties, due to the fishy odor and the perceptible discoloration.[44] Other researchers tend to agree that with 3% or less infected grains the appearance and palatability of bread, cookies, and chapaties are unaffected.[45–47] Infection levels above 5%, however, are not acceptable.[45,46] Samples with 5% infected grains can be used to produce satisfactory products if they are first washed, or samples with 10% infected grains can be used if they are first washed and steeped.[48] However, Medina[47] found that even with washing and steeping, 7% infection affected bread-making characteristics. It should be remembered that in the commercial production of flour some infected grains may be eliminated as a consequence of their low weight when air blowers are used during cleaning. In addition, infected grain lots can be mixed with less infected or noninfected lots.

There is no indication of toxic effects when Karnal bunt-infected wheat is fed to rats, chicks, or monkeys in various short-term toxicological studies.[38,46,49] Feeding bunted grains to goats, does, however, decrease the bacterial and protozoal population in the rumen and depress the formation of ammonia and volatile fatty acids.[50]

There are no known mycotoxins or ergot alkaloids present in Karnal bunt-infected wheat.[38] Nevertheless, wheat damaged by *T. indica* can be invaded by several secondary pathogens including *Aspergillus flavus*, which can be positive for toxigenic aflatoxins.[51,52]

1-4 CAUSAL ORGANISM

Mature teliospores are brown to dark brown, spherical or subspherical to oval, 22–42 $\times$ 25–40 μm in diameter, with the average 35.5 μm and the mode at 36μm[1] (Figure 1-4). Teliospore size is influenced by environmental factors.[53] Mixed with the teliospores are numerous large yellowish or subhyaline sterile cells corresponding to undeveloped teliospores. These are rounded or angular and smaller in size than the teliospores, with comparatively thinner walls.[1,30]

The teliospore has three walls or layers: endosporium, episporium, and perisporium.[54] The endosporium is thick and lamellate, whereas the episporium is adorned with thick truncate projections.[55] The perisporium is delicately fibrous and very fragile. If the teliospores are air-dried, the perisporium collapses around the episporium, often rupturing and exposing the truncate projections of the episporium. In contrast, when teliospores

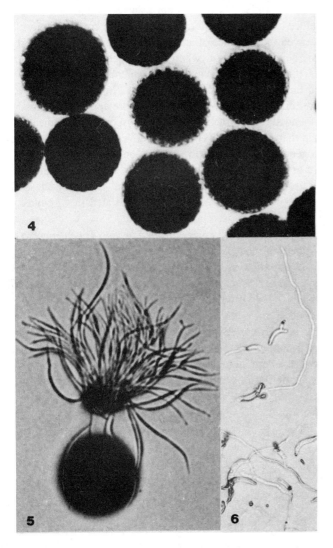

Figure 1-4 *Tilletia indica* teliospores.

Figure 1-5 A germinating *Tilletia indica* teliospore.

Figure 1-6 *Tilletia indica* secondary sporidia germination in potato dextrose agar.

are hydrated, the perisporium expands and a relatively smooth sphere is observed.[56] On the surface of the perisporium there are rodlets with an average length of 240 nm.[57]

Primary sporidia have mean lengths and widths ranging from 64.4 to 78.8 μm and 1.6 to 1.8 μm respectively; and the secondary sporidia have mean lengths and widths ranging from 11.9 to 13.0 μm and 2.00 to 2.03 μm respectively.[58] The pathogens from India and Mexico can not be differentiated for teliospore diameter, or primary and secondary sporidia characteristics.[58]

T. indica is heterothallic with bipolar incompatibility controlled by either four alleles at one locus on a homologous chromosome[59,60] or seven alleles.[61]

The placement of the Karnal bunt fungus in *Tilletia* or *Neovossia* has been debated since the organism was first discovered. Karnal bunt was originally classified as *Tilletia indica* Mitra by Mitra,[1] but Mundkur[28] transferred *T. indica* to *Neovossia indica* (Mitra) Mundkur, because of its unbranched, rather long promycelium with its whorl of nonfusing primary sporidia at the apex. In 1953, Fischer[62] again transferred the fungus to *T. indica* which the Commonwealth Mycological Institute adopted as the correct nomenclature.[63] Khanna and Payak[55] observed the presence of an apiculus in immature teliospores, which is a characteristic of the genus *Neovossia*, whereas its absence is typical of the genus *Tilletia*.[64] However, as the teliospores mature, this apiculus disintegrates, so Khanna and Payak[55] suggested that the fungus could be a species in transition between *Tilletia* and *Neovossia*, but determined that *Neovossia* was the more appropriate classification. Krishna and Singh[65] found three characteristics of teliospores that indicated a closer relationship to *Neovossia* than to *Tilletia*: (1) terminal development of teliospores from the end cell of the sporogenous hyphae, (2) an apiculus on teliospores, and (3) a nonseptate promycelium with a greater number of sporidia. Duran[4] maintained that occasional fragmentary appendages are common on teliospores of several *Tilletia* species.[64] In addition, the failure of primary sporidia to fuse has been demonstrated in a number of *Tilletia* species,[66] so this characteristic is not peculiar to *Neovossia*.[4]

The pathogen can be distinguished from *T. caries, T. foetida*, and *T. controversa* by its large teliospores,[67,68] and by the presence of an apiculus which arises from the episporium of young teliospores.[55] Also, *T. indica*, unlike *T. caries*, does not systemically infect the wheat plant from an infected seed,[27,34] and it survives in a warmer climate than *T. caries, T. foetida*, or *T. controversa*.[67–69] Attempts to hybridize *T. indica* and *T. caries* have been unsuccessful.[70]

The life cycle of *Tilletia barclayana* (Bref.) Sacc. & Syd. [synonym *Neovossia horrida* (Tak.) Padwick and A. Khan] (Kernel smut of rice) is very similar to that of *T. indica* and for this reason there are a large number of similarities between the two pathogens in culture techniques, inoculation techniques, and effectiveness of seed treatments.[27]

1-5 DISEASE CYCLE

1-5-1 Teliospore Germination

Freshly collected teliospores are incapable of germination and are considered to have a period of dormancy.[67,71,72] Dormancy in teliospores can be broken by direct exposure of bunted grains to the sun for a period of 14 days with the maximum temperature range between 40°C and 43°C.[73] Teliospores germinate, though poorly, when at least five to nine months old.[71] One- and two-year-old teliospores show the highest germination, whereas older teliospore collections show decreasing rates of germination.[72,74] The maximum length of teliospore survival at 0, 3, and 6 in. below the soil is 45, 39, and 27 months respectively.[75] Teliospore germination is significantly but not completely reduced when teliospores are ingested by Leghorn chickens and grasshoppers or passed through the intestinal tract of a cow.[76]

In all the studies conducted with teliospore germination under laboratory conditions the percentage germination has been characteristically low and variable.[72,77] Various techniques have been tried to obtain good teliospore germination. Holton[70] and Bansal et al.[72] suggested that free water was essential for teliospore germination as presoaking teliospores enhanced germination. Mathur and Ram,[74] however, obtained good germination by dusting dry teliospores on soil extract agar and water agar.

Presoaking treatments with different chemicals have been tried to stimulate teliospore germination with varying degrees of success. These include alcohols, aldehydes, alkalis, amino acids, antibiotics, fatty acids, growth regulators, inorganic acids, natural extracts, organic solvents, oxidizing agents, phenols, salts, sugars, and vitamins, in different concentrations.[72,77–83]

The optimal temperature for teliospore germination is 15°C to 22°C with extremely low germination at 2°C and total inhibition after prolonged exposure to 35°C (in darkness).[84] The temperature range 15°C to 20°C is optimal for both Indian and Mexican isolates but germination of Mexican isolates at 25°C is slightly higher than for Indian isolates.[85] Light has a stimulatory effect on germination, with artificial daylight better than near ultraviolet light, but no stimulation is obtained by prolonging illumination from 12 to 24 hours.[84] Teliospore germination occurs equally over a range of pH 6.0–9.5 but is strongly inhibited below pH 4.5 or above pH 10.0.[86]

Short periods (one to three weeks) of freezing at -5°C and desiccation do not affect teliospore germination.[86] However, if the freezing period is longer and of a lower temperature (-18°C), teliospore germination is significantly reduced.[84]

Each teliospore normally germinates with a stout promycelium bearing 32 to 128 filiform primary sporidia at its tip[28] (Figure 1-5). Occasionally the promycelium is unusually long or so much suppressed that the cluster of primary sporidia appears to arise from the teliospore itself.[70] In addition, the promycelium can be branched (false branching),[67,87] or occasionally two or even three promycelia arise from the same teliospore, with each one capable of bearing primary sporidia.[88]

When a teliospore germinates, the single diploid nucleus undergoes meiosis. Subsequent and rapid mitoses give rise to large numbers of haploid nuclei, which migrate from the hypobasidium into the promycelium and then to the sporidia. Primary sporidia formed at the promycelial tip each receive one haploid nucleus, but further mitoses produce septate sporidia with two or four nuclei.[89] These primary sporidia germinate either directly, giving rise to lateral and terminal mononucleate hyphae from the various cells, or indirectly to form sterigmata from which secondary, mononucleate falcate sporidia are forcibly discharged.[60] Krishna and Singh,[90] however, reported that primary sporidia gave rise to small, lateral, or terminal protuberances (not sterigmata), which elongated and developed into allantoid, hyaline secondary sporidia. Secondary sporidia germinate directly or by repetition[60] (Figure 1-6).

1-5-2 Sporidia Production

Secondary sporidia production occurs over a range of pH 4–9 but is greater below pH 6, and occurs more rapidly at temperatures between 20°C and 25°C but at 5°C or 30°C is significantly inhibited. Large numbers of sporidia are produced at 10°C and 15°C but at a slower rate.[27]

In the laboratory the length of time from promycelium emergence to secondary sporidia production of *T. indica* varies considerably. This may be a safety mechanism of the fungus to ensure that at least a proportion of the secondary sporidia are produced under temperature and moisture conditions favorable for their dispersal. Alternatively, under field conditions some factor stimulates a more uniform teliospore germination and production of secondary sporidia when the wheat crop is heading.[88] For example, extracts of soil, wheat grain, and cow dung at 5000 ppm have been shown to increase teliospore germination.[79] The latter theory seems more likely, although published studies have not confirmed what stimulates teliospore germination and secondary sporidia production in the field at the appropriate time.

1-5-3 Infection

Only those teliospores on or very near the soil surface can initiate disease development.[86] Sporidia are released at night, during periods of higher relative humidity and somewhat cooler air temperatures.[91] The primary sporidia or secondary sporidia are carried to the wheat spike either by air currents or by splashing water[29,34,92] (Figure 1-7). Preliminary investigations indicate that secondary sporidia may be more important for infection than the primary sporidia.[27] Contrary to earlier reports where penetration of ovary was thought to be direct,[93] recent evidences indicate that germ tubes arising from secondary sporidia penetrate through stomatal openings of glume, lemma, and/or palea. Hyphae enter the base of the ovary after first growing intercellularly through the glume, lemma, and/or palea, and possibly the rachis to the subovarian tissue and then through funiculus to ovary.[94] Wheat plants are most susceptible when the spikes emerge from the boot, but infection can take place through anthesis.[27,95] The disease progresses systemically to other florets within the spikelet initially infected and then to adjacent spikelets, including those on alternate sides of the rachis.[27,96–98]

Infection starts at the embryo end of the grain and spreads along the grain suture,[94] with the teliospores developing in extensive shallow cavities resulting from dissolution of the middle layers of the pericarp.[99]

Although teliospores partially surround the embryo, there seems to be some confusion as to whether the embryo is infected. Mitra[1] believed the embryo tissue is infected but not completely whereas, Munjal and Chatrath[94] believed the embryo is not infected. Later Rai and Singh[40] stated that in severely attacked seeds the embryo as well as the endosperm may be damaged as indicated by the production of abnormal seedlings by infected seeds. Warham[43] found that 94% of completely bunted seeds retain viable embryos irrespective of the age of the seed. This indicates that the embryo may not be

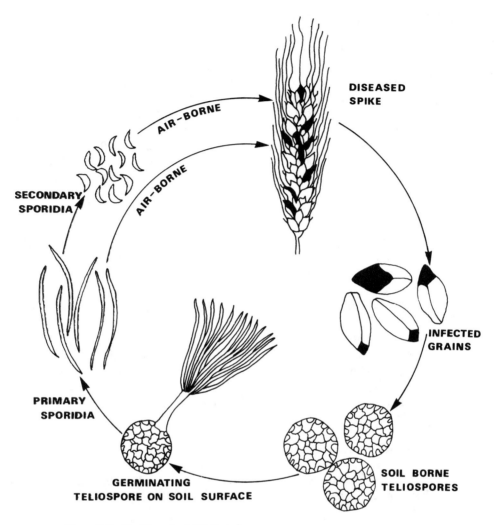

Figure 1-7 The life cycle of *Tilletia indica*.

infected but confirmation is needed with histological studies of bunted grains.

1-5-4 Dispersal

Rarely are teliospores dispersed before harvest, since the bunted grains are completely or partly concealed by the glumes and the teliospores are produced under the pericarp, which normally does not rupture until after harvest. Teliospores can be dispersed during threshing when they become deposited in the soil, lodge in the machinery, or adhere to the surface of healthy grains as an external contaminant. Wind dispersal to adjacent fields can also occur.[29,67,100,101] If the wheat stubble is burned after harvest, teliospores can become airborne and carried long distances.[102]

1-6 INOCULATION AND DISEASE RATING

1-6-1 Culture Techniques

Various techniques have been used to culture *T. indica* and maintain cultures in a sporulating condition.[103,104] The teliospores are best separated from the bunted grains by agitation and then surface sterilized with a dilute solution of sodium hypochlorite.[27] Although *T. indica* can be cultured on a number of different media,[100,104-106] in practice teliospore germination tests are best conducted in water agar with a pH 6–9, and secondary sporidia production is best on potato dextrose agar with a pH 4–6. In both cases a temperature of 20°C and alternate 12-hr periods of light and dark give good results. Three- to six-week old potato dextrose agar plates appear to give the greatest number of allantoid secondary sporidia which germinate rapidly. Older cultures not only produce less sporidia but are less reliable for inoculations.[27]

    T. indica cultures are usually white, powdery, brittle, crustaceous, umbonate colonies with dendritic margins giving rise to numerous secondary sporidia,[105] but can vary from dark to light in color, brittle to leathery in texture, and sporidial to mycelial in growth habit.[70]

    Sporidia can also be produced in a number of liquid media including potato extract,[59] potato dextrose broth,[27,42,107] potato dextrose broth supplemented with yeast, soil extract,[15] and sucrose.[27] Sporidia produced in liquid cultures are characteristically filiform secondary sporidia and can be distinguished from allantoid secondary sporidia produced on potato dextrose agar plates. A number of studies on inoculation techniques have used inoculum produced in liquid cultures.[15,42,59,107] However, Warham[27] found liquid cultures less effective as inoculum for Karnal bunt inoculations of wheat compared to potato dextrose agar cultures.

1-6-2 Inoculation Techniques

There are several artificial inoculation techniques that have been used to screen germplasm for Karnal bunt resistance:

1. Boot inoculation technique—a water suspension of secondary sporidia is injected with a hypodermic syringe into the boot growth stage. The ideal plant growth stage is the awn emergence stage,[15,27,42,95,107] and reliable infection can be obtained with low secondary sporidia concentrations (1,000 to 10,000 secondary sporidia per ml)[27,95] and low humidity.[27,95,100]

2. Spray inoculation technique—a water suspension of secondary sporidia is sprayed at various growth stages between heading and anthesis. A range of growth stages between heading and anthesis can be inoculated with this technique,[27,42,95] but high secondary sporidia concentrations (50,000 secondary sporidia per ml) and 48 hr in a humidity chamber are required.[27,95]

3. Moore's method—spikes are inoculated with a sporidial suspension under partial vacuum in a glass inoculating chamber. By this method it is possible to inoculate up to 30 spikes per hr.[108]

4. Dropper method—the floret is opened by hand with forceps and a droplet of sporidia suspension is added with a dropper. The ideal growth stage for this technique is anthesis.[100]

5. Go-go method—the central floret of individual spikelets together with the awns is removed and the inoculum syringed into the individual florets. This technique was based on the dropper method with clipping the florets, in a similar way to the go-go method of emasculation, facilitating the position of the inoculum in the florets.[109,110]

6. Cotton wool inoculation technique—small pieces of cotton wool saturated in a water suspension of sporidia are placed either inside the floret or between the floret and the rachis. Even at high secondary sporidia concentrations (100,000 sporidia/ml) this technique gives very low infection.[27,95]

The most reliable and efficient inoculation techniques are the boot and spray techniques. In the greenhouse, where humidity and temperature can be controlled, both techniques are comparable, but if anything, the spray inoculation has several advantages in that it is a more rapid screening technique and more closely mimics field infection, thereby screening for morphological resistance and to a lesser extent physiological resistance. In the field, there is insufficient humidity in dry areas for a spray inoculation technique; therefore, the boot inoculation technique is more reliable. Nevertheless, the limitations of the boot inoculation technique should be recognized in that it is a severe method that screens only for physiological resistance as inoculum is applied before heading when morphological resistance may not be fully expressed.[27,95]

1-6-3 Disease Rating

The amount of Karnal bunt infection present in a wheat crop cannot be determined until the crop is harvested and threshed. Individual grains can then be graded according to severity of infection on a scale ranging from 0 to 5 (0 for zero infection and 5 for a completely bunted grain; Figure 1-8). The overall percentage infection is calculated by dividing the number of infected grains by the total number of grains harvested.[111]

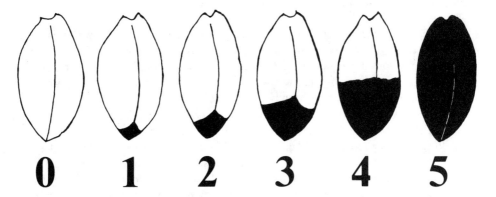

Figure 1-8   Scale of *Tilletia indica* infection severity in *Triticum aestivum*.

1-7 EPIDEMIOLOGY

High humidity and low temperatures at flowering time are conducive to the development of the disease.[29,112-116] In the Himalayan foothills the disease appears not to occur if temperatures at flowering of the wheat crop are above 23°C or below 10°C, or if the relative humidity is outside the range 54–89%.[117] A soil temperature of 17°C to 21°C is also optimum for the disease.[112] The factor most likely to be limiting in the field is moisture, both on the soil surface for teliospore germination and for infection on the spike by secondary sporidia.[27] The incidence of the disease is increased when there is little fluctuation in relative humidity during the week after inoculations.[118] Although rainfall is necessary, high rainfall alone at flowering does not always lead to the occurrence of the disease, indicating a combined role of weather factors.[117]

1-8 DISEASE MANAGEMENT

1-8-1 Regulatory

The certification standards in India for basic and certified seed in areas where the disease is not found are zero incidence, and where the disease is found, 0.1% and 0.5% incidence for basic and certified seed, respectively.[119] In either case, the seed is visually examined and contamination of healthy grains by teliospores during threshing and cleaning operations is not taken into consideration.[101] Therefore, it has been strongly argued that a seed washing test should be applied in routine testing of wheat for the detection of *T. indica*.[120]

CIMMYT nurseries produced in Mexico are examined by APHIS prior to being imported into the United States using the centrifuge wash test. Test tubes containing seeds submerged in water are shaken for 10 min. to obtain a teliospore suspension, then centrifuged for 20 min. at 3,000 rpm and the sediment examined under a compound microscope for the presence of teliospores.[64,113]

There are no experimental data available to identify the relationship between the teliospore load per seed and the subsequent development of the disease in the field, so prescribed certification standards are arbitrary.[119]

1-8-2 Cultural

The adjustment of sowing dates to avoid high humidities at flowering is not a practical disease control method in India or Mexico because both countries have certain sharp fluctuations in weather at anthesis.[27] The Karnal bunt epidemics in the Punjab appear on the early, normal, or late sown wheat crop, according to whether high rainfall occurred in January, February, or March, respectively.[34]

A higher incidence of the disease has been found in heavily manured, irrigated clay fields where the crop has lodged, compared to unirrigated fields with little or no manure and no crop lodging.[29] Where nitrogen is applied, doses greater than 80 kg/ha have been shown to increase the severity of the disease.[121] These results suggest that a reduction in irrigation or a more timely irrigation application and a reduction in fertilization may be possible disease control methods.[122] Another suggestion was the use of crop rotations to eliminate the possibility of infection from the soil;[67,122,123] but the survival of *T. indica* teliospores from 3 to 4 years in the soil[30] and the airborne infection process of the fungus[124] would limit the benefits of crop rotation.[125] Stubble burning has been suggested as another control method since teliospores on the surface and 5 cm below the soil surface do not appear to germinate after 5 hr of stubble burning.[126] However, stubble burning also causes teliospores to become airborne and carried long distances.[102] Whether or not these telio-spores remain viable still has to be determined. It is interesting to note that the majority of farmers in the Yaqui Valley, Mexico, practice stubble burning.

It has been observed that teliospore germination is inhibited under highly saline conditions, suggesting that there is a lower disease incidence on highly saline soils.[126] In the Yaqui Valley, Mexico there is less disease incidence on the west side of the valley where there is a predominance of saline soils, but farmers in the area tend to grow less wheat.[27]

Both methyl bromide (Metho-O-Gas, 570Kg/ha) and metham (Vapam, 1,100 l/ha) fumigations in wet and dry soils respectively, have been found effective in reducing ger-mination of teliospores present in soil.[127] This may be feasible for small areas of produc-tion, but is impractical on a large scale.

Farmer surveys conducted in the Yaqui Valley found that disease incidence is not significantly influenced by type of farm, soil type, land preparation, origin of seed, irri-gation, nitrogen fertilizer, weed control, or crop rotations. This indicates that although management practices may have some effect on the incidence of the disease, they are not the primary factor. The coincidence of rainfall and/or high humidity at flowering time appears to be of more importance than agronomic practices.[27]

1-8-3 Chemical

Seed treatments.   The efficacy of different compounds on wheat seed infected with Karnal bunt has been evaluated by screening teliospore germination.[27,128,129] Although some fungicides reduce teliospore germination substantially, they do not eradicate the disease when applied to infected seed.[128] This may be because teliospores in point infec-tions are well protected by the pericarp and therefore the fungicide cannot reach the teliospores. If this is correct, volatile compounds with a strong fumigant action should be more effective. However, chloropicrin and sulphur dioxide fumigations do not reduce teliospore germination more than 80% and are phytotoxic to seeds at rates lower than those that inhibit teliospore germination.[130]

Many fungicides that are very effective against *T. caries*[131] have little or no effect on *T. indica*. The reason for this difference may be that when fungicides are tested against *T. caries*, the usual procedure is to apply fungicide to seed contaminated with teliospores and measure the proportion of plants that become infected. In this case the fungicide has the opportunity to act on germination and on the delicate fungal structures involved in infection. The ability to kill the dormant teliospores is not a necessary criterion of effec-tiveness. This is the requirement for fungicides effective against *T. indica*.

Even if seed treatments were to achieve complete eradication of *T. indica* on in-fected seed samples, they would not, by themselves, provide control of the disease since the disease is primarily soilborne,[132] and teliospores may survive for several years in the soil.[75]

With the possible exception of mercurial fungicides none of the available fungicides are fungitoxic (i.e., capable of killing teliospores when applied to infected seed), but merely fungistatic (i.e., inhibit teliospore germination only when in contact with the teliospores). This means that there is no chemical seed treatment that can guarantee that wheat seed is not carrying viable *T. indica* teliospores.[27] Also, systemic fungicides applied to seed do not persist long enough to inhibit floral infection.[132]

Decontamination is possible by dry heat but requires high temperatures and long treatment times[130] which reduce seed germination.

Fungicide sprays.   Another chemical disease control method is the application of foliar fungicide sprays against the airborne sporidia that infect the wheat spike emerging from the boot. A number of fungicides applied as foliar sprays have been investigated for their effectiveness in controlling the incidence of Karnal bunt.[126,132–136] Although carbendazim (Bavistin) and mancozeb (Dithane M45) were effective each time they were tested,[126,132–135] no other fungicides were consistently effective.

Contact foliar fungicides [e.g., mancozeb (Manzate 200) or copper hydroxide] must be applied to the emerged spikes before sporidia are released from soilborne teliospores, whereas systemic fungicides [e.g., propiconazole (Tilt) and etaconazole (Vangard)] are best applied 72 hr after inoculation.[132]

The high application rates of systemic fungicides needed to control *T. indica*, when applied to wheat at late growth stages, can result in a high concentration of residue in the grain at harvest.[132] This means that systemic fungicides may be impractical. When contact fungicides are applied they do not redistribute systemically within the spike, so the residual chemical is probably removed as the grain is threshed.[132]

1-8-4 Disease Resistance

All the commercial wheat cultivars grown in India since Karnal bunt was first reported have been susceptible to the disease, although they do differ in the degree of susceptibility.[137,138] A few durum wheat and triticale cultivars offer promise as sources of resistance.[29,139–141]

Bread wheats, durum wheats, and triticales all show the same degree of susceptibility with boot inoculations. Under natural infection, triticales have been shown to be the least susceptible, with durum wheats slightly less susceptible than bread wheats. There is a similar phenomenon with spray inoculations, indicating that the resistance of durum wheat and triticale to natural infection may be morphological rather than physiological. Spray inoculation results with rye, rye addition lines, and bread wheats with 1B/1R translocations suggest rye may be another source of morphological resistance.[142] One morphological characteristic considered as a possible source of resistance to Karnal bunt was pubescence since only pubescent wheats were grown in India prior to the introduction of Mexican cultivars and the increased distribution of the disease. Nevertheless, no correlation was found between amount of infection and pubescence in bread wheats.[142] In addition, the location of stomates on the rachis and glumes was considered, since histological studies showed that the fungus penetrates through the stomates,[94] but no correlation was found between location of stomates and infection. Other morphological characteristics of bread wheat, durum wheat, and triticale that may be responsible for the higher levels of resistance in the latter two crops should be investigated.[142] It is interesting to note that in seeking rice cultivars resistant to Kernel smut of rice (*Tilletia barclayana*) which also infects the spike before or near anthesis, it has been suggested that cultivars with a shorter period of anthesis and florets that do not open wide should be considered.[143]

No line or cultivar of bread wheat screened in India or Mexico has been claimed as immune.[22,34,139,142] In the search for physiological resistance to Karnal bunt, further investigations will have to be conducted with other species related to bread wheat. Resistance, if found, could then be incorporated into bread wheat. Gill et al.[34] suggested that resistance to Karnal bunt should be sought in *Aegilops squarrosa*. Warham et al.[111] observed that six species of *Aegilops* species (*Ae. biuncialis, Ae. columnaris, Ae. crassa, Ae. juvenalis, Ae. ovata, and Ae. speltoides*) were consistently resistant to Karnal bunt after boot inoculations in the greenhouse. Another study by Warham[142] suggested barley may be a source of physiological resistance since all cultivars screened had zero infection after both boot and spray inoculations. B. Singh[126] also indicated that chromosomes 1 and 4 of barley, and 6 and 7 of rye carried genes for resistance to Karnal bunt. Studies with Chinese Spring and Kharkov rye addition lines did not confirm chromosomes 6 and 7 of rye carried Karnal bunt resistance.[142]

1-9 CONCLUSION

When first discovered, Karnal bunt was relatively unimportant and little research was conducted. In the last decade, as a result of the increased distribution of the disease and its effect on plant quarantine regulations, the disease has aroused much concern and has stimulated many more detailed investigations. Nevertheless, there are still a large number of gaps in our knowledge of the disease that should be investigated further.

A great deal more needs to be learned about the life cycle of the fungus. Examples of areas that could be investigated further are: (1) In the field there must be some factor that triggers teliospore germination and subsequent sporidia production to coincide with the heading growth stage of the wheat crop. Is it senescing leaves of the wheat plant or fertilizer applications? Moisture and temperature alone are not sufficient; otherwise, all the teliospores present in the soil would germinate after the first irrigation was applied to the crop and the fungus would have depleted itself by the time the wheat crop was heading. (2) It is now known that secondary sporidia are released at night when the temperatures are cool and there is more moisture available for their germination on the glumes. However, in what quantity are they released and how far do they travel? (3) What happens once the sporidia alight on the wheat spike emerging from the boot and germinate to penetrate the stomata? Does the fungus systemically infect other florets of the same spikelet by growing through the rachis or does it follow some other route? (4) Investigations so far indicate 54–89% R.H. with little fluctuation during the week after inoculations is conducive to the development of the disease. A more precise idea of the duration of humidity required for infection would not only indicate the conditions under which the fungus will survive in the field but would also help perfect the spray inoculation technique further.

As far as disease management is concerned, there is no recommendation for farmers to minimize the incidence of the disease. Even if seed treatments were to achieve complete eradication of *T. indica* they would not provide control of the disease since teliospores may survive for several years in the soil. In the case of agronomic practices it is unlikely that a suitable recommendation for farmers will be obtained; changes in management that appear to have any slight effect on the disease also reduce the farmers' yield. The losses from the disease are normally minimal, and therefore farmers are unlikely to accept a recommendation that reduces their wheat yield. Also, although agronomic practices may have some effect on the incidence of the disease they are not the primary factor. The coincidence of rainfall and/or high humidity at flowering time appears to be of more importance than agronomic practices. Preliminary studies have shown some fungicides applied to the wheat crop at heading are effective in controlling the disease, but they still require much research and evaluation before one will be approved for use by farmers. In addition, they are likely to be too costly for farmers to use. The use of resistant cultivars would be a more reliable disease control method. Until resistant cultivars have been identified and made available to the farmers there is no cost effective disease control method available that can be used.

In the future, more emphasis should be given to the resistant cultivars. A large number of lines have been screened but so far no lines have been identified as immune. This is partly because large numbers of bread wheats have been screened in the field with the boot inoculation technique, which is a severe method that only screens for physiological resistance. It now seems apparent that physiological resistance in bread wheats will not be found. More screening of germplasm should be carried out with the spray inoculation technique, which more closely mimics field infection thereby screening for morphological resistance. Spray inoculation results indicate that Karnal bunt resistance in durum wheats and triticales observed under natural infection is morphological and not physiological. Pubescence and location of stomates have already been investigated and found unimportant. There is a need for continued investigation of other morphological characteristics of bread wheat, durum wheat, and triticale that may be responsible for the higher levels of resistance in the latter two crops.

If physiological resistance to Karnal bunt is to be found, further investigations will have to be conducted with other species related to bread wheat. Barley and *Aegilops* species have already been shown to be possible sources of physiological resistance but there may be others worth investigating.

Seed treatments were proposed as a means of providing assurance to regulatory agents that the seed was not carrying viable teliospores. However, despite all the seed treatment investigations there is still no seed treatment for quarantine purposes that can guarantee that wheat is not carrying viable *T. indica* teliospores. Research could be continued to determine the effectiveness of any new fungicide available on the market, but in light of those investigations conducted so far, the chances of finding a chemical that will kill the teliospores and hence satisfy the quarantine regulations is remote. Thus, if seed is going to be distributed from countries where the disease prevails then it has to be produced in locations where the disease is not found. Otherwise, there can be no guarantee that the teliospores are not being transported on the seed.

In summary, research on Karnal bunt should concentrate on the production of resistant cultivars and investigations on the life cycle. *T. indica* provides an excellent example of how a disease can affect plant quarantine regulations. In the future, as more germplasm and seed are exchanged around the world, there will have to be more emphasis on seed health, to ensure only clean seed is shipped. If not, the benefits of crop improvement programs will be limited by serious restrictions on germplasm distribution.

1-10 REFERENCES

1. Mitra, M. "A New Bunt on Wheat in India." *Ann. Appl. Biol.* 18 (1931): 178.

2. Agarwal, V. K., Verma, H. S., and Khetarpal, R. K. "Occurrence of Partial Bunt on Triticale." *FAO Plant Prot. Bull.* 25 (1977): 210.

3. Commonwealth Mycological Institute. *Distribution Maps Plant Disease, No. 173,* 3rd ed. Kew, England: Commonwealth Agricultural Bureau, 1974.

4. Duran, R. "Further Aspects of Teliospore Germination in North American Smut Fungi II." *Can. J. Bot.* 50 (1972): 2569.

5. Anonymous. "U.S. Quarantine of Mexican Wheat: Dilemma for Researchers and Exporters." *Diversity* 5 (1983): 13.

6. Howard, A., and Howard, G. L. C. *Wheat in India I. Production, Varieties and Improvement.* Calcutta, India: Thacker Spink and Co., 1909.

7. McRae, W. "Report of the Imperial Mycologist." *Sci. Rep. Imperial Inst. Agric. Res., Pusa, India, 1932–1933,* (1934) 134.

8. Mundkur, B. B. "Some Rare and New Smuts from India." *Indian J. Agric. Sci.* 14 (1944): 49.

9. Mundkur, B. B. "Studies in Indian Cereal Smuts, V. Mode of Transmission of the Karnal Bunt of Wheat." *Indian J. Agric. Sci.* 8 (1943): 54.

10. Munjal, R. L. "Status of Karnal Bunt (*Neovossia indica*) of Wheat in Northern India During 1968–1969 and 1969–1970." *Indian J. Mycol. Plant Pathol.* 5 (1975): 185.

11. Singh, D. V,. Srivastava, K. D., and Joshi, L. M. "Present Status of Karnal Bunt of Wheat in Relation to its Distribution and Varietal Susceptibility." *Indian Phytopathol.* 38 (1985): 507.

12. Wheeler, W. H., Hunt, J., and Sutter, P. X. *List of Intercepted Plant Pests, 1949,* United States Department of Agriculture: Science Research Associates, Bureau of Entomology and Plant Quarantine, 1951, p. 77.

13. Anonymous. *List of Intercepted Plant Pests 1955.* Washington, D.C.: Science Research Associates, Bureau of Entomology, 1956, p. 63.

14. Locke, C. M., and Watson, A. J. "Foreign Plant Diseases Intercepted in Quarantine Inspection." *Plant Dis. Rep.* 39 (1955): 518.

15. Aujla, S. S., Grewal, A. S., Gill, K. S., and Sharma, I. "A Screening Technique for Karnal Bunt Disease of Wheat." *Crop Improv.* 7 (1980): 145.

16. Lambat, A. K., Ram, N., Mukewar, P. M., Majumdar, A., Indra, R., Kaur, P., Varshney, J. L., Agarwal, P. C., Khetarpal, R. K., and Dev, U. "International Spread of Karnal Bunt of Wheat." *Phytopathol. Mediterr.* 22, (1983): 213.

17. Nath, R., Lambat, A. K., Mukewar, P. M., and Rani, I. "Interceptions of Pathogenic Fungi on Imported Seed and Planting Material." *Indian Phytopathol.* 34 (1981): 282.

18. Chechet, S. M. "Okchranyaem Rastitel'nye Bogatstva" (The Protection of Plant Resources), *Zashch. Rast.* 8 (1980): 14, [Ru].

19. Lira, M. "The Karnal Bunt Situation in North-west Mexico," in *Karnal Bunt Disease of Wheat—Proceedings of a Conference* (Mexico: CIMMYT—Centro Internacional de Mejoramiento de Maiz y Trigo, April 1984), 24.

20. Delgado, S. "Mexican Phytosanitary Policy in Relation to Karnal Bunt," in *Karnal Bunt Disease of Wheat—Proceedings of a Conference* (Mexico: CIMMYT, Centro Internacional de Mejoramiento de Maiz y Trigo, April 1984), 27.

21. Boratynski, T. N., Matsumoto, T. T., and Bonde, M. R. "Interceptions of *Tilletia indica* at the California-Mexico Border in Mexican Railroad Boxcars." *Phytopathology* 75 (1985): 1339.

22. Hoffmann, J. A. "Karnal Bunt of Wheat." *Phytopathology* 73 (1983): 782.

23. Prescott, J. M. "Fungicide Applications to Germplasm Destined for CIMMYT'S International Nurseries," in *Karnal Bunt Disease of Wheat—Proceedings of a Conference* (Mexico: CIMMYT—Centro Internacional de Mejoramiento de Maiz y Trigo, April 1984), 18.

24. Kahn, R. P., and Hopper, B. "United States and Canadian Quarantine Policies," in *Karnal Bunt Disease of Wheat—Proceedings of a Conference* (Mexico: CIMMYT—Centro Internacional de Mejoramiento de Maiz y Trigo, April 1984), 28.

25. Anonymous. Action plan on Karnal bunt, *Neovossia indica* (Mitra) Mundkur. Washington, D. C.: United States Department of Agriculture, 1983, p. 43.

26. Aujla, S. S., Sharma, I., Gill, K. S., and Grewal, A. S. "Prevalence of Karnal Bunt in Punjab as Influenced by Varietal Susceptibility and Meteorological Factors." *Plant Dis. Res.* 1 (1986): 51.

27. Warham, E. J. "Studies on Karnal Bunt Disease of Wheat" (Ph.D. thesis, University of Wales, Aberystwyth, 1987), 220.

28. Mundkur, B. B. "A Second Contribution towards a Knowledge of Indian Ustilaginales." *Trans. Br. Mycol. Soc.* 24 (1940): 312.

29. Bedi, S. K. S., Sikka, M. R., and Mundkur, B. B. "Transmission of Wheat Bunt Due to *Neovossia indica* (Mitra) Mundkur." *Indian Phytopathol.* 2 (1949): 20.

30. Pal, B. P. *Wheat.* New Delhi: Indian Council of Agricultural Research, 1966, p. 230.

31. Mitra, M. "Studies on the Stinking Smut or Bunt of Wheat in India." *Indian J. Agric. Sci.* 7 (1937): 459.

32. Gill, K. S. "Present Status and Future Strategy for Development of Wheat Varieties Resistant to Karnal Bunt (*Neovossia indica*) Disease." *Seeds Farms* (October 1979): 33.

33. Joshi, L. M., Singh, D. V., Srivastava, K. D., and Wilcoxson, R. D. "Karnal Bunt: A Minor Disease that is Now a Threat to Wheat." *Bot. Rev.* 49 (1983): 39.

34. Gill, K. S., Randhawa, A. S., Aujla, S. S., Dhaliwal, H. S., Grewal, A. S., and Sharma, I. "Breeding Wheat Varieties Resistant to Karnal Bunt." *Crop Improv.* 8 (1981): 73.

35. Bedi, P. S., and Meeta, M. "Effect of 'Karnal' Bunt on Weight and Germination of Wheat Grains and Subsequent Metabolism of Seedlings." *Indian Phytopathol.* 34 (1981): 114.

36. Bedi, P. S., Meeta, M., and Dhiman, J. S. "Effect of Karnal Bunt of Wheat on Weight and Quality of the Grains." *Indian Phytopathol.* 34 (1981): 330.

37. Rai, R. C., and Singh, A. "Estimation of Loss of Wheat Grain Weight Due to Karnal Bunt Infection." *Indian J. Mycol. Plant Pathol.* 12 (1982): 102.

38. Bhat, R. V., Deosthale, Y. G., Roy, D. N., Vijayaraghavan, M., and Tulpule, P. G. "Nutritional and Toxicological Evaluation of 'Karnal' Bunt affected Wheat." *Indian J. Exp. Biol.* 18 (1980): 1333.

39. Bansal, R., Singh, D. V., and Joshi, L. M. "Effect of Karnal-Bunt Pathogen (*Neovossia indica* (Mitra) Mundkur) on Weight and Viability of Wheat Seed." *Indian J. Agric. Sci.* 54 (1984): 663.

40. Rai, R. C., and Singh, A. "A Note on the Viability of Wheat Seeds Infected with Karnal Bunt." *Seed Res.* 6 (1978): 188.

41. Singh, D. "A Note on the Effect of Karnal Bunt Infection on the Vigour of Wheat Seed." *Seed Res.* 8 (1980): 81.

42. Singh, R. A., and Krishna, A. "Susceptible Stage for Inoculation and Effect of Karnal Bunt on Viability of Wheat Seed." *Indian Phytopathol.* 35 (1982): 54.

43. Warham, E. J. "Effect of *Tilletia indica* Infection on Viability, Germination and Vigor of Wheat Seed." *Plant Dis.* 74 (1990): 130.

44. Mehdi, V., Joshi, L. M., and Abrol, Y. P. "Studies on Chapati Quality VI. Effect of Wheat Grains with Bunts on the Quality of Chapaties." *Bull. Grain Technol.* 11 (1973): 195.

45. Sekhon, K. S., Gupta, S. K., Bakhshi, A. K., and Gill, K. S. "Effect of Karnal Bunt on Chapati Making Properties of Wheat Grains." *Crop Improv.* 7 (1980): 147.

46. Sekhon, K. S., Saxena, A. K., Randhawa, S. K., and Gill, K. S. "Effect of Karnal Bunt Disease on Quality Characteristics of Wheat." *Bull. Grain Technol.* 18 (1980): 208.

47. Medina, C. L. "Efecto de Diferentes Niveles de Infección con Carbón Parcial en la Calidad de Trigo y las Características Organolépticas del Pan." (B. S. Thesis, Technical Institute of Sonora, Cuidad Obregon, Sonora, Mexico, 1985), 63.

48. Sekhon, K. S., Randhawa, S. K., Saxena, A. K., and Gill, K. S. "Effect of Washing/Steeping on the Acceptability of Karnal Bunt Infected Wheat for Bread, Cookie and Chapati Making." *J. Food Sci. Technol.* 18 (1981): 1.

49. Bhat, R. V., Rao, B., Roy, D. N., Vijayaraghavan, M., and Tulpule, P. G. "Toxicological Evaluation of Karnal Bunt of Wheat." *J. Food Safety* 5 (1983): 105.

50. Sharma, C. P., Singla, S. K., Sareen, V. K., Singh, S., and Bhatia, J. S. "Note on the Effect of Feeding 'Karnal' Bunt-Infected Wheat on Metabolism in the Rumen of Goat." *Indian J. Anim. Sci.* 52 (1982): 603.

51. Bedi, P. S., Singh, P. P., and Sohi, H. S. "Detection of Aflatoxin-Producing Isolates of *Aspergillus flavus* from the Wheat Grains Infected with 'Karnal' Bunt." *Indian J. Ecol.* 8 (1981): 304.

52. Singh, P. P., and Bedi, P. S. "Aflatoxin Producing Potential of Toxigenic Cultures of *Aspergillus flavus* Isolated from 'Karnal' Bunt Infected Wheat Grains." *Indian Phytopathol.* 37 (1984): 520.

53. Bansal, R., Singh, D. V., and Joshi, L. M. "Comparative Morphological Studies in Teliospores of *Neovossia indica.*" *Indian Phytopathol.* 37 (1984): 355.

54. Khanna, A., Payak, M. M., and Mehta, S. C. "Teliospore Morphology of Some Smut Fungi, I. Electron Microscopy." *Mycologia* 58 (1966): 562.

55. Khanna, A., and Payak, M. M. "Teliospore Morphology of Some Smut Fungi, II. Light Microscopy." *Mycologia* 60 (1968): 655.

56. Gardner, J. S., Allen, J. V., Hess, W. M., and Tripathi, R. K. "Sheath Structure of *Tilletia indica* Teliospores." *Mycologia* 75 (1983): 333.

57. Gardner, J. S., Hess, W. M., and Tripathi, R. K. "Surface Rodlets of *Tilletia indica* Teliospores." *J. Bacteriol.* 154 (1983): 502.

58. Peterson, G. L., Bonde, M. R., Dowler, W. M., and Royer, M. H. "Morphological Comparisons of *Tilletia indica* (Mitra) from India and Mexico." *Phytopathology* 74 (1984): 757.

59. Duran, R., and Cromarty, R. "*Tilletia indica*, a Heterothallic Wheat Bunt Fungus with Multiple Alleles Controlling Incompatibility." *Phytopathology* 67 (1977): 812.

60. Fuentes, G., "Biology of *Tilletia indica* (Mitra)" (M.S. Thesis, Washington State University, 1984), 66.

61. Krishna, A., and Singh, R. A. "Multiple Alleles Controlling the Incompatibility in *Neovossia indica.*" *Indian Phytopathol.* 36 (1983): 746.

62. Fischer, G. W. *Manual of North American Smut Fungi.* New York: Ronald Press Co., 1953, p. 343.

63. Commonwealth Mycological Institute, *Descriptions of Pathogenic Fungi and Bacteria. No. 748, Tilletia indica.* Kew, England: Commonwealth Agricultural Bureau, 1983.

64. Duran, R., and Fischer, G. W. *The Genus Tilletia.* Pullman: Washington State University Press, 1961, 138.

65. Krishna, A., and Singh, R. A. "Taxonomy of Karnal Bunt Fungus: Evidence in Support of Genus *Neovossia.*" *Indian Phytopathol.* 35 (1982): 544.

66. Fischer, G. W., and Holton, C. S. *Biology and Control of Smut Fungi.* New York: Ronald Press Co., 1957, p. 622.

67. Mitra, M. "Stinking Smut (bunt) of Wheat with Special Reference to *Tilletia indica* Mitra." *Indian J. Agric. Sci.* 5 (1935): 1.

68. Zillinsky, F. J. *Common Diseases of Small Grain Cereals: A Guide to Identification.* Mexico: CIMMYT—Centro Internacional de Mejoramiento de Maiz y Trigo, 1983, p. 141.

69. Hoffmann, J. A. "Bunt of Wheat." *Plant Dis.* 66 (1982): 979.

70. Holton, C. S. "Observations on *Neovossia indica.*" *Indian Phytopathol.* 2 (1949): 1.

71. McRae, W. "Report of the Imperial Mycologist." *Sci. Rep. Imperial Inst. Agric. Res., Pusa, India, 1930–1931,* 1932, 73.

72. Bansal, R., Singh, D. V., and Joshi, L. M. "Germination of Teliospores of Karnal Bunt of Wheat." *Seed Res.* 11 (1983): 258.

73. Aujla, S. S., Sharma, I., and Singh, B. B. "Method of Teliospore Germination and Breaking of Dormancy in *Neovossia indica.*" *Indian Phytopathol.* 39 (1986): 574.

74. Mathur, S. C., and Ram, S. "Longevity of Chlamydospores of *Neovossia indica* (Mitra) Mundkur." *Sci. Cult.* 29 (1963): 411.

75. Krishna, A., and Singh, R. A. "Longevity of Teliospores of *Neovossia indica* Causing Karnal Bunt of Wheat." *Indian J. Mycol. Plant Pathol.* 13 (1983): 97.

76. Smilanick, J. L., Dupler, M., Wiese, K., Hoffmann, J. A., Clark, D., and Dobson, D. "Germination of Teliospores of Karnal-, Dwarf- and Common-Bunt Fungi after Ingestion by Animals." *Plant Dis.* 70 (1986): 242.

77. Singh, R. S., Singh, K. P., Pal, B. P., and Chaube, H. S. "Note on Effect of Aldehydes and Fatty Acids on Germination of Chlamydospores of Karnal Bunt Fungus." *Patnagar J. Res.* 2 (1977): 238.

78. Krishna, A., and Singh, R. A. "Effect of some Organic Compounds on the Germination of Teliospores of *Neovossia indica.*" *Indian Phytopathol.* 32 (1979): 167.

79. Krishna, A., and Singh, R. A. "Effect of Physical Factors and Chemicals on the Teliospore Germination of *Neovossia indica.*" *Indian Phytopathol.* 35 (1982): 448.

80. Krishna, A., and Singh, R. A. "Effect of some Organic Compounds on Teliospore Germination and Screening of Fungicides against *Neovossia indica.*" *Indian Phytopathol.* 36 (1983): 233.

81. Krishna, A., and Singh, R. A. "Enhancing the Germination of Teliospores of *Neovossia indica* with Chemicals." *Indian J. Mycol. Plant Pathol.* 13 (1983): 103.

82. Gupta, R. P., and Singh, A. "Effect of Certain Plant Extracts and Chemicals on Teliospore Germination of *Neovossia indica." Indian J. Mycol. Plant Pathol.* 13 (1983): 116.

83. Bansal, R., Singh, D. V., and Joshi, L. M. "Effect of Liquid Nitrogen on Germination of Teliospores of Karnal Bunt." *Indian Phytopathol.* 37 (1984): 368.

84. Zhang, Z., Lange, L., and Mathur, S. B. "Teliospore Survival and Plant Quarantine Significance of *Tilletia indica* (causal agent of Karnal bunt)—Particularly in Relation to China." *Eur. Plant Prot. Bull.* 14 (1984): 119.

85. Royer, M. H., and Rytter, J. L. "A Comparison of Temperature and Photoperiod Requirements for *Tilletia indica* Teliospore Germination." *Phytopathology* 74 (1984): 758.

86. Smilanick, J. L., Hoffmann, J. A., and Royer, M. H. "Effect of Temperature, pH, Light and Desiccation on Teliospore Germination of *Tilletia indica." Phytopathology* 75 (1985): 1428.

87. Krishna, A., and Singh, R. A. "Aberrations in the Teliospore Germination of *Neovossia indica." Indian Phytopathol.* 34 (1981): 260.

88. Warham, E. J. "Teliospore Germination Patterns in *Tilletia indica." Trans. Br. Mycol. Soc.* 90, (1988), 318.

89. Fuentes, G., and Duran, R. "*Tilletia indica:* Cytology and Teliospore Formation in Vitro and in Immature kernels." *Can. J. Bot.* 64 (1986): 1712.

90. Krishna, A., and Singh, R. A. "Cytology of Teliospore Germination and Development in *Neovossia indica* the Incitant of Karnal Bunt of Wheat." *Indian Phytopathol.* 36 (1983): 115.

91. Prescott, J. M. "Sporidia Trapping Studies of Karnal Bunt," in *Proceedings of the Fifth Biennial Smut Worker's Workshop* (Mexico: CIMMYT—Centro Internacional de Mejoramiento de Maiz y Trigo, April 1986) p. 22.

92. Mundkur, B. B. "Karnal Bunt, an Air-Borne Disease." *Curr. Sci.* 12 (1943): 230.

93. Munjal, R. L., and Chatrath, M. S. "Studies on Mode of Infection of *Neovossia indica* Incitant of Karnal Bunt of Wheat." *J. Nucl. Agric. Biol.* 5 (1976): 40.

94. Goates, B. J. "Histology of Infection of Wheat by *Tilletia indica*, the Karnal Bunt Pathogen," *Phytopathology* 78 (1988), 1434.

95. Warham E. J. "A Comparison of Inoculation Techniques for Assessment of Germplasm Susceptibility to Karnal Bunt (*Tilletia indica*) Disease of Wheat," *Ann. Appl. Biol.* 116 (1990): 43.

96. Dhaliwal, H. S., Gill, K. S., Randhawa, A. S., and Sharma, S. K. "Systematic Spread of Karnal Bunt (*Neovossia indica* (Mitra) Mundkur) Disease of Wheat." *Wheat Information Service 56.* (Gurdaspur, India: Punjab Agricultural University, Regional Research Station, March 1983), p. 24.

97. Dhaliwal, H. S., Randhawa, A. S., Chand, K., and Singh, D. "Primary Infection and Further Development of Karnal Bunt of Wheat." *Indian J. Agric. Sci.* 53 (1983): 239.

98. Bedi, P. S., and Dhiman, J. S. "Spread of *Neovossia indica* in a Wheat Ear." *Indian Phytopathol.* 37 (1984): 335.

99. Roberson, R. W., Luttrell, E. S., and Cashion, N. L. "Formation of Exogenous Terminal Teliospores in *Neovossia indica* (*Tilletiaceae*)." *Mycol. Soc. Am. Newsletter,* 36 (1985): 37.

100. Chona, B. L., Munjal, R. L., and Adlakha, K. L. "A Method for Screening Wheat Plants for Resistance to *Neovossia indica." Indian Phytopathol.* 14 (1961): 99.

101. Khetarpal, R. K., and Agarwal, V. K. "Studies on Some Aspects of Black-Point and Karnal Bunt Diseases of Triticale." *Indian Phytopathol.* 32 (1979): 292.

102. Matsumoto, T. T. "Searching for *Tilletia indica* in Southwest U.S.A.," in *Proceedings of the Fifth Biennial Smut Workers' Workshop* (Mexico: CIMMYT—Centro Internacional de Mejoramiento de Maiz y Trigo, April 1986), p. 8.

103. Dhiman, J. S., and Bedi, P. S. "A Technique for the Isolation of *Neovossia indica*—the Causal Organism of Karnal Bunt of Wheat." *Indian Phytopathol.* 36 (1983): 767.

104. Munjal, R. L. "Technique for Keeping the Cultures of *Neovossia indica* in Sporulating Condition." *Indian Phytopathol.* 27 (1974): 248.

105. Ramamoorthy, C. S., and Mundkur, B. B. "*Neovossia indica* in Culture." *Curr. Sci.* 8 (1944): 49.

106. Singh, M., and Singh, A. "Effect of Culture Media and Antimicrobial Agents on Growth and Sporulation of *Neovossia indica*." *Indian J. Mycol. Plant Pathol.* 16 (1986): 331.

107. Krishna, A., and Singh, R. A. "Method of Artificial Inoculation and Reaction of Wheat Cultivars to Karnal Bunt." *Indian J. Mycol. Plant Pathol.* 13 (1983): 124.

108. Moore, M. B. "A Method for Inoculating Wheat and Barley with Loose Smuts." *Phytopathology* 26 (1936): 397.

109. Aujla, S. S., Grewal, A. S., Gill, K. S., and Sharma, I. "Artificial Creation of Karnal Bunt Disease of Wheat. *Cereal Res. Commun.* 10 (1982): 171.

110. Aujla, S. S., Grewal, A. S., and Sharma, I. "Relative Efficiency of Karnal Bunt Inoculation Techniques." *Indian J. Mycol. Plant Pathol.* 13 (1983): 99.

111. Warham, E. J., Mujeeb-Kazi, A., and Rosas, V. "Karnal Bunt (*Tilletia indica*) Resistance Screening of *Aegilops* Species and their Practical Utilization for *Triticum aestivum* Improvement." *Can. J. Plant Pathol.* 8 (1986): 65.

112. Aujla, S. S., Sharma, Y. R., Chand, K., and Sawney, S. S. "Influence of Weather Factors on the Incidence and Epidemiology of Karnal Bunt Disease of Wheat in the Punjab." *Indian J. Ecol.* 4 (1977): 71.

113. European and Mediterranean Plant Protection Organization (EPPO). *Data Sheets on Quarantine Organisms, List A1, Tilletia indica* (Mitra). Paris: EPPO, 1979.

114. Khetarpal, R. K., Agarwal, V. K., and Chauhun, K. P. S. "Studies on the Influence of Weather Conditions on the Incidence of Black-Point and Karnal Bunt of Triticale." *Seed Res.* 8 (1980): 108.

115. Joshi, L. M., Singh, D. V., and Srivastava, K. D. "Present Status of Karnal Bunt in India." *Indian Phytopathol.* 33 (1980): 147.

116. Aujla, S. S., Sharma, I., and Gill, K. S. "Effect of Time and Method of Inoculation on Karnal Bunt Development." *Indian Phytopathol.* 39 (1986): 230.

117. Singh, A., and Prasad, R. "Date of Sowing and Meteorological Factors in Relation to Occurrence of Karnal Bunt of Wheat in U.P. Tarai." *Indian J. Mycol. Plant Pathol.* 8 (1978): 2.

118. Dhiman, J. S., Bedi, P. S., and Mavi, H. S. "Relationship among Temperature, Humidity and Incidence of Karnal Bunt of Wheat." *Indian J. Ecol.* 11 (1984): 134.

119. Agarwal, V. K. "Quality Seed Production at Pantnagar, India." *Seed Sci. Technol.* 11 (1983): 1071.

120. Agrawal, K., Yadav, V., Singh, T., and Singh, D. "Occurrence and Detection of Karnal Bunt of Wheat in Rajasthan." *Indian J. Mycol Plant Pathol.* 16 (1986): 290.

121. Aujla, S. S., Gill, K. S., Sharma, Y. R., Singh, D., and Nanda, G. S. "Effect of Date of Sowing and Level of Nitrogen on the Incidence of Karnal Bunt of Wheat." *Indian J. Ecol.* 8 (1981): 175.

122. Singh, D. V., Srivastava, K. D., and Joshi, L. M. "Prevent Losses from Karnal Bunt." *Indian Farming* (February 1979), 7.

123. Singh, B., and Mathur, S. C. "Bunts of Wheat." *Agric. Anim. Husbandry in India* 3 (1953): 10.

124. Krishna, A., and Singh, R. A. "Investigations on the Disease Cycle of Karnal Bunt of Wheat." *Indian J. Mycol. Plant Pathol.* 12 (1982): 124.

125. Warham, E. J. "Karnal Bunt Disease of Wheat: A Literature Review." *Trop. Pest Management,* 32 (1986): 229.

126. Singh, B. B. "Studies on Variability, Epidemiology and Biology of *Neovossia indica* (Mitra), Mundkur Causing Karnal Bunt of Wheat" (Ph.D. thesis, Punjab Agricultural University, Ludhiana, India, 1986), 97.

127. Smilanick, J. L., and Prescott, J. M. "Effect of Soil Fumigation with Methyl Bromide, Metham and Formaldehyde on Germination of Teliospores of *Tilletia indica.*" *Phytopathology* 76 (1986): 1060.

128. Rai, R. C., and Singh, A. "Effect of Chemical Seed Treatment on Seed-Borne Teliospores of Karnal Bunt of Wheat." *Seed Res.* 7 (1979): 186.

129. Fuentes, S., Torres, E., and Garcia, C. "Chemical Seed Treatment for Partial Bunt of Wheat." *Phytopathology* 73 (1983): 122.

130. Smilanick, J. L., Hoffmann, J. A., Secrest, L. R., and Wiese, K. "Evaluation of Chemical and Physical Treatments to Prevent Germination of *Tilletia indica* Teliospores." *Plant Dis.* 72 (1988): 46.

131. Hoffmann, J. A., and Walder, J. T. "Chemical Seed Treatments for Controlling Seedborne and Soilborne Common Bunt of Wheat." *Plant Dis.* 65 (1981): 256.

132. Smilanick, J. L., Hoffmann, J. A., Cashion, N. L., and Prescott, J. M. "Evaluation of Seed and Foliar Fungicides for Control of Karnal Bunt of Wheat." *Plant Dis.* 71 (1987): 94.

133. Krishna, A., and Singh, R. A. "Evaluation of Fungicides for the Control of Karnal Bunt of Wheat." *Pesticides* 16 (1982): 7.

134. Singh, A., and Prasad, R. "Control of Karnal Bunt of Wheat by a Spray of Fungicides." *Indian J. Mycol. Plant Pathol.* 10 (1980): i.

135. Singh, A., Tewari, A. N., and Rai, R. C. "Control of Karnal Bunt of Wheat by a Spray of Fungicides." *Indian Phytopathol.* 38 (1985): 104.

136. Singh, D. V., Srivastava, K. D., Joshi, L. M., and Verma, B. R. "Evaluation of Some Fungicides for the Control of Karnal Bunt of Wheat." *Indian Phytopathol.* 38 (1985): 571.

137. Bedi, P. S. "Occurrence and Prevalence of Karnal Bunt of Wheat in the Punjab." *Indian Phytopathol.* 33 (1980): 249.

138. Gautam, P. L., Pal, S., Malik, S. K., and Saini, D. P. "Note on Reaction of Promising Wheat Strains to Karnal Bunt." *Pantnagar J. Res.* 2 (1977): 228.

139. Joshi, L. M., Renfro, B. L., Saari, E. E., Wilcoxson, R. D., and Raychaudhuri, S. P. "Rust and Smut Diseases of Wheat in India." *Plant Dis. Rep.* 54 (1970): 391.

140. Meeta, M., Dhiman, J.S., Bedi, P. S., and Kang, M. S., "Incidence and Pattern of 'Karnal' Bunt Symptoms on some Triticale Varieties under Adaptive Research Trial in the Punjab." *Indian J. Mycol. Plant Pathol.* 10 (1980): Lxxxiv.

141. Bekele, G. T., Skovmand, B., Gilchrist, L. I., and Warham, E. J. "Screening Triticale for Resistance to Certain Diseases Occurring in Mexico," in *Genetics and Breeding of Triticale* (Clermont-Ferrand, France: EUCARPIA meeting, 2–5 July, 1985, INRA, Paris, 1985), p. 559.

142. Warham, E. J. "Screening for Karnal Bunt (*T. indica*) Resistance in Wheat, Triticale, Rye and Barley." *Can. J. Plant Pathol.* 10 (1988): 57.

143. Whitney, N. G., and Frederiksen, R. A. *Kernel Smut of Rice.* Texas A&M University Report MP 1231 (1975), 12.

2

WHEAT RUSTS

S. NAGARAJAN

Indian Council of Agricultural Research
Krishi Bhawan, New Delhi

S. K. NAYAR AND S. C. BHARDWAJ

IARI, Regional Station, Shimla, India

2-1 INTRODUCTION

Wheat cultivation is concentrated in the cooler regions of the world and it enjoys a good crop husbandry. The three *Triticum* species, namely, *T. aestivum* (bread wheat), *T. dicoccum* (einkorn), and *T. durum* (durum) are the most commonly cultivated species. Wheat withstands erratic rainfall, water stress, and restricted nutrient supply, and can yield satisfactorily even in problem soils. Comparatively, yields are stable in a wide range of environments, yet epidemics of rusts and other diseases can cause great loss.

2-2 WHEAT RUSTS

2-2-1 Genus *Puccinia* and Cereals

The genus *Puccinia* infects a number of crops belonging to the family Gramineae and others. Occasionally the same pathogen infects more than one species, as is evident from Table 2-1.

There are three types of wheat rust diseases that occur almost universally. The detailed symptomatology (Figures 2-1–2-3) and other characteristics of the pathogen and the disease are summarized in Table 2-2.

TABLE 2-1: DISEASES CAUSED BY *PUCCINIA* SPP. ON WINTER CEREALS

S. No.	Crop	Disease	Causal Agent
1.	Wheat (*Triticum* spp.)	Black (stem) rust	*Puccinia graminis* Pers. f. sp. *tritici* Eriks. and Henn.
		Brown (leaf) rust	*P. recondita* Rob. ex Desm. f. sp. *tritici* Eriks. (Syn. *P. triticina, P. rubigo-vera* var. *tritici*)
		Yellow (stripe) rust	*P. striiformis* West. f. sp. *tritici* (Syn. *P. glumarum*)
2.	Barley (*Hordeum vulgare*)	Black rust	*P. graminis* Pers. f. sp. *tritici* Eriks. and Henn.
		Leaf rust	*P. hordei* Otth. (Syn. *P. anomala* Rostr.)
		Yellow rust	*P. striiformis* West. f. sp. *hordei*
3.	Oat (*Avena sativa*)	Black rust	*P. graminis* f. sp. *avenae* Eriks. and Henn.
		Crown rust	*P. coronata* Cda. f. sp. *avenae* Fraser and Ledingham
4.	Rye (*Secale cereale*)	Stem rust	*P. graminis* Pers. f. sp. *secalis*
		Brown rust	*P. recondita* Rob. ex Desm. f. sp. *secalis* (=*P. dispersa*)

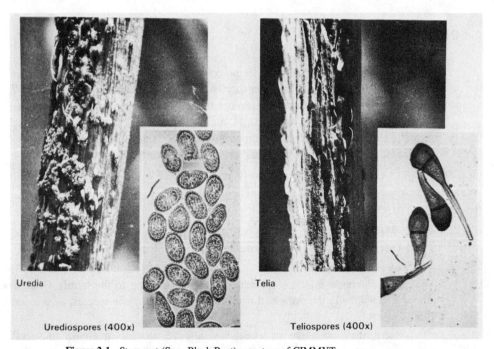

Figure 2-1 Stem rust (Syn. Black Rust), courtesy of CIMMYT.

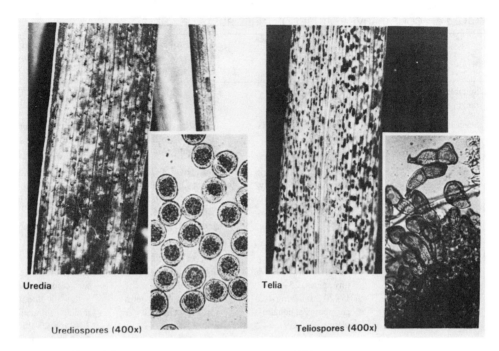

Figure 2-2   Leaf rust (Syn. Brown Rust), courtesy of CIMMYT.

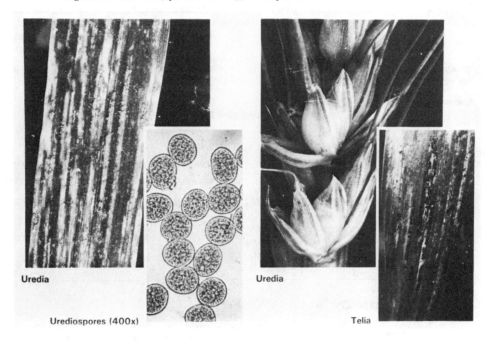

Figure 2-3   Stripe rust (Syn. Yellow Rust), courtesy of CIMMYT.

TABLE 2-2:   COMPARATIVE FEATURES OF THREE RUSTS OF WHEAT

Differentiating characters	Black rust (*Puccinia graminis* f. sp. *tritici*)	Brown rust (*Puccinia recondita* f. sp. *tritici*)	Yellow rust (*Puccinia striiformis* f. sp. *tritici*)
Suitable temperature range	25°C	20°C	10–15°C
Severity	More severe on stalks than on leaf sheaths, leaves, ear heads	Infects leaves exclusively, rarely on leaf sheaths and stalks	Infects leaves and when severe, leaf sheaths, stalks, and ears also
Urediopustule	Large elongated, dark brown, bursting early, throwing epidermal fringes in the process	Small, oval, light brown, do not run together, burst early with mild displacement of epidermis, found chiefly on upper surface of leaves	Arranged in rows of lemon yellow colored sori, epidermal rupture not visible
Urediospore	Brown, oval, thick wall, tiny spines, 25–30 × 17–20 μm having 4 germpores at equatorial plane	Brown, spherical, minutely echinulated, 16–28 μm, 7–10 germ pores dispersed all over	Spherical to ovate, spore wall colorless, contains yellow oil globule, minute echinulation 23–35 × 20–25 μm, 6–16 germ pores dispersed all over
Teliopustule	Black, found on all aerial parts, mainly stem	Pustules do not burst epidermis, found chiefly on leaf surface and rarely elsewhere	Sori are flattened and dull black, chiefly on lower surface and do not burst open easily
Teliospore (two celled)	40–46 × 15–20 μm	35–63 × 12–20 μm	35–63 x 12–20 μm
Alternate hosts	Species of *Berberis*, *Mahonia, Mahoberberis*	*Isopyrum fumarioides*, *Thalictrum* spp., *Clematis* spp., *Anchusa* spp.	Not known ·
Collateral hosts	*Bromus coloratus, B. carinatus, B. japonicus, B. mollis, B. patulus, Hordeum distichon, H. murinum, H. stenostachys, Lolium perenne, Hilaria jamesii, Aegilops squarrosa, A. ventricosa, A. trinecilis*	*Aeqilops* spp.	*Bromus catharticus, B. japonicus, Hordeum murinum, Muehlenbergia hugelii*

2-2-2 Disease Cycle

Stem rust.   *P. graminis tritici* produces teliospores late in the season. These require a dormancy period before germinating into a four-celled promycelium bearing single-celled globose basidiospores, on short sterigmata. Prior to germination, fusion of nuclei takes place in the teliospores. This diploid nucleus undergoes meiosis, forming four daughter haploid nuclei that migrate one each into the basidiospores. The basidiospores are of two different strains, namely + and −, each being self-incompatible.

The basidiospores are windborne, and if deposited on young barberry leaves, germinate and infect the leaf at 12–21 °C. Spermogonia are produced near the upper epidermis after a few days of infection. Spermatia are produced on stalks and all spermatia from a single spermogonium contain either + or − factor depending on the source of basidiospores. The mycelium producing spermatia also produce receptive hyphae carrying the same factor. While feeding repeatedly on the fragrant nectar, insects transfer the spermatia soaked in nectar on to the receptive hyphae of another factor. The hyphae protrude out through the ostiole, permitting easier fertilization. The spermatial contents pass into the receptive hyphae through a pore by dissolving the wall at the place of contact. The dikaryotic mycelium formed thereafter proliferates and forms aecial cups rupturing the lower epidermis. These cups contain chains of aeciospores in palisade layers borne on short stalks. The aeciospores are ejected from the lower side of leaves in high humidity and thus differ in the mode of spore liberation when compared to the urediospores (Figure 2-4).

Aeciospores are incapable of infecting *Berberis*. Mature aeciospores get blown off and infect wheat to produce crops of urediospores. About 83 species of *Berberis* are known to be infected by the stem rust pathogen. Of these *B. aristata, B. coriaria, B. lycium,* and *B. vulgaris* are the important ones. More than six species of *Mahonia* and *Mahoberberis neubertii* (a cross between *Mahonia* and *Berberis*) also act as alternate hosts to this pathogen.

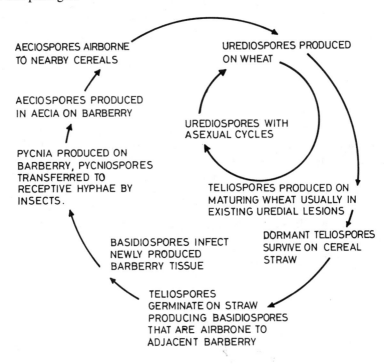

Figure 2-4 Disease cycle of *Puccinia graminis tritici*.

Mehta[1,2] showed that under Indian conditions, alternate hosts are not functional. The survival and perpetuation of the pathogen, therefore, occur through repeated urediocycles. The nonsynchronization of vulnerable tender barberry leaves when the basidiospores are available drastically curtails the role of alternate host under Indian conditions. In fact, alternate hosts are of no consequence in the recurrence of stem rust in India. So far four aecial stages have been recorded on *Berberis* but are connected with the hosts other than wheat, as indicated below:

1. An aecial stage commonly present in Shimla has been connected with *Agropyron semicostatum* Nees and is in fact *P. graminis* f. sp. *Agropyrii*.[3]

2. Aecial stage on *Berberis jaeschkeana* C.K. Schn. has been connected with *Poa nemoralis* L. and named as *P. poaenemoralis* Otth.[4]

3. The third aecial stage on *Berberis* in Shimla is connected with *Brachypodium sylvaticum* (Huds.) Beauv. being *P. brachypodii*.[5]

4. *Aecidium montanum* Butler, which is frequently observed in Shimla, Kumaon Hills, and Nepal, is not related to *P. graminis tritici*.

Many collateral hosts such as *Bromus patulus, B. coloratus, B. carinatus, B. mollis, B. japonicus, Hordeum distichum, H. murinum, H. stenostachys, Lolium perenne, Hilaria jamesii, Aegilops squarrosa, A. ventricosa, A. trinecilis* can serve as off season hosts in the annual perpetuation of the stem rust pathogen.

Brown rust.   Jackson and Mains[6] demonstrated *Thalictrum* spp. as the alternate host for *P. recondita tritici*. Later *Isopyrum fumarioides* in Siberia,[7] *Clematis* spp. in Italy,[8] and *Anchusa* spp. in Portugal[9] were also found to serve as alternate hosts. The life cycle of *P. recondita tritici* is like that of *P. graminis tritici*, which was explained earlier. An aecial stage found on *Thalictrum javanicum* in India is that of *Puccinia persistens* and therefore does not infect wheat. A few species of *Aegilops* also become infected by *P. recondita tritici*.

Yellow rust.   Although heteroecious, there is still no known alternate host for the yellow rust pathogen. In India, the primary inoculum for the Indo-Gangetic plain comes each year from the array of Himalayan hills. In the temperate countries, volunteer plants or the "green bridge" and collateral hosts help in the annual perpetuation of the pathogen. The role of collateral hosts like *Agropyron semicostatum, Bromus catharticus, B. japonicus, Hordeum murinum,* and *Muehlenbergia hugelii* is less clear in the annual recurrence of the disease. In the absence of a functional alternate host, perpetuation of the pathogen is accomplished by repeated urediocycles. When temperatures are low, infections remain latent for an unusually long time and burst open suddenly when favorable conditions return.

2-3 INTERNATIONAL SPREAD

Cereal rusts particularly the wheat rusts, are notorious international travelers. The pathogen propagules each year spread hundreds and thousands of kilometers. In many cases, the source area and target of the organism are distinct and the spread of the pathogen is accelerated by the coincidence of certain favorable weather patterns, direction of winds, inoculum load, plant age, and type of host variety at the target area. Christensen[10] estimated that a moderate infection of black rust will produce 4×10^{12} urediospores/day/hectare. That of brown and yellow rust is said to be of the order of 3.2×10,[13] and $0.6 - 2.0 \times 10^{13}$, respectively. When an adequate infection occurs over a large area, an unbelievable load of urediospores are generated.

Stepanov[11] reported a minimum wind speed of 0.51 to 0.75 m/sec necessary to remove urediospores of *P. recondita tritici*. More spores are removed when the wind is turbulent. The threshold wind speed needed to dislodge the spore varies for other rusts. Wind-initiated leaf rubbing that precedes rain results in increased spore concentration in the air. Vertical transport of these spores occurs when convective currents arise over the infected area. As the source of primary inoculum of stem rust in India and Kenya is a group of hills, urediospores that reach the air current are horizontally carried over long distances before being deposited.

There is evidence from North America, Europe, India, Southern Africa, and Australia regarding the long distance dispersal and deposition of urediospores of *Puccinia* spp. During such transport, the spore concentration in the air gets diluted with increasing distance. That aside, the atmospheric variation affects the viability of the spores in transit.

The stem rust urediospores are aerodynamically suited for being carried aloft in the air for horizontal spread. They are less sensitive to the UV radiation that is so high at the higher altitudes. During the spore trappings done in 1923 using aircraft, Stakman and co-workers[12] trapped several cereal rust spores on Vaseline-coated slides. Urediospores have been trapped at heights exceeding 4,000 m, and in the high passages of interior Himalaya, linking two valleys.[13]

Given that the terminal velocity of *P. graminis tritici* urediospores is around 1.06 cm/sec, it takes a lot of time to settle down through sedimentation, by which time they are transported over long distances. Urediospores often lose viability before deposition but occasionally an instant washdown of viable urediospores is accomplished by falling raindrops. Rain samplers have been designed and procedures for analyzing the samples have been standardized.[14] Following this procedure urediospores of wheat rusts have been trapped in rain water in the United States and India.[15,16] Such a deposition over the target area assures successful infection. Long-distance transport of rust urediospores as occurs in different countries has been exemplified here.

2-3-1 Australia and New Zealand

Wheat-growing areas within Australia are separated by long distances. Postharvest rains in the eastern part of the country facilitate the survival of pathogens on volunteer plants and *Agropyron scabrum* Beauv. (rough wheat grass). Western Australia, the other wheat area, is separated from the east by 1,300 km of desert, yet an occasional exchange of inoculum between these regions occurs. Stem rust strains move north and south and east and west over the Australian continent in which aerial transport of spores plays a major part. Similarities in the pathogenicity of stem rust strains with those of southern parts of Africa and that of Australia, identical biochemical patterns, and connecting winds from Australia indicate long-distance dispersal and deposition of viable urediospores across

5,000 km of ocean. Studies spanning a period of 40 years revealed the long-distance spread of *P. graminis tritici* urediospores. Strain 21-ANZ, 2 spread between New South Wales and South Victoria, over a distance of 1,450 km. Earlier observations suggested that west to east movements are rare, however, the epidemic of 1973–74 was a result of these movements.

Data from New Zealand clearly reflect a resemblance of race flora to eastern Australia. It is further evident from the fact that no stem rust infection occurs in winter, thereby meaning that the pathogen is eliminated in the southern island of New Zealand. The early maturing crop in Queensland provides inoculum for the North island, whereas south-eastern Australia serves as source of urediospores for southern New Zealand. This information can be further validated by the occurrence of a new strain in northern New Zealand undoubtedly detected earlier in Queensland. Such virulences spread across the 2,000 km distance of ocean from Australia to New Zealand.[17]

2-3-2 United States

In the United States, 80% of the wheat area is winter wheat, and the rest consists of spring types. The barberry eradication program and release of resistant varieties curtailed the spread of *P. graminis tritici*, but even then serious epidemics swept the United States in 1916, 1923, and thereafter. The fact that the barberry eradication program in all the wheat-growing states was nearly complete and yet the stem rust was occurring led to investigations on the long-distance dispersal of the pathogen. Stakman and his co-workers[12] clearly demonstrated the northerly movement of the stem rust pathogen from its overwintering sources in southern Texas and northern Mexico. From this overwintering area, the windborne inoculum sweeps through the United States, crossing the Rockies and infecting the fields in the Dakotas, Minnesota, and across the Canadian border. The differences in latitude, altitude, presence of susceptible wheat or grasses throughout the year, lack of topographical barrier, mass flow of the wind from north to south and vice-versa all ensure infection of *P. graminis tritici* from Mexico to the Prairie, covering a distance of 4,000 km. Gene deployment, multilines, and other management practices have been found useful in curtailing such a spread.

Favorable winds carry innumerable spores back into the southern United States and Mexico during autumn, therefore, the movement of the stem rust is bidirectional, i. e., from source to target and back to source.[18] The role of meteorological factors in the long-distance dispersal is evident from the spread of race 15B of *P. graminis tritici* that swept across the United States in a single year. Deposition of the primary inoculum of *P. graminis tritici* over the northern part of the United States occurs along with rain[15] and local climatic factors decide the further course of the epidemic. To minimize crop losses, nearly 370 different cultivars are annually grown in the United States.[19]

2-3-3 South Asia

Berberis/Thalictrum spp., the alternate hosts for black and brown rust, are nonfunctional. In the Indo-Gangetic plains during the warm postharvest summer months, the rust pathogens are eliminated due to the high temperature. There is no local inoculum left in the Indo-Gangetic plain to cause fresh infections during the succeeding wheat season. The pathogen retreats back to the Himalayas during the May–November period when summer crops of wheat and 'green bridges' are available. The Nilgiri and Palney hills of southern India where wheat is available around the year above 2,200 m altitude serves as the source area for the disease for the southern and central Indian crop.[1,20] Recent studies, however, reveal that stem rust urediospores spread primarily from the Nilgiri and Palney hills of southern India. It is believed that the Himalayas play a very minor role in the recurrence of epidemics for northern India. Inoculum that is available in the South Indian hills are carried further north by tropical cyclones and are deposited along with rains over one-month-old central Indian wheat crops. Such a deposition, which occurs during November as a consequence of tropical cyclones, is the one capable of epidemic initiation. Spread of stem rust from central India to other parts of northern India is favored by the repeated passage of "Western Disturbances" linking both areas. The associated winter rainfall permits nationwide appearance and buildup of stem rust.

A set of three weather situations that favor a northerly movement of the stem rust urediospore have been identified as the Indian Stem Rust Rules.[21] The occurrence of a tropical cyclone (ISR-1) can be monitored from weather satellites[22] and accordingly, a disease appearance prediction system has been developed and validated.[23,24]

Spread of brown rust from the southern foci to central India is identical with that of stem rust, and the appearance of leaf rust can also be predicted on the basis of Indian Stem Rust Rules.

Wheat-growing areas are relatively cooler over northwest India and Pakistan; hence, yellow rust of wheat (*P. striiformis tritici*) is more important. During both 1976 and 1980 serious epidemics of yellow rust on wheat in Pakistan caused food deficits. *P. striiformis tritici* is not a regular long-distance spreader, and from the oversummering areas in Hindu Kush, the Sulaiman ranges and in the Himalayas, the pathogen spreads up to a few hundred kilometers. Epidemic size and the terminal disease severity are greatly dependent on the number and frequency of western disturbances, a phenomenon that can be monitored by measuring the cumulative percentage areas under cloud, using weather satellite images. The suspected deviation in wheat crop health can be validated by appropriate analysis of the landsat images.[25,26] An integrated approach for an efficient management of the food resources of South Asian nations using ground truth, weather information, models on disease severity prediction, and satellite-based crop health data has been presented for appropriate adoption.[27]

2-4 MODELS IN CEREAL DISEASES STUDY

Roelfs[28] developed a model for downwind dispersal and deposition of *P. graminis* and *P. recondita* spores from a point source. Through multiple regression analysis, Burleigh[29,30] developed linear equations for predicting wheat leaf rust severities and crop loss. Of the various biological and meteorological parameters that he used in the stepwise multiple regression, only a few were of significance. Nagarajan and Joshi[31] developed a simple linear model that predicted with reasonable accuracy stem rust severity over a 7-day period.

$$Y = -29.3733 + 1.820\ x1 + 1.7735\ x2 + 0.2516\ x3$$

Where Y is the expected mean disease severity, x1 is mean disease severity 7 days before, x2 is mean weekly maximum temperature, and x3 is mean maximum relative humidity for the prediction period. The accuracy of this equation in stem rust severity prediction was validated by Karki et al.[32] through multilocation testing. Validation is the process of comparing model results to comparable real world data.

In the United States development of leaf rust is successfully predicted using the multiple regression equations developed by Eversmeyer and Burleigh[33] and Burleigh et al,[30] whereas cumulative degree days is the basis of yellow rust prediction in the western states of the United States.[34]

EPIPRE is a computerized, supervised control system for wheat pests and diseases. Essentially, EPIPRE does nothing more than to apply the principles of population dynamics and computer science in order to benefit individual farmers. The system is exclusively for wheat, both spring and winter types. EPIPRE handles six major fungal diseases and two insect pests. The model recommends whether plant protection operations are to be executed and in no case recommends more than two sprays of triadimefon. It also takes into account that it is not to interfere with the nontarget beneficial insects and predators. Thus, EPIPRE is one of the most successful models that was field validated for many years in the Netherlands. It has now gained acceptance in many of the European countries interested in integrated pest and disease management in wheat.[35]

2-5 IDENTIFICATION OF RACES/VIRULENCES

Stakman and Piemeisel[36] introduced the concept of race in the wheat stem rust system. Stakman and Levine[37] designated the physiological races and differentiated them on a set of 12 differentials. These differentials were selected from genetically diverse *Triticum* species on which reactions were temperature insensitive, now known as the International Differentials. Occasionally supplementary or additional differentials were used to characterize biotypes that are not otherwise detectable based on the standard differentials. Host pathogen interaction is recorded as resistant (0, 0;, 1, and 2), susceptible (3, 4), or heterogeneous, as given below:

R	0	Immune	No infection
	0;	Nearly immune	No uredia, but hypersensitive flecks present
	1	Very resistant	Uredia minute, surrounded by distinct necrotic areas
	2	Moderately resistant	Uredia small to medium, surrounded by chlorotic or necrotic border
S	3	Moderately susceptible	Uredia medium in size, coalescence infrequent, chlorotic areas may be present
	4	Very susceptible	Uredia large, coalescing, no necrosis but chlorosis may be present
X		Heterogeneous	Size of pustules variable, may contain all above types

Each race is thus characterized by a set of reactions. The designated race number does not quantify the reaction and is to be reckoned with the master table to ascertain the reactions of the type culture. The number given to each set of reactions is merely a serial number, the earlier detected getting a low number in comparison to the one isolated later.

Similarly, brown[38] and yellow rust races[39] are identified on a set of standard differentials.

With the advancements in the field of host-pathogen interaction, and the realized utility of the gene-for-gene hypothesis of Flor,[40] a need arose to update the race analysis procedure. The standard race analysis procedure does not indicate the specific resistance genes that accord resistance and those that are susceptible to the isolate. Watson and Luig[41] developed a new system for identification of stem rust virulences by using lines with known resistance genes. Johnson et al.[42] developed a system on parallel lines for the identification of virulences of yellow rust. Following this, Roelfs and McVey[43] and Nagarajan et al.[44,45] updated procedures for identification of rust virulences to suit various national requirements. Yet the basis of virulence identification remains the same, that is, based on host pathogen differential interaction. Entries in the differential set have been replaced by lines with known gene(s) or near isogenic lines or lines with gene combinations in addition to the condensed standard differentials. This enlarged procedure permits the identification of virulences and at the same time indicates the genes that accord seedling resistance against those cultures. This information has direct implications for the crop improvement program and hence the changed procedures are being followed in many nations.

Postulation on the probable host resistance genes are made by comparison of the seedling reaction matrix of the unknown host with that of the lines with known specific genes. Matrices can be generated by the appropriate selection of virulences that differ for pathogenicity. Computer programs have been developed for efficient and bulk comparison of seedling data and postulating probable host resistance genes.

2-6 DISEASE MANAGEMENT

2-6-1 Cultural

Excessive use of nitrogenous fertilizers should be avoided as these enhance foliar growth and make plants more prone to rust infection, whereas on the other hand, phosphatic fertilizers hasten maturity and hence reduce the chances of infection. Wheat rust epidemics can be checkmated by quick varietal changes along the path of the pathogen dispersal. A rapid change in cultivars averted or delayed the possible epiphytotic of 1976 in India, while in adjoining Pakistan, where cultivars were not changed, a severe leaf rust epidemic occurred.[46] Diversity in cultivars grown on a farm and mixed cultivation of different varieties and/or crops will reduce the multiplication of inoculum resulting in decreased development of rust.

Destruction of collateral hosts/alternate hosts and volunteer wheat plants/green bridges several weeks before planting wheat will reduce the initial inoculum and delay the epiphytotic. Barberry eradication was visualized as a means of breaking the stem rust cycle in order to maximize grain production. Such an eradication done in the United States drastically delayed disease onset, reduced initial inoculum level, decreased the number of physiologic races, and reduced the frequency of stem rust epidemics. Coupled with better varieties and plant protection operations, the barberry eradication substantially reduced the frequency of stem rust in the United States.[47]

2-6-2 Host Resistance

Various disease management approaches are in practice using vertical resistance[48] such as multiline, varietal mixture, gene cycling, adding resistance genes, and using slow rusting or adult plant resistance.[49] Releasing and cultivating a number of varieties with differing resistance base creates a diverse mosaic pattern. Such a pattern minimizes the chances of widespread epidemic.

Alien sources have been tapped as possible donors in the search for genes that would provide lasting resistance; but not all the alien genes provided the expected level of lasting resistance. Because the locating, identifying, transferring, and combining of race specific genes is easier, so far such resistance genes have been widely used. On the contrary, the rate reducing type of resistance, although it produces less terminal disease severity and lesser crop loss, is difficult to incorporate in varietal improvement attempts. Various approaches for the incorporation of such a resistance and selection procedures have been critically studied.[50,51] The genetic basis of rate reducing resistance can be due to adult plant resistance, temperature sensitive genes, polygenic resistance, and so on. However, Vertifolia effect arising out of willful selection for vertical resistance can result in the erosion of rate reducing resistance. An appropriate blend of both vertical and horizontal resistance may lead to varieties that can withstand the test of time.

2-6-3 Biological

In his experiments with rust fungi, such as *P. graminis* f. sp. *tritici, P. striiformis, P. recondita,* and *P. graminis* f. sp. *avenae,* Hassebrauk[52] isolated *Verticillium albo-minimum, V. malthousei,* and *Cephalosporium lefroyi* from the sori of rusts. *Sphaerellopsis filum* (perfect state *Eudarluca filum*), an ubiquitous hyperparasite, is associated with *P. coronata, P. graminis, P. recondita, P. striiformis, P. sorghi,* and species of *Cronartium.* In Kenya, *S. filum* was found in about 99% of the rust sori. *Gonatobotrys simplex* and *Verticillium lecanii* are also known to infect *P. graminis* f. sp. *tritici. Aphanocladium album* is also pathogenic on *P. coronata, P. hordei, P. graminis* f. sp: *avenae,* and *P. recondita* f. sp. *tritici,* by preventing production of urediospores, and so only teliospores are produced. Forrer[53,54] found increased formation of teliospores in *P. graminis* f. sp. *tritici* after leaf application of the metabolic products of *A. album* at the rate of 0.24 mg/cm^2. The biological control needs special attention as not much work has been done in this area.

2-6-4 Chemical

From the early days of chemical control using sulphur, much advancement has been made, up to the present era of systemic fungicides, in reducing the number of sprays needed to keep disease below the economic threshold. Generally, systemic fungicides are better protectants than eradicants, because subsequent to the establishment of parasitism, the pathogen is less vulnerable.[55,56] In western European nations and a few others where wheat productivity levels are very high, the dictating grain prices permit chemical control for winter wheats even in situations where crop loss is just around 2%[57] whereas in India, even in the highly productive northwestern regions where average yield is 3 T/ha, chemical control is not popular.

Of the systemics, triadimefon, triarimol, and fenapanil are the best eradicants. An appropriately timed single spray of triadimefon provides quite a lasting control against brown and yellow rust of wheat.

2-7 SUMMARY

The wheat rusts continue to be the main concern in increasing crop production despite several scientific advancements of this and earlier centuries. Causes of variability in the pathogen and how to minimize it are getting considerable attention. The new systemic fungicides that are effective against wheat rusts have added a new element for effective disease management. Various varietal diversification approaches and farm level and regional level strategies have come out of sound epidemiological investigations. The time tested varietal mixers, timely sowing of resistant varieties, and optimal fungicide spray wherever necessary are the strategies that will be continued for proper disease management.

2-8 REFERENCES

1. Mehta, K. C. "Annual Recurrence of Rusts on Wheat in India." *Presidential address (Sec. of Botany) Proc. 16th Ind. Sci. Congr.*, 199, 1929.

2. Mehta, K. C. "Further Studies on Cereal Rusts in India." *Sci., Monogr. No. 14*, Imperial Council Agr. Res., India, 1, 1940.

3. Prasada, R. "Discovery of Uredo-stage Connected with the Aecidia so Commonly Found on Species of *Berberis* in Simla Hills." *Indian J. Agr. Sci.* 17 (1947): 137.

4. Joshi, L. M., and Payak, M. M. "A *Berberis* Aecidium in Lahaul Valley Western Himalayas." *Mycologia* 55 (1963): 247.

5. Payak, M. M. "*Berberis* as the Aecial Host of *Puccinia brachypodii* in Simla Hills (India)." *Phytopathol. Z.* 52 (1965): 49.

6. Jackson, H. S., and Mains, E. B. "Aecial Stage of the Orange Leaf Rust of Wheat, *Puccinia triticina* Eriks." *J Agric. Res.* 22 (1921): 151.

7. Chester, K. S. "The Nature and Prevention of Cereal Rusts as Exemplified in the Leaf Rust of Wheat." *Chronica Botanica*, Waltham, Mass. 1946.

8. Sibilia, C. "La forma Ecidica Della Ruggine Bruna Delle Foglie di Grano *Puccinia recondita* Rob. ex. Desm. in Italia." *Boll. Stn. Patol. Veg. Rome* 18 (1960): 1.

9. de'Oliveira, B., and Samborski, D. J. "Aecial Stage of *Puccinia recondita* on Rannunculaceae and Boraginaceae in Portugal." *Proc. Cereal Rusts Conf.* 66 (1966): 133.

10. Christensen, J. J. "Long Term Dissemination of Plant Pathogens." in *Aerobiology*, No.17, ed. F. R. Moullon (Washington, D.C.: America Association for the Advancement of Science, 1942), 78.

11. Stepanov, K. "Dissemination of Infective Diseases of Plants by Air Currents." *Lenin Acad. Agric. Sci., SII* 8 (1935): 7.

12. Stakman, E. C., Henry, A. W., Curran, G. C., and Christopher, W. N. "Spores in the Upper Air." *J. Agric. Res.*, 24 (1923): 599.

13. Nagarajan, S., and Joshi, L. M. "Presence of Wheat Rust Urediospores over Rohtang Pass (3,954 m) in the Interior Himalayas." *Cereal Rusts Bull.* 3 (1975): 35.

14. Rowell, J. B., and Romig, R. W. "Detection of Urediospores of Wheat Rusts in Spring Rains." *Phytopathology* 56 (1966): 807.

15. Roelfs, A. P., Rowell, J. B., and Romig, R. W. "Samples for Monitoring Cereal Rust Urediospores in Rain." *Phytopathology* 60 (1970): 187.

16. Nagarajan, S., Singh, H., Joshi, L. M., and Saari, E E. "Prediction of *Puccinia graminis* f. sp. *tritici* on Wheat in India by Trapping the Urediospores in Rain Samples." *Phytoparasitica* 5 (1977): 104.

17. McEwan, J. M. "The Source of Stem Rust Infecting New Zealand Wheat Crops." *N.Z.J. Agric. Res.* 9 (1969): 536.

18. Roelfs, A. P. "Gradients in Horizontal Dispersal of Cereal Rust Urediospores." *Phytopathology* 62 (1972): 70.

19. Briggle, L. W., Strauss, S. L., Hamilton, D. F., and Howse, G. H. "Distribution of the Varieties and Classes of Wheat in the United States in 1979." *U.S. Agric. Res. Serv., Stat. Bull.* 676 (1982): 1.

20. Mehta, K. C. "Further Studies on Cereal Rusts in India," Part II. *Sci. Monogr. No. 18.* (Delhi: Indian Council Agric. Res., 1952).

21. Nagarajan, S., and Singh, H. "The Indian Stem Rust Rules—A Concept on the Spread of Wheat Stem Rust." *Plant Dis. Rep,* 59 (1975): 133.

22. Nagarajan, S., and Singh, H. "Satellite Television Cloud Photography as a Possible Tool to Forecast Plant Disease Spread." *Curr. Sci.* 42 (1973): 273.

23. Nagarajan, S., and Joshi, L. M. "Further Investigations on Predicting Wheat Rusts Appearance in Central and Peninsular India." *Phytopathol. Z.* 98 (1980): 84.

24. Nagarajan, S., and Singh, H. "Preliminary Studies on Forecasting Wheat Stem Rust Appearance." *Agric. Meteorol.* 17 (1976): 281.

25. Nagarajan, S. "Wheat Rust Warning Using Satellite Imagery." *Proc. Int. Conf. on Ecology.* (Netherlands: The Hague, 1974).

26. Nagarajan, S., Seidboldt, G., Kranz, J., and Saari, E. E. "Utility of Weather Satellites in Monitoring Cereal Rust Epidemics." *Z. Pfl. Krank.* 39 (1980): 296.

27. Nagarajan, S., and Ajai, "Monitoring and Mapping Long Distance Spread of Plant Pathogens," in *Techniques in Plant Disease Epidemiology*, eds. J. Kranz and J. Rotem (Berlin: Berlag, 1986).

28. Roelfs, A. P. "Gradients in the Horizontal Dispersal of Cereal Rust Urediospores." *Phytopathology* 62 (1972): 70.

29. Burleigh, J. R., Roelfs, A. P., and Eversmeyer, M. G. "Estimating Damage to Wheat Caused by *Puccinia recondita tritici*." *Phytopathology* 62 (1972): 944.

30. Burleigh, J. R., Eversmeyer, M. G., and Roelfs, A. P. "Development of Linear Equations for Predicting Wheat Leaf Rust." *Phytopathology* 62 (1972): 947.

31. Nagarajan, S., and Joshi, L. M. "A Linear Model for a Seven-Day Forecast of Stem Rust Severity." *Indian Phytopathol.* 31 (1978): 500.

32. Karki, C. B., Pande, S., Thombre, S. B., Joshi, L. M., and Nagarajan, S. "Evaluation of a Linear Model to Predict Stem Rust Severity." *Cereal Rusts Bull.* 7 (1979): 3.

33. Eversmeyer, M. G., and Burleigh, J. R. "A Method of Predicting Epidemic Development of Wheat Leaf Rust." *Phytopathology* 60 (1970): 805.

34. Coakley, S. M., and Line, R. F. "Quantitative Relationships between Climatic Variables and Stripe Rust Epidemics on Winter Wheat." *Phytopathology* 71 (1981): 461.

35. Zadoks, J. C. "EPIPRE, a Computer Based Scheme for Pest and Disease Control in Wheat," in *Cereal Production*, ed. R. J. Gallaghev (London: Butterworths, 1984), 354.

36. Stakman, E. C., and Piemeisel, F. J. "Biological Forms of *Puccinia graminis* on Cereals and Grasses." *J. Agr. Res.* 10 (1917): 429.

37. Stakman, E. C., and Levine, M. N. "The Determination of Biologic Forms of *Puccinia graminis* on *Triticum* species." *Minn. Agri. Expt., Tech. Bull.* 8 (1922): 10.

38. Johnston, C. O., and Mains, E. B. "Studies on Physiological Specialization in *Puccinia triticina.*" *USDA Tech. Bull.* 313 (1932).

39. Gassner, G. and Straib, W. "Zur Frage der Konstanz des Infektions Typus von *Puccinia triticina* Eriks." *Phytopathol. Z.* 4 (1932): 57.

40. Flor, H. H. "Inheritance of Pathogenicity of *Melampsora lini.*" *Phytopathology* 32 (1942): 653.

41. Watson, I. A., and Luig, N. H. "The Classification of *Puccinia graminis* var. *tritici* in Relation to Breeding Resistant Varieties." *Proc. Linn. Soc. N.S.W.* 880 (1963): 235.

42. Johnson, R., Stubbs, R. W., Fuchs, E., and Chamberlain, N. H. "Nomenclature for Physiologic Races of *Puccinia striiformis* Infecting Wheat." *Trans. Brit. Mycol. Soc.* 58 (1972): 475.

43. Roelfs, A. P., and McVey, D. V. "Races of *Puccinia graminis* f. sp. *tritici* in the USA during 1973." *Plant Dis. Rep.* 58 (1974): 608.

44. Nagarajan, S., Nayar, S. K., Bahadur, P. "The Proposed Brown Rust of Wheat (*Puccinia recondita* f. sp. *tritici*) Virulence Monitoring System." *Res. Bull. No. 1, IARI*, R.S. Simla, India, 1981.

45. Nagarajan, S., Nayar, S. K., Bahadur, P. and Kumar, J. "Wheat Pathology and Wheat Improvement." *IARI, Reg. Stn., Flowerdale, Simla, India*, 12, 1986.

46. Nagarajan, S., and Joshi, L. M. "Epidemiology in the Indian Subcontinent," in *The Cereal Rusts* vol. II, eds. A. P. Roelfs and W. R. Bushnell (New York: Academic Press, 1985), 371.

47. Roelfs, A. P. "Effects of Barberry Eradication on Stem Rust in the United States." *Plant Dis.* 66 (1982): 177.

48. Vanderplank, J. E. *Plant Diseases: Epidemics and Control.* New York: Acad. Press, 1963, p. 349.

49. Mundt, C. C., and Browning, J. A. "Development of Crown Rust Epidemics in Genetically Diverse Oat Populations: Effect of Genotype Unit Area." *Phytopathology* 75 (1985): 607.

50. Parlevliet, J. E. "Components of Resistance that Reduce the Rate of Epidemic Development." *Ann. Rev. Phytopathol.* 17 (1979): 203.

51. Lamberti, F., Maller, J. M., and Vander Graaf, N. A., eds. *Durable Resistance in Crops.* New York: Plenum, 1983.

52. Hassebrauk, K., "Pilzliche Parasiten der Getreideroste II. Mitteilung." *Phytopathol. Z.* 10 (1937): 464.

53. Forrer, H. R. "Der Einfluss von stoffwechselproducten des Mycoparasiten *Aphanocladium album* auf die Teleutosporenbildung von Rostpilzen," *Phytopathol. Z.* 88 (1977) 306.

54. Forrer, H. R. "Der Einfluss Von Stoffwech-selprodukten des Mycoparasiten *Aphanocladium album* auf die sporenbildung von Getreiderostpilzen." (Dissertation, ETH, Zurich, 1977).

55. Rowell, J. B. "Control of Leaf Rust in Spring Wheat by Seed Treatment, with 4-*n*-butyl-1,2,4-triazole." *Phytopathology* 66 (1976): 1129.

56. Rowell, J. B. "Evaluation of Chemicals for Rust Control." in *The Cereal Rusts* Vol. II, eds. A. P. Roelfs and W. R. Bushnell. (New York: Academic Press) 1985 p. 561.

57. Rathwell, W. G., and Skidmore, A. M. "Recent Advances in the Chemical Control of Cereal Rust Diseases." *Outlook Agric.* 11 (1982): 37.

3

BARLEY YELLOW DWARF

R. T. PLUMB

Plant Pathology Department
AFRC Institute of Arable Crops Research, Rothamsted Experimental Station,
Harpenden, Herts, AL5 2JQ, UK.

3-1 INTRODUCTION

Although there is no supporting evidence, it seems likely that the viruses that cause the disease now known as barley yellow dwarf (BYD) have been present for as long as Gramineae has been an extensive plant family and aphids have fed in the phloem of different members of the family. Indeed, the complex interactions between aphids and the BYD virus suggest that they have coevolved into a mutually beneficial association.[1] As knowledge of the causal agents of disease has increased, so has the ability to associate specific agents with specific diseases. Therefore, although descriptions such as red leaf disease of oats[2-5] are probably the consequence of infection by what we now know as BYD virus, there can be no certainty of this. The recognition of the virus etiology of red leaf did not immediately eliminate the use of the earlier name.[6-7] Bruehl[7] summarized reports describing symptoms that were, in retrospect, manifestations of what we would now call barley yellow dwarf.

The first recognition of BYD as an aphid transmitted, virus-induced disease was by Oswald & Houston,[8] who realized that a yellowing condition of barley first observed over a wide area of California and accompanied by moderate to severe stunting probably had the same cause as a stunting and chlorosis of wheat and a reddening of oats that appeared a week later. As no doubt had often happened in the past, the immediate explanation for the cause was "the result of some set of peculiar environmental conditions." However, evidence accumulated that this was not the case and attempts to associate the symptom with a fungus root rot were unsuccessful. Aphids—especially the greenbug, *Toxoptera*

(Schizaphis) graminum—were also implicated as the cause but the lack of coincidence of symptoms with heavy aphid infestations and the failure to collect greenbug in the affected area led to the conclusion that the aphids must be acting as virus vectors. Subsequent transmission experiments confirmed this.

Although barley yellow dwarf virus (BYDV) is now the accepted name of the agent that causes the disease BYD, several alternative names, in addition to oat red leaf, have been suggested, notably cereal yellow dwarf and cereal yellow dwarf virus,[9-13] and *Hordeumvirus nanescens*.[14] Since this early work, knowledge of the causal agent and techniques for virus diagnosis, identification, and characterization have improved greatly. Consequently, there is clear evidence for at least five distinct isolates of BYDV, which can be divided into two distinct groups. The importance of these distinctions and whether there is a need to identify the two virus groups as distinct viruses is discussed below. For simplicity, and often because no reference had been made to a particular isolate, the terms BYD and BYDV will be used when describing the disease and its causal agents except where an isolate is specified and it is relevant to the discussion.

3-2 DISTRIBUTION

It is interesting, but potentially misleading, to look at the distribution of BYD as given by successive Commonwealth Mycological Institute distribution maps of plant diseases, brought up to date by the author (Figure 3-1). Although there may have been a dramatic spread in the incidence of the disease, the large increase in the area in which infection is

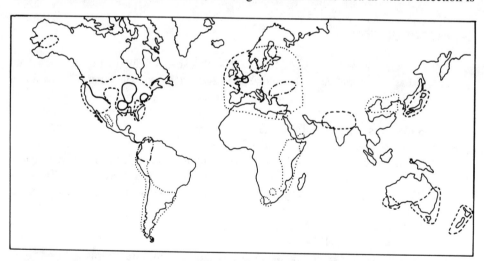

New areas found to be infected with BYDV

———	**1956**
- - - - -	**1963**
–·–·–	**1969**
············	**1987**

Figure 3-1 Map of the distribution of barley yellow dwarf virus in 1956, 1963, 1969, and 1987.

known to occur is probably principally due to improved methods of diagnosis. However, the extension of cereal cultivation, especially of wheat, into new areas, and the intensification of cereal growing, may have revealed the presence of the virus, which was already widely distributed but inapparent in local grasses. Thus, five years after the first recognition of the virus etiology of BYD, infection had been reported only from the United States, Canada, and the Netherlands. By 1963, infection had been reported widely in the United States, Canada, and Mexico, in most countries of western Europe, in Australia, New Zealand, Jordan, Egypt, India, Pakistan, and Japan. Further reports in the late 1960s identified BYDV in Colombia, the first report from South America, and parts of the USSR. At CIMMYT sponsored meetings in 1983[15] and 1987[16] it was clear that there were few places where cereals were grown that BYDV had not been reported. Infection has been confirmed in Ecuador, Peru, Bolivia, Brazil, Chile, Paraguay, and Uruguay, in South America; Spain, Portugal, Hungary, German Democratic Republic, Poland, and Greece, in Europe; Syria, Iran, Turkey, Israel, and Lebanon, in the Middle East; Morocco, Tunisia, Libya, Ethiopia, Kenya, Tanzania, Burundi, Mozambique, Republic of South Africa, and Zimbabwe, in Africa; and in the People's Republic of China.

Most of these national reports are of infection in grain crops in which symptoms of infection are usually most obvious and damage most readily quantifiable. Infection of agricultural or wild species of grass is rarely obvious and there seems little doubt that infection is also widespread in them, often in areas where cereals are not cultivated.

3-3 HOST RANGE

Wheat (*Triticum aestivum, T. durum*), barley (*Hordeum vulgare*), oats (*Avena sativa*), maize (*Zea mays*), rice (*Oryza sativa*), rye (*Secale cereale*), and sorghum (*Sorghum vulgare*) are all susceptible to BYDV infection and suffer different degrees of damage, depending on interactions between isolate of virus and host cultivar. The agriculturally important grasses in the genera *Lolium, Poa, Bromus, Festuca, Phleum,* and *Cynodon* and also susceptible. Bruehl[7] summarizes host range data, and since then the literature on susceptible host is voluminous. Bruehl[7] also records 20–30 grass species in which no symptoms were produced and no virus recovered by aphid feeding. However, there has been no systematic testing of a wide host range using modern, sensitive, diagnostic methods. Although it could be argued that hosts of BYDV are so widespread that there is a ubiquitous reservoir in most regions where Gramineae grow, Watson and Gibbs[17] showed that a greater proportion of festucoid than nonfestucoid grasses were susceptible to infection, and also pointed out that festucoid grasses were usually over-represented in host range studies as they predominate in north temperate regions where most plant virologists work. However, tests in controlled conditions in the glasshouse or laboratory do not necessarily reflect field occurrence. In transmission tests aphids are usually confined on hosts on which, out of sheer necessity, they may then feed and transmit BYDV. In crops or wild species feeding preferences of the appropriate aphid vectors may have a great influence on whether or not a species is infected.[18]

3-4 PROPERTIES OF BYDVs

The identification of a virus as the cause of BYD and its distribution and host range outlined above was largely based on the assumption that the causal agent is a single entity. However, after the virus etiology was established, studies quickly demonstrated great variability. Isolates of BYDV were found that differed in the relative damage they caused to small grain cereals,[19] in their host range and the symptoms they caused,[10,20] and especially their efficiency of transmission by different aphid species.[21-23] Subsequently studies, notably by Rochow,[24-26] clarified these differences and described five variants or isolates (RPV, RMV, MAV, PAV, and SGV) that were initially classified by their biological properties[26] and have subsequently been shown to be serologically distinct,[27-31] and to differ in the number and electrophoretic mobility of double-stranded RNAs detectable in infected plants.[32] Most recently, nucleic acid probes have been used to detect isolates of BYDV and appear to recognize the same isolates as serological methods and also support the separation of BYDV isolates into two groups.[33-34] Ultrastructural studies of infected plants showed differences between isolates[35] and barley cultivars containing the Yd_2 gene,[36] which confers tolerance to isolates of BYDV in one group were severely damaged by those in the other.[37] Therefore, the case for considering the two groups as distinct viruses seems overwhelming.[38] Nevertheless, the long accepted use of the term BYDV to cover all isolates causing yellow dwarf and the current practice of distinguishing strains when referring to them suggests that no radical deviation from current nomenclature is required. However, recent results by Martin and D'Arcy,[39] who compared a range of luteoviruses by serology and nucleic acid hybridization, suggests that the RPV isolate of BYDV and the beet western yellows virus should be considered strains of one virus. The grouping suggested by Rochow and Carmichael[40] has been substantiated by all subsequent tests, so it is suggested, in agreement with Rochow[41] and Gill and Chong,[35] that BYDV isolates that are MAV-, PAV-, and SGV-like are designated BYDV-1, and isolates that are RPV- and RMV-like, BYDV-2 (Table 3-1). The use of BYDV-MAV and BYDV-RPV as suggested by Waterhouse et al.[38] seems likely to cause more confusion than clarification.

TABLE 3-1:   STRAINS OF BYDV AND THEIR RELATEDNESS

Group 1	Group 2
BYDV 1	BYDV 2
MAV	RPV
PAV	RMV
SGV	

3-5 ISOLATE NOMENCLATURE

Four aphid vectors were used to distinguish the five isolates of BYDV reported from New York State: *Rhopalosiphum padi*, *Macrosiphum* (now *Sitobion*) *avenae*, *R. maidis*, and *Schizaphis graminum*. The basis for the distinction was the relative efficiency of transmission of the five isolates by these vectors (Table 3-2), and the designation of strains by the initial letters of their most efficient vector or vector(s) is clear.

TABLE 3-2: EFFICIENCY OF TRANSMISSION (% INFECTION) OF STRAINS OF BYDV BY DIFFERENT APHID VECTORS[a]

	RP[b]	MA	RM	SG
PAV[c]	100	77	3	30
RPV	100	0	0	40
MAV	3	100	0	1
RMV	11	6	90	6
PAV	100	72	0	31
RPV	100	0	0	53
MAV	6	97	0	3
RMV	14	0	97	14
SGV	6	0	0	75

Source: Adapted from Rochow[25] and Johnson & Rochow.[189]

[a]10 aphids/plant, 109–111 test plants.

[b]RP = *Rhopalosiphum padi*; MA = *Macrosiphum* (now *Sitobion*) *avenae*; RM = *Rhopalosiphum maidis*; SG = *Schizaphis graminum*.

[c]Mean of two PAV isolates.

However, in testing virus isolates by this method, care should be taken to avoid inadvertent bias. Clones or biotypes of aphids may differ in their efficiency of transmission,[42–43] temperature may affect differently the ability of aphid species to acquire and transmit BYDV,[25] and the host preference of vectors may affect transmission. In the UK, *R. maidis* reproduces well on barley but feeds only reluctantly on other cereals. In Canada, Gill[43] put BYDV isolates in three groups based on vector efficiency and included *Metopolophium dirhodum* as a test species. In the UK, three isolates serologically closely related to the MAV, RPV, and PAV isolates of Rochow were also tested with five aphid species (Table 3-3). The coincidence of these results with those of Rochow, even where close serological relatedness has been demonstrated, is not very good, especially as the results in Table 3-2 are based on transmission rates given by 10 aphids/plant and those in Table 3-3 by single aphids. It is possible that the Group 2 isolates from Canada are mixtures, but it is impossible to confirm this. Despite these differences the biological

TABLE 3-3: EFFICIENCY[a] OF TRANSMISSION OF BYDV ISOLATES FROM CANADA AND UNITED KINGDOM

Isolate	Rochow type[c]	Vectors[e]					
		RP	MA	RM	SG	MD	MF
Group 1[b]	MAV	1	86	2	1	6	NT
Group 2	PAV??	28	33	52	36	30	NT
Group 3	RMV??	0	0	33	13	0	NT
G[d]	PAV	72	20	16	NT	26	2
F	MAV	9	55	2	NT	38	20
R 568	RPV	66	0	13	NT	21	9

[a]Single aphid transmissions.

[b]Groups 1, 2, and 3 (Canada, Gill[43]) mean transmission from 9, 5, and 2 isolates respectively.

[c]Based on transmission for Canadian isolates and serology for UK isolates.

[d]Isolates from UK

[e]RP = *Rhopalosiphum padi*; MA = *Macrosiphum (Sitobion) avenae*; RM = *R. maidis*; SG = *Schizaphis graminum*. MD = *Metopolophium dirhodum*; MF = *M. festucae*.

characteristics are recognizably similar, but clearly serological characterization must be complemented by biological tests.

Thus, although the five biologically well defined isolates fit well into two distinct groups, it is far from certain that the full variability of BYDV is yet known. The difference in vector transmission is one example of variation that is of great epidemiological importance but is not apparently always associated with serological differences.[44] Perhaps nucleotide sequencing will demonstrate why such differences exist and may result in the production of probes to isolates of much more narrowly defined properties than can currently be distinguished. In the People's Republic of China several isolates have been described and, following Rochow's system, designated by their principal vectors.[45–47] Isolates could be divided into five groups.[46] The largest single group appeared by vector transmission and serology to be MAV-like isolates, others were like RPV and RMV, whereas a fourth group was similar but apparently not identical to PAV. The fifth and largest group reacted to none of the antisera used, which may have been because they were not infected but equally they may have been infected with serologically distinct strains. Transmission results with isolates from other regions of China were not compared serologically[45] but isolates appeared similar to MAV and RPV. However, others were difficult to relate to known BYDV isolates, and some may have been mixtures. Nevertheless, a greater range of isolates of BYDV may be present in China than is currently known from elsewhere.

3-6 APHID VECTORS

BYDV is only transmitted by aphids. There are reports of transmission through seed,[48–49] which have not been substantiated, and frit fly (*Oscinella frit*) has been claimed as a vector[50] but this also requires supporting evidence. Approximately 20 aphid species have been shown to be vectors[51–52] but nowhere do they all occur together and usually three or four species predominate in any one region. Those occurring most frequently are *Rhopalosiphum maidis, R. padi, Sitobion avenae, Schizaphis graminum,* and *Metopolophium dirhodum,* and in Australasia, *Sitobion miscanthi. S. avenae* and *M. dirhodum* are common in the cooler temperate regions or at higher elevations in lower latitudes, although *S. avenae* has extended its range and is now common in maritime and continental climates, whereas *R. maidis* and *S. graminum* are more common in warm regions or areas with a continental climate. By contrast, *R. padi* is very widespread in all cereal-growing regions.[53] The discrepancy between the species reported as vectors may be because of misidentification or confusion among virologists over changes of name required by aphid taxonomists. The only difference between the lists of vectors given by Jedlinski[52] and A'Brook[51] are the vectors given by Pei and Hsu[54] from China for millet red leaf virus, a synonym of BYDV. For detailed references and synonyms the reader is referred to A'Brook[51] and Jedlinski[52] but for completeness the species given as vectors by A'Brook[51] are listed in Table 3-4 with the additional report of transmission in South Africa[55] by *Diuraphis noxia,* the Russian wheat aphid. Vectors such as *Aulacorthum circumflexum* and *Myzus persicae* are not normally cereal feeders and are of no importance in spreading BYDV in crops.

TABLE 3-4: APHID SPECIES REPORTED AS VECTORS OF BYDV (FROM A' BROOK[51])

Anoecia corni (Fabricus)
Aphis glycines (Matsumura)
Aulacorthum (Neomyzus) circumflexum (Buckton)
Cavariella salicicola (Matsumura)
Ceruraphis eriophori (Walker)
Diuraphis noxia (Mordwilko)
Hyalopterus pruni (Geoffroy)
Melanaphis saccahari (Zehntner)
Metopolophium albidum (Hille Ris Lambers)
M. dirhodum (Walker)
M. festucae (Theobald)
M. frisicum (Hille Ris Lambers)
Myzus persicae (Sulzer)
Rhopalosiphum insertum (Walker)
R. maidis (Fitch)
R. padi (Linnaeus)
R. poae (Gillette)
R. rufiabdominalis (Sasaki)
Schizaphis graminum (Rondani)
Sipha elegans (del Guercio)
Sitobion avenae (Fabricius)
S. fragariae (Walker)
S. miscanthi (Takahashi)
Tuberocephalus momonis (Matsumura)

3-7 TRANSMISSION

BYDV isolates are transmitted in the persistent or circulative manner and acquisition and infection feeding periods of 48 hr or more are required for maximum rates of transmission. Once acquired, virus is retained for many days, often the rest of the vector's life. BYDV is largely confined to the phloem of infected plants and thus the aphids that transmit BYDV must feed to acquire virus and usually multiply on susceptible hosts. Virus is taken into the alimentary canal from where it is transported through the epithelial cells of the hindgut to the hemocoel.[56] Virus is also eliminated from the aphid in honeydew. Once in the hemocoel, virus particles are then carried in the hemolymph which bathes all the aphid's organs. For the virus to be transmitted it must reach the accessory salivary gland and be expelled, when the aphid next feeds, into the phloem of a virus-free plant through the salivary duct. The process of transport from hindgut to hemocoel is thought to be by endocytosis, and a model for the process has been developed by Gildow.[57] This envisages several stages: (1) virus attachment to the cell apical plasmalemma, (2) endocytosis of the particle into a coated pit which may bud off to give a coated vesicle and may fuse with others, (3) transport to the basal plasmalemma with which the vesicle fuses, releasing the virus particle to the hemocoel. The transport of BYDV particles from the hemocoel into the salivary gland is also described by Gildow[57,58] and is believed to involve diffusion through the basal lamina of the accessory gland and specific binding to the membrane. Virus may then be transferred to the salivary gland lumen via tubular vesicles produced by membrane invaginations. The specific binding to the accessory gland membrane is probably the site of vector specificity,[1,59] mediated through an interaction between the virus capsid protein and the accessory gland membrane. Such a mechanism is consistent with the transcapsidation theory of the transmission of BYDV isolates by normally poor vectors[40] and provides an explanation for the interactions observed between MAV and PAV isolates in transmission experiments.[60]

Despite this close association between BYDV isolates and their vectors there is no evidence that BYDV multiplies inside aphids.[61] Transmission efficiency is related to the length of acquistion feeds[61-62] and although virus is retained when nymphs molt, and the ability to infect plants can be retained throughout life from a single acquisition feed,[20] frequency of transmission generally declines as the interval between acquisition and infection feeds increases.

3-8 MIXED INFECTIONS AND DEPENDENT TRANSMISSION

Aphid transmission can be a reliable diagnostic tool but can also be misleading as hosts can be simultaneously infected by more than one isolate. When this happens the two infecting isolates are often but not always from different BYDV groups and, by a process described as heterologous encapsidation or more specifically transcapsidation,[63] in which the nucleic acid of one isolate becomes enclosed by the protein of the other, transmission can occur by aphids that either never or rarely transmit. The phenomenon, described in detail by Rochow,[63] has been studied most for the isolates MAV and RPV and their transmission by *R. padi*. *R. padi* rarely transmits MAV from plants infected with this isolate alone (Table 3-2), but does so frequently from plants infected mixedly by MAV and RPV isolates. However, the phenomenon does not work in reverse; *M. (S). avenae* does not transmit RPV from mixed RPV and MAV infections. An RMV isolate can also function as does RPV,[64] and RPV and RMV can interact together. The most important criterion for transcapsidation is the simultaneous replication of the two viruses. Thus, it only results from dual infection of a susceptible plant host, but such infections are not uncommon in field crops.

3-9 PROPERTIES OF BYDV PARTICLES AND RELATIONSHIPS TO OTHER VIRUSES

BYDV is a member of the luteovirus group and the MAV isolate is the type member.[65] As serological techniques are applied to more members of the luteovirus group it has become apparent that most are serologically related.[39, 66-69] The different BYDV isolates not only have different degrees of relationship with each other but also with other luteoviruses; most recent serological and nucleic acid relationships are given in D'Arcy et al.[68] and Martin and D'Arcy.[39]

All isolates of BYDV have many common properties. The virus particles are isometric, about 25–30 nm in diameter; the measured diameter differs with different methods of preparation for electron microscopy. Each particle has one coat polypeptide of approximately 24×10^3 daltons molecular weight and contains one molecule of positive sense ssRNA with a molecular weight of approximately $1.85-2.0 \times 10^6$ daltons.[70] Particles have a sedimentation coefficient (s_{20w}) of 115–127s.

Preparations of dsRNA from oats cv. Coast Black infected with different isolates of BYDV showed that five dsRNAs were present, designated ds 1–5, ranging in size from $0.5–3.8 \times 10^6$ daltons.[32] Two major and two or three minor classes of dsRNA were isolated from plants infected with each isolate. Isolates MAV, PAV, and SGV gave all five classes, whereas isolates RPV and RMV were associated with only four bands, thus supporting the separation of isolates into two groups. The largest, dsRNA -1, had a molecular weight of $3.6–3.8 \times 10^6$ daltons, approximately double the value for genomic RNA and is presumed to be the replicative form. The smaller dsRNAs 2–5, with molecular weights of $2.00–0.5 \times 10^6$ daltons are thought to represent subgenomic RNA. Although there were differences in mobility between dsRNAs 1 and 2 from isolates of the two groups, it is uncertain whether these are real differences; much more consistent is the absence of dsRNA-5 from plants infected by group 2 isolates.

When cDNA clones were prepared from the RNA of RPV and PAV it was found that some regions of the genome lacked any homology[33] and probes could discriminate between them. Clones have been prepared to an Australian PAV isolate[71–72] and to MAV isolates from the United States[34] and from Europe.[73] The RNA of a PAV serotype from Australia has been completely sequenced, apart from the 5′-terminal base.[74] Five open reading frames have been recognised and the genes for the coat protein[75] and the RNA-dependent RNA polymerase have been identified.

3-10 DETECTION AND DIAGNOSIS

3-10-1 Symptoms

The most obvious manifestation of BYDV infection, but often not diagnostic, is the symptoms caused by BYDV. Nevertheless, in many regions this is the only diagnostic method available, sometimes supported by the reproduction of symptoms in virus-free test plants by aphid transmission. However, symptoms produced in field crops and those on test plants in a glasshouse are often very different.

There are also differences in the test plants used which are usually chosen from cultivars known to give consistantly clear symptoms. In North America, oat cvs Clintland or Clintland 64 and barley cv. Black Hulless are the usual test plants, whereas in Europe oat cvs Blenda or Maris Tabard are often the preferred choice. None of these cultivars have been grown commercially for many years and supplies are usually multiplied by those needing them for testing.

On field-grown oats, wheat, and barley two characteristic symptoms are usually, but not always, present on infected plants: (1) discoloration, and (2) stunting (Figures 3-2–3-4). The discoloration, which starts at the tip and spreads down the margins of leaves, ranges from yellow, typically in barley but also shown in wheat, to bright reddish purple, usually seen in oats but often mixed with yellow in wheat. Variants include diffuse light and dark green stripes which are sometimes seen on winter barley. Discoloration is often but not always accompanied by stunting and, in extreme cases, plants can be killed. Stunting is usually greatest in barley and oats and least in wheat. However, the expression of symptoms depends on many variables, the isolate of virus, the age of the plant when infected, the host cultivar, and weather conditions, especially light and temperature.

Figure 3-2   Symptoms of barley yellow dwarf in barley.

Figure 3-3   Symptoms of barley yellow dwarf in wheat.

Figure 3-4 Symptoms of barley yellow dwarf in oats.

Thus, in winter-sown crops, especially wheat, infection can occur soon after emergence but symptoms of infection do not appear until six months later in the following spring. This can lead to difficulties in determining when plants are infected. A further characteristic of infection seen in all small grain cereals but most obviously in wheat is the upright leaf habit. Because symptoms usually take longer to develop in wheat than other cereals, the flag leaf is often the first leaf to show discoloration. Virus infection blocks translocation and the leaves accumulate carbohydrate; consequently they become discolored, stiff, and upright rather than bending over as in virus-free plants. As a result, in infected wheat a spiky appearance often accompanies discolored leaves. A further symptom on oat ears is blasting, in which distinctive blind florets are often numerous. Irrespective of the type of symptom their distribution is often characteristic, frequently in roughly circular patches associated with infection by a single aphid vector and secondary spread from that source.

Despite these difficulties symptoms have been widely used to express severity of BYD. For barley, Schaller, Rasmusson, and Qualset[76] devised a 0–4 scale, and Schaller and Qualset[77] developed an expanded scale of 0–9 for use with wheat and oats. On this scale, discoloration, dwarfing, vigor, floret sterility, and early maturity are combined to define the different ratings.

On other susceptible hosts symptoms are usually even less definitive than on small grain cereals. On infected maize the margins of leaves are often a dark red[78] but in some cultivars flecking and striping are often produced, sometimes much sooner than leaf discoloration.[79] On grasses there is a range of discoloration, but usually seen only at a late stage of maturity. Consequently, in grazed swards, or those cut frequently, symptoms are rare; on crops left for hay or in wild grasses symptoms are more often apparent.

In glasshouse grown test plants some of the field symptoms also appear but, depending on conditions, different symptoms are often present. In oats, for example, water soaking is often characteristic and leaves become brown. Leaf distortions and serrations of the edge are also more frequently seen in glasshouse plants than crops, and for some virus isolates, the tip of the leaf that is growing most rapidly when infection occurs fails to expand properly and the top 3–4 cm quickly fall off. In these "strap leaves" it appears that the parenchymatous tissue has failed to develop. It is only in sap extracts from these leaves that the author has unequivocally seen particles in the electron microscope by conventional leaf dip preparation, presumably because the ratio of virus-containing phloem to virus-free parenchyma was large in these abnormal leaves.

With experience, symptoms in field crops can be useful for diagnosis, and in the glasshouse where conditions are more controlled, and other pests and diseases excluded, diagnosis is reliable. However, no observer can distinguish isolates by symptoms, although very stunted and dwarfed plants of barley and oats are most likely to be caused by PAV-like isolates. In the glasshouse, temperature but especially light changes through the year. In low light conditions, discoloration caused by the least damaging isolates is often difficult to see or absent. At temperatures above 25°C symptoms are also less obvious than between 15–20°C. Ideally plants should be grown in controlled environments with high light intensity and at a temperature between 15–20°C. In these conditions virus symptoms become obvious, either as discoloration or stunting, in relation to uninoculated control plants, in 10–21 days.

3-10-2 Internal Changes

As BYDV is largely confined to the phloem it is in this tissue that most disruption of tissues is seen,[80] with degeneration and, in severe cases, necrosis. Esau[81] also reported gum deposits in the xylem of BYDV-infected oats and barley, and the accumulation of a sugary exudate in the mesophyll of oats, which sometimes exuded onto the leaf surface. The thickness of leaves was also increased, contributing to their stiff, upright appearance, and McKinney, Specht, and Stanton[82] examined the distribution of the pigments that give the distinct color characteristics of infected plants.

Clearly, if phloem function is disrupted carbohydrates will not be translocated so efficiently, and Orlob[83] found more carbohydrates, especially starch and reducing sugars, in the leaves; in the roots total nitrogen was increased.[84] Jensen[85] described physiological factors affected by BYDV infection and showed a 400% increase in soluble carbohydrate content in leaves, but not culms or ears.

Gill and Chong[86–87] compared the ultrastructural differences caused by infection with different BYDV isolates. They found differences between PAV, RPV, and MAV isolates that supported the separation of BYDV isolates into two groups. The site at which virus particles were first observed and the presence and accumulation of filaments were two distinctive features. With RPV, but not MAV or PAV infection, mitochondria were disrupted. However, such changes are determined by the time after infection that the material is examined and the combination of isolate and host, especially the presence in the barley host of the Yd_2 gene.[88]

3-10-3 Serology

To distinguish virus isolates less subjectively than by symptoms, serology and cDNA probes,[33,89] are necessary. Nevertheless, for more than 20 years the separation of BYDV isolates was based on biological characters, especially aphid transmission. In laboratories with access to reliable serological tests aphid transmission has largely been replaced by serology;[90] however, using the methods described by Rochow[26, 29] aphid transmission is still very useful. As shown in Tables 3-2 and 3-3 and by Lister and Sward[44] there can be substantial differences in vector transmission of serologically similar isolates. Therefore, supporting serological confirmation of isolate identity is always desirable.

The MAV isolate of BYDV was purified by Rochow and Brakke[91] and subsequently preparations of a PAV and an RPV isolate were produced,[92] but for all isolates yields of virus were very small. Subsequently, purification has become relatively routine although yields of virus are never very large; D'Arcy[93] has summarized the purification methods used. At first, because of the very low concentration of BYDV in plants, serological tests could be done only by using partially purified virus preparations and this was cumbersome. Early methods used to detect serological differences between the isolates included infectivity neutralization,[94-95] double-diffusion in agar, and latex agglutination.[96-97] It was not until the development of enzyme-linked systems such as ELISA (enzyme-linked immunosorbent assay) for plant viruses[98] that the rapid serological diagnosis of BYDV isolates from plant sap became possible.[26-27] These methods were simpler and more readily detected mixed infections than did vector transmission.[99] However, diagnosis of isolates is limited by the antisera that are available. In North America five antisera are required as all five isolates are present. However, in Europe it appears[100] that RMV and SGV isolates do not occur and only three antisera are required to detect BYDV isolates. However, there is less knowledge of BYDV isolates in southern Europe where *R. maidis* and *S. graminum* are more common than in the north and northwest where most work has been done.

All the serological methods described above have been largely based on polyclonal antisera usually produced in rabbits. Consequently, relatively large quantities of BYDV are required for immunization, and the low virus content of BYDV in plants means that to obtain sufficient purified virus it must often be accumulated. In addition there is a finite supply of antibodies, although this is often not too serious a problem, and there is the inherent variability of antisera produced from different animals. The production of monoclonal antibodies can overcome some of these problems and offers the potential of producing limitless quantities of a consistent quality. Monoclonal antibodies of BYDV isolates have been produced in the United States[29-30] and in the UK.[31,101] In our hands an indirect ELISA system works best with the initial coating with a polyclonal antiserum followed by a second monoclonal antibody to detect the virus isolate giving the best discrimination.

3-10-4 cDNA Probes

cDNA probes have been made to cloned sequences from BYDV isolates[33-34,73] and dot blot assays using ^{32}P-labeled probes readily detected BYDV. Cloned probes have some of the same advantages as monoclonal antibodies, in that only small amounts of purified virus are necessary for their production and once developed they can be propagated indefinitely. However, while detection relies on radioactive labeling, the use of probes will be restricted to laboratories in which the appropriate safety precautions can be taken. Nonradioactive probes, based on photobiotin, have been made[89] but as yet are not widely available and the prolonged procedure for their use makes them currently less acceptable than the ELISA system.

3-10-5 Immunospecific-Electron Microscopic (ISEM) Methods

The combination of serology and electron microcopy, first applied to plant viruses by Derrick,[102] was developed by use with BYDV by Paliwal[103] using polyclonal antibodies and by Forde[104] with monoclonals. Although it is unlikely to replace ELISA as a routine diagnostic method, it is especially useful for confirming borderline results. The appearance of particles is usually sufficiently distinctive to be conclusive (Figure 3-5).

3-10-6 Detecting BYDV in Aphids

Central to all studies of BYDV epidemiology is the role of aphid populations in the transmission and spread of the virus. Aphid populations and their infectivity have been incorporated in forecasting systems for BYDV[105-106] but the tests of infectivity have relied on transmission of BYDV to plants, supported by serological confirmation and identification of isolates. This process takes up to 3–4 weeks and although guidance can still be given to farmers in time for them to take appropriate action, a simpler, quicker system that detected virus directly in aphids would clearly be advantageous. The possible sero-

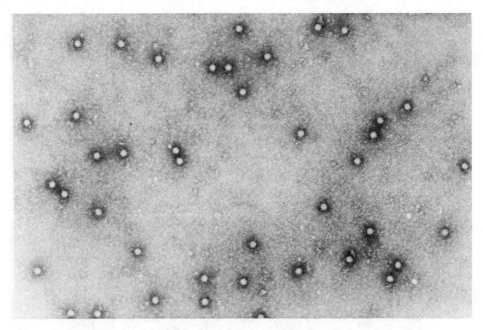

Figure 3-5 Particles of BYDV stained in phosphotungstic acid after trapping by antiserum on electron microscope grids. Magnification × 80,000.

logical methods available were reviewed by Harrison[107] but to these methods should be added the use of cDNA probes.[108] An ELISA with a fluorogenic substrate and polyclonal antisera was used by Torrance et al.[109] in the UK to compare transmission by one group of aphids, mainly *R. padi*, with the detection of BYDV in a comparable group. The method was sensitive enough to detect BYDV in single aphids, and generally a greater proportion of aphids tested by ELISA were found to contain virus than transmitted BYDV to test plants. However, the agreement was not consistent enough for the serological to replace the biological method of detecting infective vectors. One reason for this is the switch, in the UK, from parthenogenetic to sexual forms during the autumn—the most important period of migration. BYDV was detected in male and sexual female (gynoparae) *R. padi* that are biologically unlikely to transmit BYDV because their preferred host, *Prunus padus*, is not susceptible. This is especially a problem in southern Britain where *R. padi* is the dominant vector species in autumn. Earlier work by Denechere, Cante, and Lapierre[110] used a conventional ELISA to detect BYDV in a group of 80 aphids after a 48–60 hr acquisition feed. After 5 days acquisition BYDV could be detected in groups of as few as 5 aphids and it was estimated that virus concentration in each aphid was 1.4 ng or 2×10^6 particles.

Such direct methods of detection may be of great value in assessing virus in spring and summer populations that are entirely parthenogenetic, or where sexual forms of the aphids do not occur. Despite the encouraging results with fluorescent ELISA[109] the detecting equipment is expensive and a system of improving conventional chromogenic ELISA detection would be preferable. Torrance[111] has described an enzyme amplification system for increasing the sensitivity of detection of BYDV. This method not only increased the sensitivity but also decreased the time necessary for the assay to 2 hr. However, some results with polyclonal antisera and plant sap gave unacceptably high nonspecific values. Tests on individual parthenogenetic aphids detected BYDV in 80% compared with an expected transmission rate of 90%.

ISEM methods have also been used to detect BYDV in aphids.[112–113] The numbers of particles seen showed good agreement with virus transmission, and particle numbers declined if aphids were given no further access to an infected source. One problem associated with the detection of BYDV isolates in aphids is that to obtain enough extract to test for isolates, too great a dilution is necessary. Therefore, for practical purposes, only presence or absence can be determined, but as the aphid species is known little information would be lost if only the isolate(s) that the aphid transmits efficiently were included in the screen.

3-11 EFFECTS OF BYDV ISOLATES ON YIELD

With such a wide distribution, such a wide range of isolates of BYDV, and such different abilities to detect and characterize BYDV isolates in laboratories around the world, it is not surprising that there is rather little detailed evidence of the damage caused by BYDV in field crops. It is also difficult to disentangle the various components of the disease spectrum that infects most crops; consequently, much of the evidence for damage by BYDV given in the literature is potential damage, as measured in glasshouse tests or partially controlled environments, where other pathogens and pests can be largely excluded, or effectively controlled. Nevertheless, Pike[114] summarized much of the data that is available.

Isolates differ in the damage they cause and Rochow[25,91] has examined the characteristic effects of the five described isolates; SGV, RMV, and RPV were relatively weakly virulent, i. e., causing relatively little damage; MAV was moderately virulent, and PAV, strongly virulent. Results from Europe for RPV, MAV, and PAV in glasshouse transmissions would generally support these results but in inoculation tests with field-grown winter barley all isolates caused substantial yield loss when infection was in the autumn, although only the PAV isolate killed plants; however, when inoculated in spring, ranking of damage was as described by Rochow and as observed in glasshouse tests (Table 3-5). However, inoculating spring-sown cultivars before tillering with the same three isolates (Table 3-6) caused much more damage than to the winter barley inoculated at approxi-

TABLE 3-5:   THE EFFECTS OF PAV-, RPV-, AND MAV-LIKE ISOLATES OF BYDV ON WINTER BARLEY CV. IGRI INOCULATED IN THE AUTUMN OR SPRING

Inoculation time and isolate	Yield (g/plant)	1000 grain wt (g)	Height (cm)	Shoot Number
Autumn				
PAV	0.00	0.00	0.00	0.00
MAV	5.87e[a]	31.08c	44.8c	17.8a
RPV	1.31f	24.61d	49.0c	7.1e
Spring				
PAV	7.66d	36.30b	72.2b	8.6d
MAV	8.37c	39.06b	67.6b	11.8c
RPV	10.95b	38.94b	88.6a	13.1bc
Uninoculated	14.41a	48.47a	93.0a	13.5b

Source: Data from G. M. Herrera.[37]
[a]Values in each column followed by different letters are significantly different at $P = 0.005$ as determined by the Student-Newman test.

TABLE 3-6:   THE EFFECTS OF PAV-, MAV-, AND RPV-LIKE ISOLATES OF BYDV ON SPRING BARLEYS CVS ATLAS 68 (+ YD$_2$) AND ATLAS 57 (− YD$_2$)

Cultivar	Isolate	Yield (g/plant)	1000 grain wt (g)	Height (cm)	Shoot Number
Atlas 68					
	PAV	19.13b[a]	44.19b	108.9a	13.4b
	MAV	36.93a	49.07ab	118.6a	18.9ab
	RPV	16.37b	40.62c	106.6a	12.2b
Uninoculated		39.37a	52.11a	115.9a	19.4a
Atlas 57					
	PAV	0.00	0.00	0.0	0.0
	MAV	11.21b	45.83b	106.4a	6.7b
	RPV	13.16a	49.47a	122.9a	12.2b
Uninoculated		40.16a	49.47a	122.9a	22.4a

Source: Data from G. M. Herrera.[37]
[a]Values in each column followed by different letters are significantly different at $P = 0.005$ as determined by the Student-Newman test.

mately the same time but a more advanced growth stage (Table 3-5).

Although these results approximate to those that would be obtained from field-grown plants, they should still be treated with caution. The treatments were distinct, with no interaction between infected and healthy plants as would happen in the field, and all plants were infected simultaneously, a most unlikely natural occurrence. However, these results do include some that are generally applicable. For example, damage decreases as plant age when inoculated increases.[115]

In experiments, inoculation is usually with more than one aphid/plant to ensure infection. This is most unlikely to occur naturally and evidence for a dosage effect is equivocal. Smith; Burnett and Gill; and Boulton and Catherall[116–118] reported that increasing the number of aphids inoculating BYDV increased symptom severity and yield loss. These results were not related to virus content, although there was the presumption that virus content did increase with the number of inoculating aphids. Using ELISA to assess virus content and comparing 2 and 10 inoculating aphids, Skaria, Lister, and Foster[119] found no differences in virus content, symptoms, or weight of plants up to 30 days after inoculation. Clearly, there is the possibility that large numbers of aphids infecting a young seedling will cause direct feeding damage, but although such results are important in determining appropriate methods of screening germplasm for resistance, they are of academic interest in field infections. Also, different virus content in plants of plant parts had little or no effect on their efficiency as sources of virus for aphid acquisition.[120] Generally, oats and barley are more severely damaged than wheat. Thus, although no results comparable with those in Tables 3-5 and 3-6 for barley are available for wheat, it is unlikely that the PAV isolate would have killed wheat plants and the losses caused by the MAV and RPV isolates would have been much less.

The more recent survey of yield losses due to BYDV is that of Pike[114] but reports from countries or regions are given in note 15. Summaries of the most recent estimates of yield loss in different regions are given in the report of the 1987 CIMMYT Workshop on Barley Yellow Dwarf.[16] Few reports have been confident enough to put a monetary value to yield loss, but Duffus[121] estimated the annual lost yield of barley in the United States as worth $6.3 million for 1951–1960, based on a 1.6% decrease in yield, and $40 million has been estimated as the lost value in Australia.[18] In the UK a 5–10% loss was suggested by Doodson and Saunders,[115] which is worth approximately £100–200 million ($165–330 million) per year. However, averages clearly mean little and other reports from around the world give losses in yield up to 100%. In South America a 20–30% loss in wheat yield is reported, and during the period 1974–1976 losses ranged from 18–45%. In Chile there was a 10% loss in wheat yield in irrigated areas in 1986–87, but during the period 1975–1977, 10–60% losses were recorded. Other Andean countries report losses of up to 30% and here BYDV is the most important disease of cereals.

In the United States, losses have been decreased by the successive introduction of resistant cultivars and the benefits of growing resistant cultivars such as CM67 have been estimated as a saving of 19–60% of yield in California. Elsewhere in the US, losses differed with crop and time of infection. In most regions there is some loss in every year, with estimates of 10% of the annual wheat lost in Indiana and $1.8 million loss of wheat in Montana in 1980. In Canada, BYDV is present but not damaging in every year, with an epidemic likely on average one year in four. In 1986, up to 60% of oats and barley was lost and little wheat is now autumn sown because BYDV infection greatly increases the risk of winter kill.

In Europe losses have been reported on all cereals, including maize and rice. Early infection of rice in Italy caused almost total crop loss. BYDV has caused losses in wheat of 30–40% in Morocco and 9% of barley in Syria. In Iran, a 3–7% loss in wheat is reported. In central, eastern, and southern Africa BYDV is always present and sporadically epidemic. In Kenya in 1987, yield losses of barley were 27% and of wheat, 47%. In Asia, losses in the People's Republic of China can be severe, especially of wheat.

As most of these reports indicate, knowledge of losses is patchy and usually confined to experiments or measurements on small grain cereals. However, while infection of maize can be damaging[16,79,122] the problem has only relatively recently been recognized and little data are available. Rice is infected in Europe, particularly in northern Italy under the disease name of Rice Giallume, and is a serious problem; infection also occurs in Spain. However, possibly the most extensive infection is present in wild and cultivated perennial grasses. While these have prompted much interest as reservoirs of BYDV, comparatively little has been done regarding the direct effects of infection. This is partly because symptoms are usually seen only on crops left for hay or seed, but also because of the difficulty of measuring effects on a crop in which productivity often relates to milk and meat production rather than crop yield. Nevertheless, dry matter yield losses of 20% have been reported from simulated swards.[123–125] One of the difficulties in measuring the effects of BYDV is that infection sometimes stimulates tillering while stunting the whole plant.[126] Perhaps surprisingly, when the effect of a PAV-like isolate on 54 varieties of *Lolium* spp. (ryegrass) was tested, the yield of some was increased while others were unchanged or decreased and Catherall[124] speculated that it might be easier to breed for the yield increase response than for resistance or tolerance. However, even in varieties in which yield was increased, root growth was decreased, which may in the long-term prove disadvantageous. In addition to an effect on yield, there is a potential effect on quality.

3-11-1 Interactions of BYDV with Other Pathogens

In field crops the infection of plants by a single pathogen is probably the exception rather than the rule. In some cases infection by one pathogen can predispose the plant to infection by another. Infection by BYDV causes soluble carbohydrates to accumulate in leaves but decreases nitrogen.[71] Increases in the extent of powdery mildew (Erysiphe graminis)[127] infection on BYDV-infected barley and oats, and of *Septoria avenae* infection on oats[128] have been reported. In the powdery mildew/BYDV interaction the damage by BYDV was so severe that there was little opportunity to determine if joint infection increased yield loss. However, *Septoria avenae* did cause more yield loss on BYDV-infected, than on virus-free, oat plants. Increased infection of BYDV-infected winter barley by leaf blotch (*Rhynchosporium secalis*) is also evident in UK crops, patches of virus-infected plants in spring often being most evident by the necrosis on the leaves caused by leaf blotch infection. On the other hand, in controlled environment experiments,[129] BYDV infection substantially decreased infection by leaf blotch and net blotch (*Pyrenophora teres*). The latter results were explicable by known effects of an increase in water soluble carbohydrate increasing resistance to leaf blotch, but no interaction between water soluble carbohydrate and net blotch infection is known. The field observations on BYDV/leaf blotch interactions thus appear anomalous. Other examples of BYDV infection decreasing subsequent fungal infection are for oat crown rust (*Puccinia coronata* f.sp. *avenae*) and wheat brown rust (*Puccinia recondita* f.sp. *tritici*). In most of these experiments BYDV was established on plants before subsequent fungal infection. In experiments on interactions between the root-infecting take-all fungus (*Gaeumannomyces graminis* var. *tritici*) and BYDV on wheat,[130] plants were first infected by take-all and in the field the combined effects of the two pathogens were greater than the additive effects of each alone. Both take-all and BYD severely decrease root growth, and the growth of the take-all fungus was increased in BYDV-infected compared with virus-free plants.

Von Wechmar[131–132] has described the problems of separating viruses that either interact or cause confusing symptoms in cereals, but there are few detailed examples of interaction between BYDV and other viruses. In *Lolium* spp. Catherall[133] showed that mixed infection of BYDV and the mite-transmitted ryegrass mosaic virus rarely caused more damage than either virus alone, although joint infection differentially affected the competitiveness of different *Lolium* spp. in swards.

3-12 EPIDEMIOLOGY

Interactions between more than 20 potential aphid vectors, at least 5 distinct virus isolates, and a host range spanning most of the Gramineae and including 5 of the most extensively grown crops in the world—wheat, maize, rice, barley, and oats—as well as widespread grasses, both cultivated and wild, mean that the epidemiology of BYDV is complex and that no simple descriptions are possible. However, there are some unifying features and general principles.

Because BYDV is transmitted only by aphids, a knowledge of aphid populations, biology, and dispersal is essential to any understanding of BYD epidemiology. A characteristic of some of the aphid vectors of BYDV is that they are host alternating, feeding on Gramineae during the summer but completing their sexual cycle on woody perennials, immune to BYDV, overwinter. This is especially important for *R. padi, R. insertum,* and *M. dirhodum.* Other species such as *S. avenae* and *S. graminum* also have a sexual cycle but this is completed on Gramineae and thus the virus acquisition and transmission cycle is not broken. The sexual cycle is a common feature of aphid populations in temperate climates,[134] but in some regions, notably western Europe, both sexual and parthenogenetic populations overwinter. As BYDV is confined to Gramineae and does not pass through the egg of vectors, the epidemiology of BYD is greatly influenced by the relative survival rates of these two populations.[135] In the warm temperate, tropical, and subtropical regions reproduction may be almost entirely parthenogenetic (*R. maidis*) but in hot dry conditions aphids may be killed or reproduce very slowly.[53] Carter et al.[136] have reviewed the biology of cereal aphids in the UK, and Eastop[53,134] has given summaries of the important features of the most common vectors of BYDV.

3-12-1 Aphid Populations

Although infection and spread of BYDV requires aphids, there is not always a simple relationship between aphid numbers on the crop or in the air and BYDV infection. Thus, although when no aphids are present no virus can be spread, it is difficult to be certain that aphids are absent and impossible to know, from a single assessment, whether aphids had been present earlier. Counts of aphids on crops or caught in traps need to be interpreted in relation to the isolates of virus present, the infectivity of the aphids, the growth stage of the host and, perhaps most difficult of all, factors influencing future development and spread of aphid populations. Even counting aphids is far from straightforward. The distribution of populations in fields is not uniform so sampling procedures must be devised that take this into account. Aphids are not always obliging enough to make themselves visible. Some, such as *R. rufiabdominalis, R. insertum,* and *Anoecia corni,* feed on roots, others, such as *R. padi,* prefer the base of the plant and stems. Other species are more readily seen; *M. dirhodum* is usually found on leaves, whereas *S. avenae* rapidly moves to ears as they develop. On maize, populations of aphids can exceed several hundred but often cannot be seen until leaves sheathing the ear are pulled back. Some aphids are very discontinuous in their distribution; *R. maidis,* for example, can reach large numbers (in excess of 100) on single barley plants but be absent on neighbors.

Because plants are damaged by BYDV most if infected when young, it is at these early stages that aphid populations are critical and most difficult to record. Waiting until populations have increased sufficiently to be readily visible may be too late to prevent extensive virus spread. As in all monitoring systems, there is a balance between the effort required to monitor populations and the efficiency and accuracy of interpretation of the results. This is especially a problem when even the presence of aphids can be misleading. Rochow[137] reported large populations of *M. (S.) avenae* on oats but little virus infection because *M. (S.) avenae* was an inefficient vector of the predominant virus strain. Although counts of aphids on crops will continue to be made and the results used either as part of an assessment of risk from BYDV infection or for other reasons, there is a need for more objective methods of assessing aphid populations and the risk associated with their transmission of BYDV. In western Europe, Taylor[138] has described a monitoring system that uses suction traps to sample migrant aphid populations. Results from these suction traps, which sample air at 12.2 m, have enormously improved knowledge of aphid populations and their fluctuations from year to year and region to region. A similar scheme has been established in the western region of the United States.[139] More localized trapping systems have been based on yellow water pan traps[140] or sticky, yellow impaction traps.[141]

As a result of data from such trapping systems the characteristic migration patterns of the principal BYDV vectors can be demonstrated.[106] The pattern of migrant populations of *M. dirhodum, R. padi,* and *S. avenae* in the UK is given in Figure 3-6. A comparison of *R. padi* populations of 2 years at sites ranging from the Mediterranean coast to northern Scotland is given by Robert[140] and shows that the distribution, frequency, and timing of *R. padi* flights differ substantially from place to place and between years. Of particular importance to BYDV epidemiology is the relative contribution of different vector species to the total of potential vectors at different times of the year. Thus, in July most of the potential BYDV vectors are *S. avenae*; however, in autumn virtually all the aphids were *R. padi,* and only one or two *S. avenae* were caught (Figure 3-6). Without any further evidence of vector infectivity it is clear that autumn-sown crops (September–October) are unlikely to be infected by MAV isolates and RPV isolates are unlikely to be a problem on those crops sown in spring. However, two exceptionally mild winters in Europe 1988–89 and 1989–90 have demonstrated that MAV can spread in winter-sown crops. Figure 3-6 also clearly demonstrates why the worldwide observation of greater infection of early

Numbers of cereal aphids in suction trap samples throughout the year

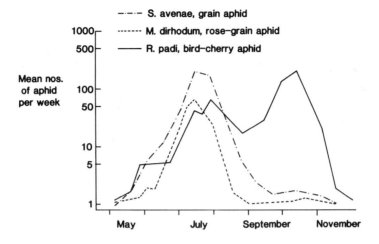

Figure 3-6 Numbers of cereal aphid vectors of BYDV in suction trap samples throughout the year in UK (mean of 20 years).

autumn and late-sown spring crops occurs. The recent change, through much of Europe,
to autumn- rather the spring-sown crops and the earlier sowing of autumn crops to enhance potential yield and to allow autumn sowing to be completed, has resulted in much
more infection and damage than was seen with the "traditionally" sown crops of mid-
October, which emerged after aphid flight had stopped.

3-12-2 Aphid Infectivity

The trapping systems referred to, as valuable as they are, still provide only entomological
and potential virus transmission information. The interpretation and use of such data
requires knowledge of the virus transmitting ability or infectivity of each aphid. A distinction vital in BYDV epidemiology is that between a viruliferous aphid—one that contains a
virus, and an infective aphid—one that contains and transmits a virus. Plumb et al.[105,142–143]
have described methods of assessing infectivity by trapping aphids alive and allowing
them to feed on test seedlings. The proportion of each BYDV vector species transmitting
each week is integrated with the numbers of the species caught in the nearest 12.2 m
suction trap to give an "infectivity index." This has been used as a predictive system in
eastern England, where a threshold value of 50 has been established, but its extension to
other regions requires further validation of the index by field trials and these are currently
under way. Recent modifications to this scheme[144] have shown that by removing males
from the potential vector total, better correlation can be achieved between the index and
virus infection. Because this system takes time, direct methods of detecting BYDV in
aphids have been tried and these are described above.

In northern and northwestern Europe the production of sexual forms, especially of
R. padi, begins at the end of August. Subsequently, the proportion of males in the population increases until approximately equal numbers of males and females are present.[145] This

change is associated, as might be expected, with a sharp decrease in the proportion of the total population that are parthenogenetic females. However, other than a measure of reproduction, or feeding preference, there is no way of separating gynoparae from parthenogenetic females. Apart from the implications of these results for trapping methods, they also suggest that the greater infection of early autumn-sown crops with BYDV is not just due to longer exposure to more aphids but that the population contains a larger proportion of aphids that can feed and multiply on Gramineae and thus spread BYDV than if sown later.

For aphids with a sexual cycle, BYDV transmission is interrupted and this clearly decreases the risk of infection. In regions with milder climates and less extremes of day length than northern Europe, the sexual cycle is absent or rare. In these circumstances survival is on Gramineae and thus virus acquisition and transmission is perpetuated. In southeastern Australia, which has a Mediterranean climate and where the sexual cycle is absent, Smith and Plumb[146] found that 60% of migrant vectors were infective compared with usually less than 5% during the autumn migration of *R. padi* in Britain.[106]

3-12-3 Survival and Dispersal

Successful overwintering as eggs is not assured, as very cold conditions can kill eggs, and in some parts of Europe, while sexual forms of *R. padi* are produced, hosts (*Prunus padus*) on which eggs are laid are often rare or absent.[135] Therefore, some regions may be infested from sources in more favorable climates some distance away. In continental regions the northward progression of milder conditions in summer is often accompanied by weather conditions that favor aphid dispersal. Low-level jet winds have been implicated in long-distance dispersal of aphids transmitting BYDV in North America[147–148] and similar conditions appear to exist in Asia (G. H. Zhou, personal communication). In Europe a similar progression of seasons occur but there is no good evidence for long-distance dispersal from the Mediterranean to more northern regions. There is, however, a suggestion that MAV isolates may be introduced to Sweden by migrations of *S. avenae* from Denmark and northern Germany. An outbreak of *M. dirhodum* in Europe in 1979 provided an opportunity to follow an infestation of aphids over large distances, and all the evidence suggested that the very large populations that occurred progressively further north were the result of the development of local populations, not migration from further south.[149] Weather systems in western Europe usually move from west to east across the path of the seasonal gradient and thus appear to decrease the chance of northward migration.

Where climates are cool and wet, perennial grasses provide a ubiquitous reservoir of infection, harboring aphids as viviparae or eggs overwinter, depending on species, from where they migrate to newly emerged spring-sown cereal crops, or longer established autumn-sown crops. Observational evidence suggests that such spring migrants preferentially alight on plants that are still in distinct rows, the fields presenting a mosaic of green and brown to migrating aphids; uniformly green areas, such as autumn-sown crops or early-sown spring crops, attract fewer aphids. Thus, as well as being young and readily infected and damaged by BYDV, late-sown spring crops are more frequently colonized. In the UK, in 1976 through 1979, barley sown in March was always much less infected by BYDV and bore fewer aphids than crops sown 6 weeks later.[150]

In northern Europe the migration of aphids from grass is relatively short and subsequently populations develop on the rapidly growing cereal crop until predation, parasitism, or crop ripening causes numbers to decline. Most species then move back from cereals to grasses (*S. avenae*) or to their primary host (*M. dirhodum*) in July and August and stay there until the following spring. As already noted. *R. padi* has a further migration in autumn (Figure 3-6) that is, depending on region, partly a sexual generation returning to its primary, egg-laying host, but that also contains other cereal colonizing forms. In some regions maize can be colonized at this time and the aphids migrate from maize later in the year, prolonging the migration period and increasing the risk of infection on autumn-sown crops.[151]

In hot, dry conditions, summer drought is the principal limitation to aphid survival. Little green tissue is normally available over summer in these regions, but aphids may survive in moister areas, or on irrigated crops, especially maize, which can provide a bridge over the dry period.[152] In northern India the only recorded occurrence of serious BYDV was associated with growing out of season crops. When this stopped, so did the BYDV problem.[153]

In addition to the carryover of BYDV and vectors on crops, they can also be perpetuated between crops on stubble regrowth or the volunteers resulting from grain shed during harvest. This problem is worst in climates where there is sufficient rain after harvest to stimulate germination and regrowth and when it does occur it can be a very effective source of virus even when ploughed in before the next crop is drilled. Ploughing inverts soil, stubble, and volunteers, but does not immediately kill them. If the following crop is sown soon after cultivation then, in some circumstances, aphids can move directly from the dying, buried plants onto the newly emerged seedlings. This is perhaps more of a problem when grass swards are cultivated as it is often difficult, especially in wet conditions, to kill the grass by cultivation before sowing cereals.[143] An alternative approach is to kill the green plant cover with a broad spectrum herbicide before cultivating and sowing the next crop.

The effectiveness of the overwintering or oversummering host as a source of BYDV is not uniform. In Britain, while there was no apparent difference in the distribution of BYDV isolates in perennial grasses, the isolates acquired by aphids and carried to nearby cereals reflected the survival of the vectors and thus differed between the west (mild winters) and the east (cool/cold winters) of the country.[154] Therefore, although an examination of the isolates of BYDV in perennial grasses may be interesting, it may not reflect accurately the risk of infection in adjacent cereals.

The epidemiology of some isolates of virus is distinct. The distribution of the damaging rice giallume isolate of BYDV, which appears to be related to the PAV isolate,[155] is correlated with the perennial grass weed *Leersia oryzoides*, which is usually the only grass weed in rice paddies in Italy. *L. oryzoides* is a host of *R. padi* and emerges before rice, thus acting as a focus for initial infection by migrant *R. padi* which then spread virus to the surrounding crop via apterous offspring of the initial migrants.[156]

The relative importance of primary infection, introduced by migrant aphids, and secondary infection, as a result of within crop multiplication and spread of aphid vectors, is critical in the control of BYDV. Where the proportion of infective migrants is large, as in regions where aphids can survive year round on Gramineae, primary infection may be extensive and secondary spread of little consequence. Where few migrants are infective, often where sexual generations are important, then primary infection may be of little consequence and most crop loss is associated with spread within the crop. Clearly, insecticidal control is a more realistic prospect in the latter rather than the former areas.

Patterns of dispersal within crops are also influenced by aphids' behavior. In Britain, patches of infection are usually associated with autumn infection by *R. padi*, whereas spring infection by *S. avenae* usually results in single apparently random infections; a distribution also recorded elsewhere.[157]

3-13 CONTROL

There are, broadly, three possible options for controlling BYDV infection:

1. Crop husbandry, largely by changing sowing dates to avoid the risk of infection.
2. Killing aphid vectors or preventing them feeding on susceptible hosts.
3. Breeding, using resistant/tolerant cultivars.

Which method is chosen depends on many local circumstances, not the least of which is the economics of growing the susceptible crop. Thus in many regions of the world, although insecticides may be very effective, the growers cannot afford them, have no effective means of applying them, and have a very low yield against which to offset the costs of treatment. As yields increase, treatments, including the use of insecticide, also become more economically justifiable. Thus, in the high yielding, high input, highly efficient farming of much of western Europe, insecticides have been used for many years. Here the main interest is in devising accurate forecasting schemes to guide the use of insecticides. Elsewhere, resistance or avoidance may be alternatives, or linked, to decrease infection. Nevertheless, other practices, especially filling in gaps in the crop sequence by growing out of season crops under irrigation, or changing sowing dates to increase yield potential, increase the risk of virus infection and damage. In these circumstances national, local, and personal judgments have to made on the advisability of the proposed action and what can be done to mitigate the risk.

3-13-1 Crop Husbandry

There are frequent reports of interactions between sowing dates and BYDV infection, although few experiments designed to examine such interaction have been done. Work in the UK to validate forecasting systems examined five different autumn-sowing dates in 2 consecutive years (Table 3-7). As well as demonstrating the decline in infection with the later sowing dates, the results also show the marked difference there can be between 2 consecutive years. However, in both years a sowing date of around October 1 would have largely avoided infection and insecticide treatments would not have been required. The British experience is not unique, and a date for sowing after which infection is unlikely has been reported from many regions. In Italy, October 15–30 is the critical time, in Montana, United States, September 10, and in New Zealand, where the term "earliest safe sowing dates" was first used, this has been determined for different regions.[158] In spring it is usually the latest sown crops that suffer most damage, as this increases the risk of young, very susceptible seedlings being exposed to aphids carrying virus. In the spring

TABLE 3-7: WINTER BARLEY, SOWING DATES, BYDV INFECTION, AND APHID INCIDENCE

1985–86			1986–87		
Sowing date	Aphids/plant	Virus (%)	Sowing date	Aphids/plant	Virus (%)
13 Sept.	1.65	21.2	12 Sept.	0.33	<0.1
23 Sept.	0.59	4.5	22 Sept.	0.17	<0.1
2 Oct.	0.07	3.0	1 Oct.	0.18	0
11 Oct.	0.05	0.5	10 Oct.	0.01	0
23 Oct.	0	0	24 Oct.	0.01	0

there is usually no conflict between sowing date, potential yield, and BYDV infection, as early sowing increases yield potential and decreases virus infection.

Ensuring clean seed beds, especially for autumn-sown crops, and avoiding carryover of virus on volunteers and on a previous grass crop, are sensible precautions to avoid virus infection. In some regions the disposal of surplus straw after harvest is a physical and environmental problem. Much of it is burnt but in many areas this will be banned. In Britain, on winter barley, there was more BYDV when straw disposal by burning, baling, or chopping and incorporating was followed by ploughing than when direct drilled. On winter wheat there was more BYDV on plots direct drilled and heavily treated with methiocarb to control slugs.[159] These effects may have been because the treatments decreased populations of aphid predators.

A further curious interaction of crop husbandry with BYDV has been reported from the United States,[160] where taking land out of wheat cultivation to decrease surpluses, the "payment in kind" scheme, led many farmers to plant a cereal crop in the spring to prevent soil erosion. As grain could not be harvested from these areas and they were not treated to contain disease or vector populations, they became large reservoirs of inoculum and increased incidence of BYDV has been associated with the scheme. The introduction in the European Economic Community from autumn 1988 of the "set-aside" program may have similar consequences.

3-13-2 Insecticides and Other Chemicals

The aphid vectors of BYDV are not difficult to kill and although resistance to insecticides has been reported in some species,[53] this does not appear to lessen the effectiveness of their control where insecticides are used regularly. No evidence was obtained for insecticide resistance in field populations of *S. avenae, M. dirhodum,* or *R. padi* in Britain where insecticides are frequently used on cereal crops to control BYDV and prevent direct damage.[161] However, when these results were obtained there was little evidence, in most of the UK, that aphids could survive the winter as viviparous forms. Thus, killing aphids with insecticide pre-empted their death by natural causes. In the recent mild winters, aphids have readily survived the winter, thus, selection for insensitivity is more likely and this problem should be re-examined. Consequently, to minimize the risk of insensitivity developing and to avoid damage to nontarget, often beneficial, organisms it is important to know if and when chemicals should be applied.

Pesticides have been most widely used in Europe to control BYDV, and it is instructive to examine why this should be so. European cereal farming is a high input, high output system with support giving guaranteed prices of about £100/ton, substantially more than world prices. National average yields are approximately 5–6 t/ha and the best farms would expect to achieve 8 t/ha for winter barley and 10 t/ha for winter wheat. Most cereals are now sown in the autumn and sown early to maximize yield potential. Consequently, BYDV is recognized as a serious constraint to yields and some regions are particularly at risk. In these areas a routine application of an insecticide, usually a synthetic pyrethroid, in late October or early November is recommended for all crops sown in September; crops sown even earlier may also require treatment at emergence. This advice recognizes the serious losses that can occur and how little, in relation to potential yield loss, is the cost of an almost completely effective insecticide treatment. Currently, products cost £5–7/ha and even if application costs are assumed to double this, a yield increase of only 0.2 t/ha is sufficient to pay for the treatment; such increases (2% on a 10 t crop) are difficult to measure experimentally. Therefore, many farmers prefer to "insure" against BYDV by using insecticide routinely. As profit margins for cereal growers decline, such an approach becomes less sensible and accurate forecasting systems are required. Such systems already exist in Britain, where they are used in areas in which BYDV is not thought to be a problem every year,[105–106, 143] in France[162], and they are being developed in northern Italy[163] and the northwestern United States.[164]

The timing of insecticide application is critical, especially if, as is clearly desirable, only one application is made. In the UK and northern France the optimum timing is about the end of October and the beginning of November. This coincides with the end of the autumn aphid migration so no reinvasion is likely after treatment. Perhaps surprisingly, such a treatment gives almost complete control of BYDV, even on crops exposed to potential vectors for 6 weeks. The small proportion of infective migrants in autumn and the consequent little primary infection, as previously described, is probably the reason for the treatment's success. However, it also implies that most spread within the crop occurs in November and December when the weather is becoming increasingly cold. If sprays are applied later the yield benefit often sharply declines, but in mild winters yield benefits can be obtained from sprays applied into the following year.

Because of their effects on nontarget organisms, the risk of developing resistance, and the justifiable environmental concerns at the use of insecticides on millions of hectares of cereals, alternative materials have been sought. Two approaches have been the development of aphid alarm pheromone derivative[165] and antifeedants.[166]

The antifeedants, derived from the plants *Polygonum hydropiper* or *Ajuga remota*, when applied three times during the autumn aphid migration, decreased BYDV incidence and increased yield,[167] and the most active compound, formed from (E)-ß-farnesene and diundecylacetylenedicarboxylate, also decreased virus infection. Although such products are not commercially useful at present, their further development could lead to materials as effective as current insecticides but that are environmentally more acceptable.

3-13-3 Breeding

Unconscious selection for resistance to BYDV had probably been done for many years before the disease and its causal agent were recognized. Yellow, stunted plants with sterile florets are unlikely to be selected as candidates for propagation by a primitive farmer, even less by a plant breeder, and Nature itself will have weeded out many such plants by natural selection. Therefore, the point from which efforts to find better varieties started just over 30 years ago was almost certainly the result of some selection. Qualset et al.[168] summarized the genetics of host plant resistance to BYDV and considered some of the problems facing plant breeding in identifying and utilizing the genes that are beneficial.

Barley. Most success in identifying specific genes conferring resistance to BYDV has been achieved in barley in which a major gene conferring tolerance has been identified. Schaller[167] summarized the genetic background and the recognition of the Yd_2 gene[36] that conditions resistance, which is located on chromosome 3.[169] Extensive testing at breeding centers throughout the world has confirmed that the Yd_2 gene does confer tolerance, but significant reversals have occurred. Nevertheless, in California, where most of the early selection work was done, five tolerant cultivars have been released and comprise the majority of the acreage. Discoloration and stunting do occur but yield losses are minimal and there is no evidence of any decline in effectiveness of the Yd_2 gene. Other tolerant cultivars have been released but do not appear to contain Yd_2.[170] In tests reported by Schaller and Qualset[77] Yd_2 containing lines outyielded their near isogenic lines without Yd_2 by 35–45%.

All but one of the cultivars released carrying Yd_2 are spring types. However, the first winter barley, cv. Vixen, known to carry Yd_2 that suffered little yield loss when infected by unspecified isolates of BYDV[171] was recommended for use in the UK in 1987. More detailed investigations of this cultivar, using defined isolates of BYDV present in the UK, gave interesting results (Table 3-8) that may account for some of the reversals reported by Schaller.[167] When inoculated with MAV and PAV isolates in autumn under conditions in which susceptible cv. Igri was killed by PAV and its yield decreased by 90% by MAV (Table 3-5), Vixen showed no yield loss. However, some components of yield were affected, which suggests that a compensating effect existed in the experiment which may not be repeated in the field, where consequently there may be a small yield loss. When inoculated with an RPV isolate, Vixen yielded no better than Igri (Tables 3-5 and 3-8). Inoculation in spring followed a similar pattern although damage was less than with the autumn inocula-

TABLE 3-8: THE EFFECTS OF PAV-, MAV-, AND RPV-LIKE ISOLATES OF BYDV ON WINTER BARLEY CV. VIXEN, CONTAINING THE YD$_2$ GENE, INOCULATED IN THE AUTUMN OR SPRING

Inoculation time and isolate	Yield (g/plant)	1000 grain wt (g)	Height (cm)	Shoot Number
Autumn				
PAV	15.87a[a]	41.57b	94.8b	15.4a
MAV	15.93a	42.63ab	94.7b	15.4a
RPV	1.25d	32.56b	52.1c	4.7c
Spring				
PAV	12.61b	34.80c	91.2b	15.0a
MAV	15.70a	45.12a	100.7a	14.6ab
RPV	10.18c	33.30c	94.5b	14.3ab
Uninoculated	15.62a	45.26a	104.0a	13.4b

Source: Data from G. M. Herrera.[37]

[a]Values in each column followed by different letters are significantly different at $P = 0.005$ as determined by the Student-Newman test.

tion. The experiment was repeated with Atlas 57 (-Yd_2) and Atlas 68 (+Yd_2) and a similar pattern was seen although the relative damage caused by the isolates on Atlas 57 differed from that on Igri (Tables 3-5, 3-6), but again Atlas 68 showed no benefit over the Atlas 57 when infected by the RPV isolate. Thus, it is tempting to suggest that the Yd_2 gene is effective only against Group 1 isolates, PAV, MAV and SGV, of BYDV. Such a result would be consistent with other evidence that the two BYDV groups are distinct viruses and therefore resistance against one would not confer resistance against another.

The early selection work was largely against undefined isolates. If this differential susceptibility of Yd_2 gene containing barley is confirmed, it will have far-reaching effects on the deployment of cultivars with Yd_2 and will require much more detailed knowledge of the spectrum of BYDV isolates present in barley-growing areas.

Wheat.   Major genes conferring resistance to BYDV are not known in wheat, although it is believed that they exist,[168] and resistance has been detected.[172] Resistance appears to be evident from several sources and thus may, if they can be combined, give better resistance than is currently available. Tola and Kronstad,[172] summarizing work up to 1984, concluded that no good resistance was present but cvs Yamhill, Novisad, and Anza all showed promising characteristics. They suggested that a recurrent selection program could increase resistance to BYDV, but so far no cultivars offering the same degree of protection that is available in barley have been produced.

The wild grass *Thinopyrum intermedium* has been demonstrated as the source of a chromosome that decreases the concentration of BYDV in lines of wheat,[173] and Zhong 4, an octoploid hybrid between wheat and *T. intermedium*, has been shown to have good resistance to BYDV in the field.[174–175] Larkin at al.[176] have also described possible sources of resistance in other wild grass species.

Oats.   Jedlinski[177] has summarized the progress that has been made in developing resistant oat cultivars in the United States and elsewhere since the severe epidemic in 1959. More than 10 cultivars have been released, each giving good yields even when BYDV infection was widespread. Good sources of tolerance have been identified in other *Avena* spp. and this has been incorporated in several breeding programs. However, most breeding work has been done in spring oats, and much less progress has been made in producing resistant winter lines.

Other crops

Triticale.   BYDV infects triticale (a wheat × rye hybrid) and can be damaging. BYDV-resistant lines have been identified[178] but none are agronomically acceptable. Nevertheless, if the apparent resistance to BYDV available in rye can be satisfactorily incorporated in triticale this may be a route for improving wheat resistance, possibly via embryo culture. Recent work on both spring and winter triticale and methods of evaluating its response to BYDV are described by Collin et al.[179]

Maize.   Although maize cultivars are susceptible to BYDV, only relatively recently[122] has the extent of infection been realized. Consequently, the selection of resistant lines is less far advanced for the small grain cereals, although existing maize hybrids in Italy are already tolerant of BYDV infection.[180]

Rice.   In northern Italy a resistance breeding program began in 1974 and selections have been made[181-182] and are now widely grown. It appears that one incompletely dominant gene governs resistance.[183]

Grasses.   Catherall[124] has examined the reactions of lines of many *Lolium* spp. to BYDV isolates in Britain and found a range of reactions. However, breeding for resistance to BYDV is not a high priority area for agriculture grasses.

Future prospects for resistance.   Recombinant DNA technology has opened up the prospect of increasing genetic diversity in crop plants by their transformation. However, no suitable systems yet exist for regenerating small grain cereal plants. It seems probable that such techniques will soon be developed and thus will make available genetic information currently either unavailable or difficult to transfer to crops susceptible to BYDV. This is likely to help efforts that have been made to transfer the Yd_2 gene from barley to wheat.[184-185]

No mention has yet been made of decreasing BYDV infection through resistance to the aphid vector. There are differences in host reaction to aphid vectors,[186-189] but there is no evidence that this contributes substantially to differences in infection of different cultivars.

3-14 CONCLUSION

In the short term, 5–10 years, there seems little prospect of the importance of BYDV or the damage that it causes decreasing. Indeed, the probable continued expansion of wheat throughout the world is likely to increase losses related to BYDV. In new areas of cultivation the losses may be initially very severe, probably associated with the multiplication of aphid vectors unrestricted by predators and parasites, whose establishment always lags behind that of the pest.

In regions of low-input, low-yield farming, and where primary infection is extensive and environmental or economic considerations prevent the use of pesticides, then crop husbandry, especially the manipulation of sowing dates, and host resistance or tolerance are clearly the only methods available. As more detailed knowledge emerges of host genotype x virus isolate interactions, the deployment of cultivars showing resistance or tolerance, especially barleys, can be more rationally planned than is presently possible. Molecular techniques may speed the process of breeding for resistance but it seems unlikely that this will show as commercially acceptable cultivars by the end of the century, and resistance to BYDV is only one of a whole range of characters that need to be present for a cultivar to be acceptable.

There is a continuing need to characterize BYDV isolates, especially their distribution. Now that effective serological systems are available, the use of the technique will spread as simple diagnostic kits are developed.

In high input agriculture, pesticides are likely to remain important but their use will decrease as tolerant cultivars are introduced, as forecasting systems become more accurate, and as the profit margins on cereals decrease and insecticides become a more expensive insurance treatment. In some regions a more extensive agricultural system may develop including crop sowing dates planned to avoid infection. It will be some time before the "yellow plague"[121] of cereals disappears.

3-15 REFERENCES

1. Gildow, F. E. "Barley Yellow Dwarf Virus-Aphid Vector Interactions Associated with Virus Transmission and Vector Specificity," in *World Perspectives on Barley Yellow Dwarf*, ed. P. A. Burnett (Mexico: CIMMYT, 1990), 111.

2. Barrus, M. F. "Red Leaf and Blast of Oats." *Plant Dis. Rep.* 21 (1937): 359.

3. McKinney, H. H. "Infectious and Noninfectious Pigment Disorders in the Small Grains, with Particular Reference to the Southeastern Areas." *Plant Dis. Rep.* 34 (1950): 151.

4. Johnston, C. O. "The 'Red Leaf' Disease of Oats Still a Mystery in Kansas." *Oat Nat. Newslt.* 1 (1951): 20.

5. Moore, M. B. "The Cause and Transmission of Blue Dwarf and Red Leaf in Oats." *Phytopathology* 42 (1952): 471.

6. Rademacher, van B., and Schwarz, R. "Die Rotblättrigkeit oder Blattröte des Hafers—eine Viruskrankheit (*Hordeumvirus nanescens*)." *Z. Pflanzenkr. Pflanzenschutz* 65 (1958): 641.

7. Bruehl, G. W. "Barley Yellow Dwarf." *Ann. Phytopathol. Soc. Monograph* 1 (1961): 52.

8. Oswald, J. W., and Houston, B. R. "A New Virus Disease of Cereals Transmissible by Aphids." *Plant Dis. Rep.* 35 (1951): 471.

9. Bruehl, G. W., McKinney, H. H., and Toko, H. V. "Cereal Yellow Dwarf as an Economic Factor in Small Grain Production in Washington, 1955–58." *Plant Dis. Rep.* 43 (1959): 471.

10. Bruehl, G. W., and Toko, H. V. "A Washington Strain of the Cereal Yellow Dwarf Virus." *Plant Dis. Rep.* 39 (1955): 547.

11. Oswald, J. W., and Houston, B. R. "Barley Yellow-Dwarf, a Virus Disease of Barley, Wheat and Oats Readily Transmitted by Four Species of Aphids." *Phytopathology* 42 (1952): 15.

12. Oswald, J. W., and Houston, B. R. "Host Range and Epiphytology of the Cereal Yellow Dwarf Disease." *Phytopathology* 43 (1953); 309.

13. Toko, H. V., and Bruehl, G. W. "Strains of the Cereal Yellow-Dwarf Virus Differentiated by Means of the Apple-Grain and the English Grain Aphids." *Phytopathology* 46 (1957): 536.

14. Klinkowski, M., and Kreutzberg, G. "Vorkommen und Verbreitung von Gramineenvirosen in Europe." *Phytopathol. Z.* 32 (1958): 1.

15. Anon. *Barley Yellow Dwarf.* in *Proceedings of the Workshop* (Mexico: CIMMYT, 1984).

16. Burnett, P. A., ed., *World Perspectives on Barley Yellow Dwarf* (Mexico: CIMMYT, 1990).

17. Watson, L., and Gibbs, A. J. "Taxonomic Patterns in the Host Ranges of Viruses Among Grasses and Suggestions on Genetic Sampling for Host-Range Studies." *Ann. Appl. Biol.* 77 (1974): 23.

18. Johnston, G. R., Sward, R. J., Farrell, J. A., Greber, R. S., Guy, P. L., McEwan, J. M., and Waterhouse, P. M. "Epidemiology and Control of Barley Yellow Dwarf Viruses in Australia and New Zealand," in *World Perspectives on Barley Yellow Dwarf*, ed. P. A. Burnett (Mexico: CIMMYT, 1990), 228.

19. Allen, T. C. "Strains of the Barley Yellow Dwarf Virus." *Phytopathology* 47 (1957): 481.

20. Rochow, W. F. "Transmission of Strains of Barley Yellow Dwarf Virus by Two Aphid Species." *Phytopathology* 49 (1959): 744.

21. Rochow, W. F. "Barley Yellow Dwarf Virus and Oats in New York." *Oat Newslt.* 9 (1959): 19.

22. Slykhuis, J. T., Zillinsky, F. J., Hannah, A. E., and Richards, W. R. "Barley Yellow Dwarf Virus on Cereals in Ontario." *Plant Dis. Rep.* 43 (1959): 849.

23. Toko, H. V., and Bruehl, G. W. "Some Host and Vector Relationships of Strains of the Barley Yellow-Dwarf Virus." *Phytopathology* 49 (1959): 343.

24. Rochow, W. F. "Specialization Among Greenbugs in the Transmission of Barley Yellow Dwarf Virus." *Phytopathology* 50 (1960): 881.

25. Rochow, W. F. "Biological Properties of Four Isolates of Barley Yellow Dwarf Virus." *Phytopathology* 69 (1969): 1580.

26. Rochow, W. F. "Variants of Barley Yellow Dwarf Virus in Nature: Detection and Fluctuation during Twenty Years." *Phytopathology* 69 (1979): 655.

27. Lister, R. M., and Rochow, W. F. "Detection of Barley Yellow Dwarf Virus by Enzyme-Linked Immunosorbent Assay." *Phytopathology* 69 (1979): 649.

28. Rochow, W. F. "Identification of Barley Yellow Dwarf Viruses. Comparison of Biological and Serological Methods." *Plant Dis.* 66 (1982): 381.

29. Hsu, H. T., Aebig, J., and Rochow, W. F. "Differences among Monoclonal Antibodies to Barley Yellow Dwarf Viruses." *Phytopathology* 74 (1984): 600.

30. Diaco, R., Lister, R. M., Hill, J. H., and Durand, D. P. "Demonstration of Serological Relationships among Isolates of Barley Yellow Dwarf Virus by Using Polyclonal and Monoclonal Antibodies." *J. Gen. Virol.* 67 (1986): 353.

31. Torrance, L., Pead, M. T., Larkins, A. P., and Butcher, G. W. "Characterisation of Monoclonal Antibodies to a UK Isolate of Barley Yellow Dwarf Virus." *J. Gen. Virol.* 67 (1986): 549.

32. Gildow, F. E., Ballinger, M. E., and Rochow, W. F. "Identification of Double-Stranded RNAs Associated with Barley Yellow Dwarf Virus Infection of Oats." *Phytopathology* 73 (1983): 1570.

33. Waterhouse, P. M., Gerlach, W. L., and Miller, W. A. "Serotype-Specific and General Luteovirus Probes from Cloned cDNA Sequences of Barley Yellow Dwarf Virus." *J. Gen. Virol.* 67 (1986): 1273.

34. Barbara, D. J., Kawata, E. E., Veng, P. P., Lister, R. M., and Larkins, B. A. "Production of cDNA Clones from the MAV Isolate of Barley Yellow Dwarf," *J. Gen. Virol* 68 (1987): 2419.

35. Gill, C. C., and Chong, J. "Cytopathological Evidence for the Division of Barley Yellow Dwarf Virus Isolates into Two Subgroups." *Virology* 95 (1979): 59.

36. Rasmusson, D. C., and Schaller, C. W. "The Inheritance of Resistance in Barley to the Barley Yellow Dwarf Virus." *Agron. J.* 51 (1959): 661.

37. Herrera, G. M. "Interactions between Host Plants and British Isolates of Barley Yellow Dwarf Virus." (Ph.D. thesis, University of London, 1989).

38. Waterhouse, P. M., Gildow, F. E., and Johnstone, G. R. "Luteovirus Group." *Assoc. Appl. Biol.* Description of Plant Viruses 339, 9 pp., 1988.

39. Martin, R. R., and D'Arcy, C. J. "Relationships among Luteoviruses Based on Nucleic Acid Hybridization and Serological Studies." *gntervirology* 31 (1990): 23.

40. Rochow, W., and Carmichael, I. E. "Specificity among Barley Yellow Dwarf Viruses in Enzyme Immunosorbent Assays." *Virology* 95 (1979): 415.

41. Rochow, W. F. In *Barley Yellow Dwarf. Proceedings of the Workshop* (Mexico: CIMMYT, 1984), 204.

42. Rochow, W. F., and Eastop, V. F. "Variation within *Rhopalosiphum padi* and Transmission of Barley Yellow Dwarf Virus by Clones of Four Aphid Species." *Virology* 30 (1966): 286.

43. Gill, C. C. "Transmission of Barley Yellow Dwarf Virus Isolates from Manitoba by Five Species of Aphids." *Phytopathology* 57 (1967): 713.

44. Lister, R. M., and Sward, R. J. "Anomalies in Serological and Vector Relationships of MAV-like Isolates of Barley Yellow Dwarf Virus from Australia and USA." *Phytopathology* 78 (1988): 766.

45. Zhang, Q. F., Guan, W. N., Ren, Z. Y., and Zhu, X. S. "Transmissibility of Barley Yellow Dwarf Virus Strains from Northwestern China by Four Aphid Species." *Plant Dis.* 67 (1983): 895.

46. Zhou, G. H., Cheng, Z-M., Qian, Y-T, and Zhang, X-C. "Serological Identification of Luteoviruses of Small Grains in China. *Plant Dis.* 68 (1984): 710.

47. Zhou, G-H, and Zhang, S-X. "Identification of the Variants of Barley Yellow Dwarf Virus in China." in *World Perspectives on Barley Yellow Dwarf,* ed. P.A. Burnett (Mexico: CIMMYT, 1990), 290.

48. Szirmai, J. "Seed Transmission Experiments with Barley Yellow Dwarf Virus." *Növenytermeles* 28 (1979): 147.

49. Mills, P. R., Mercer, P. C., and McGimpsey, H. C. "Barley Yellow Dwarf Virus." *Ann. Rep. Dept. Agric. N. Ireland* 135.

50. Jess, S., and Mowat, D. J., "Transmission of Barley Yellow Dwarf Virus by Larvae of Frit Fly, *Oscinella frit* (L.) and the Effects of Sward-Killing Herbicides on Transmission." *Rec. Agric. Res.*, *Northern Ireland* 34 (1986): 57.

51. A'Brook, J. "Vectors of Barley Yellow Dwarf Virus," in *Euraphid; Rothamsted* 1980, ed. L. R. Taylor (Harpenden, UK: Rothamsted Experimental Station, 1981), 21.

52. Jedlinski, H. "Rice Root Aphid, *Rhopalosiphum rufiabdominalis,* a Vector of Barley Yellow Dwarf Virus in Illinois, and the Disease complex." *Plant Dis.* 65 (1981): 975.

53. Eastop, V. F. "The Biology of the Principal Aphid Virus Vectors," in *Plant Virus Epidemiology,* eds. R. T. Plumb and J. M. Thresh (Oxford, UK: Blackwell Scientific Publications, 1983), 115.

54. Pei, M. Y, and Hsu, H. K. "Studies on the Red-Leaf Disease of the Foxtail Millet (*Setaria italica* (L.) Beauv.). III Further Studies on the Transmission of the Millet Red-Leaf Disease Virus." *Acta Phytopathol. Sin.* 4 (1958): 87.

55. Rybicki, E. P., and von Wechmar, M. B. "Characterization of an Aphid-Transmitted Virus Disease of Small Grains." *Phytopathol. Z.* 103 (1982): 306.

56. Gildow, F. E. "Transcellular Transport of Barley Yellow Dwarf Virus into the Hemocoel of the Aphid Vector, *Rhopalosiphum padi*." *Phytopathology* 75 (1985): 292.

57. Gildow, F. E. "Virus-Membrane Interactions Involved in Circulative Transmission of Luteoviruses by Aphids," in *Current Topics in Vector Research,* ed. K. F. Harris (Berlin: Springer Verlag, 1987), 93.

58. Gildow, F. E. "Coated-Vesicle Transport of Luteoviruses through Salivary Glands of *Myzus persicae.*" *Phytopathology* 72 (1982): 1289.

59. Rochow, W. F., Foxe, M. J., and Muller, I. "A Mechanism of Vector Specificity for Circulative Aphid-Transmitted Plant Viruses." *Ann. N.Y. Acad. Sci.* 266 (1975): 293.

60. Gildow, F. E., and Rochow, W. F. "Role of Accessory Salivary Glands in Aphid Transmission of Barley Yellow Dwarf Virus." *Virology* 104 (1980): 97.

61. Paliwal, Y. C., and Sinha, R. C. "On the Mechanism of Persistence and Distribution of Barley Yellow Dwarf virus in an Aphid Vector." *Virology* 42 (1970): 668.

62. Watson, M. A., and Mulligan, T. E. "The Manner of Transmission of Some Barley Yellow-Dwarf Viruses by Different Aphid Species." *Ann. Appl. Biol.* 48 (1960): 711.

63. Rochow, W. F. "Dependent Virus Transmission from Mixed Infections," in *Aphids as Virus Vectors,* eds. K. F. Harris and K. Maramorosch (New York: Academic Press, 1977), 253.

64. Rochow, W. F. "Barley Yellow Dwarf: Dependent Virus Transmission by *Rhopalosiphum maidis* from Mixed Infections." *Phytopathology* 65 (1975): 99.

65. Rochow, W. F., and Israel H. W. "Luteovirus (Barley yellow dwarf virus) Group," in *The Atlas of Insect and Plant Viruses*, ed. K. Maramorosch (New York: Academic Press, 1977), 363.

66. Rochow, W. F., and Duffus, J. E. "Luteoviruses and Yellows Diseases," in *Handbook of Plant Virus Infections and Comparative Diagnosis*, ed. E. Kurstak (Amsterdam: Elsevier/North Holland Biomedical Press, 1981), 147.

67. Duffus, J. E., and Rochow, W. F. "Neutralization of Beet Western Yellows Virus by Antisera Against Barley Yellow Dwarf Virus. *Phytopathology* 68 (1978): 45.

68. D'Arcy, C. J., Torrance, L., and Martin, R. R. "Discrimination among Luteoviruses and Their Strains by Monoclonal Antibodies and Identification of Common Epitopes." *Phytopathology* 79 (1989): 869.

69. Duffus, J. E., Falk, B. W., and Johnstone, G. R. "Luteoviruses—One System, Many Variations," in *World Perspectives on Barley Yellow Dwarf*, ed. P. A. Burnett (Mexico: CIMMYT, 1990), 86.

70. Brakke, M. K., and Rochow, W. F. "Ribonucleic Acid of Barley Yellow Dwarf Virus." *Virology* 61 (1974): 240.

71. Gerlach, W. L., Miller, W. A., and Waterhouse, P. M. "Molecular Genetics of Barley Yellow Dwarf Virus." *Barley Yellow Dwarf Newsletter* 1 (1987): 17.

72. Gerlach, W. L., Miller, W. A., Cheng, Z. G., and Waterhouse, P. M. "Molecular Biology of Barley Yellow Dwarf Virus," in *World Perspectives on Barley Yellow Dwarf*, ed. P. A. Burnett (Mexico: CIMMYT, 1990), 105.

73. Eweida, M., and Oxelfelt, P. "Production of Cloned cDNA from a Swedish Barley Yellow Dwarf Isolate." *Ann. Appl. Biol.* 114 (1989): 61.

74. Miller, W. A., Waterhouse, P. M., and Gerlach, W. L. "Sequence and Organization of Barley Yellow Dwarf Virus Genomic RNA." *Nucleic Acids Research* 16 (1988): 6097.

75. Miller, W. A., Waterhouse, P. M., Kortt, A. A., and Gerlach, W. L. "Sequence and Identification of the Barley Yellow Dwarf Virus Coat Protein Gene." *Virology* 165 (1988): 306.

76. Schaller, C. W., Rasmusson, D. C., and Qualset, C. O. "Sources of Resistance to the Yellow Dwarf Virus in Barley." *Crop Sci.* 3 (1963): 342.

77. Schaller, C. W., and Qualset, C. O. "Breeding for Resistance to the Barley Yellow Dwarf Virus." *Proc. 3rd Int. Wheat Conf.* (Madrid, Spain, 1980), 528.

78. Stoner, W. N. "Studies of Transmission of Barley Yellow Dwarf Virus to Corn (*Zea mays*)." *Phytopathology* 55 (1965): 1078.

79. Panayotou, P. C. "Effect of Barley Yellow Dwarf on Several Varieties of Maize." *Plant Dis. Rep.* 61 (1977): 815.

80. Esau, K. "Phloem Degeneration in Gramineae Affected by the Barley Yellow Dwarf Virus." *Am. J. Bot.* 44 (1957): 245.

81. Esau, K. "Anatomic Effects of Barley Yellow Dwarf Virus and Maleic Hydrazide on Certain Gramineae." *Hilgardia* 27 (1959): 15.

82. McKinney, H. H., Specht, A. W., and Stanton, T. R. "Certain Factors Affecting Chlorophyll Degeneration and Pigmentations in Oats and Other Grasses." *Plant Dis. Rep.* 36 (1952): 450.

83. Orlob, G. B. "Studies on the Barley Yellow Dwarf Virus Disease." (Ph.D. thesis, University of Wisconsin, Madison, 1959).

84. Orlob, G. B., and Arny, D. C. "Some Metabolic Changes Accompanying Infection by Barley Yellow Dwarf Virus." *Phytopathology* 51 (1961): 768.

85. Jensen, S. G. "Metabolism and Carbohydrate Composition in Barley Yellow Dwarf Virus-Infected Wheat." *Phytopathology* 62 (1972): 587.

86. Gill, C. C., and Chong, J. "Development of the Infection in Oat Leaves Inoculated with Barley Yellow Dwarf Virus." *Virology* 66 (1975): 440.

87. Gill, C. C., and Chong, J. "Differences in Cellular Ultrastructural Alterations between Variants of BYDV." *Virology* 75 (1976): 33.

88. Herrera, G. M. "Interactions between Host Plants and British Isolates of Barley Yellow Dwarf Virus." (Ph.D., thesis, University of London, 1989).

89. Habili, N., McInnes, J. L., and Symons, R. H. "Nonradioactive, Photobiotinlabelled DNA Probes for the Routine Diagnosis of Barley Yellow Dwarf Virus." *J. Virol. Methods* 16 (1987): 225.

90. Rochow, W. F., Hu, J. S., Forster, R. L., and Hsu, H. T. "Parallel Identification of Five Luteoviruses that Cause Barley Yellow Dwarf." *Plant Dis.* 71 (1987): 272.

91. Rochow, W. F., and Brakke, M. K. "Purification of Barley Yellow Dwarf Virus." *Virology* 24 (1964): 310.

92. Rochow, W. F., Aapola, A. I. E., Brakke, M. K., and Carmichael, I. E. "Purification and Antigenicity of Three Isolates of Barley Yellow Dwarf Virus." *Virology* 46 (1971): 117.

93. D'Arcy, C. J. "Purification of Barley Yellow Dwarf Luteoviruses." *Barley Yellow Dwarf, A Proceedings of the Workshop* (Mexico: CIMMYT, 1984), 36.

94. Rochow, W. F., and Ball, E. M. "Serological Blocking of Aphid Transmission of Barley Yellow Dwarf Virus." *Virology* 33 (1967): 359.

95. Rochow, W. F. "Barley Yellow Dwarf Virus No. 32 in Descriptions of Plant Viruses." *Commonw. Mycol. Inst. Assoc. Appl. Biologists.* (Kew, Surrey, England).

96. Aapola, A. I. E. "Serological Relationships and in vivo Interactions among Isolates of Barley Yellow Dwarf Virus" (Ph.D. Thesis, Cornell University, 1968).

97. Aapola, A. I. E., and Rochow, W. F. "Immunodiffusion Tests with Three Isolates of Barley Yellow Dwarf Virus." *Phytopathology* 58 (1968): 398.

98. Clark, M. F., and Adams, A. N. "Characteristics of the Microplate Method of Enzyme-Linked Immunosorbent Assay for the Detection of Plant Viruses." *J. Gen. Virol.* 34 (1977): 475.

99. Rochow, W. F., Muller, I., Tufford, L. A., and Smith, D. M. "Identification of Luteoviruses of Small Grains from 1981 through 1984 by Two Methods." *Plant Dis.* 70 (1986): 461.

100. Plumb, R. T. "The Epidemiology of Barley Yellow Dwarf in Europe," in *World Perspectives on Barley Yellow Dwarf,* ed. P. A. Burnett (Mexico: CIMMYT, 1990), 215.

101. Pead, M. T., and Torrance, L. "Some Characteristics of Monoclonal Antibodies to a British MAV-like Isolate of Barley Yellow Dwarf Virus." *Ann. Appl. Biol.* 113 (1988): 639.

102. Derrick, K. S. "Detection and Identification of Plant Viruses by Serologically Specific Electron Microscopy." *Phytopathology* 63 (1973): 441.

103. Paliwal, Y. C. "Rapid Diagnosis of Barley Yellow Dwarf Virus in Plants Using Serologically Specific Electron Microscopy." *Phytopathol. Z.* 89 (1977): 125.

104. Forde, S. M. D. "Strain Differentiation of Barley Yellow Dwarf Virus Using Specific Mono-clonal Antibodies in Immunosorbent Electron Microscopy." *J. Virol. Methods* 23 (1989): 313.

105. Plumb, R. T. "Barley Yellow Dwarf Virus—A Global Problem," in *Plant Virus Epidemiology*, eds. R. T. Plumb and J. M. Thresh (Oxford, UK: Blackwell Scientific Publications, 1983), 185.

106. Plumb, R. T. Lennon, E. A., and Gutteridge, R. A. "Forecasting Barley Yellow Dwarf Virus by Monitoring Vector Populations and Infectivity," in *Plant Virus Epidemics: Monitoring, Modelling* and *Predicting Outbreaks*, eds. G. D. McLean, R. G. Garrett, and W. Ruesink (Sydney, Australia: Academic Press, 1986), 387.

107. Harrison, B. D. "Two Sensitive Serological Methods for Detecting Plant Viruses in Vectors and Their Suitability for Epidemiological Studies." *Poc. Br. Crop Prot. Conf.—Pests Dis.* 3 (1981): 751.

108. Boulton, M. I., Markham, P. G., and Davies, J. W. "Nucleic Acid Hybridisation Techniques for the Detection of Plant Pathogens in Insect Vectors." *Proc. Br. Crop Prot. Conf—Pests Dis.* 1 (1984): 181.

109. Torrance, L., Plumb, R. T., Lennon, E. A., and Gutteridge, R. A. "A Comparison of ELISA with Transmission Tests to Detect Barley Yellow Dwarf Virus-Carrying Aphids. In *Developments and Applications in Virus Testing,* eds. R. A. C. Jones and L. Torrance (Wellesbourne, UK: Assoc. Appl. Biol., 1986), 165.

110. Denechere, M., Cante, F., and Lapierre, H. "Detection Immunoenzymatique du Virus de la Jaunisse Nanisante de L'Orge Dans son Vecteur." *Rhopalosiphum padi* (L.) *Ann. Phytopathol.* 11 (1979): 507.

111. Torrance, L. "Use of Enzyme Amplification in an ELISA to Increase Sensitivity of Detection of Barley Yellow Dwarf Virus in Oats and in Individual Vector Aphids." *J. Virol. Methods* 15 (1987): 131.

112. Plumb, R. T., and Lennon, E. A. "Serological Diagnosis of Barley Yellow Dwarf Virus (BYDV)." *Rep. Roth. Exp. Stn., 1980 Part I,* 181 (1981).

113. Paliwal, Y. C. "Detection of Barley Yellow Dwarf Virus in Aphids by Serologically Specific Electron Microscopy." *Can. J. Bot.* 60 (1982): 179.

114. Pike, K. S. "A Review of Barley Yellow Dwarf Virus Grain Yield Losses," in *World Perspectives on Barley Yellow Dwarf*, ed. P. A. Burnett (Mexico: CIMMYT, 1990), 356.

115. Doodson, J. K., and Saunders, P. J. W. "Some Effects of Barley Yellow Dwarf Virus on Spring and Winter Cereals in Field Trials." *Ann. Appl. Biol.* 66 (1970): 361.

116. Smith, H. C. "The Effects of Aphid Numbers and Stage of Plant Growth in Determining Tolerance to BYDV in Cereals." *N.Z. J. Agric. Res.* 10 (1967): 445.

117. Burnett, P. A., and Gill, C. C. "The Response of Cereals to Increased Dosage with Barley Yellow Dwarf Virus." *Phytopathology* 66 (1976): 646.

118. Boulton, R. E., and Catherall, P. L. "The Effects of Increasing Dosage of Barley Yellow Dwarf Virus on Some Resistant and Susceptible Barleys." *Ann. Appl. Biol.* 94 (1980): 69.

119. Skaria, M., Lister, R. M., Foster, J. E., and Shaner, G. E. "Barley Yellow Dwarf Virus Content as an Index of Symptomatic Resistance in Cereals." *Phytopathology* 75 (1985): 212.

120. Pereira, A-M. N, Lister, R. M., Barbara, D. J., and Shaner, G. E. "Relative Transmissibility of Barley Yellow Dwarf Virus from Sources with Differing Virus Contents." *Phytopathology* 79 (1989): 1353.

121. Duffus, J. E. "Aphids, Viruses, and the Yellow Plague," in *Aphids as Virus Vectors*, eds. K. F. Harris, and K. Maramorosch. (New York: Academic Press 1977), 361.

122. Stoner, W. N. "Barley Yellow Dwarf Virus Infection in Maize." *Phytopathology* 67 (1977): 975.

123. Catherall, P. L. "Effects of Barley Yellow Dwarf Virus on the Growth and Yield of Single Plants and Simulated Swards of Perennial Ryegrass." *Ann. Appl. Biol.* 57 (1966): 155.

124. Catherall, P. L., and Parry, A. L. "Effects of Barley Yellow Dwarf Virus on some Varieties of Italian, Hybrid and Perennial Ryegrasses and Their Implication for Grass Breeders." *Plant Pathol.* 36 (1987): 148.

125. Latch, G. C. M. "Effects of Barley Yellow Dwarf Virus on Simulated Swards of Nui Perennial Ryegrass." *N.Z. J. Agric. Res.* 23 (1980): 373.

126. Catherall, P. L., and Wilkins, P. W. "Barley Yellow Dwarf Virus in Relation to the Breeding and Assessment of Herbage Grasses for Yield and Uniformity." *Euphytica* 26 (1977): 385.

127. Potter, L. R. "The Effects of Barley Yellow Dwarf Virus and Powdery Mildew in Oats and Barley with Single and Dual Infections." *Ann. Appl. Biol.* 94 (1980): 94.

128. Comeau, A., and Pelletier, G. J. "Predisposition to Septoria Leaf Blotch in Oats Affected by Barley Yellow Dwarf Virus." *Can. J. Plant Sci.* 56 (1976): 13.

129. Varughese, J., and Griffiths, E. "Effect of Barley Yellow Dwarf Virus on Susceptibility of Barley Cultivars to Net Blotch (*Pyrenophora teres*) and leaf blotch (*Rhynchosporium secalis*)." *Plant Pathol.* 32 (1983): 435.

130. Sward, R. J., and Kollmorgen, J. F. "The Separate and Combined Effects of Barley Yellow Dwarf Virus and Take-all Fungus (*Gaeumannomyces graminis var. tritici*) on the Growth and Yield of Wheat." *Aust. J. Agric. Res.* 37 (1986): 11.

131. Von Wechmar, M. B., and Rybicki, E. P. "Brome Mosaic Virus Infection Mimics Barley Yellow Dwarf Virus Disease Symptoms in Small Grains." *Phytopathol. Z.* 114 (1985): 332.

132. Von Wechmar, M. B. "Other Viruses Causing Barley Yellow Dwarf Virus-like Diseases in Small Grains in South Africa," in *World Perspectives on Barley Yellow Dwarf*, ed. P. A. Burnett (Mexico: CIMMYT, 1990), 73.

133. Catherall, P. L. "Effects of Barley Yellow Dwarf and Ryegrass Mosaic Viruses Alone and in Combination on the Productivity of Perennial and Italian Ryegrasses." *Plant Pathol.* 36 (1987): 73.

134. Eastop, V. F. "Worldwide Importance of Aphids as Virus Vectors," in *Aphids as Virus Vectors*, eds. K. F. Harris and K. Maramorosch. (New York: Academic Press, 1977), 3.

135. Plumb, R. T. "The Epidemiology of Barley Yellow Dwarf in Europe," in *World Perspectives on Barley Yellow Dwarf*, ed. P. A. Burnett (Mexico: CIMMYT, 1990), 215.

136. Carter, N., McLean, I. F. G., Watt, A. D., and Dixon, A. F. G. "Cereal Aphids: A Case Study and Review." *Appl. Biol.* 5 (1980): 271.

137. Rochow, W. F. "Predominating Strains of Barley Yellow Dwarf Virus in New York, Changes During Ten Years." *Plant Dis. Rep.* 51 (1967): 195.

138. Taylor, L. R. "EURAPHID: Synoptic Monitoring for Migrant Vector Aphids," in *Plant Virus Epidemiology*, eds. R. T. Plumb and J. M. Thresh. (Oxford, UK: Blackwell Scientific Publications, 1983), 133.

139. Pike, K. S., Allison, D. W., Low, G., Bishop, G. W., Halbert, S., and Johnston, R. "Cereal Aphid Vectors: A Western Regional (USA) Monitoring System," in *World Perspectives on Barley Yellow Dwarf*, ed. P. A. Burnett (Mexico: CIMMYT, 1990), 282.

140. Robert, Y. "Aphid Vector Monitoring in Europe," in *Current Topics in Vector Research,* ed. K. F. Harris. (Berlin: Springer Verlag, 3, 1987), 81.

141. Heathcote, G. D. "Aphids Caught on Sticky Traps in Eastern England from 1965 to 1973 in Relation to the Spread of Yellowing Viruses of Sugar Beet." *Bull. Entomol. Res.* 64 (1974): 609.

142. Plumb, R. T., Lennon, E. A., and Gutteridge, R. A. "Aphid Infectivity and the Infectivity Index." *Rep. Roth. Exp. Stn. 1981* (1983): 195.

143. Plumb, R. T. "A Rational Approach to the Control of Barley Yellow Dwarf Virus." *J.R. Agric. Soc. Engl.* 147 (1986): 162.

144. Kendall, D. A., and Chinn, N. E. "A Comparison of Vector Population Indices for Forecasting Barley Yellow Dwarf Virus in Autumn Sown Cereal Crops." *Ann. Appl. Biol.* 116 (1990): 87.

145. Tatchell, G. M., Plumb, R. T., and Carter, N. "Migration of Alate Morphs of the Bird Cherry Aphid (*Rhopalosiphum padi*) and Implications for the Epidemiology of Barley Yellow Dwarf Virus." *Ann. Appl. Biol.* 112 (1988): 1.

146. Smith, P. R., and Plumb, R. T. "Barley Yellow Dwarf Virus Infectivity of Cereal Aphids Trapped at Two Sites in Victoria." *Aust. J. Agric. Res.* 32 (1981): 249.

147. Wallin, J. R., Peters, D., and Johnson, L. C. "Low Level Jet Winds, Early Cereal Aphid and Barley Yellow Dwarf Virus Detection in Iowa." *Plant Dis. Rep.* 51 (1967): 527.

148. Wallin, J. R., and Loonan, D. V. "Low Level Jet Winds, Aphid Vectors, Local Weather and Barley Yellow Dwarf Virus Outbreaks." *Phytopathology* 61 (1971): 1068.

149. Dewar, A. M., Woiwod, I. P., and Choppin de Janvry, E. "Aerial Migrations of the Rose-Grain Aphid, *Metopolophium dirhodum* (Wlk.) over Europe in 1979." *Plant Pathol.* 29 (1980): 101.

150. Jenkyn, J. F., and Plumb, R. T. "Effects of Fungicides and Insecticides Applied to Spring Barley Sown on Different Dates in 1976–9." *Ann. Appl. Biol.* 102 (1983): 421.

151. Dedryver, C. A., and Robert, Y. "Ecological Role of Maize and Cereal Volunteers as Reservoirs for Gramineae Virus Transmitting Aphids," in *Proc. 3rd Conf. Virus Disease of Gramineae*, ed. R. T. Plumb. (Harpenden, UK: Rothamsted Experimental Station, 1981), 61.

152. Carroll, T. "The Status of Barley Yellow Dwarf Virus in Maize." *Barley Yellow Dwarf. A Proceedings of the Workshop.* (Mexico: CIMMYT, 1984), 120.

153. Tandon, J. P., Shoran, J., and Pant, S. K. "Status of Barley Yellow Dwarf Virus Research in Asia with Special Emphasis on India," in *World Perspectives on Barley Yellow Dwarf,* ed. P. A. Burnett. (Mexico, CIMMYT, 1990), 81.

154. Plumb, R. T. "Grass as a Reservoir of Cereal Viruses." *Ann. Phytopathol.* 9 (1977): 361.

155. Osler, R. "Caratterizzazione Biologica di un Ceppo del Virus del Nanismo Giallo dell Orzo (BYDV) Agente Causale del Giallume del Riso." *Riv. Patol. Veg.* 20 (1984): 3.

156. Osler, R. "Il Giallume del Riso: Attuali Consocenze sul Ciclo di Infezione del Virus, con Particolare Attenzione alle Piante *Oryza sativa* e *Leersia oryzoides* e all' Afiade *Rhopalosiphum padi.*" *Riso* 29 (1980): 217.

157. Rochow, W. F., Jedlinski, H., Coon, B. F., and Murphy, H. C. "Variation in Barley Yellow Dwarf of Oats in Nature." *Plant. Dis. Rep.* 49 (1965): 692.

158. Lowe, A. D. "Avoid Yellow Dwarf Virus by Late Sowing." *N.Z. Wheat Rev.* 10 (1967): 59.

159. Kendall, D. A., Smith, B. D., Chinn, N. E., and Wiltshire, C. W. "Cultivation, Straw Disposal and Barley Yellow Dwarf Virus Infection in Winter Cereals." *Proc. Br. Crop. Prot. Conf.—Pests Dis.* 3 (1986): 981.

160. Briggle, L. W. "The Barley Yellow Dwarf Research Program in the USA." *Barley Yellow Dwarf. A Proceedings of the Workshop.* (Mexico: CIMMYT, 1984), 141.

161. Stribley, M. F., Moores, G. D., Devonshire, A. L., and Sawicki, R. M. "Application of the FAO-Recommended Method for Detecting Insecticide Resistance in *Aphis fabae* Scopoli, *Sitobion avenae* (F.), *Metopolophium dirhodum* (Walker) and *Rhopalosiphum padi* (L.) (Hemiptera: Aphididae)." *Bull. Entomol. Res.* 73 (1983): 107.

162. Bayon, F., and Ayrault, J. P. "La Jaunisse Nanisante de l'Orge (J.N.O.): Methode Actuelle de Prevision des Risques en Automne." *Def. Veg.* 223 (1983): 268.

163. Peressini, S., Bianchi, G. L., and Coceano, P. G. "Infectivity Index and Barley Yellow Dwarf Virus Infection in Friuli-Venezia Giulia," in *World Perspectives on Barley Yellow Dwarf*, ed. P. A. Burnett. (Mexico: CIMMYT, 1990), 260.

164. Halbert, S., Bishop, G. W., Blackmer, J., Connelly, J., Johnston, R., Sandvol, L., and Pike, K. S. "Barley Yellow Dwarf Infectivity of *Rhopalosiphum padi* in Maize as an Estimate of Primary Inoculum Pressure in Irrigated Winter Wheat," in *World Perspectives on Barley Yellow Dwarf*, ed. P. A. Burnett. (Mexico: CIMMYT, 1990), 273.

165. Dawson, G. W., Griffiths, D. C., Pickett, J. A., Plumb, R. T., Woodcock, C. M., and Zhang, Z.-N. "Structure/Activity Studies on Aphid Alarm Pheromone Derivatives and Their Field Use against Transmission of Barley Yellow Dwarf Virus." *Pestic. Sci.* 22 (1988): 17.

166. Dawson, G. W., Griffiths, D. C., Hassanali, A., Pickett, J. A., Plumb, R. T., Pye, B. J., Smart, L. E., and Woodcock, C. M. "Antifeedants: A New Concept for Control of Barley Yellow Dwarf Virus in Winter Cereals." *Proc. Br. Crop Prot. Conf.—Pests Dis.* 3 (1986): 1001.

167. Schaller, C. W. "The Genetics of Resistance to Barley Yellow Dwarf Virus in Barley." *Barley Yellow Dwarf. A Proceedings of the Workshop.* (Mexico: CIMMYT, 1984), 93.

168. Qualset, C. O., Lorens, G. F., Ullman, D. E., and McGuire, P. E. "Genetics of Host Plant Resistance to Barley Yellow Dwarf Virus," in *World Perspectives on Barley Yellow Dwarf*, ed. P. A. Burnett. (Mexico: CIMMYT, 1990), 388.

169. Schaller, C. W., Qualset, C. O., and Rutger, J. N. "Inheritance and Linkage of the Yd_2 Gene Conditioning Resistance to the Barley Yellow Dwarf Virus Disease in Barley." *Crop Sci.* 4 (1964): 544.

170. Grafton, K. F., Poehlman, J. M., Sechler, D. T., and Sehgal, O. P. "Effect of Barley Yellow Dwarf Virus Infection on Winter Survival and Other Agronomic Traits in Barley." *Crop Sci.* 22 (1982): 596.

171. Parry, A. L., and Habgood, R. M. "Field Assessment of the Effectiveness of a Barley Yellow Dwarf Virus Resistance Gene Following Its Transference from Spring to Winter Barley." *Ann. Appl. Biol.* 108 (1986): 395.

172. Tola, J. E., and Kronstad, W. E. "The Genetics of Resistance to Barley Yellow Dwarf Virus in Wheat." *Barley Yellow Dwarf: A Proceedings of the Workshop.* (Mexico: CIMMYT, 1984), 83.

173. Brettell, R. I. S., Banks, P. M., Cauderon, Y., Chen, X., Cheng, Z. M., Larkin, P. J., and Waterhouse, P. M. "A Single Wheat Grass Chromosome Reduces the Concentration of Barley Yellow Dwarf Virus in Wheat. *Ann. Appl. Biol.* 113 (1988): 599.

174. Chi, S. Y., Yi, S. S., Chang, Y. H., Yi, K. H., and Son, F. Y. "Studies on Wheat Breeding by Distant Hybridization between Wheat and *Agropyron glaucum.*" *Scientia Agricultura, Sinica* 2 (1979): 1.

175. Zhang, Q. F., Guan, W. N., Ren, Z. Y., Zhu, X. S., and Tsai, J. H. "Transmission of Barley Yellow Dwarf Virus Strains from Northwestern China by Four Aphid Species." *Plant Dis.* 67, (1983): 895.

176. Larkin, P. J., Brettell, R. I. S., Banks, P., Appels, R., Waterhouse, P. M., Cheng, Z. M., Zhou, G. H., Xiu, Z. Y., and Chen, X. "Identification, Characterisation, and Utilization of Sources of Resistance to Barley Yellow Dwarf Virus," in *World Perspectives on Barley Yellow Dwarf*, ed. P. A. Burnett. (Mexico: CIMMYT, 1990), 415.

177. Jedlinski, H. "The Genetics of Resistance to Barley Yellow Dwarf Virus in Oats." *Barley Yellow Dwarf: A Proceedings of the Workshop.* (Mexico: CIMMYT, 1984), 101.

178. Comeau, A., and St-Pierre, C. A. "The Genetics of Resistance to Barley Yellow Dwarf Virus in Triticale." *Barley Yellow Dwarf, A Proceedings of the Workshop.* (Mexico: CIMMYT, 1984), 107.

179. Collin, J., St-Pierre, C. A., and Comeau, A. "Analysis of Genetic Resistance to Barley Yellow Dwarf Virus in Triticale and Evaluation of Various Estimators of Resistance," in *World Perspectives on Barley Yellow Dwarf,* ed. P. A. Burnett. (Mexico: CIMMYT, 1990), 404.

180. Lorenzoni, C., Bertolini, M., Loi, N., Osler, R., and Snidaro, M. "Tolerance to Barley Yellow Dwarf Virus in Maize," in *World Perspectives on Barley Yellow Dwarf,* ed. P. A. Burnett. (Mexico: CIMMYT, 1990), 401–03.

181. Moletti, M., and Osler, R. "Determinazione della Resistenza al "Giallume" delle Piu Importanti Varieta di Riso Italiane Mediante Inoculazioni Sperimentali con l'Afide *Rhopalosiphum padi.*" *Riso* 1 (1978): 33.

182. Moletti, M., Osler, R., and Baldi, G. "Valutazione della Resistenza al Giallume Mediante Inoculazioni Artificiali con l'Afida *Rhopalosphium padi* in Linee Pure ai Riso Migliorate." *Riso* 28 (1979): 53.

183. Baldi, G., Moletti, M., and Osler, R. "Inheritance of Resistance to Giallume in Rice." in *World Perspectives on Barley Yellow Dwarf,* ed. P. A. Burnett. (Mexico: CIMMYT, 1990), 429.

184. McGuire, P. E. "Status of an Attempt to Transfer the Barley Yellow Dwarf Virus Resistance Gene Yd_2 of Barley to Hexaploid Wheat." *Barley Yellow Dwarf. A Proceedings of the Workshop.* (Mexico: CIMMYT, 1984), 113.

185. McGuire, P. E., and Qualset, C. O. "Transfer of the Yd_2 Barley Yellow Dwarf Virus Resistance Gene from Barley to Wheat," *World Perspectives on Barley Yellow Dwarf.* ed. P. A. Burnett. (Mexico: CIMMYT, 1990). 476.

186. Kieckhefer, R. W., Jedlinski, H., and Brown, C. M. "Host Preferences and Reproduction of Four Cereal Aphids on 20 Avena Selections." *Crop Sci.* 20 (1980): 400.

187. Lowe, H. J. B. "Detection of Resistance to Aphids in Cereals." *Ann. Appl. Biol.* 88 (1978): 401.

188. Lowe, H. J. B. "Characteristics of Resistance to the Grain Aphid *Sitobion avenae* in Winter Wheat." *Ann. Appl. Biol.* 105 (1984): 529.

189. Johnson, R. A., and Rochow, W. F. "An Isolate of Barley Yellow Dwarf Virus Transmitted Specifically by *Schizaphis graminum.*" *Phytopathology* 62 (1972): 921.

4

BLAST OF RICE

J. C. BHATT

Scientist (Plant Pathology)
Vivekananda Parvatiya Krishi Anusandhan Shala (ICAR)
Almora, 263 601 India

R. A. SINGH

Professor, Department of Plant Pathology
G.B. Pant University of Agriculture and Technology,
Pantnagar 263 145, India

4-1 INTRODUCTION

Rice (*Oryza sativa* L.) is the major staple food for nearly one half of the world's population. It is primarily a tropical and subtropical crop, but the best grain yields are obtained in temperate regions.[1] The crop suffers from a number of diseases, among which blast caused by *Magnaporthe grisea* (Hebert), is one of the most important, causing significant losses in yield. It has been recorded in about 85 countries in the world (Figure 4-1).

Blast disease might have existed long before it was noticed and recorded. The fact that some of the grasses have this type of disease indicates its antiquity. In China, it was probably known as "rice fever disease" as early as 1637. The early records of blast are mainly from Japan and Italy. According to Goto,[2] M. Tsuchiya first recorded blast in Japan in 1704, and it was subsequently reported by S. Miayanaga in 1788, by N. Kojima in 1793, and by T. Konishi in 1809. In Italy, the disease called *brusone* was reported by Astolfi (1828), by Brugnatelli (1838), and by Gera (1846).[3] Although some workers considered *brusone* identical to blast, it may have consisted of other diseases as well, or it may have been quite different.[4] It was reported in the United States as early as 1876,[5-6] and in India in 1913.[7] Metcalf[5,6] perhaps named this disease blast for the first time in English.

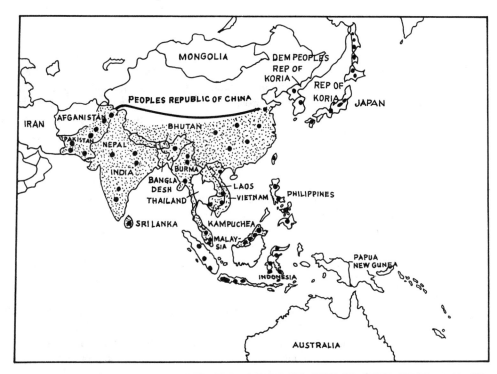

(AFTER REISSIG et al, *ILLUSTRATED GUIDE TO INTEGRATTED PEST MANAGE-MENT IN RICE IN TROPICAL ASIA,* IRRI, PHILIPPINES, 1986 WITH PERMISSION)

Figure 4-1   Geographical distribution of rice blast disease.

4-2 ECONOMIC IMPORTANCE

Several epiphytotics of the disease have been recorded in different parts of the world, resulting in serious losses in yield. In Korea, it is the only rice disease that has ever caused a serious problem.[8] The loss in yield during 1953, an epidemic year, was estimated at about 800,000 tons in Japan.[9] Yield losses were estimated at over 90% in Bicol during 1962, and at 50–60% in Leyte during 1963, in the provinces of Philippines. In India, epiphytotics have been reported in the Tanjore area of Madras state in 1919. In the hills, blast may cause more than a 65% loss in yield.[10] Its incidence on a large scale has been reported from the plains and also from peninsular India.

Several workers have attempted to correlate the percentage of panicle blast with yield losses.[11-12] Kuribayashi and Ichikawa[12] developed the formula $Y = 0.69 X + 2.8$ for estimating the losses from panicle blast, where Y is the ratio of loss and X is the percentage of panicles affected by blast. Other formulae have also been developed to calculate losses in yield.[9] In India, about 0.8% loss in yield due to blast was estimated during the 1960–61 season and the methods for estimating losses were also reviewed.[11]

4-3 SYMPTOMS

The typical symptoms appear on leaves, but the leaf sheath, rachis, nodes, and even the glumes are also attacked. The lesions on leaves appear as small, bluish, water-soaked flecks of about 1–3 mm in diameter which rapidly enlarge and may become several centimeters long and about 1 cm broad, spindle shaped. The center of the spots appears pale green or dull greyish green changing to grey, and the periphery has a dark brown band with a yellowish halo around the lesion (Figure 4-2). Lesions may coalesce and kill the entire leaves in susceptible varieties, whereas it may remain as minute, pin-sized, brown specks on the resistant variety. An early seedling infection may give the crop a burnt appearance. The spots on the leaf sheath resemble those on the leaf. The disease reduces the number of mature panicles, grain weight, and straw weight. Environmental conditions, level of resistance in cultivars, and age of the crop influence the shape and size of lesions. Nodal blast is usually noticed after heading, in which the sheath pulvinus rots and turns blackish and culm may break at the infected node. Lesions at the internode are also found in cases of severe infection. At the flower emergence the fungus attacks the peduncle, which is engirdled, and the lesion turns to brownish-black. This stage of infection is commonly referred to as rotten neck/neck rot/neck blast/panicle blast. In early neck infection, grain filling does not occur and the panicle remains erect like a dead heart caused by a stem borer. In the late infection, partial grain filling occurs. Due to the weight of such partially filled grains the base of the rachis may break down. Partial infection of the panicle also occurs. Small brown to black spots also may be observed on glumes of the heavily infected panicles.

4-4 CAUSAL ORGANISM

4-4-1 Nomenclature

There is much controversy about the name of the pathogen, as two generic names have been used—*Pyricularia* and *Dactylaria*—but Asuyama[13] concluded that the blast fungus on rice and crabgrass is reasonably included in the genus *Pyricularia*. In most of the older literature this fungus has been spelled as *Piricularia* but according to the International Rules of Botanical Nomenclature, and after Hughes'[14] publication, several researchers have used the spelling *Pyricularia*, as it predates the former one and is valid.

The specific name of the fungus on rice was first described as *Pyricularia oryzae*.[15] The fungus, which resembled *P. oryzae*, was first described in Japan by Shirai.[16]

Comparing the various characteristics of *Pyricularia oryzae* Cav. (1891) and *Pyricularia grisea* Sacc. (1880) Rossman et al.[18] concluded that there are not enough difference between these two to maintain them as different species. They are synonymous and as per rule of priority the earlier proposed name, *P. grisea*, should get the preference. Therefore, the name of the rice blast pathogen should be as follows:

PYRICULARIA GRISEA Sacc., Michelia 2 : 20 . 1880.
= *Trichothecium griseum* Cooke *in* Cooke & Ellis, Grevillea 8 : 12 . 1879, without description.

= *Trichothecium griseum* (Sacc.) Cooke *in* Ravenel, Fungi Americani Exsiccati No. 580, 1881.

= *Pyricularia oryzae* Cavara, Fungi Longobardiae Exsiccati No. 49 . 1891.

= *Dactylaria oryzae* (Cavara) Sawada, Trans. Nat. Hist. Soc. Taiwan 6 : 242. 1916.

Considering the popularity of the name *P. oryzae*, some taxonomists feel that it would be an unfortunate 'name change.' However, it may not be possible to conserve the name *P. oryzae* because there is the name 'available' for the teleomorph (perfect state), *Magnaporthe grisea* (Hebert) M. E. Barr, and as per the Sydney Code (1981) this name should be used while referring to the whole fungus in all its forms (holomorph). Therefore the correct name of rice blast fungus could be *Pyricularia* anamorph of *Magnaporthe grisea* (P. M. Kirk, CMI, Kew, Surrey, U. K.; personal communication with senior editor). In view of the popularity of the name and the fact that *P. grisea* is yet to be widely accepted by the taxonomists, in the present article pathogen of the rice blast would be referred to as *P. oryzae*.

4-4-2 Morphology

Mycelium in culture is aerial or submerged, hyaline to olievaceous and branched; conidiophores are simple or rarely branched, 2–4 septate, and emerge solitary or in clusters through stoma. The basal portion is swollen, olive to fuliginous, tapering toward the tip, with lighter color. The conidia are pyriform to obclavate, lightly pigmented, borne on simple conidiophores with truncate denticles, mostly 2-septate, rarely 1–3 septate. The size of conidia varies among different isolates collected from different hosts and environmental conditions. Conidial size ranges from 19.2 to 27.3 × 8.1 to 10.3 μm on rice and from 15.8 to 27.0 × 6.8 to 14.1 μm on other plants.[13]

Electron microscopic studies reveal that the conidial cell membrane consists of three layers, although two layers[19] or four layers[20] have also been reported in the conidial cell wall. The nucleus with nucleolus, mitochondria, endoplasmic reticulum, vacuoles, septum pore, and two kinds of dense particles, one of which might be a lysosome, have been demonstrated.[20] Laciniated or serrated epispores of the conidia have been considered characteristic of the genus *Pyricularia*.[21]

4-4-3 Cytology and Origin of Variation

There is confusion regarding the nuclear status in the fungus. Suzuki[22-23] observed the mycelia and conidiophore as multinucleate in a number of isolates of the fungus. The number of nuclei in each conidial cell varied in different isolates ranging from 1 to 12, and he considered the fungus to be "persistently heterokaryotic." Chu and Li[24] also showed the mycelium, conidia, and germ tube as multinucleate. Contrary to these, Yamasaki and Niizeki[25] reported most of the cells as uninucleate, though in certain strains 13% to 30% of the mycelial cells were multinucleate, containing 2–6 nuclei. Giatgong and Frederiksen[26-28] reported mycelial cells as uninucleate except for a few barrel-shaped cells which have as many as 4 or more nuclei. Yaegashi and Hebert[29] reported that mycelial cells are uninucleate but older cells may be multinucleate. Wherever multinucleate cells have been reported, it is suspected that some of the stained bodies in the cells may have been lipids or other bodies rather than nuclei.[30] Conidial and mycelial cells have been observed as uninucleate under an electron microscope.[19-20]

The apical cell of the conidiophores is uninucleate, which divides to send a daughter nucleus in the young conidium. Division of the nucleus in the conidium initially is followed by a septal formation between daughter nuclei. The nucleus of the apical cell of the conidium divides again, followed by a septum formation making the conidium 3-celled. Thus, 3 nuclei in a mature conidium originate from a single nucleus and are expected to be genetically identical.[31]

Conidia germinate by germ tube which in certain isolates of the pathogen may form appressoria *in vitro* and *in vivo* or in certain isolates *in vivo* only. The appressorium contains several nuclei and becomes brown and thick walled.[22] Conidia germinate by terminal and basal cells and the nucleus of the cell migrates either directly into the germ tube or one of the daughter nuclei migrates after the nuclear division.[25,28-29] Conidia germinate and produce either appressorium or barrel-shaped cells. The appressorium contain as many as 32 to 64 nuclei and the barrel-shaped cells have 2 to 6 nuclei. From these, normal uninucleate mycelia are produced.[32-33] It seems that after spore germination there is always a multinucleate stage.[32] Recent studies have shown that the nuclear behavior in hyphae, conidiophores, conidia, germ tube, appressoria, and infection pegs is uninucleate except for the occurrence of 2 nuclei in hyphal tip cells which were transitory.[31]

The somatic nuclear division in the fungus has also been the subject of great controversy. Nuclear divisions though simple mitosis,[28] atypical mitosis,[20,25] and typical mitosis[22,25] have been reported. Nuclear division by typical mitosis has recently been demonstrated beyond doubt,[34] and chromosome number is not variable as reported in literature.[35]

4-4-4 Toxins

Attempts have been made to isolate toxins from infected plant tissues and culture filtrates of the pathogen to understand their role in the pathogenesis. Tamari and Kaji[36] isolated two toxins, namely α-picolinic acid and piricularin in crystalline form from both the sources. Each toxin produced a characteristic spot on rice blades. Piricularin is also toxic to *P. oryzae*. The fungus, however, produces copper oxidase that binds the toxin and destroys its antifungal property but does not affect its toxicity to the host. It may also be detoxified by chlorogenic acid and ferulic acid, the two naturally occurring phenolics in rice plants. α-picolinic acid is another toxin, which interferes with the respiration of the host and takes part in the disease development. A third toxin, pyriculol, was isolated from *P. oryzae* cultures. The toxin inhibits the growth of rice seedlings.

In addition to these toxins, two more toxic substances have been recovered, i.e., 3, 4 dihydro 3, 4, 8 trihydroxy-1 (2H) naphthelenone, and tenuazonic acid.[37]

4-4-5 Perfect Stage

Several scientists have attempted to produce the teleomorph stage of the fungus in culture. Hebert[38] succeeded in producing the perfect stage of *Pyricularia grisea* (Cooke) Sacc. from crabgrass (*Digitaria sanguinalis* L.) and named it *Ceratosphaeria grisea* Hebert.

Later, isolates of *Pyricularia* from rice developed mature perithecia when mated with isolates of *Pyricularia* form *Eleusine coracana* Gaertn., *E. indica* (L.) Gaertn., and an isolate of *Phalaris arundinacea* L.[39] Subsequently perithecia were also produced by intra-group mating of rice isolates.[40] Studies on isolates of *P. oryzae* obtained from different rice-growing areas confirm that the fungus is heterothallic, denoted by A (+) and a (-) types. These two types are found in natural populations in many rice-growing areas, but the occurrence of perfect stage has not yet been observed in nature.[41–42]

Teleomorph stage is produced on several media in addition to Sach's medium. A short exposure of light is necessary for the formation of perithecia. Addition of sugar suppressed the formation of perithecia while acidic pH promoted their formation.[43] Plant material such as sterilized rice straw or stems of some other gramineous plants on Sach's agar are usually required for perithecial development.[44]

The morphology of a perfect state of blast fungus obtained earlier appeared to be identical to that of *C. grisea* but due to its graminicolous habit, broader perithecial neck, and presence of a sympodioconidium type imperfect state, its placing in the genus *Ceratosphaeria* is questionable. The species of *Ceratosphaeria* are lignicolous, hypersaprobes, and their perithecial peridia are carbonaceous.[45] Krause and Webster[46] erected a new genus *Magnaporthe* to accommodate a single species, *M. salvinii* (Catt.) Krause and Webster, which was described as the causal agent of stem rot of rice. Barr[45] transferred *C. grisea* to *Magnaporthe grisea*, which was further confirmed by Yaegashi and Udagawa[44] where he transferred the perfect stage of *Pyricularia* to *M. grisea* independently.

The morphology of the perithecial state produced by mating of different rice isolates was also similar to *M. grisea*.[47] Therefore, the name *Magnaporthe grisea* is retained for the perfect state of rice blast fungus. Therefore correct name of teleomorph of *P. oryzae* is as follows:

MAGNAPORTHE GRISEA (Hebert) Barr, Mycologia 69 : 954. 1977.

= *Ceratosphaeria grisea* Hebert, Phytopathology 61 : 86. 1971.

= *Phragmoporthe grisea* (Hebert) Monod, Beih, Sydowia 9 : 153. 1983.

Perithecia are often gregarious, nonstromatic, partially immersed, base spherical to subspherical, mostly 100–180 μm (80–260 μm) in diameter, 500–1,200 μm long (including neck), dark brown to black, glabrous; neck long cylindrical is often secondarily prolonged, up to 1,100 μm long, 55–160 μm in diameter, pale brown, protruding beyond the agar surface, wide-ostiolate, with walls of textura oblita, an interior lined with slender periphyses [Figure 4-3(a–d)]. Peridium is dark brown, opaque, membranaceous, pseudoparenchymatous, 8–12 μm thick, with outer layer consisting of brown angular cells measuring 4–12 × 4–8 μm. Asci are eight-spored, hyaline, cylindrical to clavate, mostly 60–90 × 10–12 μm (55–110 × 8–15 μm), unitunicate, rounded or slightly truncated above [Figure 4-3(a–d)]. Short stipitate, bases evanescent at maturity, and asci freed in the cavity; the apical ring is quite distinct, not stained by Merzer's reagent. Paraphyses are indistinct, hyaline, inflated at base, unbranched septate, and deliquescent. Ascospores are biseriate, somewhat obliquely arranged, hyaline, often guttulate, fusiform, curved, rounded at both ends, mostly 18–23 × 5–7 μm (16–25 × 4–8 μm), transversely 3-septate (sometimes 1-, 2-, or 4-septate), slightly constricted at septa, smooth-walled, at

Figure 4-2a-d.   Symptoms of rice blast disease.

maturity extruded from the ostiole in a gelatinous mass, germinating usually at each end [Figure 4-3(a–d)][4-8]

4-4-6 Variation and Pathogenic Races

The existence of pathogenic races was first noticed by Sasaki.[49] Since then the presence of races in the fungus have been characterized and reported from several countries. World distribution of races of *P. oryzae* are presented in Table 4-1.[50] The number of races identified in different countries varied depending on the number of isolates tested and differentials used. Use of different sets of differentials in different countries confused the identification and comparison of races reported from different places. A cooperative blast project between Japan and the United States was initiated in 1963 to establish an international set of differentials for identifying races of *P. oryzae* and for standardizing testing techniques. Eight rice cultivars were selected as international differentials.[51-52] Salient characters of the differentials are given by Atkins.[53] On the basis of the cooperative studies, 32 international races of the fungus were characterized in eight race groups.[51,52] International races have a prefix of the letter I, followed by the letter A, B, C, D, E, F, G, or H indicating

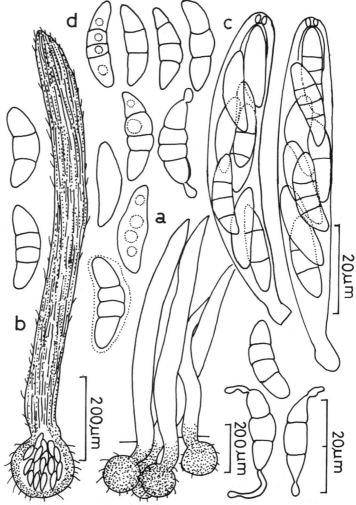

Figure 4-3   Perfect stage - *Magnaporthe grisea*, from cross of No. C10 X No. T 28. a.) Perithecia. b.) Perithecium in more detail. c.) Asci. d.) Various stages of ascospores and germinating ascospores. (After Yaegashi and Udagawa, 1978).

group. Individual races are designated by an Arabic numeral after the group letter, e.g., IA-1, IA-2, and so on.

Theoretically, 256 races were characterized using the set of eight differentials, with the addition of race group I to accommodate races that do not fit in the system.[54] A different method of classification and nomenclature of physiologic races was proposed.[55] In this method, all the 256 possible pathogenicity patterns on the eight international differentials were accommodated into 16 groups from A to P, indicating the race group and numbers indicating the physiologic races. The race groups of eight, i.e., A to H are based on susceptibility pattern, whereas the race groups from I to P are based on the resistance reaction. Susceptibility or resistance of an individual differential or sets of

TABLE 4-1: WORLD DISTRIBUTION OF RACES OF *P. ORYZAE*

Country	Race
Argentina	V$_3$-78-1, V$_3$-78-2, CU 3-78, G4-78, LP 5-78, SF 4-78 (local races), IA 2, IA 6, IB 10, II 1, IA 38, IA 9, IF 3, IB 9, ID 11, IA 126 (international races)
Bangladesh	Virulent unidentified race previously not present in the country but the disease broke out in 1980 on IR 8, Chandina, Pajam, and Pusa II.
Brazil	IA 1, IA 17, IA 65, IA 120, IB 1, IB 17, IB 39c, IC 1, IC 5, IC 30 a and b, IC 32, ID 13, IE 8, IG 2a, II 1 and II 15 (local)
Egypt	IF 1, IH 1, IJ (stable), IB 3, IC 3 (unstable)
India	IE 1, IE 2, IG 1, II 1, IE 3, IC 1, IC 13, IC 17, IC 21, ID 1, ID 6, IE 8, IB 1, IB 2, IB 10, IB 17, IB 18, IB 49, IA 11, AP 37, M$_5$ + M$_4$ (local)
Japan	A-H and 12 Japanese race groups and 7 local races, N-group in 1965, C-group in 1968, Nagain group in 1974
Peru	307 (Peru)
Philippines (IRRI)	ID 14, IH 1, IA 65, OU 243, 244, 54, BC 68, FS 66-59, Chu 66-45, TH 65-105, Fukei 73, and 100 local races
Puerto Rico	An unidentified race introduced from Texas (U.S.)
Taiwan	N 1 (local)
United States	IG 1 (Texas and Lousiana), ID 13 (Texas), IA 1-3, IB 1-6, IC 1-5, ID 1-11, IE 1, 2, IF 1, 2, IG 1 and 2, IH 1
USSR	IA, IB, IC, ID, IE, IF, IG, IH, IA 77, IC 1, IC 9, IC 17, JJ a, JF 2, JF 13, JA 165, and JC 25

For references please see Gangopadhyay and Padmanabhan.[50]

differentials taken in serial order from the list of differentials have been the criterion for classification.

In the international set, replacement of some of the differentials like Usen and Raminad Str. 3 has been suggested. Usen gives an ambiguous reaction[56] and Raminad Str. 3 does not set seeds or flower at higher latitudes.[57]

Despite this cooperative work, different sets of differentials in different countries are in use to differentiate the races (Table 4-2).[17] Some additions and/or deletions in the list of differentials are done by different workers. This trend shows that the international differentials have not satisfied the workers of different countries in race identification.

A new set of nine differentials with a single, identified resistant gene was proposed in Japan.[58] They used the Gilmour Octal system for classifying the races. The differentials are: Shin 2 (Pi-k^s), Aichi-asahi (Pi-a), Ishikari-Shiroke (Pi-i), Kanto 51 (Pi-k), Tsuyuake (Pi-m), Fukunishiki (Pi-z), Yashiromochi (Pi-ta), Pi 4 (Pi-ta^2), and Toride 1 (Pi-z^t). The races identified by this system directly indicate their virulence or avirulence to the commercial varieties cultivated in Japan.[59] This approach of using the differentials with a known resistant gene is definitely an improvement over the earlier system.

A lot of confusion exists in the literature because of classifying the intermediate lesions by some workers into susceptible types and by others into resistant types. Veeraraghavan[60] suggested that the moderate reaction of Atkins et al.[51] corresponds to type

TABLE 4-2: DIFFERENTIAL VARIETIES FOR *PYRICULARIA ORYZAE* USED IN DIFFERENT COUNTRIES[17]

Japan	United States	Taiwan	Philippines	India	Korea	Colombia
Tetep	Zenith	Kung-shan-wu-shan-ken	Kataktara DA-2	AC. 1613	Zenith	Raminad Str. 3
Tadukan	Lacrosse	Taichung 65	CI 5309	CR. 906	Ishikari-shiroke	Zenith
Usen	Caloro	Pai-kan-tao	Chokoto	Bengawan	Pi 1	NP-125
Chokoto	Sha-tiao-tsao (P)	Taichung 171	Co 25	S.M. 6	Sensho	Usen
Yakeiko	(CI 8970-P)	Chianung 242	Wagwag	Mas	Kanto 51	Dular
Kanto 51	Sha-tiao-tsao (S)	Kwangfu 1	Pai-kan-tao	Intan	Ayanishiki	Kanto 51
Ishikari-shiroke	CI 5309	Chianung 280	Peta	CR. 907	Norin 17	Sha-tiao-tsao(s)
Homare-nishiki	PI 180061	Taichung line 33	Raminad Str. 3	BJ 1	Norin 22	Caloro
Ginga	PI 201902	Kanto 51	Taichung t-c-w-c	S. 67	Norin 1	Aichi-asahi
Norin 22	Wagwag	Norin 21	Lacrosse		Tonewase	Ishikari-shiroke
Aichi-asahi	Raminad Str. 3	Sensho	Sha-tiao tsao(s)			(Napal)*
Norin 20	(Rexoro)*	Cutsugulcul	Khao-tah-haeng 17			(Bluebonnet
	(Taichung 65)*	Natala				50)*
		Kao-chio-lin-chou				
		Kaohsiung-ta-li-chen-yu				
		Taichung-ti-chio-wu-chien				

*Used as supplementary cultivar

'C' of Padmanabhan et al.[61] and the type 3 lesion[62] or C type should be considered susceptible because this lesion also has an ashy grey center and brown margin. He further
emphasized that to get more precise race identification, care should be taken to have
uninoculated controls of differentials and purity of seeds of differentials.

A large number of races have been identified in studies with isolates from different
rice-growing countries of the world. However, it is also believed that stable race flora[63] or
isolates[64] may be obtained.

The situation on the pathogenic races of blast is ambiguous as different systems and
differentials are being used in different countries. A cooperative research at the international level should be initiated to identify a good set of differentials, preferably with
known resistant gene(s) and to standardize the methodology so that the population of *P.
oryzae* throughout the world may be classified into specific race(s). The results obtained
from this endeavor will be comparable and would help in tackling the problem of blast in
different parts of the world through resistant breeding programme.

4-5 EPIDEMIOLOGY

The development of the blast in epidemic proportion is influenced by the presence of
inoculum, susceptible stage of the host, and period of favorable environmental conditions. The relation among these factors leading to development of blast and its epidemics
has been shown in Figure 4-4.

4-5-1 Disease Cycle

Inoculum sources.   The carryover of blast inoculum from one season to the
next appears not to be an important factor in tropics because conidia are present throughout the year in air,[17] but it plays an important role in the disease cycle in temperate
regions. In temperate and subtropical regions, the pathogen overwinters as mycelium and
conidia on diseased straw and seed.[65-66] Ito[67] found that spores and mycelium on the surface of straw pile died before the next spring but those within the straw pile overwintered.
Moisture, temperature, and microbial activity destroy the mycelia in straw. In hilly areas
of India, the fungus overwinters within straw pile[68] or in straw covered with winter snow,
and sporulates in April.[69] In plains, overwintering of the pathogen on seeds has been
reported, but their role in the perpetuation of the pathogen is still not certain.[69]

The pathogen has been reported from a number of cultivated and wild host species,
but this aspect is highly controversial as some workers have reported success in cross-
inoculation while others have reported failure.[13,69,70] Mackill and Bonman[71] reported that
isolates of *Pyricularia* from grassweeds *Rottboellia exaltata*, *Echinochloa colona*, and
Leersia hexandra are pathogenic to rice, and likewise some isolates from rice are pathogenic to *Brachiaria distachya*, *E. colona*, *Leptochloa chinensis*, *R. exaltata*, and *L.
hexandra*.

Disease development.   The mature conidia become airborne and lodge on the
surface of rice plants, mainly on leaves, and germinate in the presence of a thin film of
water by germ tube. Water film consisting of rain, dew, or guttation drops are essential

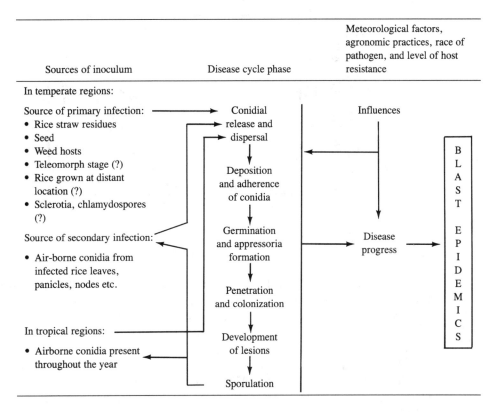

Figure 4-4 The relationships among inoculum sources, disease cycle phase, and disease development leading to blast epidemics

for germination. Germination and formation of appressoria occur between 10°C and 33°C, optimum being 25°C to 28°C for germination and between 16°C and 25°C for appressoria formation.[72] An appressorium formation starts generally 4 hr after absorption of water (Figure 4-5). The average time for appressoria formation in a population of conidia is 11 hr at optimum temperature of 24°C[73] (Figure 4-6).

The infection peg produced from the appressorium penetrates the cuticle or epidermis or enters leaf tissues through stomata. The presence of dew period plays an important role in penetration and colonization. A minimum period of 6–8 hr at 25°C was sufficient to initiate the infection, whereas at other temperatures a longer period was required[74] (Figure 4-7).

The period required to invade the host cells by conidia varies from 10 hr at 32°C to 8 hr at 28°C or 6 hr at 24°C.[75] Initiation of infection occurs from 5–7 hr at 21–27°C and 8 hr at 18°C after deposition of conidia on wet leaves, and almost all conidia complete infection at these temperatures within 18 and 21 hr, respectively.[73]

The lesions appear in 13–18 days at 9–10°C, 7–9 days at 17–18°C, 5–6 days at 24–25°C, and 4–5 days at 26–28°C.[76] Temperature does affect the growth of lesions. At 26.6°C, lesions become about twice as long in 7 days in comparison to the plants kept at 15.5°C.[77] The larger size of lesions at 26.6°C may be explained because the optimum

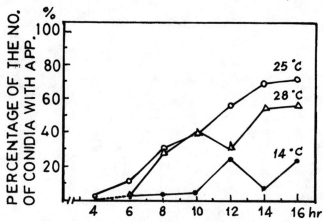

Figure 4-5 Percentage of the number of conidia of *Pyricularia oryzae* which formed appressoria with the passage of time (After Hozumi Suzuki 1969).

temperature for mycelial growth is 26–28°C. Conidiophores develop within 2–4 hr and mature within 4–6 hr if placed in water-saturated conditions[78–79] and produce conidia within 40 min. The fungus sporulates in the temperature range of 12°C to 34°C with an optimum temperature of 28°C and relative humidity over 89%, with an optimum of more than 93%.

The sporulation potential reaches a peak more quickly and decreases more rapidly at higher temperatures (a constant 32°C, and fluctuating temperatures of 32/20°C and 32/25°C) than at medium (20°C, 25°C, and 25/16°C) and lower temperatures (16°C and 20/16°C).[80] (See Figure 4-8.)

The maximum number of conidia produced on the peak day at 20°C were nearly double that at 25°C or 32°C and sevenfold that obtained at 16°C. Conidia were produced rapidly at 28°C which decreased after 9 days, whereas sporulation at 16–24°C showed a tendency to increase even after 15 days.[81] Under natural conditions sporulation and release occur simultaneously on different aged conidiophores. Spore release may commence after 4 hr, if relative humidity is beyond 95%. Release occurs more in the range of

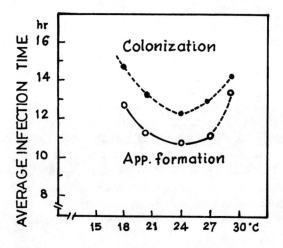

Figure 4-6 Average time for appressorium formation and colonization of a conidial population in rice blast fungus under different temperatures (After Yoshino, 1974).

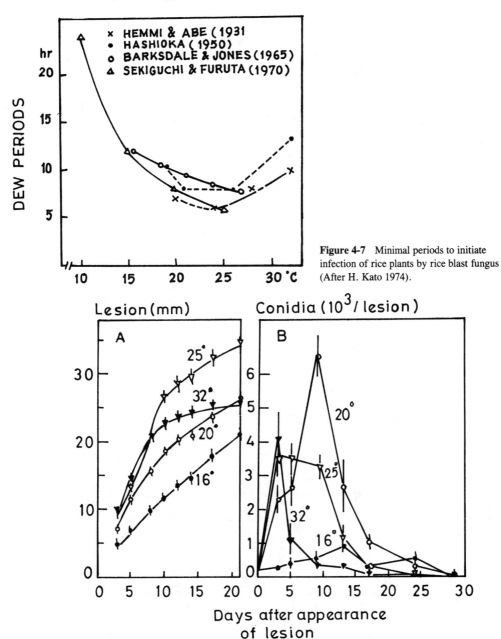

Figure 4-7 Minimal periods to initiate infection of rice plants by rice blast fungus (After H. Kato 1974).

Figure 4-8 Effect of temperature on enlargement of leaf-blast lesions and on sporulation poten *oryzae* (After Kato & Kozaka, 1974).

11–26°C than at 30°C and 35°C.[82] Water has been recognized as an agent for the release of conidia. When water droplets attach themselves to the hilum, detaching takes place[83] or they are violently discharged, although to a very short distance, by the bursting of the minute stalk cell attaching it to the conidiophore[84] or released automatically in a saturated humidity.[12] Most of the mature conidia are released immediately after splashing. A sudden increase in concentration of airborne conidia commences with the start of rain.[85] On a fair day, dew and guttation drops are important in inducing the release of conidia. The longer the dew period, the more spores are released[17] (Figure 4-9).

Most of the conidia are produced and released during the night and are dispersed to short or long distances depending on wind velocity. Conidia are dispersed for a relatively short period during the growing season in temperate regions. In Japan, the peak of the airborne population is generally during August, whereas such conidia are present in the tropics throughout the year with peaks extending from May/June to November/December[17] (Figure 4-10). Diurnal changes in the rates of sporulation, release, and dispersal of conidia are shown in Figure 4-11.[86]

Released conidia float under the canopy of rice plants and then disperse out into the air above the canopy, which are carried to short or long distances through wind, captured in rain drops, and frequently fall on leaf surface to complete the infection chain.

The disease cycle of the blast pathogen is shown in Figure 4-12.

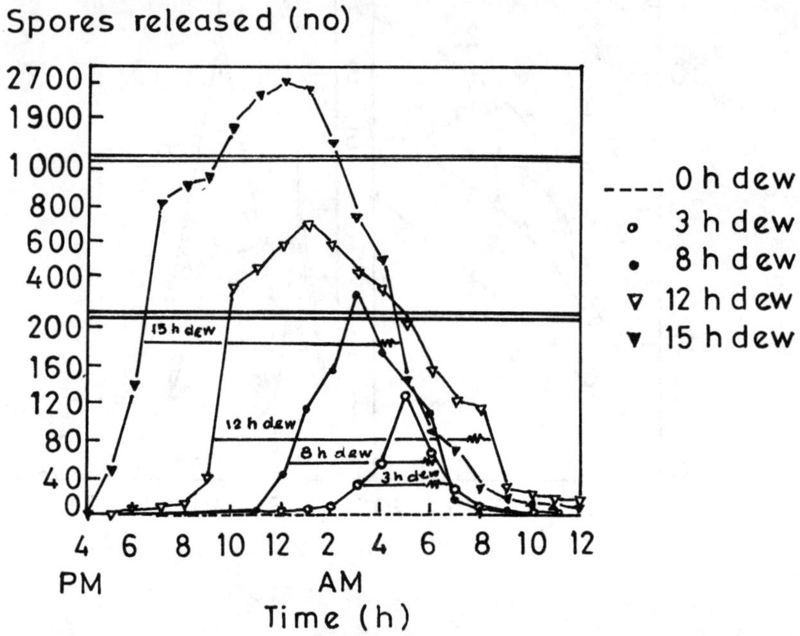

Figure 4-9   Effect of dew period on spore release of *P. oryzae* from leaf lesions in blast nursery at IRRI, 1976 dry season (IRRI, 1977).[167]

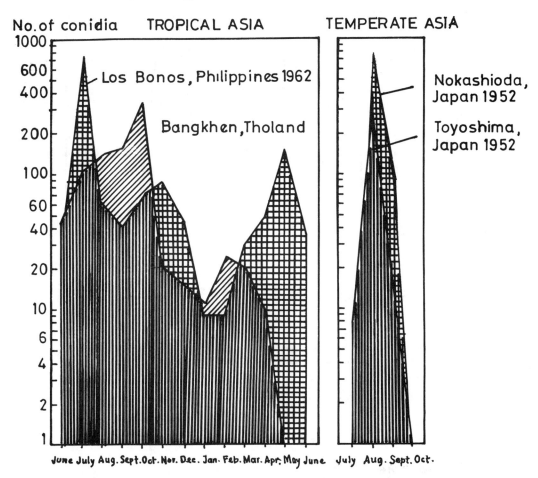

Figure 4-10   Airborne conidia of *P. oryzae* in rice fields in the tropics and temperate zone (c.f. Ou, 1985).

4-5-2 Factors Associated with Blast Epidemics

There are several reports about the relation between epidemics of rice blast and meteorological factors; nutritional level, particularly nitrogenous and silicon contents of the soil; susceptible growth stages of rice plants and level of host resistance. The most common factors that favor the epidemics are lower night temperature between 15–20°C in Uttar Pradesh hills[87] or around 20°C or less than 26°C in the plains of India,[88-89] relative humidity 90% or above, long periods of drizzling rain, more number of rainy days, longer duration of dew, cloudy weather, and slow wind movement. Higher application of nitrogenous fertilizer and soils having less silica contents increase the blast incidence. Seedling-tillering stages are more prone to foliar blast and flowering stage for neck/panicle infections. Blast causes more damage to upland rice than lowland rice probably because of less silica accumulation, favorable microclimatic conditions, and a longer dew period.

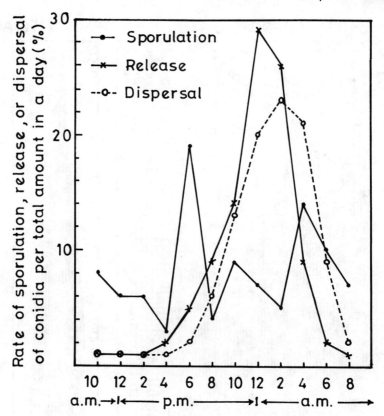

Figure 4-11 Diurnal changes in rates of sporulation, release and dispersal of conidia. Rate of sporulation based on Kuribayashi et. al. (Suzuki, 1975).

4-5-3 Blast Forecasting

Studies on rice blast forecasting have been made by several researchers[89–95] and the models have been proposed to predict blast outbreak. Information on inoculum, favorable weather conditions, and the susceptibility of host plants have been considered as the basis for blast forecasting. Critical stages of the crop for vulnerability to blast are seedling, tillering, and flowering. When the inoculum is present and the favorable environmental conditions coincide with any one of these stages, disease outbreak becomes inevitable.

Correlations of the degree of disease incidence have been made with the spores collected;[85] of blast lesions with trapped spores and wetting period of leaves;[90] with dew period and airborne spores;[94] minimum conditions of dew and temperature required for infection;[70] minimum temperature, relative humidity,[87] along with a prolonged period of dew deposition, and so on. Kingslover et al.[70] developed an equation

$$1/D = 0.265 - \frac{4.28}{T}$$

where D denotes hours of dew and T denotes temperature (degrees celcius) and indicated that this could be used to predict favorable conditions for lesion formation on about 80%

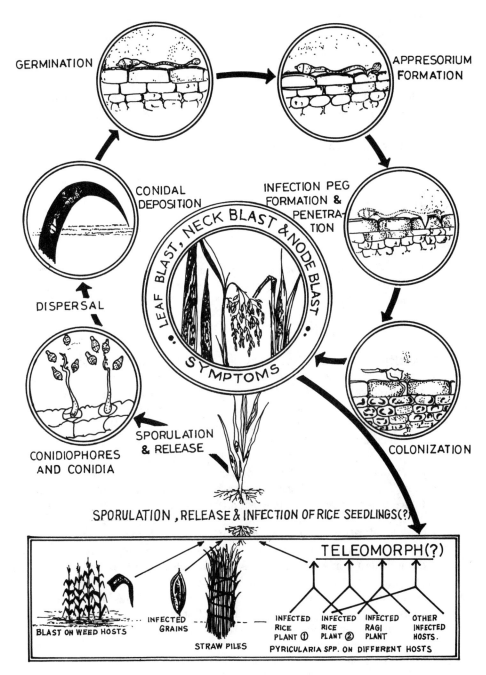

GERMINATION

APPRESORIUM
FORMATION

CONIDAL
DEPOSITION

INFECTION PEG
FORMATION &
PENETRA-
TION

LEAF BLAST, NECK BLAST & NODE BLAST

DISPERSAL

SYMPTOMS

CONIDIOPHORES
AND CONIDIA

SPORULATION
& RELEASE

COLONIZATION

SPORULATION, RELEASE & INFECTION OF RICE SEEDLINGS(?)

TELEOMORPH(?)

BLAST ON WEED HOSTS

INFECTED
GRAINS

STRAW PILES

INFECTED
RICE
PLANT ①

INFECTED
RICE
PLANT ②

INFECTED
RAGI
PLANT

OTHER
INFECTED
HOSTS.

PYRICULARIA SPP. ON DIFFERENT HOSTS

DISEASE CYCLE OF PYRICULARIA ORYZAE

Figure 4-12   Disease cycle of rice blast disease.

of the nights. Temperature below 15°C or dew periods less than 8 hr prevent infection. A fairly good forecast may be made five days ahead of the disease outbreak using dew period and airborne spores, with the equation

$$Y = 2.9 - 0.945D - 0.0098S + 0.1520D^2 + 0.004DS - 0.0000000D^2 S^2$$

where Y, D, and S denote the number of lesions per seedling, dew period in hours, and number of spores per 2.8 litres air, respectively.[94] Different methods like sheath inoculation test, number of silicated cells in flag leaf, chemical components of the rice plant, quantity of starch in rice sheath, and so on have also been used for forecasting the disease.[92,96]

4-6 CONTROL MEASURES

Use of host resistance, chemical control, manipulation of cultural practices and regulatory methods have been used to control blast. Of these, one or more than one strategies may be applicable in a given area. Strategies should be selected, developed, and modified according to the needs of farmers in a specific area, keeping in mind that the different methods when used in combination are better to manage a disease.[97]

4-6-1 Varietal Resistance

Development and use of resistant varieties is the most effective and economical method to control blast and it has long been in use by incorporating resistance into agronomically acceptable varieties. Vertical resistance (VR) and horizontal resistance[98] (HR) have been identified to reduce the losses caused by blast. Seven genetic components of slow blasting have been identified,[99] which in general is manifested by HR. True resistance and field resistance are more commonly used terms in Japan than VR and HR for rice blast.[59] Toriyama et al.[100] showed that the field resistance of Chugoku31 was controlled by a single major gene Pi-f. The field resistance of Chugoku31 also broke down in a few years.[101] Evidently, field resistance (as defined by Japanese scientists) is specific and controlled by major genes and may not differ from true resistance in longevity.[102]

Evaluating varietal resistance. HR is polygenic, generally considered stable, long-lasting, nonrace specific, recognized by decrease in disease severity, and slows down a blast epidemic after it has started, whereas VR is referred to as monogenic, recognized by absence of symptoms, and serves to prevent the occurrence of an epidemic. When favorable environmental conditions for blast are present, HR has been found ineffective, as evidenced by development of blast epidemic in Korea. However, it appears most suitable for use in subsistance crop production. Alternative disease control should be formulated for epidemic conditions, if HR is used as a disease control strategy.[97] In Japan, the resistance provided by vertical genes has not been found durable. Breakdown of resistant varieties such as IR-8, IR-20, CICA-4, CICA-8, etc., indicate that varieties screened for VR do not have durable resistance. Hence, ineffectiveness of VR imparts importance to breeding for HR.[103] On the other hand, effective and stable control with monogenic resistance is advocated for blast disease,[104] especially for intensive rice production areas such as Korea, where all of the necessary ingredients, including an excellent forecasting and race identification system, are present. A scheme of rotation of monogenes is suggested [Figure 4-13(a,b)].

Horizontal Resistance.   IRAT and IITA have emphasized the horizontal blast resistance in their breeding programs.[105-106] In the field, IRAT uses the decreasing inoculum trial for the evaluation of resistance (DITER) design,[107] in which a gradient of spores is distributed from one susceptible spreader plot.

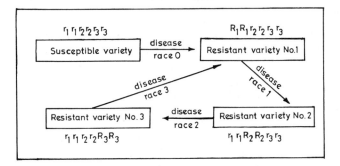

Figure 4-13a   A three-gene system of variety rotations to minimize genetic vulnerability by utilizing directed selection in the host to produce directed selection in the pathogen.

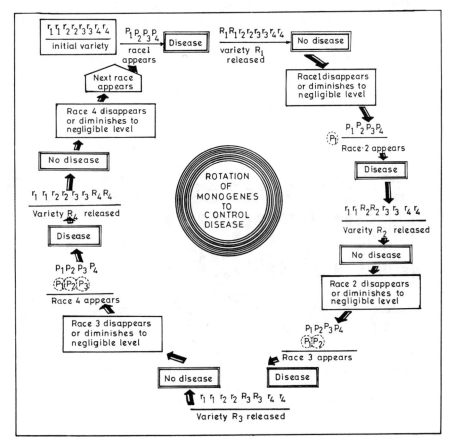

Figure 4-13b   Rotation of monogenes to control disease.

Apparent infection rate, based on a formula developed by Vander Plank, has been used at IRRI to measure HR by the equation

$$r = \frac{1}{t2 - t1} \left(\log_e \frac{x_2}{1 - x_2} - \log_e \frac{x_1}{1 - x_1} \right)$$

where rate (r) of disease increase (x) over time (t) corrected for decreasing amounts of healthy tissue ($1 - x$). The subscripts denote the beginning and end points of the time interval for which r is calculated.[108]

Vertical Resistance.   A uniform blast testing method that facilitates evaluation of the resistance of a large number of rice varieties to prevalent races in the region was adopted for the International Blast Nursery Program.[109] Layout of the uniform blast nursery is shown in Figure 4-14. This method is highly efficient for identifying resistant parents and breeding lines. Plants can be quickly and continuously screened for stability of resistance to many races around the world[110] particularly in different hot spot locations.

Screening Procedures.   The procedure for screening against blast resistance is given by Jennings et al.[110] They were of the opinion that foliar and neck resistance may vary in some varieties, hence the varieties should be screened periodically at both seedling and neck stages.

Scoring Procedure.   Various scales have been designed, based on type, color, and number of lesions as well as the amount of stunting. A Standard Evaluation System[62] has been followed in a majority of the cases for recording leaf and neck blasts. In 1988, the scale[111] has slightly modified, and is given hereunder.
Scale for leaf blast (B1) at nursery:

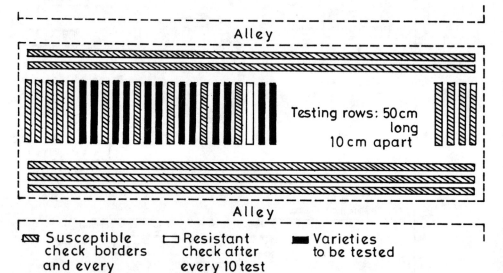

Figure 4-14   Layout of the uniform blast nursery.

0 = No lesions;

1 = small brown specks of pinpoint size or larger brown specks without sporulating center;

2 = small roundish to slightly elongated, necrotic gray spots, about 1–2 mm in diameter, with a distinct brown margin. Lesions are mostly found on the lower leaves;

3 = lesion type is the same as in scale 2, but a significant number of lesions are on the upper leaves;

4 = typical susceptible blast lesions, 3 mm or longer, infecting less than 2% of the leaf area;

5 = typical blast lesions infecting 2–10% of the leaf area;

6 = typical blast lesions infecting 11–25% of the leaf area;

7 = typical blast lesions infecting 26–50% of the leaf area;

8 = typical blast lesions infecting 51–75% of the leaf area with many leaves dead;

9 = more than 75% of leaf area affected.

For actual estimation of the disease in the field, the percentage of blast area may be taken together with predominant lesion type, the code of which is given as:

0 = no lesions;

1 = small brown specks of pinpoint size or larger brown specks without sporulating center;

3 = small, roundish to slightly elongated necrotic sporulating spots, about 1–2 mm in diameter with a distinct brown margin or yellow halo;

5 = narrow or slightly elliptical lesions, 1–2 mm in breadth, more than 3 mm long with a brown margin;

7 = broad spindle-shaped lesion with yellow, brown, or purple margin;

9 = rapidly coalescing small, whitish, grayish, or bluish lesions without distinct margins.

A scale for neck blast/panicle blast (PB) based on symptoms is:

0 = no visible lesion or lesions on only a few pedicels;

1 = lesions on several pedicels or secondary branches;

3 = lesions on a few primary branches or the middle part of panicle axis;

5 = lesion partially around the panicle base (node) or the uppermost internode or the lower part of panicle axis near the base;

7 = lesion completely around panicle base or uppermost internode or panicle axis near the base with more than 30% of filled grain;

9 = lesion completely around panicle base or uppermost internode or the panicle axis near the base with less than 30% of filled grains.

The severity of PB can be computed on the basis of number of panicles with each scale using the formula:

$$PBS = \frac{(10 \times N_1) + (20 \times N_3) + (40 \times N_5) + (70 \times N_7) + (100 \times N_9)}{\text{Total no. of panicle observed}}$$

where N_1–N_9 are the number of panicles with score 1–9.
Scoring for PB may be done using a scale
0–9, where 0 = no incidence; 1 = less than 5%; 3 = 5–10%; 5 = 11–25%;
7 = 26–50%; and 9 = more than 50%.

Potential Donors.   Varietal reaction may vary from country to country, from locality to locality, and from season to season in the same locality. Several varieties with
broad-spectrum resistance have been identified through the International Rice Blast Nursery (IRBN) which is handled as an integral part of the International Rice Testing Programme (IRTP) coordinated by the International Rice Research Institute (IRRI), Philippines. Every year the varieties and elite breeding lines are tested for blast resistance at
35–40 sites in about 20 countries. On the basis of the results, some outstanding sources of
resistance have been identified, many of which are improved plant-type breeding lines.[102]
Some potential donors for blast identified through IRTP include: Tetep, Carreon, C 46-
15, IRAT 104, IR 5533-pp 854-1, IR 1905-81-3-1, Ta-poo-cho-z, Tres Marias, Huan-sen-
goo, Camponi SML, CIAT-ICA 5, CIAT-ICA-12. Entries that are rated resistant to blast
in different countries and resistant at 75% or more locations in the IRBN conducted from
1975 onward have been identified and the details of such entries are given every year in
the IRTP reports published by IRRI. The donors combining other desirable attributes for
use in breeding programs are also selected in national and regional programs in different
countries, for making the best use of suitable resistance in different rice varieties. Sources
with multiple resistance can be used in areas where more stresses exist, in addition to
blast.[112–116]

Genetics of resistance.

Genetic studies on blast resistance were first reported
by Sasaki.[49] Subsequent work up to about 1959 was summarized by Takahashi.[117] Resistance studies indicated that the genes controlling resistance vary from one to three pairs
and that resistance is dominant in most cases. Systematic studies were undertaken only
after Goto et al.[118] established the differential system for races of *Pyricularia*. Kiyosawa
and his group used seven fungus strains in their studies, viz., p-2b (race N-2), Ken 53-33
(T-1), Ina 72 (C-3), Hoku 1 (N-1), Ken 54-20 (N-2), Ken 54-4 (N-3), and Ina 168 (N-4).
The work has been reviewed periodically.[32,119–124] Presently, more than a dozen genes imparting resistance to blast have been identified in Japan. Resistance genes identified in
different varieties and countries have been summarized by Chaudhary and Nayak (Table
4-3).[125] Blast resistance genes in about one hundred Brazilian rice varieties, using Japanese strains and a Philippine strain of *P. oryzae*, have been identified.[126]

Other genetical studies indicated that a single dominant gene Pi-1 carries resistance
to U.S. blast race 1 in Northrose and Nato and Pi-6 to race 6 in Zenith and Gulfrose;[127]
three dominant genes for races 4, 22, 24 were designated as Pi-4, Pi-22, and Pi-25 for
resistance in Japonica rices in Taiwan;[128] seven were identified from Tetep, Carreon, Ta-poo-cho-z, Pankhari 203, and Dawn against the races P06-6 (1B-4), 43 (1H-1),
and 1K 81-3 (IA-61).[129] Recently studies on mode of inheritance and genetic studies have
also been carried out by Marchetti et al.[130] and Yu et al.[131]

TABLE 4-3:   RESISTANCE GENES IDENTIFIED IN DIFFERENT VARIETIES

Origin	Variety	Genes
China	Usen, Pe Bi Hun	Pi-a, others
	Yakeo-ko, Reishiko	Pi-k, Pi-k
	To-to (short grain), Choko-to	Pi-k, Pi-a
	Hokushi Tami	Pi-k, Pi-a, Pi-m
	To-to (long grain), Taichung 65, Sha-tiao-tsao	Pi-k^s
	Oka-ine	Pi-ta
	Pai-kan-tao	Pi-ta, others
	Taichung Glu Yu 26	Pi-a, Pi-i
	Kannonsen	Pi-a, Pi-ta
India	HR-22	Pi-k ^{p} Pi-kh
	CO 25	Pi-z^t, Pi-a, others
	TKM 1	Pi-z^t, others
	Charnack	Pi-k^c, others
	CO 4	Pi-z^t
Indonesia	Tjina	Pi-b
	Tjahaja, and	Pi-b, pi-t
	Bengawan	Pi-b, Pi-t
Japan	Aichi-Asahi	Pi-a
	Ishikari Shikore	Pi-i
	Kento-51 (Toto-China)	Pi-k
	Shen 2	Pi-k^s
	Chugoku 31, St 1	Pi-f
	Fujisaka 5, Sekiyama 2, Hokuriku 12, Takasagomochi, Akishino-mochi	Pi-i
	Kusabue, Minehikari	Pi-k
	Norin 6, Shirogane, Kamenoo, Shinriki, Kinmaze-Sasashigure, Norin 17, Norin 18, Zuiho, Ishikari-Shikore, Fujisaka 5, Sekiyama 2, Hokuriku 12, Takanenishiki	Pi-k^s

TABLE 4-3: CONTINUED

Origin	Variety	Genes
Japan (Continued)	Minehikari, Tsuyuake, Sanpuku	Pi-m
	Yashiro-mochi Pi 1, k 1	Pi-ta
	Pi 4	Pi-ta^2
	Ou 244, Fukei 67	Pi-z
Korea	Doazi chall	Pi-i
	Jae Keum, Pal Tal	Pi-a, Pi-a
	Most of 86 varieties	Pi-a, Pi-i, Pi-k
Malaysia	Morak Sepilai, Kontor	Pi-z^t
	Milak Kuning	Pi-b
Pakistan	Pusur	Pi-Pk^p, Pk-a, others
	Dular	Pi-k^a, others
Philippines	Tadukan	Pi-ta and/or Pi-ta^2
	Tadukan	Pi-k
Thailand	Leuang Tawing 77-12-5, Chao Leuang 11	Pi-z^t
USA	Dawn	Pi-a, Pi-i, pi-k, others
	Zenith	Pi-z, Pi-a
	Caloro Lacrosse	Pi-k^s, Pi-k^s
	Blue Bonnet	Pi-a
USSR	Rossia No. 33	Pi-k^a
Vietnam	Te-tep	Pi-k^a, others

Source: Based on Kiyosawa,[120, 122] Ou,[32] and Chaudhary and Nayak[125]

All the 13 genes identified by Kiyosawa and his group for resistance were dominant, but some other scientists have identified recessive genes for blast resistance. One recessive gene for resistance to isolate PO6-6 was identified in IR 54[131] Three new recessive genes—Pi-n in Brazos showing resistance against race IB-54, Pi-g in Gulfrose against IG-1, and Pi-d in Lebonnet against IB-1—were identified. Close linkage between the dominant Pi-k^h gene and recessive Pi-d gene in the cultivar Lebonnet has been established.[130]

Breeding for blast resistance.   Breeding for resistance to blast has been in progress in different countries for at least 50 years. The work up to 1963 has been reviewed for Japan,[132] the United States,[133] India,[7] Taiwan,[134] and Thailand.[135] In the last workshop on rice blast, this aspect was reviewed by several workers.[57,59,136–137]

Considerable success has been achieved in developing blast resistant varieties in some countries, by using mainly vertical resistant genes in rice breeding programs. But in general, varieties with long-lasting resistance have not been developed, due to variability in pathogenicity of the fungus. As an alternative to vertical resistance, four strategies were suggested to combat the blast disease, namely, single-gene addition, pyramiding of genes, horizontal resistance, and multiline varieties.[138] In recent years, exploitation of durable resistance has also been proposed for blast control particularly for the less blast-conducive environments.[139] Durable blast resistance has been identified in certain rice cultivars [e.g., IR 20, IR 36, and IR 42 (under Tropical lowland conditions); Fu She 94, Shuang Feng 4, Xiang Ai Zao 9, Zhaoyangzao 18, Zhenluon 13, and Zhenshen 97 (under Temperate lowland conditions); Fukuton, IAC 25, Kuroka, Moroberekan, and OS 6 (under Upland conditions)] and is associated with resistance that is partial in its effect, inherited through an indefinite number of genes, and apparently race non-specific.[140]

4-6-2 Cultural Control

Seedlings raised in upland nurseries are more susceptible to the disease even after transplanting due to lower silicon content of the epidermal cells.[141] These seedlings contain more soluble nitrogen, amino acids, and amines which favor disease development, and have higher physiological activities, producing more roots and absorbing more nitrogen than seedlings grown in wet nurseries. The plants raised under flooded conditions synthesize increased amounts of phenolic compounds and total soluble and reducing sugars. Nonreducing sugar synthesis is lower in flooded conditions.[142]

Application of lesser amounts of nitrogenous fertilizer and in split doses generally reduces the disease severity. Soluble nitrogen, particularly amino acids and amines, markedly increases in plants receiving high nitrogen. The soluble nitrogen that accumulates in plants may serve as a suitable nutrient for fungus growth. Plants receiving larger amounts of nitrogen have fewer silicated epidermal cells and thus lower resistance. Application of nitrogen also reduces hemicellulose and lignin in the cell wall and weakens the plants' mechanical resistance to blast.[143]

The time of planting is also critical in the outbreak of disease.[87,144-146] By suitable adjustment of the date of sowing in each region, the coincidence of favorable blast season and the most susceptible stages of blast infection could be avoided.

4-6-3 Chemical Control

Chemicals are extensively used in Japan to control the disease. Japanese rice farmers routinely use various fungicides and methods of application for rice blast control. In the United States and Latin America, chemical control of rice diseases, especially blast, has become routine among rice farmers. These countries are highly developed agriculturally and industrially and have well-developed scientific, chemical, and extension communities that interact with farmers. Such conditions do not exist in most of the rest of the rice-growing world, where chemical control of blast is in its infancy and has the potential for rapid expansion.[97]

Several chemicals have been tested and recommended for controlling blast in different countries. Copper compounds were used in early years to control blast. Among these,

Bordeaux mixture was effectively used by Metcalf as early as 1906.[5] Various other copper compounds viz., Perenox 0.35%, Sandoz copper, Copper oxychloride, Coppesan 0.5%, and Cuprous oxide, were also reported to be effective in controlling the disease.[147-148] Later, the copper fungicides were used in mixtures with mercuric compounds like Phenyl Mercuric Acetate (PMA), which were more effective than copper alone. But fungicides of copper[89] and mercury group[149] could not become popular with farmers due to their phytotoxicity especially on high-yielding varieties.[148] Afterward the development and use of organosulphur compounds, especially Mancozeb and Zineb, became popular for blast control.

The phytotoxic effect of mercuric compounds to rice plants and the quest for more effective compounds led to the development and use of a number of antibiotics against the disease.[150] An antibiotic cephalothecin was produced by a species of *Cephalothecium*, which inhibited the growth of *P. oryzae* on rice leaves.[151] Following this, antiblastin, antimycin A, blastmycin, and blasticidin–A were developed but none of these could be used under field conditions because of their instability and/or high toxicity to fish. In 1955, blasticidin–S was developed by Fukunaga and his colleagues[150] with better efficacy to control blast. Blasticidin–S and Kasugamycin in Japan and Qingfengmycin and others in China have been used for blast control. Kasumin has also tested extensively in India and has provided good control of blast. Its spray was recommended for the control of blast on high yielding varieties because of its nontoxicity to the crop.[89]

Among the organophosphorus fungicides, Hinosan (edifenphos) and Kitazin (IBP) are the major compounds that are effective against blast and have been widely used to control the disease. In *P. oryzae*, they inhibit phospholipid N-methyltransferase,[153-154] a fungitoxic mechanism that is presumed to be the basis for their fungitoxicity and plant protective action.

Benzimidazole and related fungicides form another group of highly effective fungicides that are widely used for efficient plant disease control. The fungicides are broad-spectrum and systemic in nature. The fungitoxic moeity for the benzimidazole derivatives effective for control of disease is methyl-2 benzimidazole carbamate (MBC). Carbendazim is available in different brand names viz., Bavistin, Bengard, Derosal, Jkstein, MBC, among which Bavistin is the most popular. Benlate has also been extensively tested and reported effective for blast control. Thiophanate methyl (Topsin-M) has also given promising results for blast control. It is a thiourea-based fungicide, whose overall antifungal spectrum resembles that of benzimidazoles.

Besides the important systemic fungicides of benzimidazole group, a few more fungicides like probenazole (Oryzemate) and isoprothiolane act systemically and effectively control blast. However, the effectiveness of probenazole under Indian conditions is not as good as it is in Japan.[155]

More recently, a new group of fungicides known as melanin biosynthesis inhibitors have been introduced for disease control which have great promise as they are nonfungitoxic and control the disease at low concentrations. Members of this group block melanin biosynthesis and have been tested extensively against rice blast with practical control of the disease.[156] These fungicides are antipenetrants that act on the pathogen to prevent its entry through epidermis. Melanization of the appressorial wall appears to be necessary for the architecture and rigidity needed to support and focus the mechanical forces involved in the penetration process.[157-158] Moreover, polyketide metabolites may accumulate as a result of inhibition and may cause the cytotoxic effects, leading to antipenetrant action of fungicides.[159] The antipenetrants found effective for blast control are pyroquilon[155,160-162] and tricyclazole (EL-291).[163-164]

Apart from the effectiveness of certain fungicides for controlling blast individually, their use in combinations is desirable to minimize the chances of development of fungicide resistance in the fungus. Presently, the strategy of chemical control of blast may comprise treatment of seeds with effective fungicides, application of granular fungicides to soil, and/or foliar spray of suitable fungicides. Some fungicides, e.g., Fongorene (pyroquilon) have shown good efficacy in controlling leaf blast through seed treatment.[155,160-162,165-166] Studies and use of such fungicides may greatly help in getting profitable yield returns in epidemic areas through simple seed treatment followed by sequential application of different effective fungicides in the later stages.[166] Such a schedule would also help in minimizing the chances of development of resistant strains of *P. oryzae* to a particular chemical. Application of some granular fungicides like Coratop 5G (Pyroquilon) and Kitazin 17G (IBP) may also prove promising in coming years, as they are easier to apply as compared to conventional spray formulations.[155]

4-7 REFERENCES

1. Moomaw, J. C., and Vergara, B. S. "The Environment of Tropical Rice Production." *Proc. Symp. The Mineral Nutrition of the Rice Plant*, Baltimore, Md. Johns Hopkins Press, 1965, 3.

2. Goto, K. "History of the Blast Disease and Changes in Methods of Control." *Agric. Improvement Bureau, Ministry of Agriculture and Forestry*, Japan, 5 (1955): 1.

3. Parthasarathy, N., and Ou, S. H. "Opening Address: International Approach to the Problem of Blast," in *The Rice Blast Disease*, Proc. Symp. at IRRI, Los Banos, Philippines (Baltimore, Md.: Johns Hopkins Press, 1965), 1.

4. Padwick, G. W. *Manual of Rice Diseases*. Kew: Commonwealth Mycological Institute, 1950, p. 198.

5. Metcalf, H. "A Preliminary Report on the Blast of Rice, with Notes on Other Rice Diseases." *Bull. South Carolina Agric. Expt. St.* 121 (1906): 43.

6. Metcalf, H. "The Pathology of the Rice Plant." *Science* 25 (1907): 264.

7. Padmanabhan, S. Y. "Breeding for Blast Resistance in India," in *The Rice Blast Disease*, Proc. Symp. at IRRI, Los Banos, Philippines, (Baltimore, Md.: Johns Hopkins Press, 1965), 343.

8. Crill, P., Ham, Y. S., and Beachell, H. M. "The Rice Blast Disease in Korea and its Control with Race Prediction and Gene Rotation," in *Evolution of the Gene Rotation Concept for Rice Blast Control*, (Los Banos, Philippines: IRRI 1982), 123.

9. Goto, K. "Estimating Losses from Rice Blast in Japan," in *The Rice Blast Disease*, Proc. Symp. at IRRI, Los Banos, Philippines (Baltimore, Md.: Johns Hopkins Press, 1965), 195.

10. Bhatt, J.C. "Yield Loss in Five Rice Varieties due to Blast Disease." *J. Hill Res.* 1 (1988): 115.

11. Padmanabhan, S. Y. "Estimating Losses from Rice Blast," in *The Rice Blast Disease*, Proc. Symp. at IRRI, Los Banos, Philippines (Baltimore, Md.: Johns Hopkins Press, 1965), 203.

12. Kuribayashi, K., and Ichikawa, H. "Studies on Forecasting of the Rice Blast Disease, Special Report." *Nagano Agric. Expt. St.* 13 (1952): 229.

13. Asuyama, H. "Morphology, Taxonomy, Host Range, and Life Cycle of *Piricularia oryzae,*" in *The Rice Blast Disease*, Proc. Symp. at IRRI, Los Banos, Philippines (Baltimore, Md.: Johns Hopkins Press, 1965), 9.

14. Hughes, S. J. "Revisiones Hyphomycetum Aliquot cum Appendice de Nominibus Rejiciendis." *Can. J. Bot.* 36 (1958): 727.

15. Cavara, F. *Fungi Longobardiae Exsiccati No. 49.* 1891.

16. Shirai, M. "Notes on Plants Collected in Suruga, Totomi, Yamato and Kii." *Bot. Mag. Tokyo* 10 (1896): 111.

17. Ou, S. H. *Rice Diseases*. UK: Commonwealth Agricultural Bureaux, 1985, p. 380.

18. Rossman, A. Y., Howard, R. J., and Valent, B. "*Pyricularia grisea*, the Correct Name for the Rice Blast Fungus." *Mycologia* 82 (1990): 509.

19. Horino, O., and Akai, S. Comparison of the Electron Micrographs of Conidia of *Helminthosporium oryzae* Ito and Kurib. and *Piricularia oryzae* Cavara." *Trans. Mycol. Soc. Japan* 6 (1965): 41.

20. Wu, H. K., and Tsao, T. H. "The Ultrastructure of *Piricularia oryzae* Cav." *Bot. Bull. Acad. Sinica 8* Special No. (1967): 353.

21. Hashioka, Y., and Ikegami, H. "Fine Structure of the Rice Blast. V. Surface Structure of the Rice Blast Fungus and the Diseased Leaves." *Res. Bull.* (Faculty Agriculture, Gifu University) 28 (1969): 41.

22. Suzuki, H. "Origin of Variation in *Piricularia oryzae,*" in *The Rice Blast Disease*, Proc. Symp. at IRRI, Los Banos, Philippines (Baltimore, Md.: Johns Hopkins Press, 1965), 111.

23. Suzuki, H. "Studies on Biologic Specialization in *Pyricularia oryzae* Cav." (Tokyo: Institute of Plant Pathology, Tokyo University of Agriculture and Technology, 1967), p. 235.

24. Chu, O. M. Y., and Li, H. W. "Cytological Studies of *Piricularia oryzae* Cav. *Bot. Bull. Acad. Sinica*, Taipei 6 (1965): 116.

25. Yamasaki, Y., and Niizeki, H. "Studies on Variation of the Rice Blast Fungus *Piricularia oryzae* Cav., I. Karyological and Genetical Studies on Variation." *Bull. Nat. Inst. Agric. Sci.* D13 (1965): 231.

26. Giatgong, P., and Frederiksen, R. A. "Variation in Pathogenicity of *Piricularia oryzae.*" *Phytopathology* 57 (1967): 460.

27. Giatgong, P., and Frederiksen, R. A. "Chromosomal Number and Mitotic Division in *Piricularia oryzae.*" *Phytopathology* 58 (1968): 728.

28. Giatgong, P., and Frederiksen, R. A. "Pathogenic Variability and Cytology of Monoconidial Subcultures of *Piricularia oryzae.*" *Phytopathology* 59 (1969): 1152.

29. Yaegashi, H., and Hebert, T. T. "Perithecial Development and Nuclear Behaviour in *Pyricularia.*" *Phytopathology* 66 (1976): 122.

30. Ou, S. H. "Pathogen Variability and Host Resistance in Rice Blast Disease." *Ann. Rev. Phytopathol.* 18 (1980): 167.

31. Singh, R. A. "Studies on the Cytology, Mechanism of Variability and Mycelial Interaction in Rice Blast Fungus." (Los Banos, Philippines: IRRI, 1982), p. 32.

32. Ou, S. H. "Breeding Rice for Resistance to Blast-A Critical Review." *Proc. Rice Blast Workshop*, (Los Banos, Philippines: IRRI, 1979), 81.

33. Manibhushanrao, K., and Ou, S. H. "Karyological Studies in *Pyricularia oryzae* Cav." *Il Riso* 29 (1980): 305.

34. Row, K. V. S. R. K., Aist, J. R., and Crill, J. P. "Mitosis in the Rice Blast Fungus and Its Possible Implication for Pathogenic Variability." *Can. J. Bot.* 63 (1985): 1129.

35. Singh, R. A. "Relevance of Cytological Studies in Plant Pathology, Presidential Address." *Ann. Conf. Soc. Myc. & Pl. Pathol.* (1990): 13.

36. Tamari, K., and Kaji, J. "Biochemical Studies of the Blast Fungus (*Piricularia oryzae* Cav.), The Causative Fungus of the Blast Disease of Rice Plants. I. Studies on the Toxins Produced by Blast Fungus." *J. Agr. Chem. Soc.* Japan 28 (1954): 254.

37. Lebrun, M. H., Dutfoy, F., Gaudemar, F., Kunesch, G., and Gaudemer, A. "Detection and Quantification of the Fungal Phytotoxin Tenuazonic Acid Produced by *Pyricularia oryzae*." *Phytochemistry* 29 (1990): 3777.

38. Hebert, T. T. "The Perfect Stage of *Pyricularia grisea*." *Phytopathology* 61 (1971): 83.

39. Yaegashi, H., and Nishihara, N. "Production of the Perfect Stage in *Pyricularia* from Cereals and Grasses." *Ann. Phytopathol. Soc. Japan* 42 (1976): 511.

40. Kato, H., and Yamaguchi, T. "The Perfect State of *Pyricularia oryzae* Cav. from Rice Plants in Culture." *Ann. Phytopath. Soc. Japan* 48 (1982): 607.

41. Kato, H. "The Perfect State of *Pyricularia* Species." *Rev. Plant Prot. Res. Tokyo* 10 (1977): 20.

42. Yaegashi, H. "On the Sexuality of Blast Fungi." *Pyricularia* Spp. *Ann. Phytopathol. Soc. Japan* 43 (1977): 432.

43. Tsuda, M., Nakagawa, H., Taga, M., and Ueyama, A. "Cultural Conditions and Formation of the Perfect State of *Pyricularia oryzae* Cavara." *Trans. Mycol. Soc. Japan* 18 (1977): 161.

44. Yaegashi, H., and Udagawa, S. "Additional Note: the Perfect State of *Pryicularia grisea* and its Allies. *Can. J. Bot.* 56 (1978): 2184.

45. Barr, M. E. "*Magnaporthe, Telimeenella*, and *Hyponectria* (Physosporellaceae)" *Mycologia* 69 (1977): 952.

46. Krause, R. A., and Webster, R. K. "The Morhphology, Taxonomy, and Sexuality of the Rice Stem Rot Fungus, *Magnaporthe salvinii* (*Leptosphaeria salvinii*)." *Mycologia* 64 (1972): 103.

47. Kato, H., and Yamaguchi, T. "The Perfect Stage of *Pyricularia oryzae* from Rice Plants." *Ann. Phytopathol. Soc. Japan* 48 (1982): 607.

48. Yaegashi, H., and Udagawa, S. "The Taxonomical Identity of the Perfect State of *Pyricularia grisea* and Its Allies." *Can. J. Bot.* 56, (1978): 180.

49. Sasaki, R. "Existence of Strains in Rice Blast Fungus." *J. Plant Prot. Tokyo* 9 (1922): 631.

50. Gangopadhyay, S., and Padmanabhan, S. Y. *Breeding for Disease Resistance in Rice.* New Delhi: Oxford & IBH Publishing, 1987, p. 340.

51. Atkins, J. G., Robert, A. L., Adair, C. R., Goto, K., Kozaka, T., Yanagida, R., Yamada, M., and Matsumoto, S. "An International Set of Rice Varieties for Differentiating Races of *Piricularia oryzae*." *Phytopathology* 57 (1967): 297.

52. Goto, K., Kozaka, T., Yanagita, K., Takahashi, Y., Suzuki, H., Yamada, M., Matsumoto, S., Shindo, K., Atkins, J. G., Robert, A. L., and Adair, C. R. "U.S.–Japan Cooperative Research on the International Pathogenic Races of the Rice Blast Fungus, *Piricularia oryzae* Cav., and Their International Differentials." *Ann. Phytopath. Soc. Japan* 33 (Extra Issue) 1967: 87.

53. Atkins, J. G. "Blast Disease of Rice," in *Rice Diseases of the Americas, A Review of Literature, Agriculture Handbook* No. 448, United States Department of Agriculture 1, 1984.

54. Ling, K. C., and Ou, S. H. "Standardization of the International Race Numbers of *Pyricularia oryzae. Phytopathology* 59 (1969) 339.

55. Veeraraghavan, J. "A New Method of Classification and Nomenclature of Physiologic Races of *Pyricularia oryzae* Cav. *Int. Rice Comm. Newsl.* 24 (1975): 128.

56. Veeraraghavan, J., and Premalatha Dath. A. "Current Position of Physiologic Races of *Pyricularia oryzae* Cav. in India." *Curr. Sci.* 44 (1975): 19.

57. Padmanabhan, S. Y. "Blast Resistance in India," in *Proc. Rice Blast Workshop.* (Los Banos, Philippines: IRRI, 1979), 49.

58. Yamada, M., Kiyosawa, S., Yamaguchi, T., Hirano, T., Kobayashi, T., Kushibuchi, K., and Watanabe, S. "Proposal of a New method for Differentiating Races of *Pyricularia oryzae* Cavara in Japan." *Ann. Phytopath. Soc. Japan* 42 (1976): 216.

59. Ezuka, A. "Breeding for and Genetics of Blast Resistance in Japan," in *Proc. Rice Blast Workshop* (Los Banos, Philippines: IRRI, 1979), 27.

60. Veeraraghavan, J. "Specialisation in Pathogenicity of *Pyricularia oryzae* Cav." *Proc. Ind. Nat. Sci. Acad.* B52 (2) (1986): 300.

61. Padmanabhan, S. Y., Chakrabarti, N. K., Mathur, S. C., and Veeraraghavan, J. "Physiologic Specialisation in *Pyricularia oryzae*," *Proc. Ind. Acad. Sci.* B66 (1967): 63.

62. International Rice Research Institute. *Standard Evaluation System for Rice,* 2nd ed. IRRI (Los Banos, Philippines: IRRI, 1980), 44.

63. Veeraraghavan, J., and Premalatha Dath, A. "Host-Pathogen Equilibrium between Rice (*Oryza sativa L.*) and Its Pathogen *Pyricularia oryzae* Cav." *Curr. Sci.* 45 (1976): 333.

64. Kozaka, T., "The Nature of the Blast Fungus and Varietal Resistance in Japan" in *Proc. Rice Blast Workshop.* (Los Banos, Philippines: IRRI, 1979), 3.

65. Kuribayashi, K. "Studies on Overwintering, Primary Infection, and Control of Rice Blast Fungus, *Piricularia oryzae.*" *Ann. Phytopathol. Soc. Japan,* 2 (1928): 99.

66. Ito, S., Kuribayashi, K. "Studies on the Rice Blast Disease." Dept. Agric. Forestry, Japan, *Farm Bull.* 30 (1931): 81.

67. Ito, S. "Primary Outbreak of the Important Diseases of Rice Plants and Common Treatment for their Control." *Rep. Hokkaido Agric. Expt. Stn.* 28 (1932), 211.

68. Kapoor, A. S., and Singh, B. M. "Overwintering of *Pyricularia oryzae* in Himachal Pradesh." *Indian Phytopathol.* 30 (1977): 213.

69. Padmanabhan, S. Y. "Fungal Diseases of Rice in India." *Indian Council of Agricultural Research,* New Delhi (1974), 66.

70. Kingslover, C. H., Barksdale, T. H., and Marchetti, M. A. "Rice Blast Epidemiology." *Bulletin* 853 (1984): 29.

71. Mackill, A. O., and Bonman, J. M. "New Hosts of *Pyricularia oryzae.*" *Plant Dis.* 70 (1986): 125.

72. Suzuki, H. "Diurnal Periodicity in Spore Discharge of Rice Blast Fungus." *Rev. Plant Prot. Res.* 2 (1969): 64.

73. Yoshino, R. "Ecological Studies on the Infection in Rice Blast Epidemics. Dew, Temperature and Sequent Change of Infection Rates." *Ann. Phytopathol. Soc. Japan* 39 (1974): 186.

74. Kato, H. "Epidemiology of Rice Blast Disease." *Rev. Plant Prot. Res.* 7 (1974): 1.

75. Hashioka, Y. "Effects of Environmental Factors on Development of Causal Fungus, Infection, Disease Development, and Epidemiology" in *The Rice Blast Disease*, Proc. Symp at IRRI, Los Banos, Philippines (Baltimore, Md.: Johns Hopkins Press, 1965), 153.

76. Hemmi, T., Abe, T., Ikeya, J., and Inoue, Y. "Studies on the Rice Blast Disease. IV. Relation of the Environment to the Development of Blast Disease and Physiologic Specialization in the Rice Blast Fungus." *Mater. Rural Improv. Dept. Agric. For. Japan* 105 (1936): 145.

77. Hemmi, T., and Imura, J. "On the Relation of Air Humidity to Conidial Formation in the Rice Blast Fungus, *Piricularia oryzae*, and the Characteristics in the Germination of Conidia Produced by the Strains Showing Different Pathogenicity." *Ann. Phytopathol. Soc. Japan* 9 (1939): 147.

78. Asaga, K., Takahashi, H., and Yoshimura, S. "Observations of Sporulation by *Piricularia oryzae* by Using a Scanning Electron Microscope." *Ann. Phytopathol. Soc. Japan* 37 (1971): 372.

79. Toyoda, S., and Suzuki, N. "Histochemical Studies on the Lesions of Rice Blast Caused by *Piricularia oryzae* Cav., I. Some Observations on the Sporulation of Lesions of Different Types Occurring on Leaves of the Same Variety." *Ann. Phytopathol. Soc. Japan* 17 (1952): 1.

80. Kato, H., and Kozaka, T. "Effect of Temperature on Lesion Enlargement and Sporulation of *Piricularia oryzae* in Rice Leaves." *Phytopathology* 64 (1974): 828.

81. Henry, B. W., and Andersen, A. L. "Sporulation by *Piricularia oryzae.*" *Phytopathology* 38 (1948): 265.

82. Misawa, T., and Matsuyama, N. "Studies on Conidial Dispersal of Rice Blast Fungus." *Ann. Phytopathol. Soc. Japan* 25 (1960): 3.

83. Ono, K., and Suzuki, H. "Studies on Dissemination of Conidia of Rice Blast Fungus." *Proc. Assoc. Pl. Prot. Hokuriku*, 7 (1959): 6.

84. Ingold, C. T. "Possible Spore Discharge Mechanism in *Piricularia.*" *Trans. Brit. Mycol. Soc.* 47 (1964): 573.

85. Suzuki, H. "Studies on the Behaviour of the Rice Blast Fungus Spore and the Application for Forecasting Method of the Rice Blast Disease." *Bull. Hokuriku Agric. Expt. St.* 10 (1969): 114.

86. Suzuki, H. "Meteorological Factors in the Epidemiology of Rice Blast." *Ann. Rev. Phytopathol.* 13, (1975): 239.

87. Bhatt, J. C., and Chauhan, V. S. "Epidemiological Studies on Neck Blast of Rice in U.P. Hills." *Indian Phytopathol.* 38 (1985): 126.

88. Venkata Rao, G., and Muralidharan, K. "Effect of Meteorological Conditions on the Incidence and Progress of Blast Disease on Rice." *J. Plant Dis. Prot.* 89 (1982): 219.

89. Padmanabhan, S. Y., Chakrabarti, N. K., and Row, K.V.S.R.K. "Forecasting and Control of Rice Diseases." *Proc. Indian Acad. Sci.* 37B (1971): 423.

90. Kim, C. K., Yoshino, R., and Mogi, S. "A Trial of Estimating Number of Leaf Blast Lesions on Rice Plants on the Basis of the Number of Trapped Spores and Wetting Period of Leaves." *Ann. Phytopathol. Soc. Japan* 41 (1975): 492.

91. Kim, C. K., Kang, C. S., and Chung, B. J. "Forecasting Methods of Rice Blast Based on the Rice Plant Predisposition." *Res. Rep. ORD, Korea* 19 (1977): 145.

92. Ono, K. "Principles, Methods and Organization of Blast Disease Forecasting," in *The Rice Blast Disease*, Proc. Symp. at IRRI, Los Banos, Philippines (Baltimore, Md.: Johns Hopkins Press, 1965), 173.

93. Yoshino, R. "Influence of Temperature on the Incubation Period of *Pyricularia Oryzae* and Early Detection of Lesions by Staining with Iodine-Potassium Iodide (on rice)." *Proc. Assoc. Plant Prot. Hokuriku* 19 (1971): 11.

94. El Refaei, M. I. "Epidemiology of Rice Blast Disease in the Tropics with Special Reference to the Leaf Wetness in Relation to Disease Development." (Ph.D. thesis, Indian Agricultural Research Institute, New Delhi, 1977).

95. Kiyosawa, S. "Mathematical Studies on the Curve of Disease Increase—a Technique for Forecasting Epidemic Development." *Ann. Phytopathol. Soc. Japan* 38 (1972): 30.

96. Yamaguchi, T. "Forecasting Techniques of Rice Blast." *JARQ* 5 (1970): 26.

97. Crill, P., Ikehashi, H., and Beachell, H. M. "Rice Blast Control Strategies," in *Rice Research Strategies for the Future*. (Los Banos, Philippines: IRRI, 1982), 129.

98. Vander Plank, J. E. *Disease Resistance in Plants*. New York: Academic Press, 1968, p. 206.

99. Villareal, R. L. "The Slow Leaf Blast Infection in Rice (*Oryza sativa* L.)" (Ph.D. thesis, Pennsylvania State University, 1979), 114.

100. Toriyama, K., Yunoki, T., and Shinoda, H. "Breeding Rice Varieties for Resistance to Blast II. Inheritance of High Field Resistance of Chugoku 31." *Jap. J. Breed.* 18 (1968): 145.

101. Yunoki, T., Ezuka, A., Morinaka, T., Sakurai, Y., Shinoda, H., and Toriyama, K. "Studies on the Varietal Resistance to Rice Blast. 4. Variation of Field Resistance due to Fungus Strains." *Bull. Chugoku Agric. Exp. Stn. Ser.* E 6 (1970): 21.

102. Khush, G. S., and Virmani, S. S. "Breeding Rice for Disease Resistance," in *Progress in Plant Breeding—I*, ed. G. E. Russell. (London: Butterworths, 1985), 239.

103. Notteghem, J. L. "Breeding for Disease Resistance in Upland Rice in Africa," in *Breeding for Durable Disease and Pest Resistance*, FAO Plant Prod. & Prot. Paper 55, (1984): 107.

104. Crill, J. P., and Khush, G. S. "Effective and Stable Control of Rice Blast with Monogenic Resistance," in *Evolution of the Gene Rotation Concept for Rice Blast Control*. (Los Banos, Philippines: IRRI, 1982), 87.

105. Alluri, K., Abifarin, A. O., Zan, K., and Alam, M. S. "Varietal Improvement in Dryland Rice at the International Institute of Tropical Agriculture," in *An Overview of Upland Rice Research." Proc. Bouake, Ivory Coast Upland Rice Workshop*. (Los Banos, Philippines, IRRI, 1984), 415.

106. Bidaux, J. M. "Screening for Horizontal Resistance to Rice Blast (*Pyricularia oryzae*) in Africa," in *Rice in Africa*, eds. I. W. Buddenhagen and G. J. Persley. (New York: Academic Press, 1978), 159.

107. Institut de Recherches Agronomiques Tropicales et des Cultures Vivrieres, Memoires et travaux de l'IRAT3. *Rice Blast and its Control*. (France, 1984), 53.

108. International Rice Research Institute, Annual Report for 1979. (Los Banos, Philippines: IRRI, 1980), 538.

109. Ou, S. H. "Varietal Reactions of Rice to Blast," in *The Rice Blast Disease*, Proc. Symp. at IRRI, Los Banos, Philippines (Baltimore, Md.: Johns Hopkins Press, 1965), 223.

110. Jennings, P. R., Coffman, W. R., and Kauffman, H. E. *Rice Improvement*. Los Banos, Philippines: IRRI, 1979, p. 186.

111. International Rice Research Institute, Standard Evaluation System for Rice, 3rd ed. (Los Banos, Philippines: IRRI, 1988), 54.

112. Tandon, J. P., and Bhatt, J. C. "Potential Sources of Blast Resistance for Hilly Regions in India." *Int. Rice Res. Newsl.* 7 (1982): 7.

113. Bhatt, J. C., Garg, D. K., and Tandon, J. P. "Rices with Multiple Disease and Insect Resistance in Hilly Regions of Uttar Pradesh." *Int. Rice Res. Newsl.* 9 (1984): 11.

114. Garg, D. K., and Bhatt, J. C. "Present Status of Research on Insect-Pests and Diseases of Rice in U.P. Hills." Paper presented in Group Discussion on Hill Rices at NDRI, Karnal, 1990, 9.

115. Directorate of Rice Research, Progress Report of the All India Coordinated Rice Improvement Program, Hyderabad, Vol. 2, *Entomology and Pathology*, 1988, 396.

116. International Rice Research Institute, Annual Report for 1978. (Los Banos, Philippines: IRRI).

117. Takahashi, Y. "Genetics of Resistance to the Rice Blast Disease," in *The Rice Blast Disease*, Proc. Symp. at IRRI, Los Banos, Philippines (Baltimore, Md.: Johns Hopkins Press, 1965), 303.

118. Goto, K., Kozaka, T., Yamada, M., Matsumoto, S., Yamanaka, S., Shinada, K., Narita, T., Iwata, T., Endo, T., Shimoyama, M., Nakanishi, I., Nishioka, M., Kumanoto, Y., Kono, M., Fujikawa, T., Okadome, Z., and Tomiki, T., "Joint Work on The Races of Rice Blast Fungus, *Piricularia oryzae* (fascicule 2)." Plant Prot. Div. Minist. Agric. For. Tokyo, Japan, *Plant Insect and Disease Forecast Spec. Rept.* 18, 1964, 132.

119. Kiyosawa, S. "Genetic Studies on Host-Pathogen Relationship in Rice Blast Disease," in *Proc. Symposium on Rice Diseases and Their Control by Growing Resistant Varieties and Other Measures,*" Minist. Agric. For. Tokyo, Japan, 1967, 137.

120. Kiyosawa, S. "Gene Analysis of Blast Resistance in Exotic Varieties of Rice." JARQ 6 (1971): 8.

121. Kiyosawa, S. "Genetics of Blast Resistance," in *Rice Breeding*. (Los Banos, Philippines: IRRI 1972), 203.

122. Kiyosawa, S. "Studies on Genetics and Breeding of Blast Resistance in Rice." *Misc. Publ. Natl. Inst. Agric. Sci.* Ser. D, 1 (1974): 58.

123. Kiyosawa, S. "Pathogenic Variations of *Pyricularia oryzae* and Their Use in Genetic and Breeding Studies." *SABRAO J.* 8 (1976): 53.

124. Kiyosawa, S. "The Possible Application of Gene for Gene Concept in Blast Resistance." JARQ 14 (1980): 9.

125. Chaudhary, R. C., and Nayak, P. "Genetics and Breeding for Blast and Bacterial Blight Resistance in Rice," in *Advances in Rice Pathology,* ed. S. Kannaiyan. (Coimbatore, Tamil Nadu, India: TAU, 1987), 1.

126. Kiyosawa, S., Ling, Z. Z., and Igarashi, S. "Identification of Blast Resistance Genes in Brazilian Rice Varieties Using Japanese Strains of *Pyricularia oryzae.*" *Oryza* 24 (1987): 42.

127. Atkins, J. G., and Johnston, T. H. "Inheritance in Rice of Reaction to Races 1 and 6 of *Pyricularia oryzae.*" *Phytopathology* 55 (1965): 993.

128. Hsieh, S. C., Lin, M. H., and Liang, H. L. "Genetic Analysis in Rice, VIII. Inheritance of Resistance to Races 4, 22 and 25 of *Pyricularia oryzae.*" *Bot. Bull. Acad. Sin.* Taipei 8 (1967): 255.

129. International Rice Research Institute, Annual Report for 1981. (Los Banos, Philippines: IRRI, 1983).

130. Marchetti, M. A., Lai, X. H., and Bollich, C. N. "Inheritance of Resistance to *Pyricularia oryzae* in Rice Cultivars Grown in the United States." *Phytopathology* 77 (1987): 799.

131. Yu, Z. H., Mackill, D. J., and Bonman, J. M. "Inheritance of Resistance to Blast in Some Traditional and Improved Rice Cultivars." *Phytopathology* 77 (1987): 323.

132. Ito, R. "Breeding for Blast Resistance in Japan," in *The Rice Blast Disease*, Proc. Symp. at IRRI, Los Banos, Philippines (Baltimore, Md.: Johns Hopkins Press, 1965), 361.

133. Atkins, J. G., Bollich, C. N., Johnston, T. H., Jodon, N. E., Beachell, H. M., and Templeton, G. E. "Breeding for Blast Resistance in the United States," in *The Rice Blast Disease*, Proc. Symp. at IRRI, Los Banos, Philippines (Baltimore, Md.: Johns Hopkins Press, 1965), 333.

134. Chang, T. T., Wang, M. K., Lin, K. M., and Cheng, C. P. "Breeding for Blast Resistance in Taiwan," in *The Rice Blast Disease*, Proc. Symp. at IRRI, Los Banos, Philippines (Baltimore, Md.: Johns Hopkins Press, 1965), 371.

135. Dasananda, S. "Breeding for Blast Resistance in Thailand," in *The Rice Blast Disease*, Proc. Symp. at IRRI, Los Banos, Philippines (Baltimore, Md.: Johns Hopkins Press, 1965), 379.

136. Rosero, M. J. "Breeding for Blast Resistance at CIAT," in *Proc. Rice Blast Workshop*. (Los Banos, Philippines: IRRI, 1979), 63.

137. Ikehashi, I., and Khush, G. S. "Breeding for Blast Resistance at IRRI," in *Proc. Rice Blast Workshop*. (Los Banos, Philippines: IRRI, 1979), 69.

138. Jennings, P. R. "Concluding Remarks," in *Proc. Rice Blast Workshop*. (Los Banos, Philippines: IRRI, 1979), 217.

139. Johnson, R., and Bonman, J. M. "Durable Resistance to Blast Disease in Rice and to Yellow Rust in Wheat." Abstract *Proc. Int. Symp. Rice Research: New Frontiers*, DRR (ICAR), Hyderabad, 29, 1990.

140. Bonman, J. M., and Mackill, D. J. "Durable Resistance to Rice Blast Disease." *Oryza* 25, 1988: 103.

141. Adyanthaya, N. R., and Rangaswami, G. "The Distribution of Silica in Relation to Resistance to Blast Disease in Rice." *Madras Agric. J.* 39 (1962): 198.

142. Jayachandran-Nair, K., and Chakrabarti, N. K. "Incidence of *P. oryzae* and Biochemical Changes in Rice Plant Grown under Upland and Flooded Conditions," *Phytopathol. Z.* 98 (1980): 359.

143. Matsuyama, N. "The Effect of Ample Nitrogen Fertilizer on Cell Wall Materials and Its Significance to Rice Blast Disease." *Ann Phytopathol. Soc. Japan* 41 (1975): 56.

144. Chandramohan, J., and Palaniswamy, S. "Incidence of Blast Disease of Rice in Relation to Time of Planting." *Rice News Teller* 11 (1963): 86.

145. Bhatt, J. C. "Effect of Environmental Factors on Leaf Blast Development in Rice in U.P. Hills." *Indian Phytopathol.* (Abs), 38 (1985): 589.

146. Ou, S. H. "A Look at Worldwide Rice Blast Disease Control," *Plant Dis.* 64 (1980): 439.

147. Padmanabhan, S. Y. "Control of Rice Diseases in India." *Indian Phytopathol.* 20 (1974): 1.

148. Singh, R. A. "Recent Trend in Chemical Control of Blast and Bacterial Blight," in *Proc. National Symp. on Increasing Rice Yields in Kharif*, New Delhi: ICAR, 1978, 477.

149. Okamoto, H., Yamamoto, T., Hamaya, E., and Marks, G. C. "Studies on the Phytotoxicity of Various Organo-mercuric Compounds to Japanese and Exotic Varieties of Rice Plants and the Efficacy of these Compounds against Rice Blast When Applied in the Field." *Bull. Chugoku Agric. Expt. Sn.* 4 (1960): 225.

150. Fukunaga, K., Misato, T., Ishii, I., Asakawa, M., and Katagiri, M. "Research and Development of Antibiotics for the Rice Blast Control." *Bull. Natl. Inst. Agric. Sci.* Tokyo C-22 (1968), 94.

151. Yoshii, K. "Studies on *Cephalothecium* as a Means of Artificial Immunization of Agricultural Crops." *Ann. Phytopathol. Soc. Japan* 13 (1949): 37.

152. Misato, C. "Blasticidin-S." *Plant Prot. Assoc.* Tokyo (1961), 55.

153. Akatsuka, T., Kodama, O., and Yamada, H. "A Novel Mode of Action of Kitazin P in *Pyricularia oryzae." Agric. Biol. Chem.* 41 (1977): 2111.

154. Kodama, O., Yamashita, K., and Akatsuka, T. "Edifenphos, Inhibitor of Phosphatidylcholine Biosynthesis." *Biol. Chem.* 44 (1980): 1015.

155. Reddy, A. P. K., and Satyanarayana, K. "Evaluation of New Organic Fungicides for the Control of Blast Disease of Rice." *Pesticides* 22 (1988): 21.

156. Sisler, H. D., and Ragsdale, N. N. "Disease Control by Nonfungitoxic Compounds," in *Modern Selective Fungicides—Properties, Applications, Mechanisms of Action,* ed. H. Lyr. (UK: Longman Scientific and Technical, 1987), 337.

157. Woloshuk, C. P., Sisler, H. D., and Vigil, E. L. "Action of the Antipenetrant, Tricyclazole on Appressoria of *Pyricularia oryzae." Physiol. Plant Pathol.* 22 (1983): 245.

158. Wolkow, P. M., Sisler, H. D., and Vigil, E. L. "Effect of Inhibitors of Melanin Biosynthesis on Structure and Function of Appressoria of *Colletotrichum lindemuthianum." Physiol. Plant Pathol.* 23 (1983): 55.

159. Yamaguchi, I., Sekido, S., Seto, H., and Misato, T. "Cytotoxic Effect of 2-hydroxy-juglone, a Metabolite in the Branched Pathway of Melanin Biosynthesis in *Pyricularia oryzae." J. Pestic. Sci.* 8 (1983): 545.

160. Bandong, J. M., Torres, C. Q., Vergel, De Dios, T., Lee, Y. H., Elazegui, F. A., Merca, S. D., Estrada, B. A., Sanchez, L. M., Nuque, F. L., Mew, T. W., Rush, M. C., and Crill, J. P. "Chemical Disease Control Evaluation." (Los Banos, Philippines: IRRI, 1979), 41.

161. Singh, R. A., and Bhatt, J. C. "Effect of Seed Treatment with Systemic Fungicides on the Incidence of Foliar Blast on Rice." *Indian J. Mycol. Pl. Pathol.* 16 (1986): 249.

162. Bhatt, J. C., and Tandon, J. P. "Efficacy of the Systemic Fungicide CGA 49104 for Controlling Rice Blast." *Int. Rice Res. Newsl.* 10 (1985): 16.

163. Froyd, J. D., Guse, L. R., and Kushiro, Y. "Methods of Applying Tricyclazole for Control of *Pyricularia oryzae* on Rice." *Phytopathology* 68 (1978): 818.

164. Shiba, V., and Nagata, T. "The Mode of Action of Tricyclazole in Controlling Rice Blast." *Ann. Phytopathol. Soc. Japan* 47 (1981): 662.

165. Lewin, H. D., Mariappan, V., and Chilliah, S. "Evaluation of New Fungicides in Controlling Blast." *Int. Rice Res. Newsl.* 11 (1986): 19.

166. Bhatt, J. C. "Inheritance of Resistance to Rice Blast (*Pyricularia oryzae*) and Its Chemical Control." (Ph.D. thesis submitted to G. B. Pant University of Agriculture & Technology, Pantnagar, 1990), 158.

5

BROWN SPOT
OF RICE

N. K. CHAKRABARTI

Deep Water Rice Project, c/o Rice Research Station,
Chinsurah 712 102, Hoogly, W.B., India

SUJATA CHAUDHURI
Department of Botany, Kalyani University, Kalyani, W. B., India

5-1 INTRODUCTION

Helminthosporiosis or brown spot disease of rice is widespread and occurs in all rice-growing countries of the world (Figure 5-1). The disease is also known as sesame spot, seedling and leaf blight, and nia-yake. It causes heavy losses, particularly in the leaf spotting phase, when it can reach epidemic proportions. The disease was a major factor behind the failure of rice crops leading to the great Bengal famine of 1942−43. Its epidemic outbreak resulted in huge yield loss, extending as high as 90% in certain areas.[1] Severely infected rice grains, when cooked, tasted bitter. Even cattle refused to take it (as per discussion of senior author with cultivator from the affected area). In Punjab, losses in weight of grains ranging from 4.6% to 29% have been reported.[2]

Bui Van Ik et. al.[3] from the USSR reported that *Cochliobolus miyabeanus* attacked less developed rice plants and damage amounted to 0.5% of the yield with 1% leaf infection. In Japan, the disease is not considered destructive. In the United States also, the disease is not a serious one, but in the rice-growing pockets, the annual loss since 1965 is no less than 0.5% of the total production.[4] Herrera and Siedel[5] recorded that in Cuba seed infection decreased germination by up to 60% and caused up to 40% seedling damping-off, whereas severe leaf infection caused 40−50% reduction in grain yield.

Figure 5-1   Distribution of brown spot disease of rice.

5-2 SYMPTOMS

Symptoms appear as lesions (spots) on the coleoptile, leaf blade, leaf sheath, and glume, being most prominent on the leaf blade and glumes [(Figure 5-2(a–c)]. The lesions are brown at first, and later become typically ellipsoidal, oval to circular measuring about 0.5–22 mm × 2–5 mm. At maturity they have a light brown or grey center with a dark or reddish brown margin. Different types of brown spot syndrome are usually found to be associated with different species of *Helminthosporium* viz., brown minute spots with *H. rostratum,* elliptical brown spots with *H. oryzae,* and leaf tip blighting with *H. halodes.*[6] Larger lesions are typical of more susceptible cultivars. On the coleoptile, the spots are brown and small, whereas on the glumes they are dark brown to black in color with olivaceous velvety growth. In severe infection, the whole grain surface becomes blackened and the seeds are shriveled and discolored.

Seedlings are often heavily attacked with numerous lesions about 2.5 mm in diameter, and in such cases, leaves may dry out and ultimately die. Badly affected nurseries can often be recognized from a distance by their brownish, scorched appearance, although seedlings are usually not killed by the disease.

5-3 FUNGUS

The fungus was first described as *Helminthosporium oryzae* by Breda de Haan in 1900 and Hori et al. in 1901. In 1954, Shoemaker renamed it *Bipolaris.*[4,7] Because the conidia of this graminicolous species develop sympodially and not apically or laterally as the type

Figure 5-2 (a), (b) Symptoms of brown spot of rice.

Figure 5-2 (c) Seed infection.

species *Helminthosporium* Link ex Fries. Subramanian and Jain[8] transferred it to *Drechslera* as *D. oryzae*. Nevertheless *Bipolaris oryzae* (Breda de Haan) Shoemaker (syn. *Helminthosporium oryzae* Breda de Haan, the anamorph of *Cochliobolus miyabeanus* (Ito and Kuribayashi) Drechsler is most accepted name for the causal agent of brown leaf spot disease of rice.

The perfect stage of the fungus was found by Ito and Kuribayashi 1927, who named it *Ophiobolus miyabeanus*.[4] The fungus was considered under the new genus *Cochliobolus* by Drechsler in 1934 and confirmed by Dastur in 1942. The name *Cochliobolus miyabeanus* (Ito and Kuribayashi) Drechsler ex Dastur is in use at present for the teleomorph of *B. oryzae*.

The fungus produces hyphae that are dark brown or olivaceous and 8–15 μm or more in diameter, with sporophores arising as lateral branches measuring 150–600 × 4–8 μm without conspicuous geniculations. They are olivaceous at the base and subhyaline at the tip.

Conidia are 8–13 septate, measuring 35–170 × 11–17 μm. They are slightly curved with a bulge in the center and tapering toward the ends (Figure 5-3a,b). They

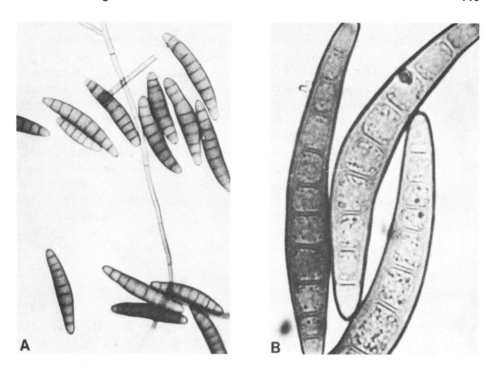

Figure 5-3 (a), (b)   Asexual spores (conidia) of *Bipolaris oryzae*.

germinate readily, producing germ tubes mostly from both end cells (bipolar) or even sometimes laterally.

The perfect stage consists of perithecia, which are globose to depressed globose, pseudoparenchymatous measuring about 560–950 × 368–377 μm. The asci are cylindrical or long fusiform, 21–36 × 142–235 μm and contain 4 to 6 filiform ascospores coiled in a helix.

The fungus can grow over a wide range of temperatures. Optimum temperature for mycelial growth and conidial germination is between 27–30°C and 25–30°C, respectively.[4] *B. oryzae* requires both light and dark periods for sporulation. However, this requirement varies with different strains of the fungus. For near-UV stimulated sporulation, the duration of exposure was more important than the dose.

Kulkarni and Ramakrishnan,[9] however, recorded that *B. oryzae* which normally did not sporulate on infected green and senescent leaves, sporulated abundantly when such leaves were detached and the water soluble components leached out.

Saltation is a common phenomenon occurring in *B. oryzae*. Chattopadhyay and Dasgupta[10] considered saltation to be due to nutritional imbalance and temperature variation. Gangopadhyay and Palit[11] considered it to be genetically controlled, being governed by a dominant factor with complementary action but with loss of pathogenicity. Lee et al.[12] reported abnormally high variation rate of *C. miyabeanus* among isolates and subcultures of the same isolate. They attributed this to perpetuation of heterokaryosis as a result of high frequency of anastomoses and nuclear divisions.

Bipolaris oryzae produces C_{25}-terpenoid phytotoxins called ophiobolin A (or cochliobolin A) and ophiobolin B (or cochliobolin B) during spore germination. Ophiobolin A, 6-epiophiobolin A, anhydroophiobolin A, 6-epianhydroophiobolin A, and ophiobolin I are produced in culture medium. All these compounds cause nonselective phytotoxicity to host and nonhost plants.[13] Ophiobolin A is most toxic. It also induces susceptibility of rice leaf to a nonpathogenic isolate of *Alternaria alternata*.[13] In vitro cytokinin secretion by the fungus that produces green islands on the rice leaves has been reported.[14] Formation of green islands, which contain more sugar, starch, and total carbohydrates than the surrounding tissues suggests a role for cytokinins in ensuring nutrition for the pathogen. The fungus also produces a number of enzymes, which include pectic enzymes, xylanase, cellulase, amylase, and protease. Nishikado in 1926 first reported the existence of different strains of the fungus.[4] Tochinai and Sakamoto[15] separated ten growth types differing in morphology, saltation, and virulence. Such strainal variations were also observed by Nawaz and Kausar[16] and Vorraurai and Giatgong.[17] However, the existence of distinct physiologic races could not be sufficiently considered by Eruoter[18] while making a comparative study of six isolates of *C. miyabeanus* from the United States.

5-4 HOST–PARASITE INTERACTION

Oku[19] made detailed studies on the mechanism of disease development and resistance, and according to him the pathogen produces toxin that kills the susceptible host cell, but in a resistant tissue, the sublethal dose of toxin increases the phenolic content of the host tissue, which is in turn oxidized into quinones by polyphenoloxidase secreted by the fungus. The quinones are further polymerized by fungal enzymes into a brown pigment which spreads into the necrotic lesions to form typical symptoms of the disease. Because of the toxicity of the accumulated oxidation products of polyphenols, the fungus cannot grow further and remains limited to the spot. Thus, phenolic substances formed after infection are related to disease resistance.

Polyphenol oxidase activity in infected tissue was found to be twice as high as the noninfected one by Chattopadhyay and Samaddar.[20] Infection of the pathogen also increases oxygen uptake and decreases photosynthetic CO_2 fixation in leaf tissue.[21]

Oku and Nakanishi[22] reported a phytoalexin-like substance in the infected tissue. Similar reports were made by Giri and Sinha.[23-24] Fungitoxicity may be induced in the susceptible rice cultivars if conidia of mildly virulent strains of *D. oryzae* are deposited on the phylloplane.[25-26]

Sridhar and Mahadevan[27] made an exhaustive review of the physiology and biochemistry of rice plants infected by *B. oryzae*. According to them, prohibitins offer the first chemical barrier of resistance to *B. oryzae* and this is influenced by nitrogen fertilization of the plants. The prohibitins may be *p*-hydroxy benzoic, salicylic, protocatechuic, cinnamic, *p*-coumaric, vanillic, ferulic, caffeic, and chlorogenic acids.

Chattopadhyay and Chakrabarti[28] investigated the relationships between anatomical characters of the rice leaves and resistance to brown spot infection and postulated that resistance is influenced by thicker epidermis and increased number of epidermal silicate bulliform cells. In a similar type of study conducted later, confirmatory results were obtained.

Resistance to brown spot was found to be positively correlated with the anatomical characteristics of rice leaves like thickness of cuticle and number of epidermal silicated bulliform cells.[28-29]

5-5 EPIDEMIOLOGY

The disease is primarily seedborne (Figure 5-4) but primary infection is likely to be severe only when the soil temperature at the sowing time is below 26°C. The fungus may survive for 2 to 3 years. Survivality period of the fungus in the seed varies under different conditions. Padmanabhan[30] demonstrated that the pathogen was viable from one year to the next growing season only on the seed. Stubbles left over in the field after the harvest have been found to harbor the pathogen; however, the importance of stubbles as the source of primary infection does not seem very significant.[31] In northern Japan, conidia on infected grains were viable for an average period of two year and mycelium for three years.[32] Under certain conditions, they may even survive in soil.

In general the survival of the pathogen is poor in unsterilized soil.[33-34] It may be due to its poor competitive saprophytic ability, highly aerobic nature, and high susceptibility to antagonism by soil isolates of *Trichoderma, Streptomyces,* and *Bacillus.*[35]

In irrigated soils in Cuba, the pathogen survived longer in infected rice straw (70 days) than on rice leaves (40 days) and longer at soil depth of 30 cm than at 15 cm.[35]

The diseased seeds need not always necessarily give rise to infected seedlings. Sometimes the coleoptile and roots are affected but due to the rapid growth of the leaves, lesions may not be formed on the subsequent leaves. Leaf spots generally arise from secondary infection by airborne conidia. The secondary spread of the disease can be very rapid and under favorable conditions for the growth of the pathogen may result in severe outbreaks. In India conidia are present in the air near rice fields throughout the off-season,[41] and in many parts of the tropics such conidia could form another source of primary infection.

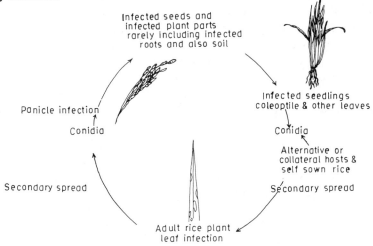

Figure 5-4   Disease cycle of brown spot of rice.

Conidia germinate, under favorable conditions within a few hours either from the apical, basal, or even intercalary cells, form the appressoria and penetrate either directly through the cuticle and epidermis or through the stomata. The first signs of infection may appear within 24 hr[37] and within a few days conidia may be produced.

In the study of relationship between conidial discharge in *B. oryzae* and weather conditions, the minimum number of trapped conidia (3–5) were observed in the month of March when maximum temperature of the day ranged from 36°C to 38°C, while maximum number of trapped conidia were observed during October–January.[38–39,41] Rise in minimum temperature was a factor for dissemination of conidia, but lowering of maximum temperature and increase in relative humidity were also required for increased dissemination of conidia of the pathogen.[40] No airborne conidia were detected at 54–60% R.H. There was significant correlation between wind velocity and presence of airborne conidia. Highest number of conidia was recorded at wind velocity of 7.0–7.5 km/h. The aeroscopic studies carried out by Chakrabarti[41] and Padmanabhan[42] showed that conidial dissemination was favored by discontinuous drizzles, cloudy weather, lowering of daily range of temperature, reduced sunshine hours, and so on.

Temperature is one factor that limits the development of brown spot. Primary infection from the grain does not occur in India if the temperature exceeds 25°C.[43] At higher temperature range, the rice plant overgrows the fungus and injury is of a much lesser degree.

A relative humidity of over 89% at 25°C is required for successful infection by conidia. In south Indian conditions severe disease between November to March is associated with high R.H. and low temperature. When night temperature is below 20°C accompanied by high R.H., outbreak follows.[46] Sunshine hours below 60 and above 110 in a fortnight reduced the intensity of brown spot.[45] Infection was more rapid in darkness than in light.

Disease is found to be higher in dry soil than in wet soil. It is more severe in rainfed than in irrigated/flooded fields.[46] Under Indian conditions plants transplanted within the middle of August developed less infection than those planted toward the end of August or afterward.[47] Similarly plants transplanted during the early part of the second crop season in December or January had a higher incidence of brown spot.[48]

The plants are most susceptible when they are in boot or in flowering stages of growth. Susceptibility of rice plant increases with age.[37,49] In the early stages only minute spots are formed and this is not affected by either soil or nutrients but after ear formation large spots are formed on the lower leaves, especially in plants grown in nutrient deficient soil.

Ono et al.[50] observed that conidial deposition was influenced by the leaf angle of the host plant under laboratory conditions. Variation in susceptibility of rice cultivars to brown spot was correlated with the variation in the leaf angles. Gangopadhyay and Chattopadhyay[51] reported that on narrower leaf angles, there was lesser conidial deposition. The leaf angle also influenced the microclimatic conditions of the rice field.

Another important factor governing disease incidence is the physiological condition of the host plant, which in turn is mainly governed by soil conditions. In Japan, plants grown in sandy, peaty, or thin soil with poor drainage and low nutrient status are most susceptible to the disease. A physiological disease of rice called "akiochi" in Japan was found to be not only associated but correlated with brown spot.[52] The more severe the akiochi the worse is the brown leaf spot, and therefore it was used as an index of akiochi.[53]

Brown spot development increases with both deficiency and excess of nitrogen. Germ tube development and spread of the mycelium in mesophyll, bundle sheath, phloem, and xylem of the leaf and collapse of the tissues was more rapid and extensive at deficient or high levels of ammonium nitrogen than at intermediate levels. Mohanty and Chakrabarti[54] reported that supply of nitrogen in slow release form was associated with lower incidence of the disease. Low phosphorus and potassium content is conducive to infection. Disease incidence was higher with leaching of nutrients from soil.[55] Lowering of oxidation-reduction protential (Eh) value is closely related to the occurrence of both brown spot and akiochi. The Eh value of the cell sap is reduced by the lack of nutrients like K, Mn, Si, Mg, and excess of P and N and also by the addition of H_2S or other reducing agents.

The presence of trace element like Mg, Mn, I, or Zn and Ca decreased susceptibility, whereas Cd and Co increased it.[44] The Fe/Mn status of the soil appeared to be important. Higher pH and low cation exchange capacity of soil was associated with higher disease incidence. Padmanabhan[1] in his analysis of the 1942−43 Bengal famine, studied the weather trends during those years and the biology of the host and pathogen and attributed the outbreak to the following factors:

1. enormous spore release due to weather conditions,

2. greater susceptibility to infection at the time of flowering and maturity; and

3. predisposing factors such as (a) leaching of nutrients from soil, (b) low potassium status, (c) imbalance of Fe/Mn (iron/manganese) ratio, (d) low nitrogen status of soil, (e) excessive soil moisture, and (f) water-logged condition.

Several weed grasses can be infected by *C. miyabeanus* artificially, and such grasses growing around rice fields have been reported as a natural host of the fungus. Chattopadhyay and Chakrabarti[56] reported *Leersia hexandra* to be an alternative host of the fungus in nature. *Echinochloa colonum* and *Arundo donux* have also been reported as collateral hosts from Central Rice Research Institute, Cuttack. Dath and Chakrabarti[57] and Subramanian and Ramakrishnan[58] reported that several wild rice varieties might also serve as collateral hosts of the pathogen. Other grass species that could be artificially infected include *Cynadon dactylon, Digitaria sanguinalis, Setaria italica,* and so on. The fungus can also cause infection in different degrees in maize, oat, barley, sorghum, sugarcane, and wheat.[4]

5-6 CONTROL

5-6-1 Seed Treatment

Nisikado in 1918 and Tisdale in 1922 started the treatment of seeds by hot water.[4] Padmanabhan[69] advised presoaking of seeds in cold water for 8 hr. followed by hot water treatment at 52 °C for 15–20 min. In most rice-growing countries it may not, however, be practical at the ordinary farmer's level.

In the 1950s mass seed treatment with organomercurials was popular. Seed treatment with Campogram (furmecyclox) followed by Dithane M-45 (mancozeb), Vitavax (carboxin), or Sicarol (pyracarbolid) was very effective in controlling seedborne *B. oryzae*. Complete control was achieved after seed treatment with Panoctine (guazatine) or RH 2161 (fenapanil)[59]

Nanda and Gangopadhyay[60] suggested seed treatment with *Bacillus subtilis* enriched soil followed by two foliar sprays of bacterial suspension when the plants were 30 and 60 days old as a prevention for primary and secondary infection of brown spot.

5-6-2 Foliar Application

Spraying of edifenphos (Hinosan) (0.1%) is widely recommended for the control of brown spot.[61] Biloxazol is also equally effective. It can be substituted for edifenphos to control brown spot in case the latter loses its effectiveness in the field due to development of tolerant strains.[62] Although edifenphos tolerant strains of *B. oryzae* have been isolated, both from laboratory and field, they are not yet a practical problem. A tolerant mutant produced in the laboratory showed possible negatively correlated cross resistance to bitertanol (Baycor) and mancozeb (Dithane M-45)[63]

Mukherjee and Bagchi[64] suggested the use of boron, copper, manganese, and zinc mixture as foliar spray for the control of airborne infection. Trivedi and Shinha[25] induced resistance to brown spot in rice cultivars by spraying conidia of wild races of *B. oryzae*. Plants developed only mild symptoms and developed a fairly high degree of resistance. Sinha and Hait[65] obtained appreciable protection against *C. miyabeanus* after spraying seedlings with chemicals known as phytoalexin inducers such as ferric chloride, nickel nitrate, sodium molybdate, DL-methionine, sodium selenite, and thioglycollic acid.

Treatment with benomyl is reported to increase the incidence of brown spot, probably by enhancing the proliferation of the pathogen.[66]

5-6-3 Varietal Resistance

Anatomical characters like thicker epidermal cuticles of the leaves, distribution of greater number of silicated bulliform cells, were observed to be associated with traditional tall *indica* cultivars like *Patnai*-23, *Bhasamanik*, and *Tilakkacharry*, which were less susceptible to brown spot disease.[67]

At the Central Rice Research Institute, Cuttack (India), after screening of 1000 varieties for 9 years against *B. oryzae*, *CH* 13, *CH* 45. *CH* 20, *T* 141, *T* 498-2A, *T* 988, *T* 2114, *T* 2118, *T* 960, *Bam* 10, *IET* 13238, *CR* 10-4025, *CR* 84-30, *JBS* 83, *JBS* 21, *JBS* 218, *JBS* 238, *J* 568, *JBS* 781, *JBS* 1510, *JBS* 1199, etc. were found resistant.[68] Padmanabhan[60] confirmed that resistance in *CH* 13 and 9515 is governed by three pairs of recessive genes indicating its horizontal nature.

Results of the field screening of 165 *japonica* and *indica* types by Omar et al.[70] showed that 26 were resistant at seedling, adult plant, and kernel stages of growth. They suggested that these varieties could be used in breeding programs.

While screening for brown spot resistance in deep water rice varieties, resistant lines selected were *CR* 98-77216, *CRRP* 34, *CR*1002, *BK NFR* 2606-3-2-1-2 and *IR*-42.[71]

Inheritance of field resistance was studied at the seedling, adult plant, and kernel stages in 6 crosses including 2 resistant (*Pi* 1 and *YNA* 282) and 2 susceptible varieties (*Giza* 171 and *Sakha* 1) by Balal et al.[72] Seedling and adult plant leaf reactions were similar and governed by 3 genes (*Her₁*, *Her₂*, and *Hes₁*). The first two were dominant for resistance and the third for susceptibility. Kernel resistance was dominant over susceptibility and controlled by 2 genes (*Hekr₁* and *Hekr₂*) carried by *Pi*1 and *YNA* 282, respectively. Seedling and adult plant leaf reactions were highly associated in all crosses, but both were independent of kernel reaction.

Eruoter[73] also concluded that a minimum of 3 major genes controlled lesion size and at least one gene controlled lesion number. Three genes conditioned differences in disease rating which included factors for both lesion size and number.

5-7 CONCLUSION

Although brown spot disease occurs in all the rice-growing countries, the intensity of the disease is generally low. However, occasional outbursts of the disease in fairly serious form may not be ruled out especially under conditions of poor fertilization and favorable environmental conditions. In developing countries, a vigilance always needs to be maintained on the occurrence of this disease and its spread.

Since new rice varieties are being bred regularly, it is necessary that constant screening of the progenies are carried out and moderately resistant types are selected so that the cultivars having horizontal resistance are ultimately recommended.

Since the control of the disease can be achieved through balanced nutrition, studies on this aspect at various agroclimatic zones are required to be carried out on a larger scale. Though the literature so far obtained does not indicate the existence of physiologic specialization in the pathogen, studies having been carried out with few isolates, more studies in this regard appear to be necessary in order to reach a definite conclusion.

5-8 REFERENCES

1. Padmanabhan, S. Y. "The Great Bengal Famine." *Ann. Rev. Phytopath.* 2 (1973): 11.

2. Bedi, K. S., and Gill, H. S. "Losses Caused by the Brown Leaf Spot Disease of Rice in Punjab." *Indian Phytopathol.* 13 (1960): 161.

3. Bui Van Ik, Chung, H. M., and Zubkov, A. F. "Estimation of the Harmfulness of Rice Diseases Taking into Account Selectivity of Plants for Pathogens," in *Ecological Aspects of Harmfulness of Cereal Crop Diseases.* All Union Research Institute of Plant Protection, 1987, 58.

4. Ou, S. H. *Rice Disease.* 2nd ed. Kew, England: CMI, 1985.

5. Herrera, L., and Siedel, D. "On the Injuries Effect of *Cochliobolus miyabeanus* (Ito and Kuribayashi) Drechslera ex Dastur in Rice Growing in Cuba." *Arch. Phytopath, Pflanzenschutz,* 14, (1978): 285.

6. Ramakrishnan, G., and Subramanian, C. L. "Studies on Helminthosporiose of Rice: Epidemiology of Brown Spot of Rice." *Madras Agric. J.* 64, (1977): 58.

7. Alcorn, J. L. "The Taxonomy of 'Helminthosporium' Species." *Ann. Rev. Phytopath.* 26 (1978): 37.

8. Subramanian, C. V., and Jain, B. L. "A Revision of Some Graminicolous *Helminthosporia.*" *Curr. Sci.* 35 (1966): 352.

9. Kulkarni, S., and Ramakrishnan, K. "Sporulation in *Drechslera oryzae.*" *IRRN* 4 (1979): 11.

10. Chattopadhyay, S. B., and Dasgupta, C. "Saltation in *Helminthosporium oryzae* Breda-de-Haan." *Indian Phytopathol.* 11 (1958): 144.

11. Gangopadhyay, S., and Palit, S. "An Additional Concept on the Saltation of *Bipolaris oryzae.*" *Indian Phytopathol.* 36 (1986): 278.

12. Lee, C. H., Hou, H. H., and Jong, S. C. "Cytology of *Helmonthosporium oryzae.*" *Mycopathologia.* 87 (1984): 23.

13. Xiao, J. Z., Tsuda, M., Doke, N. and Nishimura, S. "Phytotoxins Produced by Germinating Spores of *Bipolaris oryzae.*" *Phytopathology* 81, 1991: 58.

14. Gulati, A., and Mandahar, C. L. "Pathogenesis of Rice Leaves by *Helminthosporium oryzae:* Secretion of Cytokinin *In vitro* by fungus." *Research Bull. Punjab Univ.* 35 (1984): 115.

15. Tochinai, Y., and Sakamoto, M. "Studies on the Physiological Specialization of *Ophiobolus miyabeanus* Ito and Kuribayashi." *J. Fac. Agric. Hokkaido Univ.* 41 (1937): 1.

16. Nawaz, M. , and Kausar, A. G., "Cultural and Pathogenic Variation in *Helminthosporium oryzae.*" *Biologia,* Lahore 8 (1962): 35.

17. Vorraurai, S., and Giatgong, P. "Pathogenic Variability and Cytological Studies on *Helminthosporium Oryzae.*" *Ninth Natn. Conf. Agric. Sci.* Feb 1970, Bangkok, Thailand.

18. Eruoter, P. G. "A Comparative Study of Six Isolates of *Cochliobolus Miyzbeanus* in Rice from USA." *IRRN* 10 (1985): 23.

19. Oku, H. "Host–Parasite Relationship in *Helminthosporium* Leaf Spot Disease of Rice Plant." *Ann. Rep. Sankyo Res. Lab.* 17 (1965): 35.

20. Chattopadhyay, A. K., and Samaddar, K. R. "Comparative Physiological Changes Induced by *Helminthosporium oryzae* Infection and Ophiobolin." *Phytopathol. Z.* 98 (1980): 118.

21. Chattopadhyay, A. K., and Samaddar, K. R. "Effect of *Helminthosporium oryzae* Infection and Ophiobolin on Phenol Metabolism of Host Tissue." *Phytopathol, Z.* 98, (1980): 193.

22. Oku, H., and Nakanishi, T. "Relation of Phytoalexin Like Antifungal Substance to Resistance of the Rice Plant against *Helminthosporium* Leaf Disease." *Ann. Rep. Takamine Lab.* 14 (1962): 120.

23. Sinha, A. K., and Giri, D. N. "An Approach to Control Brown Spot of Rice with Chemicals Known as Phytoalexin Inducers." *Curr. Sci.* 48 (1979): 782.

24. Giri, D. N., and Sinha, A. K. "Correlation of Rice Varietal Reduction to Brown Spot Disease and Post Infectional Production of Fungitoxic Substances." *IRRN* 5 (1980): 6.

25. Trivedi, N., and Sinha, A. K. "Production of a Fungitoxic Substance in Rice in Response to *Drechslera* Infection" *Trans. Brit. Mycol. Soc.* 70 (1978): 57.

26. Mukhopadhyay, S., and Sinha, A. K. "Spore Germination Fluid as Inducer of Resistance in Rice Plants against Brown Spot Disease." *Trans. Brit. Mycol, Soc.* 74 (1980): 69.

27. Sridhar, R., and Mahadevan. A. "Physiology and Biochemistry of Rice Plants Infected by *Pyricularia oryzae, Helminthosporium oryzae, Xanthomonas oryzae* and *Xanthomonas translucens* f. oryzicola." *Acta Phytopathol. Acad. Sci. Hung.* 14 (1979): 49.

28. Chattopadhyay, S. B., and Chakrabarti, N. K. "Relationship between Anatomical Characters of Leaf and Resistance to Infection of *Helminthosorium oryzae* in Paddy." *Indian Phytopath.* 10 (1957): 130.

29. Nanda, H. P., and Gangopadhyay, S. "Role of Silicated Cells in Rice Leaf on Brown Spot Disease Incidence by *Bipolaris oryzae." Int. J. Trop. Plant Dis.* 2 (1984): 84.

30. Padmanabhan, S. Y. "Specialization in Pathogenicity of *Helminthosporium oryzae,"* in *Proc. 40th Indian Sci. Cong.* Part 4, Abstr. 18, 1953.

31. Chattopadhyay, S. B., and Chakrabarti, N. K. "Survival of *Helminthosporium oryzae* Breda de Haan. in the Field in Nature," in *Proc. 41st Indian Sci. Cong.* Part 3, Abstr. 123, 1954.

32. Kuribayashi, K. "Overwintering and Primary Infection of *Ophiobolus miyabeanus (Helminthosporium oryzae)* with Special Reference to the Controlling Method." *J. Pl. Prot. Tokyo* 1929: 16.

33. Luke, P. "Saprophytic Potentiality of *Helminthosporium oryzae* Breda de Haan in Soil *in situ." J. Indian Bot. Soc.* 50 (1971): 370.

34. Hiremath, P. C., and Hegde, R. K. "Role of Seed-borne Infection of *Drechslera oryzae* on the Seedling Vigour of Rice." *Seed Res.* 9 (1985): 45.

35. Kulkarni, S., Ramakrishnan, K., and Hegde, R. H. "Epidemiology and Control of Brown Leaf Spot of Rice in Karnataka. II. Survival of *Drechslera oryzae* in Nature." *Curr. Res.* 10 (1981): 67.

36. Chattopadhyay, S. B., and Dickson, J. G. "Relation of Nitrogen to Disease Development in Rice Seedling Infected with *Helminthosporium oryzae." Phytopathology* 60 (1970): 434.

37. Sato, K. "Studies on the Blight Disease of Rice Plant." *Bull. Inst. Agric Res. Tohoku. Univ.* 15, 1964, 199; 15, 1964, 239; 16, 1965, 1; 19, 1968, 64.

38. Gangopadhyay, S., and Chattopadhyay, S. B. "Vertical Dispersal of the Conidia of *Helminthosporium oryzae in vitro." Indian Phytopathol.* 26, (1973): 249.

39. Kulkarni, S., Ramakrishnan, K., and Hegde, R. H. "Epidemiology and Control of Brown Leaf Spot of Rice in Karnataka IX Atmospheric Variation in Spore Load of *Drechslera oryzae." Indian Phytopathol.* 35 (1982): 80

40. Chakrabarti, N. K. "Epidemiology of Major Fungal Diseases of Rice in India," in *Proc. Natl. Symp. on Increasing Rice Yield in Kharif.* 1978.

41. Chakrabarti, N. K. "Studies on Certain Aspects of *Helminthosporium* Disease of Paddy with Particular Reference to Conditions in West Bengal." (Ph.D. thesis, University of Calcutta, 1955).

42. Padmanabhan, S. Y. "Studies on Forecasting Outbreaks of Blast Disease of Rice I. Influence of Meteorological Factors on Blast Incidence at Cuttack." *Proc. Indian Acad. Sci.* 47 (1965): 117.

43. PANS Manual No. 3, Pest Control in Rice, Ministry of Overseas Development, Britain, London, 1970.

44. Kaur, P., and Padmanabhan, S. Y. "Control of Helminthosporium Disease of Rice with Soil Amendmants." *Curr. Sci.* 43 (1974): 78.

45. Gangopadhyay, S. "Sunshine Hours and Incidence of Brown Spot Disease Incidence in Upland Rice." *Oryza,* 18 (1978): 176.

46. Kulkarni, S., Ramakrishnan, K., and Hegde, R. K. "Dose Response in *Drechslera oryzae* (Breda de Haan) Subram. and Jain—A Casual Agent of Brown Leaf Spot of Rice." *Curr. Res.* 8 (1979): 194.

47. Chattopadhyay, S. B., and Chakrabarti, N. K. "Effect of Agronomic Practices of Susceptibility of Rice Plants to Infection of *Helminthosporium oryzae.*" *Rice News Teller* 13 (1965): 38.

48. Chakrabarti, N. K., and Mohanty, S. K. "Relationship between the Difference Dates of Transplanting and Incidence of Brown Spot Disease in High-Yielding Rice Varieties Grown in *Rabi Season* (December–May)." In *Proc. 60th Indian Sci. Congress.* Part III, 1975.

49. Padmanbhan, S. Y., and Ganguly, D. "Relation between Age of the Rice Plant and Its Susceptibility to *Helminthosporium and Blast Diseases.*" in *Proc. Indian Acad. Sci. B* 39 (1954): 44.

50. Ono, K., Suzuki, H., Onuma, M., and Inoune, E. "Data of Investigations Concerning Spore Flight of Blast Fungus." *Ann. Rep. Pl. Path. 2nd Lab.*, Natn. Hokuriku Agric. Exp. Stn., 1963.

51. Gangopadhyay, S., and Chattopadhyay, S. B, "Weather Condition and Brown Spot Disease Development of Rice in West Bengal." *Indian J. Mycol. Plant Pathol.* 4 (1975): 207.

52. Baba, I. "Nutritional Studies on the Occurrence of *Helminthosporium* Leaf Spot and Akiochi of the Rice Plant." *Bull. Natn. Inst. Agric. Sci.* Tokyo, Ser. D, 7 (1958): 1.

53. Goto, I. "Studies on the *Helminthosporium* Leaf Blight of Rice Plants." *Bull. Yamagata Univ., Agric. Sci.* 2 (1958): 237.

54. Mohanty, S. K. and Chakrabarti, N. K. "Effect of Slow Release Forms of Nitrogen Application on the Incidence of Brown Spot Disease of Rice." *Oryza* 18 (1982): 241.

55. Chattopadhyay, S. B., and Chakrabarti, N. K. "Effect of Agronomic Practices on Susceptibility of Rice Plants to Infection of *Helminthosporium oryzae.*" *Rice News Teller* 13 (1965): 38.

56. Chattopadhyay, S. B. and Chakrabarti, N. K. "Occurrence in Nature of an Alternate Host (*Leersia hexandra*) of *Helminthosporium oryzae* Breda de Haan." *Nature.* 172 (1953): 550.

57. Dath, A. P. and Chakrabarti, N. K. "Wild Rice Species, the Alternative Hosts of *Helminthosporium oryzae* Breda de Haan." *Sci. Cult.* 39 (1973): 394.

58. Subramanian, C. L., and Ramakrishnan, G. "The Occurrence of Brown Spot (*Helminthosporium oryzae* Breda de Haan) on Some Wild Species of *Oryza.*" *Curr. Sci.* 42 (1973): 327.

59. Ranganathiah, K. G., and Gowda, N. N. "Seedborne Infection of Rice by *Drechslera oryzae* and Its Control in Karnataka." *Pesticides,* 19 (1985): 44.

60. Nanda, H. P., and Gangopadhayay, S. "Control of Rice Helminthosporiose with *Bacillus subtilis* Antagonistic Towards *Bipolaris oryzae.*" *Int. J. Trop. Plant Dis.* 1 (1983): 25.

61. Lakshmanan, P., and Jagannathan, N. T. "Optimum Age of Rice for Brown Spot Control by Fungicide Spray." *IRRN* 10 (1985): 13.

62. Rao, H. N. and Lalithakumari, D. "Effect of Systemic Fungicides on *Drechslera oryzae*— The Brown Spot Pathogen of Rice." *Indian Phytopathol* 40 (1987): 168.

63. Lalithakumari, D. and Annamalai, P. "Observation of the Field Performance of Edifenphos against *Drechslera oryzae* (Brown Spot of Rice)." *ISPP Chemical Control Newsletter* 10 (1988): 26.

64. Mukherjee, S. K., and Bagchi, B. N. "Control of Secondary Airborne Infection of *Helminthosporium* Disease of Paddy." *Rice News Teller* 12 (1964): 103.

65. Sinha, A. K., and Hait, G. N "Host Sensitization as a Factor in Induction of Resistance in Rice against *Drechslera* by Seed Treatment with Phytoalexin Inducers." *Trans. Brit. Mycol. Soc.* 79 (1982): 213.

66. Taylor, D. R. "Influence of Rice Straw, Potash and the Fungicide Benomyl on Brown Spot Disease of Rice." *IRRN* 14 (1989): 26.

67. Chattopadhyay, S. B., and Chakrabarti, N. K. "Effect of Methods of Cultivation on the Incidence of Brown Spot Disease in *Aus* Paddy." *Indian Agriculturist* I. (1957): 45.

68. Chakrabarti, N. K., and Padmanabhan, S. Y. "Recent Trend in Control of Rice Diseases." *Proc. Nat. Acad. Sci.*, India, 46 (B), 1976: 137.

69. Padmanabhan, S. Y. "Control of Rice Disease in India." *Indian Phytopathol.* 27 (1974): 27.

70. Omar, R. A., Balal, M. S., El-Kazzaz, M. K., and Aidy, I. R. "Reaction of Rice Varieties to Brown Spot Disease of Rice *Cochliobolus miyabeanus* at Different Stages of Growth." *Agric. Res. Rev.* 57, (1979): 103.

71. Prasad, Y., and Singh, R. S. "Screening of Brown Spot Resistance in Deep Water Rice." *IRRN* 10 (1985): 7.

72. Balal, M. S., Omar, R. A., El-Khadem, M. M., and Aidy, I. R. "Inheritance of Resistance to the Brown Spot Disease of Rice, *Cochliobolus miyabeanus.*" *Agric. Res. Rev.* 57 (1979): 119.

73. Eruoter, P. G., "Inheritance of Sheath Coloration and of Brown Leaf Spot of Rice." *Z. Pflanzenkr. Pflanzenschutz* 96 (1986): 47.

6

RICE SHEATH BLIGHT:
The Challenge Continues

MRINAL K. DASGUPTA

Plant Pathology Laboratory, Department of Plant Protection
Palli-Siksha Bhavana (Institute of Agriculture), Visva-Bharati
SRINIKETAN 731236, West Bengal, India

6-1 INTRODUCTION

6-1-1 Historical Background and Distribution

Rice sheath blight (ShB) is regarded as an internationally important disease that is second only to and often rivals the blast disease, particularly since the introduction of high-yielding varieties (HYV) in the 1960s.[1,2] Its distribution at least in Asia is the widest of all rice diseases (Figure 6-1; Table 6-1). The disease has so far eluded any concrete and inexpensive solution for the generally poor rice farmers. Rather frequent reviews and summaries from different parts of the world are also an indication of the continuing challenge of ShB.[1-6]

6-1-2 Economic Importance and Modeling Yield Loss

Some regionwise and countrywise loss estimates are presented in Table 6-2. Attempts have been made to estimate yield losses due to ShB on various bases including vertical and horizontal spread (Table 6-3). A precise estimate of the damage in the current season is possible as these models are based on the final phases of the disease. But to achieve models of more predictive value, even including the affected host variables,[7] comparative validation of different yield loss models seems to be necessary. Also, partitioning loss due to ShB to rice from multiple pests has not been made,[8] which is in fact reality.

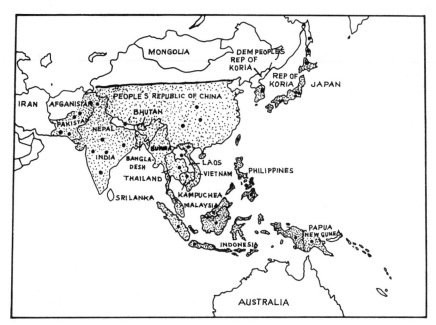

(AFTER REISSIG et al, ILLUSTRATED GUIDE TO INTEGRATTED PEST MANAGE-
MENT IN RICE IN TROPICAL ASIA , IRRI , PHILLIPINES , 1986 WITH PERMISSION)

Figure 6-1 Distribution of rice sheath blight in tropical Asia. Also recorded from Sri Lanka[2] and Iran.[21]

6-2 SYMPTOMS

The seedlings at 15 days after sowing may be rotted at the base, resulting in dead patches particularly where seedborne inoculum is high. Usually attacked at mid-tillering or during late tillering and early internode elongation (as in the southern United States), the lesions first appear on the leaf sheaths at or above water level, generally 0.5–3 cm below leaf collar in the wetland crop, or generally at or above ground level, but may be anywhere on the lower sheaths in the upland crop. Watersoaked, circular to oblong, ellipsoid or ovoid, even irregularly elongated, 3 cm × 1 cm discolorations appear. They turn into discrete lesions with pale greenish-grey to greyish-white center with narrow blackish to dark brown margin. The lesions later become fawn or off-white with brown to purple brown margin. Finally, 4–5 such lesions coalesce and girdle the whole leaf sheath, culm, boot, and flag leaf, whereby the tiller is encircled to death. A series of characteristic coppery bands may appear across the lesions, hence the name banded blight (Figure 6-2). Sclerotia appear loose among fine silvery threads of mycelium after about 6 days on or near the lesion, on or between leaf sheath and culm, within lumen of the culm, and within larger cells of sheath, and are easily detached at maturity.

Under favorable conditions, the disease may progress in three ways: (a) inward spread—from outer to inner sheaths with bleached center and irregular purple-brown border; (b) vertical spread—upward rapidly invading the lamina, loosening the sheath

TABLE 6-1: EARLY RECORDS, NOMENCLATURE, AND CHANGING STATUS OF RICE SHEATH BLIGHT

Country: Causal organism named as: Disease name proposed as	Findings and remarks	References
Japan: *Sclerotium irregulare* I. Miyake: Sheath blight (ShB)	Disease first reported	Miyake (1910)[9]
Hypochnus sasakii Shirai: ShB ShB	Causal fungus found to be same as Shirai (1906)[a] described from camphor leaves; 0.12–0.19 m ha affected	Sawada (1912)[a] Natl. Inst. Ag. Sci. (1954)[a]
Pellicularia sasakii (Shirai) Ito: Banded sheath spot	Spreading from Southeast Asia and the Pacific to temperate; serious on highly fertilized crops	Hashioka and Makino (1969)[10]
Philippines: *Rhizoctonia solani* Kühn (RS): ShB	Disease first reported; high frequency (<57%) of RS among three sclerotial pathogens on leaf sheath	Reinking (1918)[a] Palo (1926)[a] Shangzhi and Mew (1987)[11]
Japan and tropical Asia: RS: Oriental leaf and sheath blight	Second most important rice disease affecting 44% area in Japan; considered limited to Asia; total destruction of a susceptible cv not uncommon	Kozaka (1970)[a] Ou (1972)[6] Kozaka (1975)[d] Ou (1985)[2]
Japan and Taiwan: *Corticium sasakii* (Shirai) Matsumoto: ShB	Compared *R. solani* with *H. sasakii*, rejected these and *C. vagum* Berk. and Curt.	Matsumoto (1934)[a]
Taiwan: RS: ShB	90,000 ha affected, being the most important disease of the second crop of rice	Chien (1979)[b]
Sri Lanka: RS: ShB	Disease first reported	Park and Bertus (1932)[a]
Republic of Korea: RS: ShB	Of six sclerotial pathogens on rice leaf sheath, 88% due to RS AG-1	Kim and Kim (1988)[12]
PRC: RS: ShB, Sheath and culm blight	Reported from southeast China	Wei (1934)[a] Cai and Xue (1984)[13]
India: RS: Sheath spot	Probably known in Bengal since Butler (1918)[c]	Padwick (1950)[b]
ShB [Various synonyms of the teleomorph in literature: *Corticium sasakii, C. vagum, C. solani, Pellicularia sasakii, P. filamentosa, P.f.* f. sp. *sasakii*; similar situation in other countries]	First observed in 1960 in Gurdaspur (Punjab), then Uttar Pradesh; Tamil Nadu; Madhya Pradesh; endemic hot spot in Maruteru, West Godavari (Andhra Pradesh); most destructive in Kerala; epidemic in entire Punjab (1978); cv Karuna totally destructed at Central Rice Research Institute, Cuttack (Orissa) in 1980. Panicle and boot blight on IR 50 and many other cvs, widespread in Manipur.	Paracer and Chahal (1963);[b] Kohli (1966);[b] Rao and Kannaiyan (1973);[b] Gangopadhyay and Chakrabarti (1982);[4] Gangopadhyay (1983);[3] Mathur (1983);[14] Singh et al. (1988), (1989)[15,16]

TABLE 6-1: CONTINUED

Country: Causal organism named as: Disease name proposed as	Findings and remarks	References
India: *Thanatephorus cucumeris* (Frank) Donk: Banded blight and spot [syn: *Hypochnus cucumeris* Frank]	Basidiospore infection of grains proved	Saksena and Chaubey (1972)[a] Saksena (1973)[a]
USA: RS: Brown-bordered leaf and sheath spot; bordered leaf and sheath spot	Probably known since 1930s, but Ryker and Gooch's (1938)[1] description of bordered sheath spot (c.o., *R. oryzae*) is a similar but different disease	Templeton and Johnston (1969)[d] Rush (1971)[e] Atkins (1974)[e]
: ShB	Second most important disease of rice, often rivals blast; highly susceptible or totally destructed; close-planted semi-dwarf rice suffers more than twice loss than others; milling quality affected; ratoon crop severely affected; most important rice disease in southern U.S. since 1972	Lee and Rush (1983)[1] Marchetti (1983)[17] Lee and Templeton (1988)[18]
Nigeria: RS: ShB	Low incidence in upland on cv OS6; serious in Chad Basin	COPR (1976)[19] Kamel (1979)[b]
Brazil: RS: ShB	First reported in 1973; up to 39% RS in rice seed mycoflora	COPR (1976)[19] Soave et al. (1983)[20]
Iran: RS: ShB	Seedborne infection high (9.9–39.1%)	Binesh and Torabi (1985)[21]

References cited in [a] Ou,[2] [b]Gangopadhyay and Chakrabarti,[c] Hashioka and Makino,[10] [d]Lee and Rush,[1] [e]Marshall and Rush.[43]

from the culm, even causing blight of boot, flag leaf, panicle, and lodging. Such lesions are ovoid, smaller, 5–10 mm × 5 mm, which coalesce rapidly, more so on lower leaf sheaths than on upper leaves. Grains become chaffy or partially filled, particularly in the lower parts of the panicle, or the entire panicle may even be matted together;[15,16] (c) horizontal spread—laterally, the disease spreads from tiller to tiller and hill to hill apparently by physical contact in a densely crowded planting.

When conditions are less favorable, the lesions remain restricted to lower leaf sheaths or, following some advancement of disease, as subsequent opening of the plant canopy due to leaf death permits increased penetration of sunlight and reduced RH, lesions become dry, oblong, well defined, white, tan, or grey with brown to reddish brown borders. They coalesce, overlap, and form a typical cobra skin pattern.

TABLE 6-2: LOSSES DUE TO RICE SHEATH BLIGHT

Region/country	Loss estimates	References
East Asia	0.8%	IRRI (1983)[22]
Japan and Republic of Korea	5.2%	IRRI (1973)[22]
Vietnam and Philippines	25–50%	Mizuta (1956)[a]
Japan	24–38 thousand t annually lost	Natl. Inst. of Ag. Sci. (1954)[a]
	20–25% loss if disease spreads to flag leaf	Mizuta (1956)[a] Hori (1969)[a] Hori and Anraku (1971)[c]
Philippines	24% loss in susceptible varieties with heavy N and high disease pressure	IRRI (1975)[22]
India	5.2–50% loss depending on cv and cultural conditions	Kannaiyan and Prasad (1978)[b]
United States (Arkansas)	Maximum yield losses occur when infected at half internodal elongation growth stage. Highly susceptible (HS) cv Lebonnet totally lost; 22% yield loss in moderately susceptible (MS) cvs Starbonnet and Mars	Boyette and Lee (1979)[c]
Southern United States	Loss in moderately resistant cv Mars (7–15%), MS cv Starbonnet (15–25%), HS cv Lebonnet (25–50%)	Lee and Rush (1983)[1] Lee and Templeton (1988)[18]
	Up to 50% losses in susceptible cvs when all leaf sheaths and blades are affected	Lee and Templeton (1988)[18]

References cited in [a]Ou,[2] [b]Mathur,[14] [c]Gangopadhyay and Chakrabarti.[4]

6-3 ETIOLOGY

6-3-1 Identity and Nomenclature

Confusion as to the identity and nomenclature of the causal fungus (Table 6-1) is not yet over, although the situation is largely put in order. The names for the ShB pathogen generally accepted in current literature are: imperfect sclerotial stage or the anamorph as *Rhizoctonia solani* Kühn (Deuteromycotina, Deuteromycetes, Agonomycetales) and perfect stage, basidial stage, or the teleomorph of the same as *Thanatephorus cucumeris* (Frank) Donk (Basidiomycotina, Hymenomycetes, Holobasidiomycetidae, Tulasnellales, Ceratobasidiaceae). Tu and Kimbrough, (1978),[27] however, have proposed the recombined name, *T. sasakii* (Shirai) Tu and Kimbrough. The genus *Rhizoctonia*, created by De Candole,[27] in 1815 is now generally characterized as follows: (a) branching near the distal septum of cells in young, vegetative hyphae, (b) formation of a septum in the branch near the point of origin, (c) constriction of the branch, (d) dolipore septum, (e) no clamp connection, (f) no conidium, except moniloid cells, (g) sclerotium not differentiated into rind and medulla, and (h) no rhizomorph.[27] Of the three groups under *Rhizoctonia* spp. the one with multinucleate (3 per cell), larger hyphae (0.6–10 μm in diameter) having teleomorph in the genus *Thanatephorus* Donk is represented by *R. solani* (RS). Instead of Kühn's insufficient description of RS, Duggar's (1915) characterization is considered

TABLE 6-3:   EXPERIMENTAL YIELD LOSS ESTIMATES AND MODELING

Country: Author	Yield loss estimates and modeling
Taiwan: Tsai (1974)[b]	Maximum loss when inoculated at 60 DAS and at booting. At maximum of four inoculum levels, 1st and 2nd crops of cv Tainan 5 and TN1 showed 43%, 22%, 33%, and 16% yield losses, respectively
Nigeria (Chad Basin): Kamel (1979)[b]	Curvi linear relationship between disease and grain yield with r = −0.2737, r^2 = 0.0749, and regression coefficient of 0.5399 ± 0.2157 kg/plot
Philippines: Ou and Bandong (1976)[a]	With 0–50% hills inoculated, low and high N applied S and MR cvs showed 7.5–22.7%, 8.6–23.7%, 9.4–10.0%, 2.5–13.2% yield losses, respectively; with fungicidal treatment at 5%, 50%, and 100% hills infected, yield increases were 1.6%, 6.4–7.1%, and 8.9–10.1%, respectively
: Ahn and Mew (1986)[23]	Linear relation between relative lesion height (RLH) and yield under moderate disease pressure; no significant yield loss at RLH <20%, at 90% RLH, 46% yield loss possible
Taiwan: Chang (1986)[24]	Linear relation between disease and yield, panicle wt, and percentage fertility; at IRRI SES disease scores 1, 3, 5, 7, 9, yield losses in Tainan 5 and Tainung 67 are (in percent) 14.2, 26.9, 29, 32.9, 40.2 and 8.3, 18, 24.8, 49.9, respectively
India: Rajan (1987)[25]	Linear relation between yield loss and percentage diseased leaf and sheath area; at disease scores 3, 5, 7, and 9, yield losses were (in percent) 16.8, 22.9, 36.2 and 48.4, respectively
Japan: Hashiba (1984)[26]	Loss of fully ripened kernels estimated at 8.5 g/3.3 m^2 for each additional 1% disease incidence; from which the equation derived was L(yield loss in kg/10 acres) = (41.3 × −826.2) A/100, where x is RLH and A is % of affected hills. Models validated for five years

References cited in [a]Ou,[2] [b]Gangopadhyay and Chakrabarti.[4]

Figure 6-2   Rice sheath blight symptoms. Mature lesions coalescing and girdling the stem, leaf sheath, and blade, also showing vertical spread.[2]

standard, viz., wide hyphae (8–12 μm), constriction at the point of branching, and the branching of mature hyphae at right angles.[27]

6-3-2 Morhphology of RS Causing ShB of Rice

Mycelium.   Young mycelium is silvery, becoming yellow and brown with maturity, 8–12 μm broad and infrequently septate. Three types of mycelium are seen: (a) Runner mycelium—straight, creeping, trophic, noninfecting hyphae [Figure 6-3(a)] or sometimes thick and flattened; (b) Lobate mycelium—branching out from the former as short, swollen, much branched, single or multiple lobate appressoria and penetration pegs (Figure 6-4), which later becomes intracellular all through the lesions. Such mycelium can withstand desiccation. The lobate state associated with the infection process may be the counterpart of the sporogenous moniloid state;[3] (c) Moniloid mycelium—forming the sclerotia [Figure 6-3(b)] but also formed on the walls of culturewares [Figure 6-3(c)].

Sclerotia.   These are dark brown to black, 4–5 mm in diameter, spherical but flattened when pressed between leaf sheath and culm. Several sclerotia (≥5 mm) may form a larger mass in culture. Five phases of sclerotial morphogenesis have been recog-

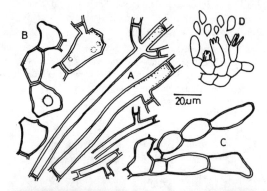

Figure 6-3 *Rhizoctonia solani*[2]
(a) Runner hyphae
(b) Moniloid cells of sclerotia
(c) Moniloid mycelium in culture
(d) Basidia and basidiospores.

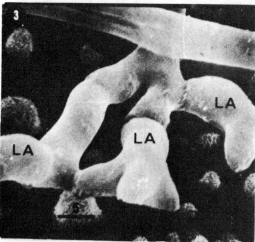

Figure 6-4 Lobate mycelium (appressorium) (LA)-each appressorium forms several lobes at the apex (bar represents 5μm).[43]

nized in RS: (a) repeated hyphal branching resulting in short, thick, lateral moniloid hyphae; (b) hyphal aggregation as clusters of thick-walled cells rich in nutrients and network formation (5 μm wide $\times$ 0.09 μm thick); (c) formation of sclerotial initial; (d) formation of white and immature sclerotium attaining maximum size in 30 hr; (e) maturation and pigmentation by melanin incrustation. Freshly formed sclerotia sink into water but become buoyant when the outermost layer becomes thick (0.51 mm) with empty cells. The evacuation takes 40 hr, whereby the living central cells attain 0.15 mm in diameter.

Basidia and basidiospores. Basidiospores 2–4, terminal in imperfect cymose or racemose clusters formed by branching of short-celled ascending hyphae. Sterigmata 2–7, usually 4 per basidium [Figure 6-3(d)]. The measurements respectively provided by Sawada (1915), Matsumoto et al. (1932) (both cited by Ou[2]) and Gangopadhyay and Chakrabarti[4] are: basidia 10–15 $\times$ 7–9 μm, 10–16 $\times$ 8–9 μm, 11–15 $\times$ 8–9 μm; sterigmata 4.5–7 $\times$ 2–3 μm, 5–8 $\times$ 2.2–2.7 μm, 7–10 $\times$ 2–3 μm; basidiospores 8–11 $\times$ 5–6.5 μm, 6–10 $\times$ 4–7 μm, 9–12 $\times$ 5–7 μm. Basidia arise in 9–19 days. Some remain probasidial, while the majority grow further with 1–2 stout sterigmata but no hymenium. Basidiospores can grow a little without nutrients but later only on yeast extract-enriched PDA. Some fertile single basidiospore isolates are homothallic. Basidial stage has been observed on upland rice appearing in white powdery or frosty layers on the healthy leaves or healthy areas adjacent to the lesions under extremely moist conditions.

6-3-3 Physiology

RS can grow within a range of 10–40 $\pm$ 1°C, or 10–37°C, preferably 20–30°C or 20–28°C, but optimum at 28°C, 30°C, 25–31°C, or 28–31°C.[2,28,29]

Geographical isolates may differ in thermosensitivity depending on local temperature regimes. Also, empty cells were proportionately more and prominent sclerotia from rice were larger, more thermotolerant, and contained more CO_2 than O_2; hence, they were more viable than those from culture.[22,30]

6-3-4 Intraspecific Variation

Based on morphological and pathogenic characters, the rice ShB pathogen was placed in sasakii (type IA) as one of the seven cultural types of Watanabe and Matsuda (1966).[27] By hyphal anastomosis, 10 anastomosis groups (AGs) have been reported in RS—all represented by the same teleomorph. Each AG represents a noninterbreeding population and a genetically independent entity. Again, within these 10 AGs of RS, by pathogenic variations[24] and by DNA–DNA homology, intraspecific groups (ISGs) (Kuninaga and Yokosawa 1982–1985)[27] have been recognized. For instance, the ShB fungus has been placed in RS AG-1, which can be divided into at least three ISGs, viz., IA, IB, and IC. AG-1-IA causes true rice sheath blight, the web blight fungus IB produces different symptoms on rice, and the sugarbeet and buckwheat fungus IC fuses well with IA and IB, but is not pathogenic on rice. Bolkan and Rebeiro,[31] however, viewed that the AGs lacked host specificity. It is hoped that molecular markers like plasmids, ds RNA mycoviruses, drug sensitivity, etc. as well as, more hopefully, symptom specificity on indicator hosts may be available in the future for precise intraspecific genetic analysis.[32]

Ecopathology.   The biological-cultural-pathogenic groups and AGs or ISGs of RS reported from various parts of the world correspond to each other. While AG-1,2,3,4 are ubiquitous, others are limited in distribution, but the knowledge is as yet incomplete in this regard. A given AG or ISG is considered the primary pathogen of a plant disease due to RS, although several AGs or ISGs may be isolated from an infected plant or a crop field.[33-35] But from a given crop field different clones of the same AG may remain in clusters.[36] Growth rate, sclerotial production, aggressiveness, and the like of different isolates derived from the hosts within and outside Poaceae including rice not only differed, but the *Pennisetum* isolate was the most virulent on rice.[37]

6-4 DISEASE CYCLE

6-4-1 Infection

Seedling infection.   The rice seeds may carry ShB inoculum and produce 4–6.6% seedling infection in India.[2,14] It is likely that the seeds on the lodged panicles catch inoculum from soil.[38] But on transplantation the infected seedlings are unable to develop disease earlier.[39] However, high frequency of seedling infection in Iran and high percentage of RS in seed mycoflora in Brazil (see Table 6-1) are alarming enough even without any information on "systemic infection," particularly since ShB is a rather recent introduction in these countries. Expansion of ecological limits of a plant pathogen is too well known.

Sheath infection and infection process in mature plant.   Notwithstanding seedborne infection, the ShB disease cycle takes place predominantly through sclerotia,[40] at least in the humid tropics (Figure 6-5). When the buoyant sclerotia tend to accumulate in undisturbed standing water at the plant-water-interface, the aerobic fungus

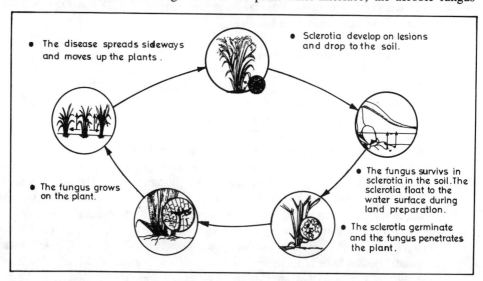

• The disease spreads sideways and moves up the plants.

• Sclerotia develop on lesions and drop to the soil.

• The fungus survivs in sclerotia in the soil.The sclerotia float to the water surface during land preparation.

• The sclerotia germinate and the fungus penetrates the plant.

• The fungus grows on the plant.

Figure 6-5   Disease cycle of rice sheath blight (role of seedborne inoculum ignored).

creeps up several centimeters in 24 hr and the primary infections are caused in wetland rice. Rain water runoff and flood irrigation permit good dispersal of floating sclerotia,[41] and consequently, provide the primary foci of infection through the stretches of rice fields. Further, with the increasing size of sclerotia or their fragments, number and size of lesions also increased.[3]

Between the two modes of penetration, the stomatal by means of lobate appressorium (LA; Figures 6-4,6-6) and the cuticular by infection cushion [IC; Figure 6-6(b)], the former is more common. A strong positive stomatropism was indicated by one or more appressoria per stoma,[42] and extensive ectotrophic growth of runner hyphae from the outer to the inner sheath surface to accomplish stomatal penetration.[43] However, a strong positive correlation ($r = 0.98$) between LA and IC[43] suggests a definite proportion between the two modes of penetration. Manner of IC formation by RS is strikingly similar on different hosts including rice and follows the classical pattern.[44] IC fails to form if physically disconnected from the sclerotia or when supplied with glucose or 3-O-methyl glucose, even on highly susceptible cultivars.[45]

Field application of SiO_2 reduced ShB (Kannaiyan and Prasad, 1978)[5] possibly due to increased silication of the cuticle and the underlying tissue presumably by preventing cuticular penetration. Thick wax deposits on cuticle also inhibit infection.[45] Further spread takes place by intra- and intercellular colonization of the host.

Pathophysiology and biochemistry of resistance.   Precise experimentation is needed to establish the relation between pectic and cellulolytic enzymes and phosphatidases, and the aggressiveness of the RS isolates on rice,[46] although a number of claims are there. Increase followed by decrease in permeability of the infected host cells and its relation with the conversion of phenols to high molecular weight compounds (Roy, 1977)[4] also need experimental evidence.

Increase in respiration and decrease in transpiration upon infection have been noted

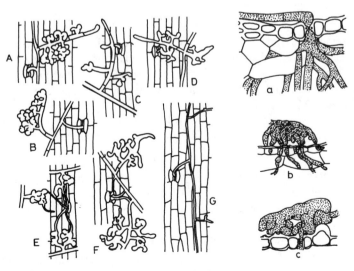

Figure 6-6 Entrance of sheath blight fungus into host tissue. (a) and (c) (cross section); (b) through cuticle cells; A to G through stomata (surface view).[2]

in susceptible cultivars (Roy, 1982).[14] Infected sheath tissues had higher dry matter and lower moisture. Reduction in total soluble sugars and starch could be correlated with increasing disease intensity,[2] perhaps by enhanced lesion expansion prompted by increased consumption by the pathogen.

Increase following infection, and N or carbendazim application, or with aging, in total phenols in highly susceptible (HS), susceptible (S), and moderately resistant (MR) cultivars (cvs) follows an identical pattern; only flavonoids increase in an MR cv at late stage of infection.[5] Induction of phenols is evidently neither rapid nor enough in the MR rice-ShB system. Ascorbic acid was lower in HS than S cv but increased in both following infection or carbendazim application (Karthikeyan et al., 1989).[5]

Both S and MR cvs had lower peroxidase; phenylalanine ammonia lyase (PAL), and tyrosine ammonia lyase (TAL) decreased in MR cv and increased in S cv. No dihydroxy-phenylalanine could be detected in inoculated or uninoculated plants. This was sought on the assumption that fungal melanins were biosynthesized homogeneously as in the animals. Only two isolates of RS with lowest virulence showed mycelial and extracellular peroxidase bands. Both S and MR cvs had lower peroxidase levels.[5] Peroxidase isozyme bands were fewer with the application of phenylacetic acid (PAA) and its derivatives in rice leaves (Ramalingam, 1984).[5] But these weak auxin-like compounds were liberated by the pathogen (Chen, 1958)[2] and might inhibit secondary root formation in rice (Wu, 1978).[4] In view of the precise role of plant peroxidases in lignification, cell differentiation, and cell wall synthesis commensurate with the usual lack of extracellular peroxidases in fungi,[47] of PAL and TAL in melanin/phenol biosynthesis as well as varied and heterogeneous origins of fungal melanins,[48] a clear picture has yet to be formulated.

Although healthy HS cv leaf sheaths had more chlorophyll which decreased on infection, the healthy MR cv contained less but the infected sheaths of MR cv could fix more [14]CO_2 than HS cv, yet vice versa in their healthy sheaths (Naidu et al., 1981).[4]

6-4-2 Dissemination

Dissemination of the ShB pathogen ought to be ordinarily limited, adjacent to the primary infection, but is perhaps aided by compensation factors when the conditions are extremely favorable during the period of infectiousness. The row-to-row spread of ShB was unexpectedly found to be faster than hill-to-hill spread,[49] or else little or no dissemination takes place and the disease is monocyclic.

6-4-3 Perennation

Dormant mycelium in crop refuse may remain active in soil for almost a year (Kannaiyan and Prasad, 1978).[4] Rice chaff and rachillae also contained viable inoculum and caused disease. Relative contributions of these sources vis-a-vis sclerotia have not been assessed.

There is considerable disagreement among different scientists regarding the viability of sclerotia (from 14 days to 2 yr) under the varying influences of the microenvironment—surface or underground, moisture and temperature regimes, soil texture, organic matter, organic amendments, green manuring, nutrient status, bacterial activity, etc.

Sclerotial population in soil has been estimated variously at $0.6-3.1 \times 10^6$ per ha or more. Viability was 42–78%, and $c.$ 49% of germinated sclerotia were pathogenic.[12] Sclerotia survived longer and accumulated more in undisturbed soils after a high incidence of ShB (Lee, 1979).[4] Sclerotial population at 0–0.6 cm layer of soil (216–701 /1) having 42–51.4% viability represented the annual production which sought to replenish the sclerotia at 0.6–7.6 cm layer, being 27–87 /1 and having 14.2–30% viability.[1] The number of sclerotia trapped floating in standing water one month after transplanting was 3.1×10^6 per ha.

Source and size of sclerotia and their fragments, melanin status and quality, membrane permeability, endogenous respiration, etc. of the sclerotia, 0_2 - CO_2 status in soil and in the sclerotia, H-ion concentration of the clay micelle, etc., have been emphasized by earlier researchers.[4] Apparently, these have not been pursued perhaps because the practical implications of the phenomenon of survival demanded more attention.

Collateral Hosts.   They may act as the primary sources of infection, provide new or additional primary foci of infection, and may contribute significantly toward annual replenishment of sclerotial population in soil. Further, atypical symptoms may be produced on some of these hosts which can indulge as surrogate hosts. The voluminous data generated in this area from India, Japan, the Philippines, and Peoples Republic of China (PRC) in particular need reexamination with regard to cross-inoculation, natural occurrence, characterization of AG or ISG involved, and their epidemiological significance.

6-5 INOCULATION, DISEASE RATING, AND BIOASSAY

Field tests are uniform and precise but seedling tests are done for mass screening. Inoculum is provided with sclerotia, hyphal tip, culture on rice grain, straw or *Typha* sp. to be inserted in a leaf sheath wrapped with paper or covered with cellophane tape (Amin, 1975),[4] to avoid drying after inoculation. Syringe inoculation (0.14–0.44 ml) was considered useful for screening.[50] Smearing with mycelial fragments in carboxymethyl cellulose suspension developed for blast which will require low inoculum may be tried for ShB.[51]

Disease rating by IRRI Standard Evaluation System (SES) has been widely adopted for the sake of convenience and uniformity of results.[51] Separate 0–9 SES scales are there for ShB to rate percentage of leaf sheaths infected and vertical spread.

For more precise evaluation of ShB, several disease rating methods have been based on the absolute number or proportion of infected sheaths in a hill.[2] Manian and Manibhusan Rao (1980)[4] developed a scale of four grades of resistance/susceptibility by normal distribution method on the basis of lesion length, percentage of infected sheaths per tiller, and percentage of infected tillers per hill. Currently, at IRRI, a scale of lesion length on detached leaves (R = 0.3 cm, S = 3 cm (IR58 as check), HS = >3 cm) and a scale of RLH (HR 20% RLH, MR 20–30%, MS 31–45%, S 46–65%, HS $\geq$65%) are followed for screening of elite lines. In addition to RLH, percentage area infected (PAI) as ratios of leaf to sheath area infection recorded 28 days after inoculation has been included.[52]

Fungicide Bioassay Techniques.   Kannaiyan and Prasad (1979)[4] have developed a poisoned bait technique for sclerotial plant pathogens, particularly RS. An 8 mm agar disc with 5 ml fungicidal solution at varying concentrations with a given number of sclerotia placed on the disc is taken on some amount of sterile soil in a 25 ml vial and is further covered by a layer of 2.5 ml sterile soil. The assembly is inoculated for 24 hr and then the sclerotia are transferred to Czapeck's agar to test viability.

Mycelial discs dipped in antibiotic/fungicide solution inoculated into leaf sheaths enabled testing fungicidal efficacy at the early phases of infection and symptom expression. Spraying the antibiotic/fungicide 8–16 hr after inoculation enabled testing curative action.[53]

6-6 EPIDEMIOLOGY

6-6-1 Factors Affecting Disease Development

Pathogen factors.   With the same inoculum density (ID), ShB and yield were not different when inoculations were made four times at 10 day intervals, but these were more with higher ID at booting or heading than with lower ID at booting.[54]

Host factors.   The characters making the high yielding varieties particularly susceptible to ShB are: dwarf, broad leaf, highly responsive to N and P resulting in high sugars and free aminoacids for the consumption of the fungus, closed canopy, congenial warm humid microclimate in the hills, closely adjacent tillers in the compact hills facilitating physical contact of inoculum, and synchronous tillering offering nearly the same physiological age of the sheaths and flag leaves.

Generally, early maturing varieties are more "resistant" than late maturing varieties under conditions less favorable for ShB. Such differences exist when they are also tall and short, respectively (Miah et al., 1979).[4] However, the differences are rendered small when planting dates are staggered so that all cultivars head at the same time or become insignificant when the disease pressure is high (IRRI, 1987).[22] But because of low temperatures in late autumn in Japan, late maturing cultivars can escape the disease.[2]

Aging.   The young (2–3 week) leaf sheaths or blades are more resistant than old (5–6 week) ones. The mature sheaths are loosened around the culm, facilitating penetration through the inner surface of the sheath. Upper sheaths and blades are also less vulnerable before heading than after. Biochemistry and physiology of differentiation-linked breakdown of resistance may be interesting to examine.[55] At flowering, the plant canopy opens, the microclimate changes, and the disease severity is usually lessened. All tillers being fertile and without compactness, longer internodes near the ground, longer panicles and high wax content, heavy silication of culm, sheath, boot, and flag leaf, compressed sheath, short flag leaf, etc. might constitute the ideotype for ShB resistance. Path analysis and factor analysis of susceptibility attributes may rank their importance.

Environmental factors

Atmospheric Factors.   Optimum disease development took place at 28°C or 30–32°C over the range of 23–35°C and 81–92% or 96–97% RH.[2-3] Apparently, the isolates do vary. Reexamination is also needed with modern tools and including factors like wetness and temperature of sheath/leaf. The disease was severe on the monsoon crop in eastern India (CRRI, 1977).[56]

While lesion development was rapid in light,[57] disease severity was maximum at 50% light intensity, medium at normal sunshine, and least in complete shading irrespective of susceptibility/resistance of cultivars. While normal sunshine produced maximum total phenols and complete shading produced physiologically weak plants, 50% light intensity enabled an increase in total and soluble sugars, aminoacids, and total N.[58]

Basidia produced at night following drizzling days remain viable longer than when produced in dry conditions when they mature less and collapse during the day (IRRI, 1973).[22]

Soil Factors.   High soil moisture adversely affected infection of seedlings but apparently was not a deterrent in wet rice. The disease became serious particularly when soil temperatures were around 20°C.

Nutritional Factors.   Similar to foliar diseases, high nitrogen directly favors only the later stages of lesion development.[52] High N and/or P increase disease and high K or PK lowers disease (CRRI, 1977).[56] There were more phenols in inoculated plants with high K than with low K (Sundaram and Prasad, 1981).[14] The lessons from the foliar diseases of rice suggest a different role for P which at higher doses can partly reverse the ill effects of high doses of N.[59] On the susceptible cv Pusa—2-21 ShB was maximum with urea, less with some slow release forms, and least with crotolidene diurea.[60] Foliar diseases of rice are, however, dose-sensitive, finally reaching the same intensity irrespective of quick or slow release forms.[61]

Salt at 0.01−1.0% reduced infection but also retarded plant growth (Endo, 1933).[2] Moderate resistance may be searched from among the salt-tolerant lines.

Biotic Factors.   The rice root nematode, *Hirschmanniella oryzae*, was shown to increase ShB and was itself promoted.[62] Whether or not an interaction is there has to be ascertained more critically.[63]

6-6-2 Forecasting

For forecasting of ShB, the factors considered most critical are the sclerotial number, distribution, floatation, viability, and pathogenicity. The methods of collection and sampling of sclerotia ($c.$ 40−400 samples for varying sclerotial density at 10% accuracy and P $= 0.05$),[4,64,65] and the knowledge of rice microclimate[66] need to be emphasized. Systematic sampling was most efficient for ShB survey.[67] For 15% accuracy, they[67] determined the sample sizes such as for percentage of infected hills (10 hills), for RLH (100 hills), for disease incidence (20 fields in 1,500 ha; single hills from five rows; every 15th hill; and 20 hills from each row).

In Japan, Hashiba et al.[68] derived a model for vertical development (VD) of ShB lesion. First model curves were based on temperature-humidity effects in stages—RLH at fixed RH, varying temperatures (T); then at fixed T, varying RHs. Second model curves were fitted for RLH and age of sheaths (2−5, 10, 15 days) after heading. The rate of VD was 1.59 cm/day. Similarly, model curve for horizontal development of ShB was developed on T, RH between hills, and sclerotia per unit area. In early maturing varieties, they obtained the relationship for disease (in terms of VD), $y = 1.62x - 32.4$, where $x =$

$0.73z - 4.13$, where z is the height of the uppermost infected part. Finally, Hashiba et al.[69] based their calculation of disease intensity as a product of vertical and horizontal disease development (HD), respectively, by (height of topmost lesion/plant ht, x), and p.c. of diseased hills (A). Thus, disease incidence (D) at maturity = RLH × HD = xA = $(1.62x - 32.4)$ A/100. The final model curve was validated for 10 years against the observed incidence of disease. Ohta et al.[70] proposed a computer simulation model for ShB forecasting in Japan.

From Korea, Kim and Min[71] estimated disease severity at maturity (y) of cv Yushin (Tongil type) as $y = 4.6x - 13.2$, where x is top lesion height (simpler than percentage of top height versus plant height or lesion index). In Texas, Belmar et al.[64] obtained linear relation ($r^2 = 0.85$, P = 0.01,) between the number of sclerotia and ShB.

Comparative validation of these models for various locations and conditions is necessary, although predictive value of these models for forecasting is limited but useful for current season crop loss estimation.[72]

More importantly, cropping systems for 34 fields had a significant effect (P = 0.05) on the preplant inoculum density (ID) of RS and ShB incidence. Mean numbers of sclerotia recovered per kg soil were 4.02, 1.43, 0.07, with average disease intensities of 5.4, 2.7, 0.4% for rice (R) - soybean (S) - R, S-S-R, pasture (P) - P - R, respectively. Significantly higher (P = 0.05) IDs and disease intensities were more in 1:1 than in 2:1 rice and soybean. Distribution of sclerotia and ShB in the fields was spatially aggregated in larger ($>1m^2$) grids and random at smaller sample spacings. Inoculum was spatially autocorrelated but disease was not. In matrix surveys, IDs ranged from 0–44 sclerotia recovered/ 440 cm^3 soil core, and disease incidence ranged from 0–26 diseased per 50 tillers.[64]

6-7 MANAGEMENT

6-7-1 Cultural Management

Cultural eradication.   Destruction or burning of crop residue could not eliminate sclerotia from soil but was suggested to be a part of integrated management.[73] Destruction of collateral hosts has to be nearly total on the fields and bunds (manual, mechanical, tarping, or chemical) in order to be effective. Two hand weedings in upland rice decreased blast but increased ShB in FRG.[74] Considering the varied effects of weeding on various soil and crop variables, detailed studies are necessary.

Cultural manipulation.   Crop rotations may be ineffective even if long because of wide host range, or they may be uneconomical or unacceptable to farmers. Large-scale shifting of rice–soybean rotation from a 1:2 to 1:1 system proved disastrous for both rice and soybean in the southern United States.[1]

Balanced NPK, split N, slow-release forms (including prilled urea, large size or large granule urea which have not been tried) may be tried or recommended. On an S cv, two sprays at 10 day intervals of borax and sulphates of Zn, Cu, and Fe each at 0.05% reduced ShB and increased yield of rice during both monsoon and winter crops.[75]

Wide spacing reduced ShB and could even counter the effects of high N, but the yield was not consistent[4,76] and close spacing could not compensate yield loss due to increased disease severity.[77] Choice of early or late maturing varieties will depend more on location and season than on any intrinsic resistance. Early transplanting, close spacing (27 × 12 cm), high N (220 kg/ha) induced more disease than 10 day late transplanting, wide spacing (27 × 15 cm), and low N (110–150 kg/ha).[78] Optimum combinations of cultural and sanitation practices have to be developed for any given rice-based agroecosystem. Further, some probable cultural management measures like deep ploughing, soil solarization, exploration of suppressive soils, and application of compost have not been attempted specifically against ShB.

6-7-2 Biological Control

Direct inoculation.   Various fungi such as *Gliocladium virens, Trichoderma viride* (TV), *T. harzianum* (TH), *T. aureoviride, T. pseudokoningii, Trichoderma spp., Aspergillus niger, A. terreus, Neurospora crassa, Penicillium spp., Laetisaria arvalis* [Aphyllophorales, corticiaceae] have been detected from soil, sclerotia, healthy or infected plant parts, or crop refuse. They can inhibit mycelial growth, suppress formation and germination of sclerotia, cause hyperparasitism and lysis, and exhaust substrates by early and rapid competitive saprophytic colonization.[2,4,79,80] *Trichoderma* population is widespread in tropical rice soils, higher in rainfed than in irrigated soils. Its population is high when ShB incidence is low, has very high competitive saprophytic ability (IRRI, 1984–1988).[22] TH releases extracellular chitinase, 1,3-β-D-glucanase, etc. and degrades the RS cell wall and may have some specificity for AGs.[81] TH and TV coil tightly around RS hyphae, and may penetrate and erode them in natural field soils.[82–83] Foliar spraying of spore suspension of *Trichoderma* spp. (Roy, 1977)[4] or application in soil[84] or standing water before planting, mass production of native or exotic *Trichoderma,*[85] improvement of their strains, etc. have to be developed before commercialization of the practice.

Nineteen fluorescent and nonfluorescent pseudomonads, 6 enterobacteria, *Bacillus subtilis, B. pumilus,* and *B. laterosporus* have been isolated from various sources including ShB RS sclerotia yielding 61% of the antagonists rendering loss of viability of 43% sclerotia in soil. Because of high sclerotial production, even this natural high antagonism may not be good enough. Seed bacterization not only affected the ShB pathogen but also promoted rice germination, caused protection form infection, and suppressed the disease of the current and subsequent crops by being established in soil[86] Of 16 bacterial isolates, while a few were active against any one of blast, ShB, and bakanae diseases, fewer were active against all three of them. Against ShB, one of the isolates reduced disease intensity to 17% against 46% in check (IRRI, 1982–1988).[22] But great caution is warranted as some of the nonfluorescent pseudomonads are known to cause rotting of grain, seedling, sheath, and foot as well as sterility in rice. Biocide formulation is a problem.

Besides, presence of plasmids and dsRNA mycoviruses of RS (as of AG-2) make them slow-growing and weakly pathogenic.[87,88] These are the parts of biocontrol weaponry on the blueprint.

Promotion of natural control.   RS is a good saprophytic colonizer but has low competitive saprophytic ability. Green manuring with legumes and nonlegumes can reduce the survival of sclerotia and saprophytic activity of RS and enhance mycostasis.[89] Sclerotia survived less in the root zone of *Sesbania aculeata.*[82]

Organic amendments like edible and nonedible oilcakes, farmyard manure, agricultural and industrial wastes, are known to reduce sclerotial viability and ShB incidence, and some of them may increase rice shoot growth.[2,4,83,90,91,92] The results were not always consistent; some of them even decreased root growth, or failed to reduce ShB incidence. For both green manuring and organic amendments, long-term field trials are needed.

6-7-3 Chemical Control

Many fungicides, other pesticides, metal chelates,[93] a series of triazines,[94] phenols and melanins,[5] plant extracts including oils, etc. have been tested for toxicity against ShB.

For seed treatment and soil application, the effective fungicides reported too often are iprodione, benomyl, and quintozene. Many other soil applicants were effective against ShB but none protected the crop from yield loss.[95] Tolclofos - methyl, effective against RS, has not apparently been tried against the RS causing rice ShB.

Foliar applicants have been sought more vigorously. Of all the fungicides in use on ShB, most workers have found iprobenfos (considered best for yield protection), iprodione, and benomyl (worked well for over 10 years with a marginal yield of 500–900 kg/ha in the southern United States[96]) to be most efficacious and economic. However, under upland and lowland conditions, iprodione and benomyl, singly or in combination effectively reduced ShB but yield was not increased.[97] The fungicides no longer approved for use include fentin hydroxide (approved for only single application in the southern United States,[1] and ferric methyl arsenate (Monzet$^{(R)}$). Organic arsines were in commercial use in Japan till around 1984. They were very effective against ShB and protective of yield loss. Others with not so consistent results, but in use, include carbendazim, thiophanate, chloroneb, chlorothalonil, derosal, edifenphos, mepronil, thiabendazole, 1,2,4-triazoles and many others.[22,98,99]

New Fungicides. Some recent fungicides are more or less specific to *Rhizoctonia* or sclerotial diseases.[100,101] Propiconazol(e) [Desmel$^{(R)}$, Tilt$^{(R)}$], a broad-spectrum triazole compound, having cytokinin-like properties and inhibiting ergosterol biosynthesis, has been used effectively against ShB. Pencyuron [Monceron$^{(R)}$] is a broad-spectrum fungicide recommended against ShB,[96,102,103] but contrary reports are also there. Diclonazine [Mongnaw$^{(R)}$] has not apparently been tried against ShB. Flutolanil [Moncut$^{(R)}$], a systemic benzanilide, with acropetal translocation in rice, is claimed to be remarkably protective and curative as foliar spray (25 μg/ml with high residual effects), foliar dusting, seed treatment (1.6–3.2 μg/g—achieving 80% control), and application in standing water.[100,104]

Combinations. Some combinations of fungicides have been tried with the hope of "synergistic" action (better called additive since interaction was not tested statistically). Thus, mancozeb + thiobencarb and iprodione + carbendazim were better than several single formulations in disease control and yield protection, while mancozeb + carbendazim tank-mix was not effective.[93,105,106,107] Fungicide—insecticide mixture formulations (pencyuron 12.5% + isoprocarb 30%) showed high synergism against ShB and brown plant hopper (BPH) in Korea under tropical conditions.[108]

Antibiotics. The antibiotics validamycin (VM, specific for *Rhizoctonia*) and polyoxin developed in Japan are in commercial use. Chingfengmeisu and Jinggangmycin, both from PRC, and aureofungin from India were claimed to be effective against ShB and other diseases. Validamycin A (VMA, derived from *Streptomyces griseus*) is particularly effective on ShB at 30 ppm. Poor efficacy of VMA against ShB as also failure to protect yield have been reported (IRRI, 1979).[22] Polyoxin (antibiotic complex from *S. cacasi* var. *ascensis*) is almost as effective as organoarsines but not phytotoxic.[10] Uyeda et al.[109] proposed that validamycin (from *S. hydroscopicus var. limoneus*) activates B-1,3-glucan synthetase by increasing the levels of AMP and Pi in mycelia. Shibata et al.[110] isolated hyphal extension inhibitors I and II from RS (the former being 5 times more toxic than the latter) with similar inhibitory spectra to that of VM. VMA can reduce inositol content and cause hyphal distortion in RS.[111]

Fungicidal Schedules.   In India, soil application of thiram and quintozene before transplanting followed by single edifenphos spray at maximum tillering proved better than any other combination.[112] In the southern United States, two fungicide applications are recommended: first, shortly after appearance of symptoms (usually after early inter-node elongation) and 10–14 days later, before heading.[1] In Korea, based on the incidence of disease over 90 days, a 21-day period of rapid progress was identified during which two fungicide sprays were recommended at 10-day intervals.[71]

Application Methods.   Some novel application methods for soil and seed treatment and chemirrigation have been attempted but none for ShB in particular.[2,113,114] Liquid N and fungicide could be tank-mixed against ShB of rice.[115] For soil application, broadcasting of fungicide (WP)—mixed sand is practiced in India.

Comments on Chemical Control of ShB.   Losses sustained after single foliar application of fungicides or lower rates are considerable;[1] hence, fungicide recommendations are to be meticulously followed in practice. Effects of fungicides in use against ShB on nontarget organisms have not been intensively studied.[75] An international long-term ShB uniform fungicide trial with comprehensive scope and objective is needed so that inconsistency of fungicide field performance can be properly statistically analyzed rather than averaging for the determination of overall performance.[1] Chemical control of diseases due to soilborne pathogens will remain difficult by soil application because of inadequate distribution of the fungicide through the required soil volume. Symplastically mobile fungicides are theoretically attractive but technically elusive.

6-7-4 Induced Resistance

Physiological induction.   Ninety percent protection of rice from ShB was claimed with preinoculation by a potato RS isolate.[116] Can a preceding crop of potato substantially reduce ShB on a succeeding crop of rice? This is a question worth asking. Phytoalexin inducers can control ShB to maximum (45%) with cycloheximide.[117] RS on pea can demethylate pisatin,[118] but no such mechanism has yet been discovered in the ShB pathogen.

Mutation breeding.   Some attempts were made without any positive result. As 96 early mutants of HS cv Pelita-I/1 showed a continuous grade of disease reaction from resistant to highly susceptible, it was rightly concluded that the chance of induction of resistance by mutation was remote.[119]

Hybridization. Crosses between IR - 930 - 14 - 14 × Colombia 1 yielded a cv showing multiple resistance against several diseases including ShB.[3] *Indica* types are more resistant than *japonica* types in Japan but have not been found useful for genetic improvement through hybridization. In the United States, a higher level of resistance is found in cultivars with short and medium grains—more like *japonica* types than with long and fine grains with short maturity as in *indica* types.[1]

Broad and narrow sense heritabilities between IR 9752-71-3.2 and IET 499 were 0.52 and 0.37, respectively. Performance of 28 diallelic F_1s (nonreciprocal crosses) showed that general combining abilities are more important than specific ones, suggesting the importance of additive gene effects. Recurrent selection was considered unlikely to improve resistance beyond a moderate level.[120]

6-7-5 Disease Resistance

Sources and screening. Screening has rarely located high resistance among the cultivars, breeding lines, or related wild species of *Oryza*. Moderate resistance was recognized rather early, such as Kataktara DA-2 (IRRI, 1972),[22] 12814/SI 26 (James, 1974).[4] As summarized in IRRI (1987),[22] of 7,614 lines, 72 have been identified as R, of which 48 belonged to upland, 17 to wild rice, and only 3 among elite lines. Only a few lines have shown 0–2 SES disease scores, such as RP1821,[119] F47 (CC147F-112-18-4-106) and HM 34-6-4-P.[121]

Different marker characters are also being sought. Seven *O. glaberrima* collections from West Africa and some non-Japanese cvs with long internodes lower down the stem and many panicles/hill showed resistance, and some taller Japanese cvs with longer panicles (upper) internodes and leaf sheaths were also relatively resistant.[54] *Oryza eichengeri*, *O. minuta* Acc.191089, and *O. rufipogon* showed resistance (RLH <20%).[122]

Screening is usually done with single isolates of RS and with varying methods of inoculation; consequently, the results are at best tentative.

Inheritance of resistance. Resistance may be dominant and monogenic,[2,45] governed by two pairs of complementary genes and modified by epistasis,[123] or, most possibly and nearly universally, incompletely dominant due to multiple genes.[120] Lack of strong genes is suggested. Wax thickness is correlated with resistance and segregates at 3:1.[45] Genetics of other susceptibility-yielding attributes need to be worked out together with locating the sources for them.

Susceptibilities to ShB and brown plant hopper (BPH) are related. IR40 has two major genes for resistance against BPH, but it is susceptible to ShB, whereby the fungus produces honeydew containing amino acids at the collar zone of the rice and BPHs are attracted to this cv (IRRI, 1983).[22] The line, IRRI 1974-11-2, showed resistance to both ShB and BPH (Gangopadhyay and Misra, 1984).[120] Six lines have shown multiple resistance to both ShB and bacterial blight.[124]

Recommendations and prospects. Yield loss differences between moderately resistant (MR—7–15%) and susceptible (S—15–25%) and highly susceptible (HS—25–50%) cvs can be substantial.[1] Thus, HS cvs such as Pusa 2-21 can be discouraged and high-yielding MR cvs such as Jaya, Triveni, Jyoti, Pankaj, IET 1991 can be encouraged, at least in endemic areas on the basis of local experience until more dependable solutions are at hand.

Borthakur and Addy[125] are optimistic, hoping that as high yield and low ShB scores are related, high ShB resistance will be achieved with breeding for high yield. However, Gangopadhyay[3] argued that since the ShB fungus is more macerative than toxicogenic, a specific or vertical resistance may be lacking in rice.

Instead of vertical and horizontal resistance, Zadoks and Schein[126] proposed five groups on the basis of cultivar (C)-pathogen (P) interaction, viz. (a) uniform, (b) P-differential, (c) C-differential, (d) strongly interactive, (e) weakly interactive. Rice–ShB relationship appears to belong to group (e). As the rice-bacterial blight system is also weakly interactive, Mew[127] suggested that aggressiveness rather than virulence could be quantifiable with respect to a group of isolates. Current screenings are based on IRRI SES scores, and some specific parameters have recently been included. The pattern of their inheritance may be clearer in future. However, since there is a limit to pyramiding of genes, the prospects of conventional breeding do not appear to be very bright.

6-7-6 Integrated Management

In Krishna delta (Andhra Pradesh, India), the combination of moderately resistant varieties, cultural management, conservation of natural enemies, and need-based use of pesticides reduced ShB and other pests, and average yield at farmers' level increased by 36% in 1985 over the base year 1981.[128] Instead of sophisticated integrated pest management (IPM) as developed in the West, concept and tactics of appropriate IPM may be applicable in ShB-endemic areas.[129,130]

6-8 CONCLUDING REMARKS

1. The world of ignorance expands as the knowledge grows, for we can know what we have to know yet we know not. For the rice sheath blight the situation is worse, since we do not know what we have to know. This reviewer could not hide his pessimism.

2. Rice ShB literature can be sorted into three types: (a) the largest number of papers producing the bulk of data, which are often inconclusive; (b) location-specific experiments from which generalizations are difficult; and (c) generalizable design of experiments which ignore location-specific answers. Yet we need voluminous and conclusive data, need to theorize as well as obtain general and location-specific solutions—all from the same hands. Plant pathologists are indeed at a disadvantage, which becomes glaring when the challenge is as formidable as the rice sheath blight has proved to be.

3. Amazing and interesting as *Rhizoctonia solani* is, in spite of its almost interminable array of miniscule differences in a continuous spectrum of variation, each such entity is consistently characterizable in terms of any given character, but they do certainly overlap when more than one character is considered. No won-

der, if it proves to be a model of a fungal genetic system and worked out in detail to act as the model genetic system in the age of microbiotechnology. Biotechnology can raise some hope *only after* basic studies on genes, their products, and the mechanism of genetic engineering applicable to this system are worked out.

4. From the knowledge gained in several rice pathogen systems including ShB as also elsewhere, the researchers are already in a difficult era marked with revolutionary thinking in resistance breeding. None other than a great visionary, Vanderplank[131] stated that race identification and study should be discontinued because the number of potential races increases geometrically as the number of available genes increases, and that races become fixed taxa. Yet old habits die hard.

5. Some new more or less specific fungicides are on extensive trials. They will remain expensive for the poor rice farmers. We can only hope that their efficacy will be consistent and the problem of resistance to them will not be insurmountable with proper management.

With these comments I conclude that the rice ShB has come to stay and the challenge continues. A challenge is an alias for an opportunity as well. If ShB is a formidable challenge, the encounter can turn out to be an epic of many battles lost but the war won.

6-9 REFERENCES

1. Lee, F. N., and Rush, M. C. "Rice Sheath Blight: A Major Rice Disease." *Plant Dis.* 67 (1983): 829.

2. Ou, S. H. *Rice Diseases*, 2nd ed. (Kew: Commonw. Mycol. Inst., 1985), 272.

3. Gangopadhyay, S. *Current Concepts on Fungal Diseases of Rice*. New Delhi: Today and Tomorrow's Publishing Co. 1963, 349 pp.

4. Gangopadhyay, S., and Chakrabarti, N. "Sheath Blight of Rice." *Rev. Plant Pathol.* 61 (1982): 451.

5. Manibhusan Rao, K., Baby, U. I., and Joe, Y. "Basic Research on Rice Sheath Blight Disease," in *Basic Research for Crop Disease Management,* ed. P. Vidhyasekaran. (Delhi: Daya Publishing House, 1990), 190, Chap. 31.

6. Ou, S. H. *Rice Diseases*. Kew: Commonw. Mycol. Inst., 1972, 256.

7. Rouse, D. I. "Use of Crop Growth-Models to Predict the Effects of Disease." *Ann. Rev. Phytopathol.* 26 (1988): 183.

8. Gangwar, S. K., Chakraborty, S., Dasgupta, M. K., and Huda, A. K. S. "Modeling Yield Loss in Indica Rice in Farmers' Fields due to Multiple Pests." *Agric. Ecosyst. Environ.* 17 (1986): 165.

9. Miyake, I. "Studien Uber die Pilze der Reispflanze in Japan." *J. Coll. Agric.* Imp. Univ. Tokyo 2 (1910): 237.

10. Hashioka, Y., and Makino, M. "*Rhizoctonia* Group Causing the Rice Sheath Spots in the Temperate and Tropical Regions with Special Reference to *Pellicularia sasakii* and *Rhizoctonia oryzae.*" *Res. Bull., Fac. Agric.*, Gifu-ken Prefectural Univ. 28 (1969): 51.

11. Shangzhi, Y., and Mew, T. W. "Sheath Blight Diseases in Tropical Rice Fields." *Rice Res. Newsl.* 12 (1987): 19.

12. Kim, W. G., and Kim, C. K. "Density of Overwintered Sclerotia in Paddy Fields in Korea, Viability of the Sclerotia and Pathogenicity of the Sclerotial Fungi." *Korean J. Plant Path.* 4 (1988): 207.

13. Cai, K., and Xue, Q. F. "Preliminary Observation of Occurrence and Spread of Sheath and Culm Blight of Rice Influenced by Precipitation and Fields' Humidity." [*Rhizoctonia solani*] *Acta Phytopath. Sinica.* 14 (1984): 186.

14. Mathur, S. C. "Fungal Diseases of Rice in India," in *Recent Advances in Plant Pathology*, eds. A. Hussain, K. Singh, B. P. Singh, and V. P. Agnihotri. (Lucknow: Print House, 1983), 368.

15. Singh, N. I., Devi, R. K. T., and Singh, Kh. U. "Occurrence of Rice Sheath Blight (ShB) *Rhizoctonia solani* Kühn on Rice Panicles in India." *Int. Rice. Res. Newsl.* 13 (1988): 29.

16. Singh, N. I., Devi, R. K. T., and Singh, Kh. U. "*Rhizoctonia solani*: An Agent of Rice Boot Blight." *Int. Rice Res. Newsl.* 14 (1989): 22.

17. Marchetti, M. A. "Potential Impact of Sheath Blight on Yield and Milling Quality of Short-Statured Rice Lines in the Southern United States." *Plant Dis.* 67 (1983): 162.

18. Lee, F. N., and Templeton, G. E. "Rice Research Overview. Disease Control." *Arkansas Farm Res.* 37 (1988): 10.

19. Centre for Overseas Pest Research (COPR). *Pest Control in Rice*, 2nd ed. (London: PANS Manual No. 3, 1976), 59.

20. Soave, J., Azzini, L. E., Villela, O. V., and Gallo, P. B. "Selection of Irrigated Rice Cultivars for Low Incidence of Spotted Seeds." *Summa Phytopathol.* 9 (1983): 179.

21. Binesh, H., and Torabi, M. "Mode of Transmission of Rice Sheath Blight through Seeds and Reaction of Rice Cultivars to the Disease." *Iranian J. Plant Pathol.* 21 (1985): 15.

22. Int. Rice Res. Inst. (IRRI). *Annual Reports*, 1972–1988.

23. Ahn, S. W., and Mew, T. W. "Relation between Rice Sheath Blight (ShB) and Yield." *Int. Rice Res. Newsl.* 11 (1986): 21.

24. Chang, Y. C. "Studies on the Effect of Disease Severity of Sheath Blight on Rice Yield." *J. Agric. Res., China* 35 (1986): 202.

25. Rajan, C. P. D. "Estimation of Yield Losses due to Sheath Blight Disease of Rice." *Indian Phytopathol.* 40 (1987): 174.

26. Hashiba, T. "Forecasting Model and Estimation of Yield Loss by Rice Sheath Blight Disease." *JARQ* 18 (1984): 92.

27. Ogoshi, A. "Ecology in Pathogenicity of Anastomosis and Inter-Specific Groups of *Rhizoctonia solani* Kühn." *Ann. Rev. Phytopathol.* 25 (1987): 125.

28. Kardin, M. K., Oniki, M., Ogoshi, A., and Sakai, R. "Effect of Air Temperature on Mycelial Growth Rate of *Rhizoctonia solani* Kühn from Indonesia and Japan." *Penilitian Pertanian* 8 (1988): 23.

29. Kim, W. G., Kim, C. K., and Yu, S. H. "Anastomosis Grouping and Cultural Characteristics of Isolates of *Rhizoctonia solani* Kühn from Sclerotia Overwintered in Paddy Fields." *Korean J. Plant Path.* 4 (1988): 136.

30. Hashiba, T., and Yamada, M. "Viability of Sclerotia Formed on Rice Plants and on Culture Media by *Rhizoctonia solani*." *Ann. Phytopathol. Soc. Japan* 47 (1981): 464.

31. Bolkan, H. A., and Rebeiro, W. R. C. "Anastomosis Groups and Pathogenicity of *Rhizoctonia solani* isolates from Brazil." *Plant Dis.* 69 (1985): 599.

32. Michelmore, R. W., and Hulbert, S. H. "Molecular Markers for Genetic Analysis of Phytopathogenic Fungi." *Ann. Rev. Phytopathol.* 25 (1987): 383.

33. Borthakur, B. K., and Addy, S. K. "Anastomosis Grouping in Isolates of *Rhizoctonia solani* Causing Rice Sheath Blight Disease." *Indian Phytopathol.* 41 (1988a): 351.

34. Tsukiboshi, T., and Sato, T. "Anastomosis Grouping of *Rhizoctonia solani* Kuhn Causing Sheath Blight in Forage Crops and Summer Blight in Herbage." *Bull. Nat. Grassl. Res. Inst. Japan* 39 (1988): 50.

35. Yu, C. M. "Brown Sheath Blight Disease of Rice—A New Disease of Rice in Taiwan." *Plant Prot. Bull. Taiwan* 25 (1983): 53.

36. Ogoshi, A., and Ui, T. "Diversity of Clones within an Anastomosis Group of *Rhizoctonia solani* Kühn in Field." *Ann. Phytopathol. Soc. Japan* 49 (1983): 239.

37. Shahjahan, A. K. M., Fabellar, N., and Mew, T. W. "Relationship between Growth Rate, Sclerotia Production and Virulence of Isolates of *Rhizoctonia solani* Kühn." *Int. Rice Res. Newsl.* 12 (1987): 28.

38. Roy, A. K. "Source of Seed-Borne Infection of Sheath Blight in Rice." *Oryza* 26 (1989): 111.

39. Naidu, V. D. "Sheath Blight Occurrence in Rice Nurseries." *Int. Rice Res. Newsl.* 8 (1983): 10.

40. Leu, L. S., and Yang, H. C. "Distribution and Survival of Sclerotia of Rice Sheath Blight Fungus, *Thanatephorus cucumeris* in Taiwan." *Ann. Phytopathol. Soc. Japan* 51 (1985): 1.

41. Lee, F. N. "Sheath Blight Sclerotia Found in Arkansas Soil." *Arkansas Farm Res.* 28 (1979): 5.

42. Manian, S., and Manibhushan Rao, K. "Histopathological Studies in the Rice Sheath Blight Disease Incited by *R. solani.*" *Z. Pflkr. Pflsch.* 89 (1982): 523.

43. Marshal, D. S., and Rush, M. C. "Relation between Infection by *R. solani* and *R. oryzae* and Disease Severity in Rice." *Phytopathology* 70 (1980): 941.

44. Doodman, R. L., and Flentje, N. T. "The Mechanism and Physiology of Plant Penetration by *Rhizoctonia solani,*" in *Rhizoctonia solani Biology and Pathology*, ed. J. R. Paremeter, Jr. (Berkeley: University of California Press, 1970): 149.

45. Marshall, D. S., and Rush, M. C. "Infection Cushion Formation on Rice Sheaths by *Rhizoctonia solani. Phytopathology* 70 (1980): 947.

46. Collmer, A., and Keen, N. T. "The Role of Pectic Enzymes in Plant Pathogenesis." *Ann. Rev. Phytopathol.* 24 (1984): 383–409.

47. Gaspar, Th., Penel, C., Thorpe, T., and Greppin, H. *Peroxidase 1970–1980, A Survey of Their Biochemical and Physiological Role in Higher Plants.* Geneva: University of Geneva Press, 1982, 324 pp.

48. Bell, A. A., and Wheeler, M. H. "Biosynthesis and Functions of Fungal Melanins." *Ann. Rev. Phytopathol.* 24 (1984): 411.

49. Mgonja, A. P., Lee, F. N., and Courtney, M. "Horizontal and Vertical Spread of Sheath Blight (ShB)." *Int. Rice Res. Newsl.* 11 (1986): 17.

50. Wasano, K., Oro, S., and Kido, Y. "The Syringe Inoculation Method for Selecting Rice Plants Resistant to Sheath Blight, *Rhizoctonia solani* Kühn." *Jap. J. Trop. Agric.* 27 (1983): 131.

51. Guochang, S., and Shuyan, S. "A New Inoculation Technique for Rice Blast." *Int. Rice Res. Newsl.* 14 (1988): 15.

52. Shahjahan, A. K. M., Fabellar, N., and Mew, T. W. "Nitrogen Level, Cultivar and *Rhizoctonia solani* Isolate Effect on Sheath Blight (ShB) Development." *Int. Rice. Res. Newsl.* 12 (1987): 27.

53. Endo, T., Matsuura, K., and Wakae, O. "Effect of Validamycin A on Infection of *Rhizoctonia solani* in Rice Sheaths." *Ann. Phytopathol. Soc. Japan* 49 (1983): 689.

54. Wasano, K., and Hirota, Y. "Varietal Resistance of Rice to Sheath Blight Disease Caused by *Rhizoctonia solani* Kühn, by the Syringe Inoculation Method." *Bull. Fac. Agric.* Saga U. 69 (1986): 49.

55. Dasgupta, M. K., and Chattopadhyay, S. B. "Variation in Susceptibility of Rice Leaves to the Brown Spot." *Indian Phytopathol.* 28 (1975): 17.

56. Central Rice Research Institute (CRRI), Cuttack, *Annual Reports,* 1977, 1979.

57. Yuno, I., Hashiba, T., and Mogi, S. "Effect of Temperature on Disease Development of Rice Sheath Blight Caused by *R. solani.*" *Proc. Assoc. Plant Prot. Hokuriku* 26 (1978): 4.

58. Chaudhuri, S., and Chakrabarti, N. K. "Effect of Different Daylight Intensities on the Susceptibility of Rice to Sheath Blight Infection and Related Biochemical Changes." *Indian Phytopathol.* 42 (1989): 308.

59. Dasgupta, M. K., and Chattopadhyay, S. B. "Effect of Different Doses of N and P on the Susceptibility of Rice to Brown Spot Caused *Helminthosporium oryzae.*" *Z. Pflkr. Pflsch.* 84 (1977): 276.

60. Roy, A. K. "Effect of Slow-Release Nitrogenous Fertilizers on the Incidence of Sheath Blight and Yield of Rice." *Oryza* 23 (1986): 198.

61. Mukhopadhyay, S. K., and Dasgupta, M. K. "Variation in Yield and Disease-Pest-Weed Situation in Lowland Rice." *Fertil. News.* 25 (1980): 30.

62. Gokulapalan, C., and Nair, M. C. "Role of Rice Root Nematode in the Severity of Sheath Blight Disease of Rice in Kerala." *Indian Phytopathol.* 39 (1986): 436.

63. Sikora, R. A., and Carter, W. W. "Nematode Interactions with Fungal and Bacterial Plant Pathogens—Fact or Fantasy," in *Vistas on Nematology,* eds. J. A. Veech and D. W. Dickson. (Hyattsville, Md.: Soc. Nematologists, 1987), 307.

64. Belmar, S. B., Jones, R. K., and Starr, J. L. "Influence of Crop Rotation on Inoculum Density of *Rhizoctonia solani* and Sheath Blight Incidence in Rice." *Phytopathology* 77 (1987): 1138.

65. Kim, C. H., and Kim, C. K. "Density and Viability of Sclerotia of Rice Sheath Blight Pathogen Overwintering in Field." *Korean J. Plant Prot.* 26 (1987): 99.

66. Hirai, G., Nakayama, N., Inano, T., Chujo, H., Takaichi, M,. and Tanaka, O. "Growth of Plants and Microclimate in Rice Population. I. Horizontal Distribution of Several Meteorological Elements in Rice Population." *Jap. J. Crop Sci.* 58 (1989): 13.

67. Koike, K., Kojima, A., and Hashiba, T. "Sample Size and Sampling Method Needed for Investigating the Disease Incidence of Rice Sheath Blight Disease." *Ann. Phytopathol. Soc. Japan* 52 (1986): 47.

68. Hashiba, T., Uchiyamada, H., and Kimura, K. "A Method to Estimate the Disease Incidence Based on the Height of the Infected Parts in Rice Sheath Blight Disease," *Ann. Phytopathol. Soc. Japan* 47 (1981): 194.

69. Hashiba, T., Koike, K., and Yamada, M. "A Method for Estimating the Yield Loss by Rice Sheath Blight Disease Based on Height of Lesions and Percentage of Affected Hills in Rice." *Ann. Phytopathol. Soc. Japan* 49 (1983): 143.

70. Ohta, K., Imai, T., Shimada, K., and Chiba, J. "A Computer Simulation Model for Rice Sheath Blight Disease I. Synthesis of Disease Model." *Ann. Rep. Soc. Plant Prot.* North Japan 39 (1988): 74.

71. Kim, C. K., and Min, H. S. "Ecological Studies on Rice Sheath Blight Caused by *Rhizoctonia solani*. II. Forecasting and Control of Rice Sheath Blight." *Korean J. Plant Prot.* 22 (1983): 21.

72. Dasgupta, M. K., and Chakrabarti, S. "Mathematical and Statistical Epidemiology." *Basic Research for Crop Disease Management*, ed. P. Vidhyasekaran. (Delhi: Daya Publishing House, 1990), 213, Chap. 20.

73. Lee, F.N., and Courtney, M. L. "Burning Rice Straw Complements Other Sheath Blight Control Tactics." *Arkansas Farm Res.* 31 (1982): 7.

74. Kurschner, E. "Effect of Weed Control on the Development of Crop Plants and Infection with Rice Blast in Upland Rice." *Mitteil. Biol. Bundes, Land. Forest.* Berlin Dahlem 245 (1988): 201.

75. Kannaiyan, S., and Prasad, N. N. "Effects of Foliar Spray of Micro-nutrients on Rice Sheath Blight Disease." *Int. Rice. Res. Newsl.* 4 (1979): 13.

76. Kannaiyan, S., and Prasad, N. N. "Effect of Spacing on the Spread of Sheath Blight Disease of Rice." *Madras Agric. J.* 70 (1983): 135.

77. Mithrasena, Y. J. P. K., and Adikary, W. P. "Effect of Plant Density on Sheath Blight (ShB) Incidence." *Int. Rice. Res. Newsl.* 11 (1986): 20.

78. Chang, K. K., Dong, S. R., and Hong, S. M. "Ecological Studies of Rice Sheath Blight Caused by *Rhizoctonia solani*. III. Cultural Method and Disease Development." *Korean J. Plant Prot.* 24 (1985): 7.

79. Mew, T. W., and Rosales, A. N. "Relationship of Soil Micro-organisms to Rice Sheath Blight Development in Irrigated and Dryland Rice Cultures." *Tech. Bull., ASPAC Food Ferti. Technol. Center* 79 (1984): 11.

80. Manibhusan Rao, K., Sreenivasaprasad, S., Baby U. I., and Joe, Y. "Susceptibility of Rice Sheath Blight Pathogen to Mycoparasites." *Curr. Sci.* 58 (1989): 515.

81. Ridout, C. J., Coley-Smith, J. R., and Lynch, J. M. "Enzyme Activity and Electrophoretic Profile of Extracellular Protein Induced in *Trichoderma* spp. by Cell Walls of *Rhizoctonia solani.*" *J. Gen. Microbiol.* 132 (1986): 2345.

82. Roy, A. K. "Studies on the Sheath Blight of Rice in Assam. *Indian Phytopathol.* 42 (1989): 308 abstract.

83. Wu, W. S., Liu, S. D., Chang, Y. C., and Tschen, J "Hyperparasitic Relationships between Antagonists and *Rhizoctonia solani.*" *Plant Prot. Bull. Taiwan* 28 (1986): 91.

84. Manian, S., and Paulsamy, S. "Biological Control of Sheath Blight Disease of Rice." *J. Biol. Control.* 1 (1987): 57.

85. Lethbridge, G. "An Industrial View of Microbial Inoculants for Crop Plants," in *Microbial Inoculation of Crop Plants,* eds. R. Campbell and R. M. Macdonald. (Oxford U.: IRL Press, 1989): 11.

86. Vasantha Devi, T., Malar Vizhi, S., Sakthivel, N., and Gnanamanickam, S. S. "Biological Control of Sheath-Blight of Rice in India with Antagonistic Bacteria." *Plant Soil.* 119 (1989): 325.

87. Hashiba, T., Homma, Y., Hykumadu, M., and Matsuda, I. "Isolation of a DNA Plasmid in the Fungus, *Rhizoctonia solani.*" *J. Gen. Microbiol.* 130 (1984): 2067.

88. Tavantzis, S. M., and Bandy, B. P. "Properties of a Mycovirus from *Rhizoctonia solani* and Its Virion-Associated RNA Polymerase." *J. Gen. Virol.* 69 (1988): 1465.

89. Manibhusan Rao, K., Baby, U. I., and Joe, Y. "Effect of Organic Amendments on the Saprophytic Survival of the Rice Sheath Blight." *Oryza* 26 (1989): 71.

90. Kannaiyan, S., and Prasad, N. N. "Effect of Organic Amendments on Seedling Infection of Rice Caused by *Rhizoctonia solani.*" *Plant Soil* 62 (1981): 131.

91. Lakshmanan, P., and Nair, M. C. "Effects of Soil Amendments on the Viability of Sclerotia of *Rhizoctonia solani* in Soil." *Madras Agric. J.* 71 (1984): 526.

92. Kannaiyan, S., and Prasad, N. N. "Influence of Certain Green Manures on Seedling Infection of Rice due to *Rhizoctonia solani* Kühn." *Madras Agric. J.* 70 (1983): 809.

93. Rao, D. S., Ganorkar, M. C., Rao, B. L. S., and John, V. T. "Potential Fungitoxicity of Some Transition Metal Chelates Derived from Dehydroacetic Acid on *Rhizoctonia solani*, the Causal Organism of Sheath Blight of Rice Plants." *Natnl. Acad. Sci. Letters* 1 (1978): 402.

94. Konno, S., Sagi, M., Kimura, C., Kikuchi, J., Yamanaka, H., Fujita, F., Yamada, Y., and Adachi, M. "Studies on Astriazine Derivatives. XI. Synthesis of 5-aryl-1,2,4-triazine Derivatives and Their Anti-fungal Activities." *Yakugaku Zassni* 108 (1988): 142.

95. Telan, I. F., and Lapis, D. B. "Greenhouse Trials of Seed Dress Method for Controlling Sheath Blight (ShB)." *Int. Rice Res. Newsl* 13 (1988): 21.

96. Groth, D. E., and Rush, M. C. "New Fungicides to Control Sheath Blight of Rice." *Louisiana Agric.* 31 (1988): 8.

97. Telan, I. F., and Lapis, D. B. "Soil Incorporation of Fungicides to Control Sheath Blight (ShB)." *Int. Rice Res. Newsl.* 13 (1988): 23.

98. Reddy, A. P. K., Bhaktabatsalam, G., and Jonn, V. T. "Sheath Blight of Rice: Relationship between Disease Severity and Yield." *Pesticides* 15 (1981): 11.

99. Suryadai, Y., and Kadir, T. S. "Field Evaluation of Fungicides to Control Rice Sheath Blight (ShB)." *Int. Rice Res. Newsl.* 14 (1989): 35.

100. Kurono, H. "Steps to Moncut, a New Systemic Fungicide." *Japan Pestic. Informn.* 46 (1985): 6.

101. Sugiyama, M. "Rice Sheath Blight and Chemical Control in Japan." *Japan Pestic. Informn.* 52 (1988): 9.

102. Mithrasena, Y. J. P. K., Wickramasinghe, D. L., and Adikari, W. P. "Fungicidal Control of Rice Sheath Blight (ShB)." *Int. Rice Res. Newsl.* 12 (1987): 26.

103. Lee, F. N., and Courtney, M. L. "Foliar Fungicide Testing for Rice Sheath Blight Control." *Arkansas Farm Res.* 30 (1981): 11.

104. Hirooka, T., Miyagi, Y., Araki, F., and Kunoh, H. "Biological Mode of Action of Flutolanil in Its Systemic Control of Sheath Blight." *Phytopathology* 79 (1989): 1091.

105. Singh, U. D., and Sethunathan, N. "Individual and Combined Effects of Certain Pesticides on *Rhizoctonia solani*, Sheath Blight Pathogen of Rice." *J. Phytopathol.* 119 (1987): 240.

106. Thangasamy, T. A., and Rangaswamy, M. "Fungicide Timing to Control Rice Sheath Blight (ShB)." *Int. Rice Res. Newsl.* 14 (1989): 24.

107. Torabi, M., and Binesh, H. "Effect of Some Fungicides on Growth of *Rhizoctonia solani*, Causal Organism of Sheath Blight Disease of Rice on Media and in the Field." *Entomol. Phytopathol. Appl.* 54 (1987): 51.

108. Song, B. H., Jeong, Y. H., Kang, C. S., and Park, H. M. "Stability and Efficacy of Mixed Pesticides to Control Sheath Blight and Brown Planthopper." *Res. Rep. Rural Dev. Adm., Korea Repub., Plant Env., Mycol. Farm Prod. Utilization* 29 (1987): 266.

109. Uyeda, M. Suzuki, K., Tsuruta, H., and Shibata, M. "Effects of Validamycin on Glucan Synthesis by Cell-free Extracts from *Rhizoctonia solani.*" *Agric. Biol. Chem.* 52 (1988): 2607.

110. Shibata, M., Kido, Y., Honda, Y., and Shimizu, K. "Hyphal Extension Inhibitors I and II with Similar Inhibitory Spectra to Validamycin, Isolated from Hyphae of *Rhizoctonia solani.*" *Agric. Biol. Chem.* 53 (189): 869.

111. Robson, G. D., Kuhn, P. J., and Trinci, A. P. J. "Effect of Validamycin A on the Inositol Content and Branching of *Rhizoctonia cerealis* and Other Fungi." *J. Gen. Microbiol.* 135 (1989): 739.

112. Dev, V. P. S. "Sheath Blight Control with Soil Fungicides." *Int. Rice Res. Newsl.* 5 (1980): 14.

113. Kloepper, J. W., and Schroth, M. N. "Development of a Powder Formulation of Rhizobacteria for Inoculation of Potato Seed Pieces." *Phytopathology* 71 (1981): 590.

114. Rollett, A. C., Roberts, D. M., Malcom, A. J., and Wainwright, A. "Techniques for the Application of Pencyuron to Control Black Scurf (*R. solani*) on Potatoes." *Monograph—Brit. Crop Prot. Council* 39 (1988): 363.

115. Lee, F. N., Wells, B. R., Courtney, M. L., and Shockley, P. A. "Liquid Nitrogen-Foliar Fungicide Combinations in Rice Evaluated." *Arkansas Farm Res.* 32 (1983): 5.

116. Kalaiselvi, K., Sreenivasaprasad, S., and Manibhushan Rao, K. "Acquired Resistance of Rice Leaves to *Rhizoctonia solani.*" *Int. Rice Res. Newsl.* 11 (1986): 16.

117. Sarkar, M. L., and Sinha, A. K. "Use of Phytoalexin-Inducing Chemicals to Control Rice Sheath Blight (ShB)." *Int. Rice. Res. Newsl.* 14 (1989): 23.

118. Delserone, L. M., and VanEtten, H. D. "Demethylation of Pisatin by Three Fungal Pathogens of *Pisum sativum.*" *Phytopathology* 77 (1987): 116.

119. Gangopadhyay, S., and Padmanabhan, S. Y. *Breeding for Disease Resistance in Rice.* New Delhi: Oxford & IBH, 1987, 340 pp.

120. Sha, X. Y., and Zhu, L. H. "Resistance to Sheath Blight (ShB) in China." *Int. Rice Res. Newsl.* 14 (1989): 14.

121. Majumder, N. D., Ansari, M. M., and Mandal, A. B. "Reaction of Rice Germplasm to Sheath Blight (ShB)." *Int. Rice. Res. Newsl.* 14 (1989): 8.

122. Amante, A. D., Pena, R de la, Sitch, L. A., Leung, H., and Mew, T. W. "Sheath Blight (ShB) Resistance in Wild Rice." *Int. Rice Res. Newsl.* 15 (1990): 5.

123. Goita, M. "Inheritance of Resistance to Sheath Blight in Long Grain Rice." *Diss. Abstr. Int., Sci. & Eng.* 46 (1985): 1808B.

124. Raj, R. B., Tayaba Wahab, Rao, G. V., Rao, A. S., and Reddy, T. C. V. "Evaluation of Rice Cultures against Bacterial Leaf Blight and Sheath Blight Disease." *Indian Phytopathol.* 40 (1987): 397.

125. Borthakur, B. K., and Addy, A. K. "Screening of Rice (*Oryza sativa*) Germplasm for Resistance to Sheath Blight (*Rhizoctonia solani*)." *Indian J. Agric. Sci.* 38 (1988b): 537.

126. Zadoks, J. C., and Schein, R. D. *Epidemiology and Plant Disease Management.* New York: Oxford University Press, 1979, p. 427.

127. Mew, T. W. "Current Status and Future Prospects of Research of Bacterial Blight of Rice." *Ann. Rev. Phytopathol.* 25 (1987): 359.

128. Krishnaiah, K., Reddy, P. C., and Rao, C. S. "Integrated Pest Control in Rice in Krishna Delta Area of Andhra Pradesh." *Indian J. Plant Prot.* 14 (1986): 1.

129. Dasgupta, M. K. "An Appropriate Integrated Pest Management System for Rice in Tropical Asia: Concept and Tactics." *Natnl. Symp. Integrated Pest Control—Progress and Perspectives,* ed. N. Mohandas and G. Koshy (Trivandrum: Assocn. Adv. Emtom., Kerala A.U.: 1988), 19.

130. Dasgupta, M. K. "Integrated Pest Management for Rice." *Indian Fmg.* 39 (1990): 25.

131. Vanderplank, J. E. "Should the Concepts of Physiological Races Die?" in *Durable Disease Resistance in Crops,* eds. F. Lamberti, J. M. Waller, and N. A. van der Graaff. (New York: Plenum, 1983), 41.

7

BACTERIAL BLIGHT OF PADDY

S. DEVADATH

Central Rice Research Institute,
Cuttack-753 006, Orissa, India

7-1 INTRODUCTION

There is no more important crop in the world today than rice; it is the staple food of the teeming millions of Asia where population explosion is taking place rapidly. There is a pressing need to double or even treble our rice production to keep pace with the increasing population. In the quest for increasing rice production, man has resorted to intensive methods of rice cultivation involving high-yielding susceptible cultivars with reduced genetic variability, higher plant population per unit area, high doses of nitrogenous fertilizers, and staggered sowing and planting which intensified the severity of bacterial blight of rice in most of the Asian countries. Bacterial blight is widely distributed, devastating, and more intimidating than other bacterial diseases of rice. It has been reported in Japan, India, Sri Lanka, Bangladesh, Pakistan, Taiwan, Korea, Thailand, Vietnam, Philippines, Indonesia, Australia, several Latin American countries, Malaysia, West Africa, Mali, Niger, and Senegal[1] (Figure 7-1) and recently from United States.

7-2 HISTORY AND ECONOMIC IMPORTANCE

Bacterial blight was first seen by farmers in 1884 in the Fukuoka area of Japan and by 1960 it was observed in all parts of Japan except the northern island of Hokkaido.[2] The bacterial nature of this disease was established by Takahasi in 1908 and by Bokura in 1911, and the bacterium was named *Bacillus oryzae* Hori and Bokura. Ishiyama in 1922

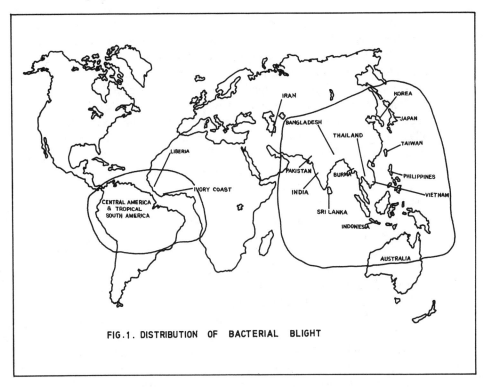

FIG.1. DISTRIBUTION OF BACTERIAL BLIGHT

Figure 7-1   Distribution of bacterial blight.

renamed the bacterium *Pseudomonas oryzae* (Uyeda and Ishiyama), adopting Migula's system. It was later transferred to *Bacterium oryzae* (Uyeda and Ishiyama) Nakata and subsequently to *Xanthomonas oryzae* (Uyeda and Ishiyama) Dowson, the name used for the last 40 years. According to the recent revision, the name *X. campestris* pv. *oryzae* (Ishiyama) Dye is currently followed.[3] The *Kresek* disease in Indonesia was studied by Reitsma and Schure[4] and they named the bacterium *Xanthomonas kresek* Schure,[5] but Goto[6] has shown it to be a severe form of bacterial blight that is occurring in various parts of the tropics.

Yield losses range from 20% to 30% in severely infected fields and may go up to 50% in rare occasions. In the tropics, the disease has been very destructive. In the Philippines and Indonesia, losses are higher than in Japan, but few figures are available.[1] Losses ranging from 6% to 60% have been reported in India.[7]

7-3 SYMPTOMS

Tiny water soaked spots at the margin of the fully developed lower leaves first appear on the seedlings, although its occurrence is rare in seedbeds. Infected leaves turn yellow, dry rapidly, and wither as the spots enlarge. The disease usually becomes noticeable in the field after the maximum tillering stage and is conspicuous at the heading stage.

A little below the tip of the leaves, water soaked lesions, usually on one side or sometimes on both sides appear at the margin. The lesions enlarge both in length and width with a wavy margin and turn straw yellow within a few days, covering the entire leaf (Figure 7-2). Water soaking is seen adjacent to the enlarging lesion. On susceptible cultivars, the lesion spreads rapidly to the lower end of the leaf sheath. As the disease portion enlarges, the infected leaves wilt and roll while the leaves are still green. Subsequently, the entire leaf may dry up (Figures 7-2, 7-3). Although the lesions appear generally at the leaf margins, the lesions may start at any point of the leaf, if injured. Milky or opaque bacterial exudate is observed in the morning hours when the relative humidity is high, and then drys up forming pale yellow, small spherical beads which are shaken off by the wind and drop into the field water (Figure 7-4). On the other straw yellow lesions, bacterial exudate is seldom noticed.

On resistant cultivars or in incompatible host–parasite relations, a brown discoloration at the advancing end of the lesion is observed, thus restricting the rapid enlargement of the lesion.

On the glumes, discolored spots surrounded by water soaking are conspicuously seen when the grains are young and green on severely infected plants. Infected grains mature earlier than the healthy grains.

Kresek or pale yellow symptoms are also noticed in the tropics. The kresek symptoms (Figures 7-5, 7-6) were first described in Indonesia.[4] In tropical countries, tips of the leaves are trimmed off before transplanting and these cut ends of the leaves serve as entry points for the bacterium. One or two weeks after transplanting, infection starts from the cut end of the leaves as green, water soaked spots, becomes greyish-green and folds up and rolls along the midrib. The rolling and withering of the entire leaf, including leaf sheath, follows.

The bacterium reaches the growing point of the young plant through the xylem vessels and infects the base of other leaves, thereby killing the young plant. When only a few leaves have wilted and can be seen floating on water, the disease is called kresek. Complete death of the entire plant is called *hama lodoh*.[4] These symptoms for the sake of simplicity are generally called kresek. These symptoms are often confused with stem borer damage. In infected young plants, when survive, the growth of tillers is arrested, and the tillers are stunted with yellowish green color. The crop shows uneven growth with missing hills.

Pale yellow leaves is another type of symptom sometimes observed. The youngest leaf is pale yellow or has a yellow or greenish-yellow broad stripe, while the older leaves are green. Although bacteria can be detected from the crown of the stem and from the internodes immediately below infected leaves, bacteria are not observed in pale yellow leaves.[6] Probably the buildup of bacterial population in the xylem vessels at the crown reaches a point at which little nutrient is available to the young leaves, thereby causing pale yellow leaves.

When leaf bits from freshly infected leaf are mounted in a drop of water and examined under a microscope, bacterial streaming from the veins is always observed. Yellowish sticky bacterial ooze is observed when the lower end of kreseked plant is cut and squeezed between the fingers. Sometimes the entire internode of the kreseked tiller shows yellow sticky bacterial mass.

Figure 7-2   Development of bacterial blight on rice leaves.

Figure 7-3   Rice crop severely affected by bacterial blight.

Figure 7-4   Bacterial exudate on the diseased leaf.

7-4 CAUSAL ORGANISM

The causal bacterium is Gram-negative, non-spore-forming, short rods with round ends measuring $1-2 \times 0.8-1$ μm with monotrichous polar flagellum of $6-8$ μm. The bacterial cells are capsulated and are joined to form an aggregate mass.   Colonies are circular, convex with entire margins, whitish yellow to straw yellow later and opaque. The bacte-

Figure 7-5   Potted plants showing kresek symptoms.

Figure 7-6 Kresek incidence in the field.

rial cells from host tissue are smaller than those grown on culture media.

The causal bacterium is a strict aerobe. The optimum temperature for growth is found to lie between 25–30°C. It fails to grow at 5°C and at a maximum temperature of 40°C. It cannot tolerate more than 3% concentration of sodium chloride. The thermal death point is between 51–53°C. Good growth is obtained at 6.5–7.5 pH levels. The isolates of the bacterium exhibit oxidative metabolism of glucose. The isolates vary to a large extent in the production of acid from carbohydrates and related carbon sources. They produce acid from xylose, glucose, fructose, galactose, sucrose, cellobiose, and starch. They fail to utilize inositol, glycerol, mannitol, and sorbitol. The bacterial isolates are negative to methyl red and Vogos-Proskaur tests. They can utilize citric, lactic, pyruvic, succinic, and fumaric acids but fail to utilize benzoic, oxalic, propionic, and tartaric acids. Asparagine is found to be inadequate as the sole source of carbon and nitrogen. The cultures are positive to catalase and negative to urease and tyrosinase activities. The isolates vary in the degree of liquefaction of gelatin. They exhibit proteolysis with alkaline reaction in milk media. Indole and acetoin are not produced and they fail to reduce nitrate to nitrite. All the isolates can produce ammonia and hydrogen sulphide, but they vary in the degree of hydrolysis of "soluble" starch and potato starch. They are found to be negative to lipolytic activity but positive to the utilization of uric acid. Nucleic acids are utilized. Lecithinase is not produced. They are weakly cellulolytic. Pectin methyl esterase is not produced but polygalacturonase is produced. The isolates of the causal bacterium vary much in the rate of utilization of different carbon and nitrogen sources.

Some minor differences are noticeable in the physiologic and biochemical characters by different workers.[6,8–14]

7-5 DISEASE CYCLE

7-5-1 Perpetuation of the Pathogen During Off-Season

Most of the reports on the perpetuation of the pathogen during off-season are based on disease reappearance in a newly planted crop or on indirect methods indicating the presence of the pathogen. The slow-growing nature of the pathogen, confusion with yellow saprophytes, and lack of selective medium impeded successful approach to ecological investigation. Though a number of methods have been devised, only few of them—needle-prick inoculation technique,[15] bacteriophage method,[16,17] streptomycin resistant mutants,[18–20] and radioimmuno assay[21] have been frequently used in ecological investigations. No selective medium has been developed for such an important pathogen. Devadath and Kelman[22] developed a better and partly selective medium which, if improved further, may serve as an excellent tool for reliably studying the ecology of the pathogen.

Survival of the pathogen in dry and growth form.   The pathogen survives in dry form and growth form. The dry form is in a dormant state and found on the diseased plants as an aggregated mass (bacterial exudate) or in xylem and xylem parenchyma of the infected tissues. The pathogen survives for longer periods at low humidity and low temperature, while it is killed in a short period at high humidity and high temperature.[12,18,23–24] When the bacterial cells in dry form reach the root zone of rice plants and grasses, they are activated into growth form and serve as primary inoculum.[25] In the growth form the pathogen survives in the rhizosphere of winter crops and perennial wild plants, especially *Leersia* spp. It is in the active state of multiplication and incites infection when it comes in contact with the susceptible host.[23] In Japan, the bacterial cells hibernating in growth form speed up their multiplication with an increase in temperature and become the primary source of infection.[25]

The longevity of the pathogen varies depending on the survival form and the environmental conditions. The pathogen in growth form can survive only when the environment is favorable, but in dry form it can survive longer even under unfavorable conditions.[2]

Survival in soil and field water.   The pathogen cannot survive for more than 2 months in soil except in the rhizosphere of *Leersia* spp.[2,26] Longevity in soil is influenced by soil pH, physico-chemical properties of the soil, soil moisture, soil temperature, and antagonistic microflora.[18,25–26] It does not survive long in field water either; survival in field water depends on temperature, phages, and antagonistic microflora.[18,27–29] Although the pathogen does not survive long in irrigation water, it can establish itself as an epiphyte on grasses along the banks of irrigation canals or on rice crop transplanted earlier.[30]

Survival in stubbles/ratoons.   In Japan, high pathogen population is found in infected stubbles after the crop is harvested but the population thereon diminishes rapidly. In warmer areas, it multiplies at the basal parts and in new roots of ratoons. But when the stubbles are ploughed, the pathogen dies in a couple of months along with the decomposition of stubbles.[31] In cooler areas, the pathogen does not survive through winter as it is killed in winter.[32]

In Thailand[33] and Sri Lanka,[34] the pathogen survives in infected stubbles and serves as foci for primary infection. In India, it perpetuates in diseased stubbles in double cropped areas,[20,35] but it cannot survive from season to season in single cropped areas.[7] Ratoons in lowlands with plenty of soil moisture constitute the source of primary inoculum in some parts of the country.[34] Survival of the diseased stubbles is greatly dependent on the soil moisture from harvest to ploughing,[34,36] which influences the survival of the pathogen.[37] The stubbles buried in the soil rot earlier than the ones on the top of the soil, exposing the pathogen earlier to the action of phages and antagonists,[12,20] but it cannot survive for more than 20 days when stubbles are ploughed down and inundated with water for a day.[18,20]

Survival in infected straw and self-sown plants.   In infected straw protected from rain, the pathogen survives till next spring in Japan, but when straw is scattered or ploughed into the soil, it dies within a month or two.[31] Infected straw, when applied in nursery soil or near an irrigation canal, serves as the source of primary inoculum.[2,38] Infected straw scattered or buried does not constitute a significant means of perpetuation from one season to another.[12,20,39] Survival in infected straw is governed by temperature,[18,40] soil moisture,[12,18,36] humidity, and sunlight.[18] Infected straw in soil of high moisture content decomposes rapidly at high temperatures and the pathogen in the disintegrating tissues soon loses its viability. Thus, in constantly humid and hot climates its chances of surviving a long break without the living host appear to be less.[18] In some countries, stacking of straw on bunds is practiced to facilitate the spider population in biological control of brown planthopper. Infected straw should not be used for this purpose. Straw is stacked and used as a cattle feed and also utilized for thatching the houses in some countries. The role of such straw in contributing the primary inoculum during rainy season is not known.

Infected self-sown rice plants in lowlands serve as a source of inoculum in some of the single cropped areas, whereas such plants in uplands and medium lands constitute the source of primary inoculum in double cropped areas.[41]

Survival in/on seed.   Presence of the pathogen in/on seeds is reported from different countries.[16,21,35,42–44] In Japan, the pathogen is not detected in dehusked rice grains,[2] but from China[45] and India[46] the presence of the pathogen in the endosperm has been reported. Hsieh and Buddenhagen[18] and Natural[47] isolated the pathogen from hull and embryo but not from endosperm. Most of the population is in the hulls while fewer are located in the brown rice. The average number of bacteria in the hull is 1.42×10^5 cells/seed and in brown rice it is 1.54×10^4 cells/seed.[47] The majority of the seedborne *X. campestris* pv. *oryzae* is in the vascular system of dead lemma and palea, with no direct connection with true seed or seedling.[18] Seed infection occurs through the vascular system and 90% of seed infection is observed immediately after harvesting.[48] In India, the pathogen is propagated from season to season through infected seeds.[49–52]

Survival of the pathogen in seeds depends on the temperature at which they are stored: under low temperature it survives longer than at high temperatures.[37,53] Percentage of seeds carrying the pathogen decreases with storage time.[2,7,18,38,54] Percentage of seed infection varies depending on the cultivar[55] and the season in which the crop is grown; wet season crop carries more seed infection than dry season crop.[18,37]

Chaffs dumped near threshing places seem to play an important role in primary infection.[56] Different seed storing practices are in vogue in different countries and even within a country, and the effect of these practices on the survival of the pathogen needs to be understood.

Survival in/on weeds and other crops.   In Japan, *Leersia sayanuka*, *L. oryzoides* var. *japonica*, *Zizania latifolia*, and *Phalaris arundinacea* are infected severely, while *Phragmites communis*, *Isachne globosa*, and *Setaria viridis* are slightly infected when inoculated.[2] *L. sayanuka*, *L. oryzoides*, and *Z. latifolia* generally show natural infection by the end of the nursery stage.[24,57] Distribution of *Leersia* spp. in general has a positive correlation with the distribution of habitually disease-occurring areas.[27]

The pathogen is found on many weeds at the time of harvesting but subsequently only on *L. sayanuka*. The pathogen population increases remarkably with the increase in temperature in the rhizomes and in newly formed roots of *L. sayanuka* and becomes the source of primary inoculum to the subsequent rice crop in Japan.[2]

In the Philippines, *Leptochloa chinensis*, *L. panacea*, *Z. aquatica* are susceptible when inoculated.[58] *L. filliformis* is also susceptible when pin-prick inoculated.[59] In India, none of the weeds in and around rice fields showed symptoms of the disease when inoculated.[12,37]

Leersia hexandra was susceptible when pin-prick or clip-inoculated.[60–63] None but Bakr and Miah[62] found natural infection on this weed. However, a number of workers[12,37,59,64,65] failed to confirm the susceptibility of this weed. *Cyperus rotundus* and *C. defformis* were also reported to be susceptible both under natural and artificial inoculation,[66] but others[59,67] failed to confirm these results. When clip-inoculated, *Paspalum scrobiculatum* developed symptoms but other methods of inoculation[59,65] did not produce symptoms on this weed. Nor could the susceptibility of *P. cummunis* reported in Japan[57] be proved.[59] *Hygroryza aristata*[62] and *Vossia euspidata*[68] in Bangladesh and *L. japonica*, *L. sayanuka*, and *Z. caudiciflora* in Korea[69] are reported to be susceptible. In Latin American countries, bacterial blight symptoms are observed on *Manisuris* sp., *Oryza latifolia*, *Brachiaria* sp., *Panicum maximum*, and a large sedge with triangular pseudostems[70] and also on *Rottboellia exaltata*, *Paspalum* spp. *Panicum* spp., *C. rotundus* and several grasses.[71] However, the susceptibility of these weeds needs confirmation.

None of the 10 rotation crops, i.e., species of *Eleusine*, *Panicum*, *Pennisetum*, *Sorghum*, and *Triticum*, are susceptible when spray-inoculated.[72] In Sri Lanka, none of the graminaceous weeds and wild rices are susceptible[34] but in India, several wild rice spp. are found susceptible both under artificial inoculation and natural conditions.[20,65,73] A few such infected spp. are even found in ponds, ditches, and irrigation canals during the off-season in both single and double cropped areas.[20,65] Pimples on carnation plants were reported[74] to be due to a strain of *X. campestris* pv. *oryzae* which differed from normal *X. campestris* pv. *oryzae* in acid production.

Admittedly, a great diversity of results exists in literature regarding the host range of the pathogen. To study the host range of the pathogen, the test plants should not be pin-prick and clip-inoculated. Both these methods injure the plants; the pathogen colonizes in injured portions and causes reddish restricted lesions which are often mistaken for the symptom. Therefore, spraying the inoculum on test plants is preferred.[65] Epiphytic survival of the pathogen on *P. scrobiculatum, L. hexandra, C. rotundus,* and *P. repens* has been reported[30] and has great epidemiological significance.

The rhizosphere of weeds and rotation crops forms an important ecological factor in the perpetuation of the pathogen.[2] It is known to survive in the rhizosphere of *Digitaria sanguinalis, Plantago major, Paspalum dictum, Cynodon dactylon,* soybean, rice,[75] wheat,[76] *Alopecurus aequalis,* and rape.[77] The pathogen surviving in other habitats is activated by the root systems of the plants and multiplies.[2] Whether the pathogen overwinters in winter crops in Japan is questioned by Tagami et al.[31]

7-6 MODE OF PRIMARY INFECTION

The hibernated bacterial cells on coming in contact with the rice seedlings are activated, multiply, and invade through wounds caused by root development at the basal part of the stem or through hydathodes on the leaf.[25] The pathogen also can enter the seedlings through stomata and multiply in the intercellular spaces of the parenchyma in the coleoptile and leaf sheath. No external symptoms appear at this stage.[78] The bacteria are carried in the "carrier seedlings" until they are exuded through the stomata, enter the vascular bundles through hydathodes, multiply in the epitheme, and then invade the vessels to incite infection.[79] It also enters through injured roots, and wounds hasten infection.[80] Flooding carries the bacterial cells present in the field water to the foliage to incite infection.[40,81] Some of the larvae and weevils in the field that feed on rice plants also help the pathogen to enter into rice plants.[82–83]

Bacteria that multiply at the crown base invade the interior from wounded parts of stem, enter the central cylinder from the crown, and block the vascular bundles of the stem at the base, often causing kresek symptoms.[34]

7-7 DISSEMINATION OF THE PATHOGEN

Infected seeds are reported to transmit the disease.[46,52,84] The bacteria may establish viable colonies on root and shoot surfaces that can eventually become primary inoculum,[85] and when such seedlings are injured at the crown, seedling infection develops.[86] Flooding,[40,82] shading at nursery, and heavy application of nitrogenous fertilizers[40] help seed transmission of the disease. The bacterial cells present in the vascular system of lemma and palea ooze out into soil water at the time of seed germination and flooding accomplishes the transmission of the pathogen from soil water to foliage to incite infection.[52]

But other researchers[4,18,53,54] reported that seed is not a source of infection. Several reasons are attributed to the failure of seed transmission. During seed soaking and germination, the pathogen oozes out and the population declines rapidly in/on seeds and cannot be recovered after 3–5 days of soaking the seed,[18,31] due to action of phages[54] and due to low number of viable cells.[53]

In my opinion, infected seeds serve as inoculum from season to season in single cropped areas. The hibernating low number of viable cells that ooze out from the germinated seedlings are transmitted to the foliage by submergence of seedlings in the seedbeds, establish themselves, multiply, and initiate infection under favorable environmental conditions. Different seed-sowing practices are followed in Asian countries. The part played by the seedborne inoculum in disease occurrence under such seed-sowing practices needs to be understood.

The pathogen is known to spread through irrigation water,[33,34,87] but the pathogen from irrigation water does not reach the leaf blade through leaf sheaths to incite infection.[2,81] Contact of rice leaves with the contaminated field water is a prerequisite for the initiation of infection in the absence of any other mode of transmission of the pathogen into the host and this is accomplished by flooding.[81] Injury on leaf sheath, crawling insects, splashes of wind-driven field water, and injured floating roots may also aid the pathogen to enter the host.

Bacterial cells released through guttation or from leaves injured during typhoonic rains[25,27,46] and those exuded from the leaves under high humidity as bacterial exudate[78] are dessiminated through wind-driven rains and by mechanical contact of the leaves. The amount of dessiminated bacterial cells varies with weather conditions; rainy winds carry greater amounts of bacteria than do dry winds.[24] Experiments carried out (unpublished) in our laboratory have indicated the dessimination of the pathogen through aerosols. Insects transmit the pathogen mechanically.[82,83,88-90] Man himself and other agricultural implements used in field operations may also transmit the pathogen from plant to plant and from field to field.

7-8 DISEASE MANAGEMENT THROUGH HOST RESISTANCE

7-8-1 Factors Associated with Resistance

Cultivars with few, short, narrow, and erect leaves suffer less disease than with luxuriant growth and spreading leaves.[91] Cultivars with glabrous leaves show negligible disease under natural field conditions (Figure 7-7) while those with hairy leaves show maximum disease.[92-93] The difference between hairy and glabrous cultivars in disease occurrence is not due to the genetic difference for resistance, but because of more hairs on hairy cultivars that help in holding more inoculum than glabrous cultivars.[93] Leaf angle also plays an important role in disease buildup under field conditions.[93]

Lesion enlargement is inhibited in incompatible host–parasite relationships and this has been attributed to a variety of factors. Quantities of reducing, nonreducing and total sugars, total nitrogen, water soluble and insoluble protein, amino nitrogen, and total phenols,[94-97] and their ratios,[58,96] inorganic elements like nitrogen, phosphorus, potassium, calcium, iron, and manganese,[98] enhanced production of different enzymes,[99-103] altered membrane permeability,[102,104-105] respiration rates,[106] phytoalexins,[107] enhanced production of antibacterial compounds,[108-109] fibrillar material in the xylem vessels,[110] and the degree of attractive activity of guttation drops from leaves[111] have been correlated with compatible or incompatible host–parasite relationships. Whether one or a combination of all these factors or their intermediate metabolic products participate in determining the compatible or incompatible host–parasite relationships is not clear.

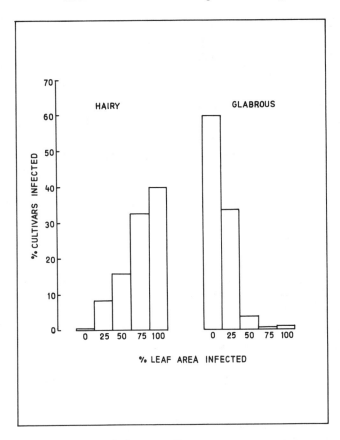

Figure 7-7   Incidence of bacterial blight on hairy and glabrous cultivars.

7-8-2 Resistance Donors

A number of cultivars have been identified as resistant in Japan, Philippines, and Indonesia and a few in India, Bangladesh, Sri Lanka, Nepal, Taiwan, and Korea. The most commonly used resistant donors are: Badshabhog, BJ 1, Dulabhog, DZ 192, Habigonj-A-IV, Babigonj A-VII, Habigonj A-VIII, Hashikalmi, IR 22, Kaoshiung 52, Lacrosse x Zenith-Nira, Malagkit SungSong, Nizersail, Pelita 1/1, Remadja, Sigadis, TKM 6, Wase Aikoku 3, W 1263, Zenith, Lead Rice, Chugoku 45, 70 x-46, 70x-38, 70x-42, 70x-43, Anak Naga 21, ARC 5942, ARC 10331, ARC 10342, ARC 10980, A-23-55-4, Bajal Dhan, Dharial, Hom Thong, Khoya Motor, Mashuri, Nagakayat, PI 209938, PI 205936, R.L. Gopher, Sel. 9210, Semora Mangga, Serendah Kuning 23, Serendah Kuning 60, and *Oryza barthii* (=*O. longistaminata*).

7-8-3 Inheritance of Disease Resistance

Two genes of Kogyoku and one of Shimotsuki confer resistance and it is dominant.[112] Nishimura and Sakaguchi,[113] using several resistant cultivars, concluded that resistance is monogenic and dominant.

Resistance to bacterial group A isolates is conditioned by two complementary dominant genes, *X-1* and *X-2*; *X-1* acted with *X-3* found in Shimotsuki while *X-2* did not act with *X-3*. Resistance in Norin 27 and Kanto 60 to bacterial group A isolates is governed by a single dominant gene and in Wase Aikoku 3, the resistance to bacterial group C isolates is controlled by single dominant gene or genes. Lesion enlargement seems to be controlled by polygenes to bacterial group A isolates, while it is controlled by major genes to bacterial group B.

Kogyoku group cultivars possess a dominant gene *Xa-1* for resistance against group I bacterial isolates.[114] The gene *Xa-1* is linked with liguleless gene *lg* and with phenol staining gene *Ph*[115] belongs to Pl linkage group (Group II) corresponding with chromosome X1.[116] Resistance in Rantai Emas group cultivars is controlled by *Xa-1* and *Xa-2* genes. *Xa-2* confers resistance against bacterial group II isolates. These two genes exhibit resistance against both I and II groups of bacterial isolates throughout the plant growth and are closely linked to each other.[114] Cultivar Pi 1, a derivative from Tadukan belonging to Rantai Emas group, show resistance against group I bacterial isolates but not to group II bacterial isolates, thus indicating that it derived only *Xa-1* but not *Xa-2* from Tadukan.[117]

Resistance in Wase Aikoku 3 to all the three groups of bacterial isolates is controlled by a single dominant gene *Xa-w* which is independent of *Xa-1* and *Xa-2*. Tkm 6 and Nakashin 120 appear to carry resistant gene(s) different from *Xa-w*.[118] Wase Aikoku 3 does not express resistance in the seedling stage while it does in older plants. Resistance in Mimigura Mochi, Kurokara, and Kuromochi to bacterial groups I, II, and III is conferred by a single dominant gene and the resistance in Kurokara and Kuromochi is controlled by *Xa-w* or if not so, another gene closely linked with *Xa-w*.[119]

Resistance in Kogyoku group cultivars to group V bacterial isolates is governed by a single dominant gene *Xa-kg* which is independent of *Xa-w* and closely linked with *Xa-1*. The resistance in Java 14 to group V bacterial isolates by three dominant genes, *Xa-1*, *Xa-w*, and *Xa-kg* or by three other genes allelic to the three genes or linked with them very closely.[120] One major gene controlled resistance of IR 28 to group I and another closely linked major gene to group V bacterial isolates. Several multiple genes and/or polygenes appeared responsible for resistance to bacterial groups II, III, and IV.[121] In IR 28, IR 29, and IR 30 two closely linked major genes controlled resistance to the bacterial groups I and V and are multiple alleles at *Xa-1* and *Xa-kg* loci respectively. These genes are designated as *Xa-1*[h] and *Xa-kg*[b], respectively.[122]

Six major dominant genes have been identified so far in Japan. Inheritance of resistance to bacterial blight in several cultivars is investigated at the International Rice Research Institute to Pxo 61 isolate belonging to pathotype I of the Philippines and six loci conferring resistance have been identified.

Wase Aikoku 3 and PI 215963 have nonallelic genes for resistance.[123] A dominant gene *Xa-4* confers resistance in Tkm 6 while a single recessive gene *xa-5* governs resistance in BJ 1. *Xa-4* and *xa-5* segregate independently of each other, cultivars with *xa-5* gene are resistant both at seedling and at adult plant stage. But cultivars with *Xa-4* are divided into two groups: cultivars resistant at all stages of crop growth and cultivars resistant at flowering but susceptible at seedling stage. The allele at *Xa-4* governing adult stage resistance is designated as *Xa-4*[b] to distinguish from *Xa-4*[a] which governs resistance at all stages of plant growth. *Xa-4*[a] and *Xa-4*[b] may be pseudoalleles at *Xa-4*.[124] A number of cultivars from India, Sri Lanka, and Indonesia and a few from Korea and Vietnam are known to possess *Xa-4*[b].[122,125,126]

Resistance in Malagkit Sunsong is conferred by a single dominant gene *Xa-6* which segregates independently of *xa-5* but is linked to *Xa-4*. All cultivars with this gene are susceptible at seedling stage and resistant at booting and flowering stages.[127] Resistance conferred by *Xa-6* is recessive at booting but dominant at flowering. This "reversal of dominance" was first observed by Olufowote et al.[128] and later confirmed by Sidhu and Khush.[127]

Resistance in DV 85 is controlled by a single recessive gene at seedling stage but two genes, one recessive and one dominant, confer resistance at flowering stage. The recessive gene is allelic to *xa-5* but the dominant gene is nonallelic to *Xa-4* and *Xa-6* and hence designated *Xa-7*, which segregates independently of *Xa-4*, *xa-5*, and *Xa-6*. A single recessive gene *xa-8*, which is nonallelic to and is independent of *Xa-4*, *xa-5*, and *Xa-6* is reported in PI 231129.[125] The resistance, both at seedling and adult stages in Khao Lay Nhay and Sateng, is due to a recessive gene *xa-9* which is nonallelic to *xa-5* and *xa-8*.[129] The dominant gene for resistance in Cas 209 is designated as *Xa-10*.

Most of the cultivars reported to be resistant in Japan and Philippines are susceptible to highly aggressive isolates of the pathogen in India. Resistance in Lacrosse x Zenith-Nira, Wase Aikoku 3, Early Prolific, BJ 1, and Tainan iku is controlled by recessive genes, but the field resistance is horizontal.[130] Digenic, trigenic, and tetragenic inhibitory ratios are postulated to H_{14}, H_{89}, and H_{146} isolates respectively in BJ 1 and Sigadis. Resistance to all isolates is assumed to be dominant but the expression is controlled by an inhibitory gene.[131] Resistance in IRRI 69/469, IRRI 70/470, BJ 1, and Lacrosse x Zenith-Nira is controlled by a dominant gene. The expression of resistance in Lacrosse x Zenith-Nira and BJ 1 is suppressed by a dominant inhibitory gene. Two dominant complementary genes appear to govern resistance in Malagkit Sungsong.[132]

In Early Prolific, Lacrosse x Zenith-Nira, Tkm 6, Sigadis, Malagkit Sungsong, and Wase Aikoku 3, resistance is dominant and monogenic and the gene conferring resistance in Malagkit Sungsong and Wase Aikoku 3 is nonallelic to the gene conferring resistance in the rest of the cultivars.[133] The resistance in BJ 1 is reported to be controlled by a polygenic system.[134-135] Using bacterial isolates that show differential reaction, Tembhurnikar and Padmanabhan[136] studied the inheritance of resistance in various cultivars. Maternal influence on inheritance of resistance is observed in some of the crosses.[137-138] Resistance is conditioned by 2–3 major genes acting in an independent, duplicate, and complementary manner while the lesion development is under control of a large number of minor genes operating as modifiers.[139] Incomplete dominance and over-dominance for the lesion development at tillering and boot leaf stage respectively is observed. The maximum number of genes or gene groups conferring resistance and exhibiting dominance at some loci are 3 and 1 for tillering and boot leaf stages, respectively.[138] No maternal influence is observed in any one of the 56 crosses.[140] Single dominant genes which are allelic to one another control resistance in Patong 32, Nam Sagui, Lua Ngu, and Nam Sakouy. The resistance in BJ 1, DV 85, PI 231129 is controlled by single recessive gene at flowering stage which seems to be allelic.[129]

Resistance to Sri Lankan isolate in Wase Aikoku 3 is conferred by two dominant genes *Xa-a* and *Xa-k*; in PI 209938 and Zenith also two dominant genes *Xa-p* and *Xa-i* confer resistance; R.L. Gopher has only *Xa-i* but Blue Bonnet/Rexark has incompletely dominant gene *Xa-b*. *Xa-a*, *Xa-k* as well as *Xa-p* and *Xa-i* show additive effect. In some, cultivar resistance is conferred by minor genes.

Resistance in BJ 1 and Tkm 6 is conditioned by a single recessive and by a single dominant gene respectively.[141] The genes reported from various countries are identified by utilizing domestic bacterial isolates alone and hence, the results are not comparable. When different isolates are used for genetic analysis, different gene/genes for resistance is/are detected even from one and the same cultivar. Admittedly the number of genes reported for resistance even in a given cultivar varied considerably from country to country and investigator to investigator within the same country.

7-8-4 Breeding for Disease Resistance

Induced mutations through thermal neutrons, gamma rays, chemicals, and the combination of physical and chemical mutagens have been tried for developing bacterial blight resistant mutants mostly from high-yielding susceptible cultivars.[142-143] Breeding for bacterial blight resistance started in Japan in the 1920s. Kono 35 was selected as the resistant individual from Shinriki in 1926.[117] Zensho was developed from Shiga Sekitori 11 by 1932. Zensho 26, a strain of Zensho, was crossed with Jukoku and Kanto 53 and from these crosses, Hoyoku, Kokumasari, Shiranui, Oyoda were developed by 1964. Utilizing Hoyoku, a number of resistant cultivars were developed. From Shobei, Kogyoku was developed by 1932 and from Kogyoku, Koganemaru and Hagareshirazu were developed in 1938 and 1965 respectively. Utilizing Norin 27, Asakaze, Hayatomo, and Nishikaze were developed in 1957, 1964, and 1967 respectively. Sakaguchi[144] utilized lead rice and Wase Aikoku 3 resistant to all the three bacterial groups and developed 70x-38, 70x-42 and 70x-46 (from Lead rice), Chugoku 45 and Chukei 314 (from Wase Aikoku 3). At the International Rice Research Institute, more than 100 cultivars are used as sources of resistance to bacterial blight. Single and multiple crosses are made and mass crossing and screening permits recombination of different genes for resistance in the progenies.[145]

Concentrated efforts have been made to develop durable resistant cultivars in India and a number of fixed lines are being evaluated in multilocations to identify high-yielding broad spectrum resistant lines. But the so-called resistant cultivars released in India (Figure 7-8) failed to show durable resistance. BJ 1, Tkm 6, IR 22, IR 20, Malagkit Sungsong, Early Prolific, Wase Aikoku 3, Lacrosse x Zenith-Nira, DV 85, Sayaphal, and Cemposelak have been utilized in resistance breeding. The IRRI cultivars released in India also could not show the durable resistance in different regions of the country.

In Indonesia, the resistant cultivars Sigadis, Ramadja, Djelita, Syntha, Dewi Tara, Arimbi, Batara, Bengawan, Pelita 1/1, and Pelita 1/2 are developed utilizing the sources of resistance from Benong, Baiang, Sigadis, and Syntha. All the IRRI cultivars released in Indonesia with PB numbers are reported to be resistant.[146]

The resistant donors used in Asian countries do not show satisfactory levels of resistance, although their resistance is satisfactory in eastern and southern countries. Hence, there is an urgent need for seeking better resistant donors and/or pyramiding the genes of such donors in different gene combinations in order to develop suitable resistant cultivars for South Asia. The immune wild rice strain *O. barthii* (= *O. longistaminata*) identified, is an excellent source for breeding durable resistant cultivars in southern Asia in general and India in particular. Excellent progress has been made to incorporate the immunity of this wild rice strain into *O. sativa* background in our laboratory and at the

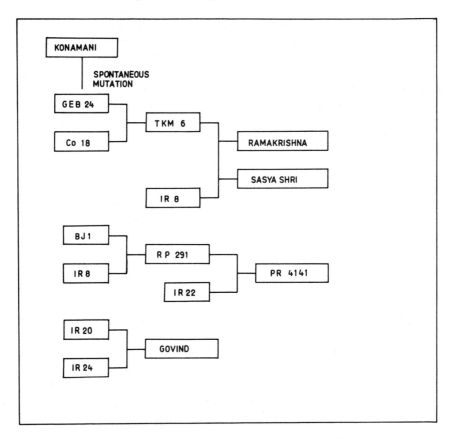

Figure 7-8   Pedigree of bacterial blight-resistant cultivars released in India.

International Rice Research Institute (Dr. G. S. Khush, personal communication) through the F_1s supplied by us.

To rely on vertical resistance alone involves the risk of breakdown of resistance. Therefore, horizontal resistance is vital in a breeding program as it will not be broken down by the shift of pathotypes of the pathogen. Nakashin 120, Chinsurah Boro II, Pelita 1/1, IR 29, Jelita, Dora, Remadja, IR 20 and IR 22 in Indonesia, Kogyoku, Shimotsuki, Nipponbare followed by Kunihikari, Yaeho, Satominori, Asominori, and Gomashirazu in Japan are reported to possess a high degree of horizontal resistance.[147-148]

7-9 DISEASE MANAGEMENT THROUGH CHEMICALS

7-9-1 Disease Forecasting

Disease forecasting is very essential to use need-based chemical application. Several methods for forecasting bacterial blight are reported, but the validity of some is questionable.

By natural infection.   Inspecting the seedlings toward the end of the nursery stage in 15 cm^2 at the entrance, middle, and near the outlet of the irrigation water in nursery to detect the primary infection and to assess subsequent disease development is suggested. Many resistant and susceptible cultivars are grown under heavy nitrogenous fertilization in the forecasting field and the incidence and severity of the disease are assessed periodically; rice leaves are wounded by needles periodically and the disease development on wounded leaves is estimated and utilized to forecast the disease, and occurrence of the disease on collateral host such as *L. sayanuka* is also used to forecast the disease occurrence.[2]

By climatic conditions.   Floods, typhoons, amount of rainfall, number of rainy days, reduced sunshine, and optimum temperature range influence the disease development and hence these factors are used to forecast the disease.[2]

By bacteriophage population.   Phage population increases much in advance of the disease and hence, this has been used to forecast the disease occurrence in several countries.[33,34,149]

In nurseries, generally, the number of phages detected are less than 30/ml. When the phage population exceeds 100/ml, seedling infection commences, and if a far greater number of phages are detected at a later stage of the nursery, severe disease outbreak in early tillering stage is forecasted. Immediately after transplanting, if 50/ml phages are detected in irrigation canals, the earliest disease occurrence is expected on *L. oryzoides*. If more than 100/ml phages are detected, it is estimated that the first disease outbreak is within 10–14 days. In mid-tillering stage, if the phage population is less than 50, more than 100 and more than 1000/ml in paddy field water, the disease is forecasted to be slight, moderate, and severe respectively. If the phages detected in irrigation canal are more than 100/ml, the disease outbreak is expected shortly in this area, and the disease is already prevailing if the phage population is more than 1,000/ml. In maximum tillering stage, if the phage population in paddy field water is less than 100, more than 500, and more than 5,000/ml, the disease occurrence is going to be slight, moderate, and severe respectively. When the phage population is regularly over 1,000/ml, the disease is starting widely in this area. When 1,000–2,000/ml phage are detected frequently in a wide area, it is considered that many secondary infection sources already exist and a high possibility of disease occurrence is forecasted.[2] Kresek incidence is expected after transplanting, if the phage population in field water of nurseries exceeds 500–1,000/ml.[34]

7-9-2 Use of Chemicals to Reduce the Disease

Seed treatment.   Steeping seeds for 12 hr in a mixed solution of 0.05% wettable Ceresan and 0.02% agrimycin followed by hot water treatment at 52–54°C for 30 min;[150] soaking the seeds for 8 hr in 0.01% Ceresan wet and 0.6 g of streptocycline (75% streptomycin + 25% chloroxytetracycline) in 5 gal of water;[151] overnight soaking of seeds in 100 ppm of streptocycline solution; steeping the seeds in water for 12 hr followed by hot water treatment;[152] soaking the seeds for 16–24 hr in decinormal nitric acid followed by washing in tap water and finally drying them well in the sun;[153] and seed treatment with triphenyltin chloride and HPMTS (2-hydroxy propyl-methane-thiosulphonate)[154] are the various methods suggested to eradicate the seedborne inoculum.

Chemical spray.   Attempts to control through chemicals had begun in Japan over a decade before its bacterial nature was established. Bordeaux mixture with or without sugar, Bordeaux mixture and copper soap mixture and copper compounds were tried without much success. After World War II copper-mercury fungicides were used.[2] Mercury fungicides considerably inhibited the lesion development when sprayed, but with lapse of time after infection, the inhibitory effect was reduced.[155] Spraying of chemicals or antibiotics at the nursery stage was helpful when the nurseries harbored severe infection or when the secondary spread of the pathogen in the field was mild.[2] Since streptomycin was more effective than copper-mercury fungicides, spraying of streptomycin solution frequently at short intervals was recommended.[156] High concentrations of streptomycin induced etiolation[157] thereby reducing the yields;[158] chloramphenicol inhibited the lesion development.[2] Dipping of seedlings in copper and zinc compounds was effective in controlling the kresek occurrence.[159]

Rabbing (i.e. burning of trash on top of soil) followed by spraying copperoxychloride is reported to control the disease. Secondary spread was arrested to a large extent by seven sprays of streptocycline in a tolerant cultivar while it was checked to a limited extent in a highly susceptible cultivar.[151] Chlorination of field water was found to decrease the disease,[160-161] but these results could not be confirmed.[162]

Phenazine or its 5-oxide at 150–200 ppm effectively controlled the disease,[163-164] but spraying phenazine after the entry of the pathogen does not give satisfactory control[165] and it cannot translocate in rice plants.[157] Cellocidin, chloramphenicol, phenylmercury acetate are effective in reducing the disease, and synthetic organic bactericides such as nickel dithiocarbamate, dithianone, phenazine, and phenazine *N*-oxide are also recommended for the control of bacterial blight.[166]

Sprays of Agrimycin-500,[167] Agrimycin-100,[168] Sankel,[169] and streptocycline, along with copper compounds were reported to be effective, but the efficacy of these chemicals in India was not confirmed.[165]

Very low doses of 1-methylthiosemicarbazide when root-fed showed preventive as well as curative effect but it was neither effective when sprayed nor inhibitive in vitro.[170] When applied to irrigation water, the compound was converted into three substances including 2-amino-1,3,4-thiadiazole, an antibacterial substance that is readily translocated into rice plants.[171] Spraying of techlofthalam was useful compared with soil application and it translocated readily and reduced bacterial population.[172] Though probenazole (Oryzemate) was not inhibitory to the pathogen, it induced resistance in rice plants through host-mediation and is easily translocated in rice plants when applied to soil.[173]

As the invasion of the pathogen occurs almost throughout rice-growing season, bactericides must be applied at the most appropriate time. Spraying at the later stages of nursery, at the initiation of the disease at maximum tillering stage, and at early stage of secondary spread of the disease at boot leaf stage is suggested,[174] and spraying of chemicals in high concentrations at the critical time, just before the manifestation of the disease, is more successful than spraying frequently in dilute formulations.[2]

Variation in the sensitivity of the causal bacterial isolates to chemicals, and existence or development of drug-resistant strains in nature, pose serious problems in formulating fool-proof chemical control.[175-178]

7-10 REFERENCES

1. Ou, S. H. *Rice Diseases,* 2nd ed. (Kew, UK: Commonwealth Mycological Institute, 1985), p. 380.

2. Tagami, Y., and Mizukami, T. "Historical Review of the Researches on Bacterial Leaf Blight of Rice Caused by *Xanthomonas oryzae* (Uyeda *et* Ishiyama) Dowson." *Special Rep. Plant Dis., and Insect Pest Forecasting Serv.*, Minist. Agric. For., Japan 10 (1962): 112.

3. Dye, D. W., Bradbury, J. F., Goto, M., Hayward, A. C., Lelliot, R. A., and Schroth, M. N. "International Standards for Naming Pathovars of Phytopathogenic Bacteria and List of Pathover Names and Pathotype Strains." *Rev. Plant Pathol.* 59 (1980): 153.

4. Reitsma, J., and Schure, P. S. J. " 'Kresek', a Bacterial Disease of Rice." *Contr. Centr. Res. Inst. Agric. Bogor* 117, 1950: 17pp.

5. Schure, P. S. J. "Attempts to Control the Kresek Disease of Rice by Chemical Treatment of the Seedlings." *Contr. Centr. Res. Inst. Agric. Bogor* 136 (1953): 17pp.

6. Goto, M. " 'Kresek' and Pale Yellow Leaf, Systemic Symptoms of Bacterial Leaf Blight of Rice Caused by *Xanthomonas oryzae* (Uyeda and Ishiyama) Dowson." *Plant Dis. Reptr.* 48 (1964): 858.

7. Srivastava, D. N. "Epidemiology and Control of Bacterial Blight of Rice in India," in *Proc. Symp. Rice Diseases and Their Control by Growing Resistant Varieties and Other Measures.* (Agric. For. Fish. Res. Counc., Minist. Agric. For., Tokyo, Japan, 1967), 11.

8. Ishiyama, S. "Studies on Bacterial Leaf Blight of Rice." *Rep. Agric. Exp. Stn. Tokyo* 45 (1922): 231. [English Abstr. in *Jap. J. Bot* 1 (1922): 21.

9. Yoshimura, S., and Tahara, K. "Morphology of Bacterial Leaf Blight Organism Under (Electron) Microscope." *Ann. Phytopathol. Soc. Japan* 26 (1960): 61.

10. Muko, H., and Isaka, M. "Re-examination of Some Physiological Characteristics of *Xanthomonas oryzae* (Uyeda *et* Ishiyama) Dowson." *Ann. Phytopathol. Soc. Japan* 29 (1964): 29.

11. Shekhawat, G. S., and Srivastava, D. N., "Variability in Indian Isolates of *Xanthomonas oryzae* (Uyeda and Ishiyama) Dowson, the Incitant of Bacterial Leaf Blight of Rice." *Ann. Phytopathol. Soc. Japan* 30 (1968): 289.

12. Devadath, S. "Studies on *Xanthomonas oryzae* (Causal Organism of Bacterial Blight) Occurring on Rice." (Ph.D. thesis, Utkal University, Bhubaneswar, India, 1969).

13. Hifni, H. R., Nishiyama, K., and Ezuka, A. "Bacteriological Characteristics of Some Isolates of *Xanthomonas oryzae* Different in their Pathogenicity and Locality." *Contr. Centr. Res. Inst. Agric. Bogor* 16 (1975): 1.

14. Reddy, O. R., and Ou, S. H. "Characterization of *Xanthomonas oryzae* (Uyeda and Ishiyama) Dowson, the Bacterial Blight Pathogen of Rice." *Ann. Phytopathol. Soc. Japan* 42 (1976): 124.

15. Mizukami, T., and Seki, M. "Studies on the Bacterial Leaf Blight of Rice Plant. On the Distribution of *Bacterium oryzae* (Uyeda *et* Ishiyama) Nakata upon the Rice Plants (Preliminary report)." *Kyushu Agric. Res.* 15 (1955): 57.

16. Wakimoto, S. "The Determination of the Presence of *Xanthomonas oryzae* by the Phage Technique." *Sci. Bull. Facul. Agric. Kyushu Univ* 14 (1954): 495.

17. Wakimoto, S., and Yoshii, H. "Quantitative Determination of the Population of a Bacteria by Phage Technique." *Sci. Bull. Facul. Agric. Kyushu Univ.* 15 (1955): 161.

18. Hsieh, S. P. Y., and Buddenhagen, I. W. "Survival of Tropical *Xanthomonas oryzae* in Relation to Substrate, Temperature, and Humidity." *Phytopathology* 65 (1975): 513.

19. Devadath, S., Premalatha Dath, A., and Rao, C. S. "Estimation of the Population of *Xanthomonas oryzae* Introduced into Natural Eco-systems." *Curr. Sci.* 45 (1976): 236.

20. Thri Murty, V. S., Devadath, S., and Rao, C. S. "Ecology of *Xanthomonas campestris* pv. *oryzae*, the Incitant of Bacterial Blight of Rice." *Indian J. Agric. Sci.* 52 (1982): 524.

21. Dong, Y. D., So, B. H., and Meng, X. J. "Radioimmuno Assay for Detecting the Pathogen of Rice Bacterial Leaf Blight." *Application of Atomic Energy in Agriculture* 3 (1981): 47.

22. Devadath, S., and Kelman, A. "A Better and Semi-selective Medium for *Xanthomonas oryzae* and *X. translucens* f. sp. *oryzicola*," in *Proc. First Natl. Symp. Plant Bacterial Diseases* [Sri Venkateswara University, Tirupati, India, July 14 to 16, 1980, 33 (abstract)].

23. Wakimoto, S. "Considerations on the Overwintering and the Infection Mechanisms of *Xanthomonas oryzae*." *Plant Protect. Japan* 10 (1956): 421.

24. Yoshimura, S. "Diagnostic and Ecological Studies of Rice Bacterial Leaf Blight, Caused by *Xanthomonas oryzae* (Uyeda *et* Ishiyama) Dowson." *Bull. Hokuriku Natl. Agric. Exp. Stn.* 5 (1963): 27.

25. Mizukami, T. "Studies on the Ecological Properties of *Xanthomonas oryzae* (Uyeda *et* Ishiyama) Dowson, the Causal Organism of Bacterial Leaf Blight of Rice Plant. *Agric. Bull. Saga Univ.* 13 (1961): 85.

26. Thri Murty, V. S., and Devadath, S. "Survival of *Xanthomonas campestris* pv. *oryzae* in Different Soils" *Indian Phytopathol.* 35 (1982): 32.

27. Goto, M. "Ecology of Phage-Bacteria Interaction of *Xanthomonas oryzae* (Uyeda *et* Ishiyama) Dowson." *Bull. Facul. Agric. Shizuoka Univ.* 19 (1969): 31.

28. Uematsu, T., and Wakimoto, S. "Biological and Ecological Studies of *Bdellovibrio*. I. Isolation, Morphology and Parasitism of *Bdellovibrio*." *Ann. Phytopathol. Soc. Japan* 36 (1970): 48.

29. Thri Murty, V. S., and Devadath, S. "Survival of *Xanthomonas campestris* pv. *oryzae* and Its Phage in Field Water at Different Temperatures." *Indian Phytopathol.* 35 (1982): 25.

30. Thri Murty, V. S., and Devadath, S. "Studies on Epiphytic Survival of *Xanthomonas campestris* pv. *oryzae* on Some Graminaceous Weeds." *Indian Phytopathol.* 34 (1981): 279.

31. Tagami, Y., Kuhara, S., Kurita, T., Fujii, H., Sekiya, N., Yoshimura, S., Sato, T., and Watanabe, B. "Epidemiological Studies on the Bacterial Leaf Blight of Rice, *Xanthomonas oryzae* (Uyeda *et* Ishiy.) Dowson. I. The Overwintering of the Pathogen." *Bull. Kyushu Agric. Exp. Stn.* 9 (1963): 89.

32. Yoshimura, S. "Bacterial Leaf Blight Disease of Rice in Hokuriku Area." *Plant Protect. Japan* 13 (1959): 395.

33. Tabei, H., and Eamchit, S. "Infection Source of the Bacterial Leaf Blight of Rice in Thailand." *JARQ* 8 (1974): 123.

34. Watanabe, Y. "Ecological Studies on Kresek Phase of Bacterial Leaf Blight of Rice." *Bull. Tokai-Kinki Natl. Agric. Exp. Stn.* 28 (1975): 50.

35. Srivastava, D. N., and Rao, Y. P. "Epidemic of Bacterial Blight Disease of Rice in North India." *Indian Phytopathol.* 16 (1963): 393.

36. Chattopadhyay, S. B., and Mukherjee, N. "Survival of *Xanthomonas oryzae* (Uyeda and Ishiyama) Dowson, the Incitant of Bacterial Leaf Blight of Rice in Soil, Stubbles, Dead and Live Tissues in Field." *Riso* 23 (1974): 309.

37. Thri Murty, V. S. "Studies on the Ecology of *Xanthomonas oryzae*, the Incitant of Bacterial Blight of Rice." (Ph.D. thesis, Ravishankar University, Raipur, India, 1977).

38. Choi, Y. C., Cho, Y. S., and Chung, B. J. " 'Kresek' Disease in Korea. II. Effect of Pathotype of Pathogens and the Use of Infected Straw on the Development of Kresek." *Korean J. Plant Protect.* 17 (1978): 23.

39. Wakimoto, S., and Tamari, K. "Studies on the Overwintering of *Xanthomonas oryzae* in Dried State (Preliminary report)." *Proc. Assoc. Plant Protect. Kyushu* 2 (1956): 107.

40. Isaka, M. "Studies on the Forecasting of Bacterial leaf Blight of Rice Plants Caused by *Xanthomonas oryzae* (Uyeda *et* Ishiyama) Dowson, with Special Reference to the Development and the Utilization of Bacterial Exudation Method." *Spec. Bull. Fukui Agric. Exp. Stn.* 4 (1973): 165.

41. Thri Murty, V. S., and Devadath, S. "Role of Volunteer Rice Seedlings in the Perpetuation of *Xanthomonas campestris* pv. *oryzae.*" *Oryza* 18 (1981): 59.

42. Nakazawa, M., and Kato K. "Distribution of Pathogenic Bacteria in Diseased Rice Plants." *Proc. Assoc. Plant Protect. Kansai* 1 (1958): 12.

43. Supriaman, T., and Tantera, D. M. "Detection of *Xanthomonas oryzae* in Rice Seed Samples." *Contr. Centr. Res. Inst. Agric. Indonesia* 1 (1972): 14.

44. Hsieh, S. P. Y., Buddenhagen, I. W., and Kauffman, H. E. "An Improved Method for Detecting the Presence of *Xanthomonas oryzae* in Rice Seed." *Phytopathology* 64 (1974): 273.

45. Fang, C. T., Lin, C. F., and Chu, C. L. "A Preliminary Study on the Disease Cycle of the Bacterial Leaf Blight of Rice." *Acta Phytopathol. Sinica* 2 (1956): 173.

46. Srivastava, D. N., and Rao, Y. P. "Seed Transmission and Epidemiology of the Bacterial Disease of Rice in North India." *Indian Phytopathol.* 17 (1964): 77.

47. Natural, M. P. "Survival of Streptomycin Resistant Isolates of *Xanthomonas oryzae* in Rice Seeds." (M.S. thesis, University of Philippines, Los Banos, Philippines, 1975).

48. Durgapal, J. C., Baleshwar Singh, and Pandey, K. R. "Mode of Infection of Rice Seeds by *Xanthomonas oryzae.*" *Indian J. Agric. Sci.* 50 (1980): 624.

49. Chakravarti, B. P., and Rangarajan, M. "Study on Bacterial Blight of Rice: Association of *Xanthomonas oryzae* with Seeds and Varietal Reactions." *Oryza* 5 (1968): 20.

50. Singh, R. A., and Rao, M. H. S. "A Simple Technique for Detecting *Xanthomonas oryzae* in Rice Seeds." *Seed Sci. Technol.* 5 (1977): 123.

51. Rao, Y. P., and Srivastava, D. N. "Application of Phages in Investigation of Epidemiology of Bacterial Blight Disease of Rice," in *Proc. Symp. Epidemiology, Forecasting and Control of Plant Diseases* (Univ. Lucknow, Lucknow, India, January 18 to 20), *Indian Natl. Sci. Acad.* 46 (1978): 314.

52. Thri Murty, V. S., and Devadath, S. "Role of Seed in the Survival and Transmission of *Xanthomonas oryzae,*" in *Proc. First Natl. Symp. Plant Bacterial Diseases* [Sri Venkateswara University, Tirupati, India, July 14 to 16, 1980, 27 (abstract)].

53. Eamchit, S., and Ou, S. H. "Some Studies on the Transmission of Bacterial Blight of Rice Through Seed." *Phil. Agr.* 54 (1970): 33.

54. Kauffman, H. E., and Reddy, A. P. K. "Seed Transmission Studies of *Xanthomonas oryzae* in Rice." *Phytopathology* 65 (1975): 663.

55. Mukherjee, N. "Some Aspects of Seed Transmission of Bacterial Plant Pathogens." *Seeds Farms* 3 (1977): 41.

56. Devadath, S., and Premalatha Dath, A. "Infected Chaff as a Source of Inoculum of *Xanthomonas campestris* cv. *oryzae* to the Rice Crop." *Z. Pflanzenkr. Pflanzenschutz* 92 (1985): 485.

57. Goto, K., Fukatsu, R., and Ohata, K. "Bacterial Leaf Blight of Rice and Its Infection to Wild Grasses." *Ann. Phytopathol. Soc. Japan* 17 (1953): 187.

58. International Rice Research Institute (IRRI), Annual Report for 1967, Los Banos, Philippines, 1967.

59. Dalmacio, S. C., and Exconde, O. R. "Host Range of *Xanthomonas oryzae* in the Philippines." *Phil. Agr.* 51 (1967): 283.

60. Rao, P. S., and Kauffman, H. E. "A New Indian Host of *Xanthomonas oryzae*, Incitant of Bacterial Leaf Blight of Rice." *Curr. Sci.* 40 (1971): 271.

61. Reddy, P. R., and Nayak, P. "A New Host for Bacterial Leaf Blight Pathogen of Rice." *Curr. Sci.* 43 (1974): 116.

62. Bakr, M. A., and Miah, S. A. "New Weed Hosts of *Xanthomonas oryzae* (Uyeda and Ishiyama) Dowson in Bangladesh." *Int. Rice Commn. Newsl.* 24 (1975): 16.

63. Chu, C. L., and Chien, C. C. "*Leersia hexandra*—A Host Plant of Bacterial Leaf Blight." *J. Agric. Res. China* 24 (1975): 20.

64. Fang, C. T., Ren, H. C., Chen, T. Y., Chu, Y. K., Faan, H. C., and Wu, S. C. "A Comparison of the Rice Bacterial Leaf Blight Organism with the Bacterial Leaf Streak Organisms of Rice and *Leersia hexandra* Swartz." *Acta Phytopathol. Sinica* 3 (1957): 99.

65. Devadath, S., Premalatha Dath, A., and Padmanabhan, S. Y. "Wild Rice Plants as Possible Source of Bacterial Blight Inoculum of Cultivated Rice." *Curr. Sci.* 43 (1974): 350.

66. Chattopadhyay, S. B., and Mukherjee, N. "Occurrence in Nature of Collateral Hosts (*Cyperus rotundus* and *C. defformis*) of *Xanthomonas oryzae*, Incitant of Bacterial Blight of Rice." *Curr. Sci.* 37 (1968): 441.

67. Silva, J. P. "Some Studies on Bacterial Leaf Blight Disease of Rice in the Philippines." Saturday Seminar, International Rice Research Institute, Los Banos, Philippines, 1968, June 15.

68. Miah, S. A. "Recent Research Results on Rice Diseases in Bangladesh." International Rice Research Institute, Los Banos, Philippines, April 23 to 27, 1973.

69. Lee, K. H. "Recent Research on Rice Diseases in Korea." International Rice Research Conference, International Rice Research Institute, Los Banos, Philippines, April 23 to 27, 1973.

70. Muko, H., Kusaba, T., Watanabe, M., and Tabei, H. "Several Factors Related to the Occurrence of Bacterial Leaf Blight Disease of Rice." *Proc. Assoc. Plant Protect. Kanto-Tosan* 4 (1957): 7.

71. Lozano, J. C. "Identification of Bacterial Leaf Blight in Rice, Caused by *Xanthomonas oryzae*, in America." *Plant Dis. Reptr.* 61 (1977): 644.

72. Central Rice Research Institute (CRRI), Annual Report for 1971, CRRI, Orissa, Cuttack, Orrissa, India, 1972.

73. Mohan, S. K., and Rao, Y. P. "Bacterial Blight and Leaf Streak Diseases of Rice—Their Epidemiology and Control," in *Current Trends in Plant Pathology*, eds. S. P. Raychaudhuri and J. P. Verma. Lucknow, India: Lucknow University. 1974, 119–123.

74. Thomas, W. D., and Dickens, L. E. "Carnation Pimple." *Plant Dis. Reptr.* 37 (1953): 634.

75. Duan, Y. J., Wang, Y. X., and Lu, X. D. "Studies on the Weed Carriers of Bacterial Leaf Blight of Rice." *Acta Phytophylacica Sinica* 6 (1979): 19.

76. Mohiuddin, M. S., Rao, Y. P., Mohan, S. K., and Verma, J. P. "Survival of *Xanthomonas oryzae*, the Incitant of Bacterial Blight of Rice in the Rhizosphere of Wheat." *Sci. Cult.* 43 (1977): 124.

77. Wakimoto, S. "Overwintering of *Xanthomonas oryzae* in Soil." *Agric. Hortic.* 37 (1956): 1413.

78. Tabei, H. "Anatomical Studies of Rice Plant Affected with Bacterial Leaf Blight, with Special Reference to Stomatal Infection at the Coleoptile and the Foliage Leaf Sheath of Rice Seedling." *Ann. Phytopathol. Soc. Japan* 33 (1967): 12.

79. Tabei, H., and Muko, H. "Anatomical Studies of Rice Plant Leaves Affected with Bacterial Leaf Blight (*Xanthomonas oryzae*) in Particular Reference to the Structure of Water Exudation System." *Bull. Natl. Inst. Agric. Sci. Japan Ser.* C 11 (1960): 37.

80. Zaragoza, B. A., and Mew, T. W. "Relationship of Root Injury to the 'Kresek' Phase of Bacterial Blight of Rice." *Plant Dis. Reptr.* 63 (1979): 1007.

81. Premalatha Dath, A. "Factors Influencing the Development of Bacterial Blight of Rice Incited by *Xanthomonas oryzae*." (Ph.D. thesis, Ravishankar University, Raipur, India, 1974).

82. Devadath, S., and Rao, P. S. P. "Indications of Insect Transmission of *Xanthomonas oryzae*." *The Rice Pathology Newsl.* 1 (1975): 13.

83. Noda, T., Sato, A., and Sato, Z. "Occurrence of Kresek Symptoms of Bacterial Leaf Blight of Rice Seedlings Injured by Rice Plant Weevil, *Echinocnemus squameus* Billberg." *Ann. Phytopathol. Soc. Japan* 47 (1981): 84.

84. Singh, D. V., Banerjee, A. K., Rai, M., and Srivastava, S. S. L. "Survival of *Xanthomonas oryzae* in Infected Paddy Seeds in Plains of Uttar Pradesh." *Indian Phytopathol.* 33 (1980) 601.

85. Buddenhagen, I. W. "Report of the Consultancy in Pathology." All India Coordinated Rice Improvement Project, Progress Report 3, 2. 1, Indian Council of Agricultural Research, New Delhi, 1969.

86. Durgapal, J. C., Baleshwar Singh, and Pandey, K. R. "Prospects of Applying 'Crown inoculation' in Investigation of Epidemiology of Bacterial Blight of Rice." *Indian Phytopathol.* 33 (1980): 565.

87. Inoue, Y., Goto, K., and Ohata, K. "Overwintering and Mode of Infection of Leaf Blight Bacteria of Rice Plant." *Bull. Div. Plant Breed. Tokai-Kinki Natl. Agric. Exp. Stn.* 4 (1957): 74.

88. Goto, K., Inoue, Y., Fukatsu, R., and Ohata, K. "Field Observations on the Outbreak and Fluctuation of Severity of Bacterial Leaf Blight of Rice Plant." *Bull. Div. Plant Breed. Tokai-Kinki Natl. Agric. Exp. Stn.* 2 (1955): 53.

89. Mohiuddin, M. S., Rao, Y. P., Mohan, S. K., and Verma, J. P. "Role of *Leptocorisa acuta* Thun. in the Spread of Bacterial Blight of Rice." *Curr. Sci.* 45 (1976): 426.

90. Thri Murty, V. S., and Devadath, S. "Studies on the Transmission and Survival of *Xanthomonas campestris* pv. *oryzae* through Insects." *Indian Phytopathol.* 34 (1981): 162.

91. Kiryu, T., and Mizuta, H. "On the Relationship between Habits of Rice Plant and Varietal Resistance against Bacterial Leaf Blight 1. *Kyushu Agric. Res.* 15 (1955): 54.

92. Premalatha Dath, A., Padmanabhan, S. Y., and Devadath, S. "The Relation between Certain Host Characters and Bacterial Blight Incidence in Rice." *Proc. Indian Acad. Sci.* 85B (1977): 301.

93. Raju Philip., and Devadath, S. "Relation between Some Leaf Characters and the Occurrence of Bacterial Blight of Rice." *Riso* 29 (1980): 317.

94. Misawa, T., and Miyazaki, E. "Studies on the Leaf Blight of Rice Plant (1) Alteration of Contents of Carbohydrates, Nitrogenous and Phosphorus Compounds in the Diseased Leaves." *Ann. Phytopathol. Soc. Japan* 38 (1972): 375.

95. Purushothaman, D. "Phenolic Changes in Rice Varieties Infected by *Xanthomonas oryzae.*" *Acta Phytopathol. Hung.* 9 (1974): 65.

96. Moses, G. J., Rao, Y. P., and Siddiq, E. A. "Studies on Physiological Changes in Relation to Resistance-Susceptibility Reaction of Rice Plants to Bacterial Blight." *Indian Phytopathol.* 28 (1976): 508.

97. Rao, N.S.R.K., and Nayudu, M. V. "Changes in the Organic Constituents of Rice Leaves Infected by *Xanthomonas oryzae. Phytopathol. Z.* 94 (1979): 357.

98. Raju, Philip. "Studies on the Host–Parasite Relationships in *Xanthomonas oryzae*," (Ph.D. thesis, Ravishankar University, Raipur, India, 1977).

99. Purushothaman, D. "Phenylalanine Ammonialyase and Aromatic Amino Acids in Rice Varieties Infected with *Xanthomonas oryzae.*" *Phytopathol. Z.* 80 (1974): 171.

100. Miyazaki, E., Yamanaka, S., and Misawa, T. "Studies on the Bacterial Leaf Blight of Rice. II. A Comparison of Hydrolytic Enzyme Activity between Diseased and Healthy Tissue." *Ann. Phytopathol. Soc. Japan* 42 (1976): 21.

101. Akutsu, M., and Watanabe, M. "Studies on the Physiological Changes in Rice Plants Infected with *Xanthomonas oryzae.* II. Changes in the Activities of Peroxidase in the Infected Leaves." *Ann. Phytopathol. Soc. Japan* 44 (1978): 499.

102. Rao, N.S.R.K., and Nayudu, M. V. "Enzymological and Permeability Changes Involved in Bacterial Leaf Blight of Rice." *Phytopathol. Z.* 96 (1979): 77.

103. Mohanty, S. K., Reddy, P. R., and Sridhar, R. "Phenylalanine and Tyrosine Ammonia Lyases in Bacterial Leaf Blight Syndrome of Rice." *Z. Pflanzenkr. Pflanzenschutz* 89 (1982): 422.

104. Watanabe, M., Samejima, S., Hayashi, N., and Hosokawa, E. "Studies on the Physiological Changes in the Rice Plants Infected with *Xanthomonas campestris* pv. *oryzae.* IV. Behaviour of ^{14}C-Photosynthetic Assimilates in Infected Rice Plants." *Ann. Phytopathol. Soc. Japan* 46 (1980): 656.

105. Kohno, Y., Watanabe, M., and Hosokawa, D. "Studies on the Physiological Changes in the Rice Plants Infected with *Xanthomonas campestris* pv. *oryzae.* V. Permeability Changes in Infected Rice Plants." *Ann. Phytopathol. Soc. Japan* 47 (1981): 555.

106. Watanabe, M., and Asaumi, T. "Studies on the Physiological Changes in the Rice Plants Infected with *Xanthomonas campestris* pv. *oryzae.* I. Relationship between the Increase in Respiratory Rate and Multiplication of the Bacteria in the infected Leaves." *Ann. Phytopathol. Soc. Japan* 41 (1975): 364.

107. Uehara, K. "On the Phytoalexin Produced by the Results of the Interaction between the Rice Plant and the Leaf-Blight Bacterium (*Xanthomonas oryzae*)." *Ann. Phytopathol. Soc. Japan* 25 (1960): 149.

108. Nakanishi, K., and Watanabe, M. "Studies on the Mechanisms of Resistance of Rice Plants Against *Xanthomonas oryzae.* III. Relationship between the Rate of Production of Antibacterial Substances and of Multiplication of Pathogenic Bacteria in Infected Leaves of Resistant and Susceptible Varieties." *Ann. Phytopathol. Soc. Japan* 43 (1977): 265.

109. Nakanishi, K., and Watanabe, M. "Studies on the Mechanisms of Resistance of Rice Plants against *Xanthomonas oryzae*. IV. Extraction and Partial Purification of Antibacterial Substances from Infected Leaves." *Ann. Phytopathol. Soc. Japan* 43 (1977): 449.

110. Horino, O. "Ultrastructural Histopathology of Rice Leaves Infected with *Xanthomonas campestris* pv. *oryzae* on Kogyoku Group Rice Varieties with Different Levels of Resistance at the Seedling Stage." *Ann. Phytopathol. Soc. Japan* 47 (1981): 501.

111. Fang, T. Y., and Kuo, T. T. "Bacterial Leaf Blight of Rice Plant. VI. Chemotactic Responses of *Xanthomonas oryzae* to Water Droplets Exuded from Water Pores on the Leaf of Rice Plants." *Bot. Bull. Academica Sinica* 16 (1975): 126.

112. Kariya, K., and Washio, O. "Effect of the Selection during Early Segregating Generations for Bacterial Leaf Blight Resistance in Rice." *Chugoku Agric. Res.* 5 (1956): 39.

113. Nishimura, Y., and Sakaguchi, S. "Inheritance of Resistance in Rice to Bacterial Leaf Blight, *Bacterium oryzae* (Uyeda *et* Ishiyama) Nakata (Abstr.)." *Japan J. Breed.* 9 (1959): 58.

114. Sakaguchi, S. "Linkage Studies on the Resistance to Bacterial Leaf Blight, *Xanthomonas oryzae* (Uyeda *et* Ishiyama) Dowson, in Rice." *Bull. Natl. Inst. Agric. Sci. Japan Ser.* D 16 (1967): 1

115. Nishimura, Y. "Studies on the Reciprocal Translocations in Rice and Barley" *Bull. Natl. Inst. Agric. Sci. Japan Ser.* D 9 (1961): 171

116. Iwata, N., and Omura, T. "Linkage Analysis by Reciprocal Translocation Method in Rice (abstr.). *Japan J. Breed.* 21 (Suppl. 1) (1971): 16.

117. Toriyama, K. "Breeding for Resistance to Major Rice Diseases in Japan," *Rice Breeding*, International Rice Research Institute, (Los Banos, Philippines, 1972), 253.

118. Ezuka, A., Horino, O., Toriyama, K., Shinoda, H., and Morinaka, T. "Inheritance of Resistance of Rice Variety Wase Aikoku 3 to *Xanthomonas oryzae*." *Bull. Tokai-Kinki Natl. Agric. Exp. Stn.* 28, (1975): 124.

119. Yamada, T., Horino, O., and Samoto, S. "Studies on Genetics and Breeding of Resistance to Bacterial Leaf Blight in Rice. 3. Inheritance of Resistance of Newly Discovered Wase Aikoku Group Varieties in Japanese Native Varieties to Bacterial Groups I, II and III of *Xanthomonas oryzae* (Uyeda *et* Ishiyama) Dowson." *Ann. Phytopathol. Soc. Japan* 45 (1979): 321.

120. Ogawa, T., Morinaka, T., Fujii, K., and Kimura, T. "Inheritance of Resistance of Rice Varieties of Kogyoku and Java 14 to Bacterial Group V of *Xanthomonas oryzae*." *Ann. Phytopathol. Soc. Japan* 44 (1978): 137.

121. Yamada, T., Horino, O., and Samoto, S. "Studies on Genetics and Breeding of Resistance to Bacterial Leaf Blight of Rice. IV. Inheritance of Resistance of IR 28 to Bacterial Groups I, II, III, IV and V of *Xanthomonas oryzae* (Uyeda *et* Ishiyama) Dowson from Japan." *Japan J. Breed.* 29 (1979): 279

122. Yamada, T., and Horino, O. "Studies on Genetics and Breeding of Resistance to Bacterial Leaf Blight of Rice. V. The Multiple Alleles Resistant to Bacterial Groups I and V of *Xanthomonas campestris* pv. *oryzae* of Japan in the Varieties IR 28, IR 29 and IR 30." *Japan J. Breed.* 31 (1981): 423.

123. Murty, V. V. S., Khush, G. S., and Jensen, N. F. "Inheritance of Resistance to Bacterial Leaf Blight, *Xanthomonas oryzae* (Uyeda *et* Ishiyama) Dowson in Rice. I. Allelic Relationships of Resistance Genes in Donor Varieties." *Japan J. Breed.* 23 (1973): 325.

124. Librojo, V., Kauffman, H. E., and Khush, G. S. "Genetic Analysis of Bacterial Blight Resistance in Four Varieties of Rice." *SABRAO J.* 8 (1976): 105.

125. Sidhu, G. S., Khush, G. S., and Mew, T. W. "Genetic Analysis of Bacterial Blight Resistance in Seventy Four Cultivars of Rice, *Oryza sativa* L. *Theor. Appl. Genet.* 54 (1978): 105.

126. Sidhu, G. S., Khush, G. S., and Mew, T. W. "Genetic Analysis of Resistance to Bacterial Blight in Seventy Cultivars of Rice, *Oryza sativa* L., from Indonesia." *Crop Improv.* 6 (1979): 19.

127. Sidhu, G. S., and Khush, G. S. "Dominance Reversal of Bacterial Blight Resistance Gene in Some Rice Varieties." *Phytopathology* 68 (1978): 461.

128. Olufowote, J. O., Khush, G. S., and Kauffman, H. E. "Inheritance of Bacterial Blight Resistance in Rice." *Phytopathology* 67 (1977): 772.

129. Sidhu, G. S. "Inheritance of Resistance to Bacterial Blight in Rice," in Group Meeting on *Bacterial Blight of Rice*, Central Rice Research Institute, Cuttack, India, December 8 to 10, 1980.

130. Padmanabhan, S. Y., Mathur, S. C., Devadath, S., Row, K.V.S.R.K., and Misra, R. K. "Inheritance of Disease Resistance in Rice." Final Technical Report U.S. PL-480 Project, Central Rice Research Institute, Cuttack, (undated), Indian Council of Agricultural Research (mimeographed).

131. Jayaraj, D., Seshu, D. V., and Shastry, S. V. S. "Genetics of Resistance to Bacterial Leaf Blight in Rice." *Indian J. Genet.* 32 (1972): 77.

132. Moses, G. J., Rao, Y. P., and Siddiq, E.A. "Inheritance of Resistance to Bacterial Leaf Blight in Rice." *Indian J. Genet.,* 34 (1974): 271.

133. Singh, D. P., and Nanda, J. S. "Allelism of Bacterial Leaf Blight Resgenes in Rice." *Indian J. Genet.* 37 (1977): 335.

134. Nagaraju, M., Reddy, P. R., and Balakrishna Rao, M. J. "Genetics of Resistance to Bacterial Leaf Blight in Rice." *SABRAO J.* 9 (1977): 21.

135. Tembhurnikar, S. T., and Padmanabhan, S. Y. "Inheritance of Resistance to Bacterial Leaf Blight of Rice in a Cross BJ 1 x Taichung Native 1." *Oryza* 18 (1981): 22.

136. Tembhurnikar, S. T., and Padmanabhan, S. Y. "Inheritance of Resistance to Bacterial Blight in Rice." *Oryza* 18 (1981): 153.

137. Nayak, P., Ratho, S. N., and Misra, R. N. "Maternal Influence on Bacterial Leaf Blight Reaction in Rice." *Curr. Sci.* 44 (1975): 744.

138. Ratho, S. N., Nayak, P., Misra, R. N., and Padmanabhan, S. Y. "Diallel Analysis of Bacterial Leaf Blight Resistance in Rice. I. Nature of Gene Action." *Riso* 25 (1976): 65.

139. Nayak, P., Ghosh, A. K., Ratho, S. N., and Misra, R. N. "Inheritance of Bacterial Leaf Blight Resistance in Rice." *Oryza* 14 (1977): 32.

140. Reddy, O. R, and Rao, A. V. "Diallel Analysis of Resistance to Bacterial Blight in Rice." *Riso* 27 (1978): 137.

141. Hsu, T. H., Teng, Y. C., and Chiu, S. M. "Breeding and Genetic Studies of Bacterial Leaf Blight (*Xanthomonas oryzae*) Resistant Varieties of *indica* Type of Rice. II. Inheritance of Resistance to Bacterial Leaf Blight Disease in Rice Variety BJ 1." *J. Agric. Res. China* 25 (1976): 269.

142. Ismachin, K. M., and Mugiono, O. "Selection for Bacterial Leaf Blight (*Xanthomonas oryzae*) and Sheath Blight (*Rhizoctonia oryzae*) Resistant Mutants in a Collection of Early Rice Mutants." *Induced Mutations Against Plant Diseases*, IAEA, Vienna, 1977, 199.

143. Nayak, P., Padmanabhan, S. Y., and Mishra, R. N. "Variability in Bacterial Leaf Blight Reaction in Rice due to Mutation and Hybridization." *Riso* 27 (1978): 311.

144. Sakaguchi, I. "Breeding of Rice Strains with High Resistance to Bacterial Leaf Blight." *Bull. Natl. Inst. Agric. Sci. Japan Ser.* D 28 (1977): 89.

145. Mew, T. W., and Khush, G. S. "Breeding for Bacterial Blight Resistance in Rice," in *Proc. Fifth Int. Conf. Plant Path. Bact. Cali.* Colombia 1981, 504.

146. Harahap, Z., Siregar, H., and Siwi, B. H. "Breeding Rice Varieties for Indonesia." International Rice Research Institute, *Rice Breeding*, Los Banos, Philippines, 1972, 1941.

147. Yoshida, H., and Yasugi, M. "Resistance of Rice Varieties Commonly Cultivated in Tottori Prefecture to Bacterial Leaf Blight." *Bull. Tottori Agric. Exp. Str.* 17 (1977): 7.

148. Yamada, T., Horino, O., and Samoto,S. "Studies on Genetics and Breeding of Resistance to Bacterial Leaf Blight in Rice. II. Newly Discovered Wase Aikoku Group Varieties Native to Japan." *Japan J. Breed.* 29 (1979): 191.

149. Tagami, Y., Kuhara, S., Kurita, T., and Sekiya, N. "Relation between the Population of *Xanthomonas oryzae* Phage in Paddy Field Water and the Occurrence of Bacterial Leaf Blight." *Proc. Assoc. Plant Protect. Kyushu* 4 (1958): 63.

150. Srivastava, D. N., and Rao, Y. P. "Paddy Farmers Should Beware of Bacterial Blight— Measures for Preventing the Disease." Indian Farming 14 no. 6 (1964): 32.

151. Jain, S. S., Reddy, P. R., and Padmanabhan, S. Y. "Control of Leaf Blight of Rice, Caused by *Xanthomonas oryzae* (Uyeda and Ishiyama) Dowson." *Bull. Indian Phytopathol. Soc.* 3 (1966): 85.

152. Sinha, S. K., and Nene, Y. L. "Eradication of the Seed-borne Inoculum of *Xanthomonas oryzae* by Hot Water Treatment of Paddy Seeds." *Plant Dis. Reptr.* 51 (1967): 882.

153. Subramoney, A., and Abraham, A. "A New Method of Seed Treatment Against Bacterial Leaf Blight." *Int. Rice Commn. Newsl.* 18 (1969): 33.

154. Singh, R. A., and Rao, M. H. S. "Evaluation of Several Chemical Treatments for Eradicating *Xanthomonas oryzae* from Rice Seeds." *Seed Sci. Technol.* 10 (1982): 119.

155. Muzukami, T., and Seki, M. "On the Effects of Some Fungicides upon the Infection and the Development of Lesions of the Bacterial Leaf Blight of Rice Plant." *Kyushu Agric. Res.* 14 (1954): 209.

156. Seki, M., and Mizukami, T. "Application of Antibiotics Against Bacterial Leaf Blight of Rice Plant." *Kyushu Agric. Res.* 17 (1956): 98.

157. Devadath, S. "Upward Translocation of Antibiotics and Their Role in Controlling Bacterial Blight of Rice." *Riso* 22 (1973): 33.

158. Yoshimura, S., Tahara, K., and Aoyagi, K. "On the Control of Bacterial Leaf Blight Disease of Rice by Antibiotics." *Proc. Assoc. Plant Protect. Hokuriku* 9 (1961): 24.

159. International Rice Research Institute (IRRI), Annual Report for 1978, Los Banos, Philippines, 1979.

160. Padmanabhan, S. Y., and Jain, S. S. "Effect of Chlorination of Water on Control of Bacterial Leaf Blight of Rice, Caused by *Xanthomonas oryzae* (Uyeda and Ishiyama) Dowson." *Curr. Sci.* 35 (1966): 610.

161. Chand, T., Singh, N., Singh, H., and Thind, B. S. "Field Efficacy of Stable Bleaching Powder to Control Bacterial Blight of Rice." *Int. Rice Res. Newsl.* 4 no. 4 (1979): 12.

162. Palaniswami, A., and Ahmad, N. F. "Bleaching Powder for Bacterial Blight Control." *Int. Rice Res. Newsl.* 4 no. 3 (1979): 16.

163. Sekizawa, Y., Watanabe, T., and Oda, M. "Effects of Phenazine against Rice Leaf Blight Bacterium, and Its Biochemical Mechanism." *Ann. Phytopathol. Soc. Japan* 30 (1965): 145.

164. Oda, M., Sekizawa, Y., and Watanabe, T. "Phenazines as Disinfectants against Bacterial Leaf Blight of the Rice Plant." *Appl. Microbiol.* 14 (1966): 365.

165. Devadath, S., and Premalatha Dath, A. "Screening of Antibiotics and Chemicals against Bacterial Blight of Rice." *Oryza* 7 (1970): 33.

166. Fukunaga, K. "Antibiotics and New Fungicides for Control of Rice Diseases," in *Proc. Symp. Plant Diseases in the Pacific*, XIth Pacific Sci. Congr. Tokyo, August 25 to 27, 1966, 170.

167. Krishnappa, K., and Singh, R. A. "Chemical Control of Bacterial Leaf Blight of Rice." *Mysore J. Agric. Sci.* 11 (1977): 530.

168. Singh, R. A., Das, B., Ahmed, K. M., and Pal, V. "Chemical Control of Bacterial Leaf Blight of Rice." *Trop. Pest Management* 26 (1980): 21.

169. Mukherjee, G. G., Santra, A., and Mukherjee, S. K. "Efficacy of Sankel and New Sankel against Bacterial Blight of Rice *Xanthomonas oryzae* (Uyeda and Ishiyama) Dowson in Field Tests in West Bengal, India." *Int. Rice Res. Newsl.* 1 no. 1 (1976): 18.

170. Ohmori, K., Nakagawa, T., Suzuki, T., Koike, K., Baba, T., Ishida, S., and Misato, T. "Activity of Thiosemicarbazides for the Control of Bacterial Leaf Blight of Rice." *J. Pestic. Sci.* 1 (1976): 95.

171. Ohmori, K., Suzuki, T., Ishida, S., and Misato, T. "Conversion of 1-Methylthiosemicarbazide into Antibacterial Substances to *Xanthomonas oryzae* in the Flooded Soil System." *J. Pestic. Sci.* 1 (1976): 295.

172. Nakagami, K., Tanaka, H., Yamaoka, K., and Tsujino, Y. "Population of *Xanthomonas oryzae* and Concentration of Techlofthalam in Guttation Droplets on Rice Leaf Sprayed with Techlofthalam." *J. Pestic. Sci.* 5 (1980): 607.

173. Ohashi, T. "Oryzemate: Novel Systemic Anti-bacterial and Anti-fungal Agrochemical." *Japan Pestic. Inf.* 37 (1980): 37.

174. Yoshimura, S., and Tagami, Y. "Forecasting and Control of Bacterial Leaf Blight of Rice in Japan," in *Proc. Symp. Rice Diseases and Their Control by Growing Resistant Varieties and Other Measures*, Agric. For. Fish. Res. Counc., Minist. Agric. For., Tokyo, Japan, 1967, 25.

175. Wakimoti, S., and Mukoo, H. "Natural Occurrences of Streptomycin Resistant *Xanthomonas oryzae*, the Causal Bacteria of Leaf Blight Disease of Rice." *Ann. Phytopathol. Soc. Japan* 28 (1963): 153.

176. Shekhawat, G. S., and Srivastava, D. N. "Variability in Indian Isolates of *Xanthomonas oryzae* (Uyeda and Ishiyama) Dowson, the Incitant of Bacterial Leaf Blight of Rice." *Ann. Phytopathol. Soc. Japan* 34 (1968): 289.

177. Devadath, S. "Sensitivity of *Xanthomonas oryzae* Isolates to Antibiotics *in vitro*." *Indian J. Microbiol.* 11 (1971): 53.

178. Cho, W. C., and Shim, J. W. "Studies on the Chemical Resistance of Phytopathogenic Bacteria. II. Selective Effect of Chemical Resistance on the Rice Bacterial Leaf Blight Pathogen, *Xanthomonas oryzae* (Uyeda and Ishiyama) Dowson to Agrepto." *Korean J. Plant Protect.* 16 (1977): 229.

8

RICE TUNGRO

S. MUKHOPADHYAY

Plant Virus Research Centre, Department of Plant Pathology
Bidhan Chandra Krishi Viswavidyalaya, Kalyani-741 235, West Bengal

8-1 INTRODUCTION

Tungro is a Philippine word. Etymologically it means "degenerated growth." The rice crops in the Philippines suffering from the degenerated growth until the middle of the twentieth century were regarded as having Tungro disease without any reference to its etiology.[1] Later, Rivera and Ou[2] established the viral nature of this disease in the Philippines. They found that the disease could be transmitted by *Nephotettix virescens* Distant. Ling[3] observed that the virus could also be transmitted by *N. nigropictus* Stal and *Recilia dorsalis* Motsch.

Similar degenerative disease of rice was recorded in different southern and southeast Asian countries, China, and Italy.[4-5] Because of its unknown etiology, this disease was recorded by different local names in different countries, like "penyakit merah" in Malayasia,[6] "yellow orange leaf" in Indonesia. Later on, the association of a common causal virus was claimed and a regional map of the distribution of rice tungro virus was visualized. It has been claimed further that the *rice waika* found in Japan is closely related to tungro. Figure 8-1 illustrates the up-to-date knowledge on the regional distribution of tungro virus.

Although no scientific assessment of crop losses due to tungro disease in different countries is available, an estimate shows that serious damage took place in the past in different countries from time to time (Table 8-1) due to this or similar disease.[9-10]

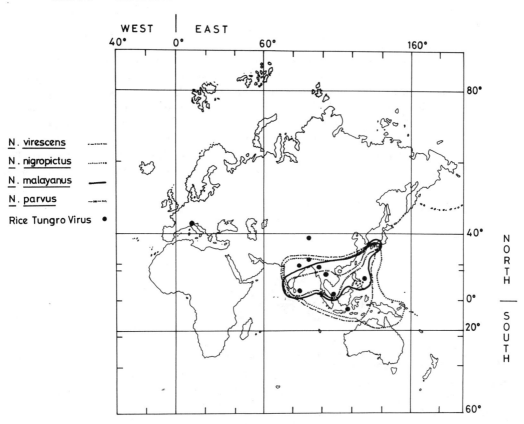

Figure 8-1   Geographical distribution of rice tungro disease and vectors.

TABLE 8-1:   MAJOR OUTBREAKS OF RICE TUNGRO AND SIMILAR DISEASES[9]

Year	Country	Area damaged ($\times$ 100 ha)
1965	Thailand	50
1966	Thailand	660
1969	Bangladesh	10–15%
1969	India (Eastern)	—
1969	Malayasia	21
1970	Philippines	3.5
1970–72	Indonesia	5
1971	Philippines	7.87
1971	Philippines	456
1972–74	Indonesia	50
1980	Indonesia	16
1981*	India (West Bangal)	213.4
1982–83	Malayasia	50
1983–84	Indonesia	25

*Mukhopadhyay[10]

8-2 SYMPTOMATOLOGY

The characteristic of rice tungro virus (RTV) disease are stunting of the plants and discoloration of the leaves (Figure 8-2). The extent of stunting usually depends on the variety and age of the plant at the time of infection. Early infection causes more stunting. The type of discoloration mostly depends on the variety. Some varieties show prominent vein clearing, while in others it usually remains obscure. In some varieties, various shades of yellow predominate, while in others orange coloration becomes more common. There may also be small, rusty, necrotic spots in the discolored areas of the older leaves. The discoloration symptoms occasionally may get suppressed due to the application of inorganic nitrogen fertilizers.[11] Similar foliar symptoms are often found in plants infested with mealy bugs or termites.

Tungro affects both vegetative and reproductive activity of rice plants. It causes stunting; reduces tillering, number and length of panicles and number of spikelets; delays

Figure 8-2 Stunting and discoloration of foliage in RTV-infected plants.

maturation; makes incomplete grain filling; and reduces grain yield, grain weight and grain starch content.[12-13] RTV infection also results in accumulation of starch in the leaves which is often used for routine diagnostic purposes.[14]

8-3 PATHOGEN

Tungro is a composite disease caused by two morphologically unrelated viruses: rice tungro bacilliform virus (RTBV) and rice tungro spherical virus (RTSV) (Fig. 8-3).[15-17] RTBV has a bacilliform capsid 130 x 30 nm made up of a single species of coat protein of MW 36 K and a single molecule of circular double-stranded DNA of 8.3 kbp. RTBV is considered to be a member of the newly described Badna virus group. RTSV has a isomatric capsid, 30 mm in diameter, comprising two to three polypepetide species and a single species of polyadenylated single-stranded RNA of about 10 kb.[67] Spherical particles are coupled either along the sides or at the end with bacilliform particles and association remained even after density gradient centrifugation.[17-18] Omura et al.[15] found no serological relationship between RTBV and RTSV; however, a recent study of Mishra et al.[18] indicates some antigenic relationship between two types of particles as both could be trapped and decorated by RTSV antiserum. A virus of another rice disease reported from

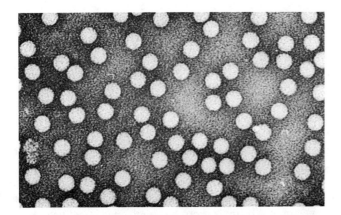

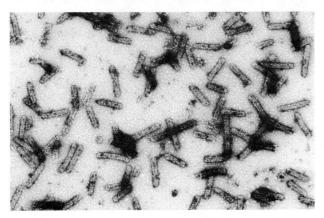

Figure 8-3 Transmission electron micrograph of RTV components from purified preparations: RTSV (upper) and RTBV (lower) (courtesy Dr. H. Koganezawa, IRRI, Los Banos, Philippines)

Japan, rice waika virus (RWV), has been found to be not only morphologically similar but also serologically identical to RTSV.[19]

Both RTBV and RTSV can multiply independently. Both viruses are transmitted by the green leafhopper and other leafhopper species. RTSV is transmitted independently, while RTBV requires RTSV or a RTSV-related "helper" factor for transmission by green leafhopper.[20–21]

Generally plants infected with both viruses show severe tungro symptoms, including yellowing and stunting of plants, while RTBV-infected plants show mild stunting and yellowing. RTSV causes no clear symptoms on most rice cultivars and enhances the symptoms caused by RTBV infection.[22]

Earlier observations regarding the existence of variability in RTV[23–24] have lost their signification after the demonstration of two types of viruses in tungro disease. In view of the identification of RTSV and RTBV, the actual nature of the strains of RTV needs to be reinvestigated.

8-4 TRANSMISSION

Rice tungro is transmitted only by rice green leafhoppers.[25] *Nephotettix virescens* Distant has been identified as the principal vector.[5] *N. nigropictus* Stal can also transmit the virus to a limited extent. Both the adults and nymphs can transmit the virus but the adults are more efficient. All the vectors found in nature are not capable of transmitting the RTV. The extent of the occurrence of active transmitters may depend on the agroecological conditions of the location.[5] There are also claims that hybrids of *N. virescens* and *N. nigropictus* transmit RTV[26] but the incidence of such hybrids in nature has been refuted by several workers. In addition to the above two species, *N. malayanus* Ishihara and Kawase, *N.parvus* Ishihara and Kawase, and *Recilia dorsalis* Motschulsky have been demonstrated to be capable of transmitting RTV.[25,27] Hibino,[21] however, could not find any transmission by *R. dorsalis*; on the other hand, he found transmission by *N. cinticeps*.

The transmission parameters have been investigated mostly with *N. virescens*. The acquistion threshold access period varies from 5 min to 30 min.[25] The acquisition access between 3–5 days can, however, provide maximum virus uptake.[25] Preacquisition fasting improves the efficiency of transmission.[28] The inoculation threshold access period varies from 7–30 min.[25] There is also a claim that a single probe by an infective insect can cause infection.[25]

After the acquisition of the virus, the leafhoppers may immediately become viruliferous and the virus does not need any incubation period in the vector to become infective. Serial transmissions conducted by different workers showed gradual loss of infectivity of the vectors.[28] Contradictory reports are available on the retention of the virus in the vectors. It has been reported that a viruliferous leafhopper can retain the virus from 2–6 days.[25] The period of retention again may be increased with the decrease of temperature up to a certain limit.[29] Thus, the transmission characteristics of rice tungro resembles nonpersistent virus with certain reservations. The relationship has been termed as "transitory."[30]

After the identification of the association of two different viruses in tungro disease, the transmission characteristics of the concerned viruses were reinvestigated. The minimum acquisition access period for RTSV was found to be only 10 min.[31] After an acquisition access of 10 min in RTSV-infected plant, RTBV could be acquired in an access period of at least 30 min by *N. virescens*; for efficient transmission of both the viruses, however, an acquisition access for a period longer than 4 hr was required.

Hibino[31] further showed that when a *N. virescens* is allowed to feed on a rice plant infected both by RTSV and RTBV, it can acquire and transmit both the viruses together or individually; the transmission of RTSV alone, however, occurred at a low rate. But leafhoppers, when fed on plants infected with RTSV alone, transmit RTSV at a high rate. On the contrary, when leafhoppers are fed on plants infected with RTBV alone, no transmission could occur. When the leafhoppers are separately exposed to RTSV and RTBV and fed together on rice plants, no symptom normally develops but there may be occasional symptoms of RTSV. If the same leafhoppers first acquire RTSV then RTBV, they can transmit both the viruses together or RTBV alone. Once an insect acquires RTSV, it retains the ability to transmit RTBV for at least 7 days. The extent of infection in leafhoppers-resistant cultivars by RTBV is more than that found by RTSV. If the same leafhoppers acquire RTBV first, then RTSV, they usually fail to transmit RTBV. Although there may be cross-infection of RTSV-infected plants by RTBV and RTBV-infected plants by RTSV to the same extent, chances for cross-infection of RTBV-infected plant by RTSV are more because of the shorter acquisition and inoculation access periods of RTSV.

8-5 EPIDEMIOLOGY

The epidemics of rice tungro virus disease are the function of several interacting factors: availability of the virus, host susceptibility to the virus and the vector, growth stage of the host, vector availability, agronomic practices, and environmental conditions. The availability of the virus depends on its sources in the fields or migration of viruliferous leafhoppers from a short or long distance. The rice varieties show a wide range of susceptibility both to the virus and the vector. There is also variation in susceptibility with age of the host plant. The availability of the vectors differs with season. The availability of the virus and its uptake by the host plant through the vectors are related to the agronomic practices and environmental or meteorological conditions. Short or long-distance migration of viruliferous vectors depends on the wind speed or wind field frontal system.

8-5-1 Sources of the Virus

Rice tungro is a crop adaptive virus and it mainly infects and survives in rice plants. There are, however, reports on its availability in wild rice and a few weeds. It does not survive in seeds or soil.[25]

Weeds and wild rice.   A number of species of wild rice (e. g., *Oryza barthii, O. officinalis, O. ridlev, O. rupipogon, O. sativa*) and weeds (e. g., *Eleusine indica, Echinocloa colonum, E. crusgalli, Dactyloctenium aegypticum, Eragrostis tenella, Ischinum gogossum, Leersia hexandra, Paspalam scrobiculatum, P. distichum, Triticum aestivum, Brachiaria reptans, Pennisetum typhoides, Hemarthuria compressa, Polypogon monospeliensis, Sorghum vulgare, S. halepense, Sporobolus tremulus, Setaria* *glauca, S. verticillata, Digitaria adscensdens, Bothriochloa odorata, B. raptans, Cynodon dactylon, Rottboelia compressa,* etc.) are reported to be infected by RTV either under natural or artificially inoculated conditions.[7,8,32–33] However, most of the weeds can retain the virus only for a limited period of time.[33] The extent of infection and the retention of the virus in different weeds were found to vary with the season.[34–35] A majority of the weeds also do not support the feeding of the principal vector, *N. virescens.* Therefore, the possibilities of the weeds acting as the source of RTV in the fields seem to be remote. Wild rices, however, have such potentialities as they support the feeding of *N. virescens* on them.

Stubbles.   Ling and Palomer[36] reported that root debris or stubbles of infected rice plants may carry RTV in them. Rao and John[37] further observed that leafhoppers, when collected from infected stubbles, could transmit the virus. Anjaneyulu and John[38] suspected that rice stubbles may be the reservoirs of the virus between the crops. Kondaiah and John[39] found the presence of the virus in dried rice leaves for more than 40 days. Tarafder and Mukhopadhyay[40–41] reported that the stubbles could keep the virus in the field up to 75 days depending on the variety and crop season. When the efficiency of freshly infected stubbles in transmitting the virus was tested under field condition, 11% transmission could be obtained. But the stubbles, when chopped off and puddled in the field, failed to act as the source of the virus.[42]

Symptomless hosts.   There are several varieties of rice which on infection show very mild symptoms or no symptoms at all. Mukhopadhyay[5,43] found that IR-20, IR-30, IR-36, IR-442-58, Vijoya, Cauvery, Ratna, Bala, Krishna, etc. usually showed mild symptoms on inoculation. Among the local (tall *indica*) varieties tested, Latisail, OC-1393, Dudhshar, Ashanalaya, Laghusail, and Nagra showed mild symptoms whereas Indrasail, Rajmalati, Lakshmansail, Kalamkathi, Kathamuk, Madhumalati, Jhulur, and Dhushri remained symptomless. These long duration varieties are grown in Kharif season and mature in December. These varieties, if infected, remain unnoticed and may serve as the virus source for the next *boro* season.

8-5-2 Inter-Seasonal Cycling

RTV is mostly a crop adaptive, nonpersistent virus and lives in rice plants. Its survival mostly depends on the availability of the rice crop. In a monorice cropping system, it may survive in stubbles and multiples in ratoons. In case of a double or triple crop system, however, the virus may survive through transmission from one rice crop to another. Rice in West Bengal (India) is almost a continuous crop. It is grown in three cropping seasons: *Aus* (June to September), *Aman* or *Kharif* (July to November–December), and *Boro* (January–February to April–May).

Rice-rice cropping system is very common in this state where rice is grown as *Kharif-boro* or *Kharif-aus* sequence. In case of *Kharif-boro*, no overlapping of crops, usually occurs, particularly with respect to early sown high-yielding varieties, but overlapping of crops occurs in case of local varieties. *Boro* nurseries are usually raised before the harvesting of the *Kharif* rice. During the late *Kharif* months (December), the vector population usually declines. Vectors tend to move for a better climate from high or medium land situation having mature rice crops or stubbles to low-lying ditches where *boro*-nurseries are being raised. In case the moving insects are viruliferous, they may infect *boro* seedlings at the nursery beds.[44] Therefore, in *kharif-boro* passage of the RTV, seed-beds may play a key role as a reservoirs of infective or uninfective vectors.

Boro and *aus* crop usually overlaps. The vectors usually survive as resident population in low-lying *boro* crop or overwinter from November–December as nymphs and emerge into adults during March–April. During this period, the *boro* crop almost matures and becomes unsuitable to the emerging leafhopper population or to get fresh infection of RTV. Thus, the passage of the virus from standing *boro* to *aus* is very unlikely. The *aus* crop again is mostly broadcasted. Therefore, the question of the flight of leafhoppers from standing *boro* to *aus* seedbeds does not arise but the scope of the movement of the leafhoppers from the *boro* stubbles to the *aus* standing crop remains. During Kharif, seedbeds are usually raised during June to July and the leafhoppers tend to move from *boro* stubbles and *aus* standing crop to kharif seedbeds. During this season, rice is grown in enormously vast areas having a wide range of varieties and leafhoppers get hosts at different growth stages of the crop. In the early part of this, passage of RTV and the vector takes place from the preceding season to the seedbeds and then a secondary passage takes place from early sown crop to late sown crop. In an operational project conducted during *kharif*, 1977 to *boro*, 1978–79, viruliferous leafhoppers could be observed in standing *kharif, boro* seed beds, standing *aus* and *kharif* seedbeds in succession. By introducing the seedbed treatments, however, the incidence of viruliferous leafhoppers could be minimized and the interseasonal passage of the virus could be broken.[42] In another study, the incidence of tungro and the viruliferous leafhoppers were controlled by early ploughing of the infected kharif stubbles, vector monitoring, and successive seedbed treatments.[45] An extensive survey conducted in the rice fields of West Bengal (India) during kharif, 1981, when a tungro epidemic occurred in this state, also suggested the significance of seedbed infection in spreading the epidemics.[10,46]

8-5-3 Spread

The spread of RTV is directly dependent on the availability of viruliferous leafhoppers. In the absence of the virus, only the availability of the leafhoppers will not create any epidemic. Similarly, in the absence of the leafhoppers the presence of RTV alone can not cause any epidemic. The availability of both the virus and leafhoppers are necessary for the spread of tungro. Ling[47] estimated the relation between the proportion of plants infected in an area and the size of the virus source. He recorded more spread when the source plants were scattered. A vector was found to spread the virus from the source plant to a maximum distance of 250 m. The spread of RTV again depends on the variety of the source plants.[48–49]

Plant age adversely affects the spread of the virus.[36,48,50–53] Susceptibility of the plant to tungro decreases with the increase of plant age at the time of inoculation.[36,51] Tarafder et al.[53] observed that the effect of transplanting age on the spread of RTV is further influenced by the transplanting time. In addition, the preference of the leafhoppers for the host age is also important. Individual treatments of these factors with respect to the spread may be misleading.

Leaving aside the question of the source plant and its associated conditions, the spread of RTV is directly dependent on the vector activity. Theoretically a single infective vector can infect 288 plants/day, but practically it is much lower (i. e., 11 to 30 plants/day at 27°C and 40 plants/day at 34°C).[54] Ling et al.[55] under experimental conditions determined the mathematical relationship of the effect of vectors on infection and found that the percentage of infected seedlings increases rapidly up to 1 *N. virescens*/seedling, but only slightly when the number of vectors increased to 3/seedling. Adult insects are three times more efficient vectors than are nymphs.[56] However, the behavior of vectors usually differs under field conditions depending on the cultivar, plant age, microclimatic and meterological conditions, and lunar phases. The field spread of RTV may further be complicated by the migration of viruliferous leafhoppers from a short to a long distance. There may also be a wind field frontal system transportation of vectors.

8-6 INTEGRATED MANAGEMENT

The control of RTV needs to be considered in an integrated manner involving elimination of natural virus sources, agronomic adjustments, seedbed management, and vector control processes.

8-6-1 Elimination of Natural Sources of RTV

To eliminate the natural sources, steps should be taken to reduce the inoculum potential in the field. For this purpose, there should be field inspection for the presence of the virus in the stubbles and removal of this source by simple ploughing. Natural sources of the virus can also be minimized by raising resistant varieties to both virus and vectors. Several sources for resistance have been identified in different places and utilized to breed tungro resistant varieties. During the last 20 years several RTV-resistant high-yielding rice cultivars have been released,[57-60] and many of them are in wide cultivation. Studies on nature of tungro resistance suggest that a low level of disease in some cultivars is due to their resistance to the leafhoppers[61] while it may not be the case in few other cultivars (e. g., IR-20, IR-26, IR-36, and IR-40).[59]

8-6-2 Agronomic Adjustment

The agronomic adjustments for the management of RTV disease depend on certain concepts. First of all, it is to be kept in mind that the susceptibility of a plant declines with age and usually it does not take up any infection beyond 75 days of its age. Usually the transplanting age of the seedlings is susceptible to the virus and particularly to the vectors. The availability of residential population of the vectors at the time of transplantation will depend on the rainy days during April–May,[42,62] so a forecasting system can be developed on the basis of these interacting factors, to suggest a suitable time for transplantation in order to avoid vector pressure on the seedlings at the early stage of growth.

8-6-3 Seedbed Management

Seedbed management is the most efficient practice to break the interseasonal passage of RTV and to control the disease. Mukhopadhyay et al.[45] recommended a safe distance of

10 m from standing crop to raising seedbeds. An application of Furadan 3G (1.5 kg a. i./ha) or Foratox 10G (1.75 kg a. i./ha) immediately after seedling emergence can keep the seedbeds free from the vectors for 9–12 days. If the proper isolation distance is maintained, this application can be avoided subject to the pursuance of regular monitoring of vectors in the beds. Depending on the increase in vector population in the seedbeds, a second application of pesticides may be profitably made 5 days prior to transplantation. For this purpose, granular or emulsifiable concentrates may be used. Quinalphos and phosphamidon have been found to be useful. This treatment ensures transplantation of virus-free seedlings. The second application of pesticides at the seedbeds needs to be considered in *kharif* and not in *boro* season as the vector population starts rising in *kharif* and declining in *boro* season during this period.

8-6-4 Field Management

The field management usually involves chemical control of the vectors. Pesticidal control of RTV or vectors has been studied by a larger number of workers,[62] but hardly any experiment has been conducted on the integrated management of the disease indicating the management practices required at different field stages of cultivation. Mukhopadhyay et al.[45] observed that if vector-free seedlings are transplanted there would be hardly any leafhopper in the field immediately after transplantation. The vectors arrive from outside through their active movement, or may migrate possibly by the windfield frontal system; hence there should be a strong monitoring system for recording the arrival of vectors. The seedlings are to be kept free from vectors usually up to 45–50 days after transplantation (if 30–35 day old seedlings are transplanted). Depending on the arrival of vectors, one application of pesticides may be necessary in the field at 15–20 days after transplantation. If the recommended practices are rigorously pursued, the spread of RTV by resident vectors can be successfully controlled. In case the viruliferous leafhoppers migrate by the windfield frontal system and a sudden influx of them takes place in the field at the susceptible age of crop, then a separate strategy needs to be developed to control the immigrating viruliferous vectors.

As far as pesticides are concerned, cypermethrin among emulsifiable insecticides[63] and carbofuran among granular insecticides[64] are very efficient in management of tungro. Satapathy and Anjaneyulu[65] demonstrated that wettable powder insecticides—acephate, bendiocarb, carbaryl, isoprocarb, and carbofuran—when sprayed at 0.1% concentration, not only prevented tungro infection in artificial inoculation tests, but also efficiently managed the disease incidence under field conditions. In another study, Satapathy and Anjaneyulu[66] observed that under field conditions carbofuran as root zone application (@ 1 kg a. i./ha) was much superior to broadcast application. The effects of bendiocarb, BPMC, and isoprocarb as root zone application were almost similar to broadcast application of carbofuran.

8-7 CONCLUSION

Rice tungro is now considered as a complex disease in the etiology of which two unrelated viruses RTSV (rice tungro spherical virus) and RTBV (rice tungro bacilliform virus) are involved. RTSV does not usually produce any symptom except mild stunting in susceptible varieties but facilitates the acquisition of RTBV, which induces tungro symptoms in rice plants. A contradictory claim, however, has been made in Japan, where waika virus, which is either identical or related to RTSV, caused epidemics. In a preliminary survey conducted on the incidence of RTSV and RTBV and RTSV + RTBV in rice plants of West Bengal, RTSV could be detected widely even from apparently healthy plants, whereas RTSV + RTBV were rare. At early tillering stage, some RTSV containing plants showed recoverable symptoms. Thus, RTSV may not be a simple nonsymptom producing virus. It may have variants or strains which by themselves may cause epidemics. In any case the nature, characteristics, and variation of RTSV need to be properly elucidated.

Information so far available indicates that the chances of the spread of RTSV in nature are much more than RTBV. The minimum acquistion access period of RTSV is less than 10 min whereas an insect takes at least 30 min to acquire RTBV from a source usually provides optimum acquisition of RTBV. Insects may not like to feed on plants for such a long time under natural conditions. Moreover, RTSV is efficiently transmissible by both *N. virescens* and *N. nigropictus* irrespective of the presence of RTBV whereas RTBV is transmissible by *N. virescens* only in presence of RTSV. Thus, the survival value of RTBV seems to be very limited. If this virus is implicated to the frequent outbreak of tungro disease in different South and southeast Asian countries, it may have a separate survival and spreading system. Thus, the nature, characteristics, perpetuation, and spread of this virus need to be critically examined.

Rice green leafhoppppers, particularly *N. virescens* and *N. nigropictus*, are involved in the spread of tungro disease. The distribution of tungro coincides with the distribution of these insects at the global level. *Nephotettix virescens* is exclusively rice plant feeder, whereas *N. nigropictus* equally feeds on certain weeds. Thus, availability of rice fields is important for the incidence of these insects and the viruses as well. The abundance of these insects again depends on certain agrometeorological conditions, particularly temperature and relative humidity. In those countries, where the minimum temperature goes below 15°C, in winter months, these insects become scarce during that period. The population starts rising again with the onset of the favorable climatic conditions and availability of rice fields. Besides the seasonal growth cycle of the resident population, there may also be occasional immigration of vector from short distances by active flying along the wind direction or by passive transportation from long distances by the windfield frontal system followed by a climatic depression. In case the immigrants are viruliferous, the epidemiology of the disease becomes more complex. Although indirect evidence is available on the immigration of the leafhoppers, adequate windfield studies and the synoptic analysis of the leafhopper catches need to be conducted for confirming windfield transportation of leafhoppers.

It is true that rice green leafhoppers do fly. They have definite nocturnal flight activity, so they prefer to stick to the host plant during daytime and disperse from civil twilight to 5 hr after sunset and again 2 hr before sunrise. This dispersion again is related to the lunar periodicity. *Nephotettix nigropictus* is more sensitive to this periodicity than *Nephotettix virescens*. In a general way, both the species disperse from first quarter to full moon. This lunar phase dependent dispersal of these insects is implicated to their infestation of the rice fields and the spread of the viruses as well. These lunar phase related dispersions of *N. virescens* and *N. nigropictus* and tungro need to be critically studied for the proper understanding of the mechanics of field spread of rice tungro disease.

8-8 REFERENCES

1. Agathi, J. A., Sison, P. L., and Abalos, R. "A Progress Report of the Rice Maladies in Central Luzon with Special Reference to the Stunt or Dwarf Disease." *Philipp. J. Agric.* 12 (1941); 197.

2. Rivera, C. T., and Ou, S. H. "Leafhopper Transmission of Tungro Disease of Rice." *Plant Dis. Rept.* 49 (1965): 127.

3. Ling, K. C. "Ability of *Nephotettix apicalis* to Transmit the Rice Tungro Virus." *J. Econ. Ent.* 63 (1970): 582.

4. Xie, L. H., and Lin, J. Y. "The Occurrence of Rice Tungro Disease (spherical virus) in China." *J. Fjuian Agric. Coll.* 3 (1982): 16.

5. Mukhopadhyay, S. "Ecology of Rice Tungro Virus and Its Vectors, in *Virus Ecology*, eds. A Misra and H. Polasa. (New Delhi: South Asian Publishers, 1984), 139.

6. Ou, S. H., Rivera, C. T., Navaratnam, S. J., and Goti, K. G. "The Virus Nature of 'Penyakitmerah' Disease of Rice in Malayasia." *Plant Dis. Rept.* 49 (1965): 778.

7. Wathanakul, L., and Weerapat, P. "Virus Diseases of Rice in Thailand," in *Virus Diseases of Rice Plant*. (Baltimore, MD.: Johns Hopkins Press, 1969), 79.

8. Raychaudhuri, S. P., Mishra, M. D., and Ghosh, A. "Preliminary Note on Transmission of a Virus Disease Resembling Tungro of Rice in India and Other Virus Like Symptoms." *Plant Dis. Rept.* 51 (1967): 300.

9. Hibino, H., John, V. T., Miah, S. A., and Shagir Sama. "Present Status of Tungro Problems in Tropical Asia and the RTV Collaborative Project." *Proc. Workshop on the RTV Collaborative Project,* International Rice Research Institute, Los Banos, 1984.

10. Mukhopadhyay, S. *PVRC Survey Report I; Sudden Outbreak of Rice Tungro Virus Disease in West Bengal during 1981,* Plant Virus Research Centre, Bidhan Chandra Krishi Viswavidyalaya, West Bengal, India (Mimeo), 1982.

11. Rao, G. M., and Anjaneyulu, A. "Influence of Nitrogen Nutrition of Tungro Diseased Plants of Different Cultivars." *Oryza* 13 (1976): 75.

12. Chowdhury, A. K., and Mukhopadhyay, S. "Effects of Virus on Yield Components." *Int. Rice Comm. Newslt.* 24 (1975): 74.

13. Srinivasan, S. "Yield Loss Due to Rice Tungro Virus." *Int. Rice. Res. Newsl.* 4 (1979): 13.

14. Chowdhury, A. K., and Mukhopadhyay, S. "Effect of Rice Tungro Virus Infection on Starch, Reducing Sugar and Phosphorus Contents of Leaves of Different Varieties of Rice." *Curr. Sci.* 43 (1974): 280.

15. Omura, T., Saito, Y., Usugi, T., and Hibino, H. "Purification and Serology of Rice Tungro Spherical and Rice Tungro Bacilliform Viruses." *Ann. Phytopath. Soc. Japan* 49 (1983): 73.

16. Cabauatan, P. Q., and Hibino, H. "Isolation, Purification and Serology of Rice Tungro Bacilliform and Rice Tungro Spherical Viruses." *Plant. Dis.* 72 (1988): 526.

17. Jain, R. K., and Mishra, M. D. "On the Association of Rice Tungro Virus Components with Tungro Disease." *Curr. Sci.* 58 (1989): 457.

18. Mishra, M. D., Niazi, F. R., and Jain, R. K. "Serological Relationship of Rice Tungro Spherical Virus and Bacilliform Virus Components Associated with Rice Tungro Disease." *Cur. Sci.* 59 (1990): 228.

19. Hibino, H., and Cabauatan, P. Q. "Purification and Serology of Rice Tungro Spherical Virus." *Int. Rice Res. Newslt.* 10 (1983): 10.

20. Cabauatan, P. Q., and Hibino, H. "Transmission of Rice Tungro Bacilliform and Spherical Viruses by *Nephotettix Virescens* Distant." *Phillipp. Phytopathol.* 21 (1985): 103.

21. Hibino, H. "Transmission of Two Rice Tungro Associated Viruses and Rice Waika Virus from Doubly or Singly Infected Source Plants by Leafhopper Vectors." *Plant Dis.* 67 (1983): 774.

22. Hibino, H., Roechan, M., and Sudarisman, S. "Association of Two Types of Virus Particles with Penyakit Habang (Tungro Disease) of Rice in Indonesia." *Phytopathology* 68 (1978): 1412.

23. Anjaneyulu, A., and John, V. T. "Strains of Rice Tungro Virus." *Phytopathology* 62 (1972): 111.

24. Mukhopadhyay, S., and Bandopadhyay, B. "Incidence of Different Strains of Rice Tungro Virus in West Bengal." *Int. J. Trop. Plant Dis.* 1 (1984): 125.

25. Ling, K. C. *Rice Virus Diseases*, International Rice Research Institute, Los Banos, 1972.

26. Ling, K. C. "Hybrids of *Nephotettix impicticeps*. Ish. and *N. apicalis* (Motsch) and Their Ability to Transmit the Tungro Virus of Rice." *Bull. Ent. Res.* 58 (1968): 393.

27. Ling, K. C. *Synonymies of Insect Vectors of Rice Viruses*, International Rice Research Institute, Los Banos, 1973.

28. John, V. T. "Identification and Characterization of Tungro Virus Diseases of Rice in India." *Plant Dis. Rept.* 52 (1968): 871.

29. Ling, K. C., and Tiongco, E. R. "Effect of Temperature on the Transmission of Rice Tungro Virus by *Nephotettix virescens Philipp. Phytopathol.* 11 (1975): 46.

30. Ling, K. C., and Tiongco, E. R. "Transmission of Rice Tungro Virus at Various Temperatures; A Transitory Virus–Vector Reaction," in *Leafhopper Vectors and Plant Disease Agents*, eds. K. Maramorosch and K. F. Harris. (New York: Academic Press, 1974), 344.

31. Hibino, H. "Relations of Rice Tungro Bacilliform and Rice Tungro Spherical Viruses with Their Vector *Nephotettix virescens.*" *Ann. Phytopathol. Soc. Japan* 49 (1983): 545.

32. Rivera, C. T., Ling, K. C., and Ou, S. H. "Suspect Host Range of Rice Tungro Virus, *Phillip.*" *Phytopathol.* 5 (1969): 16.

33. Mishra, M. D., Ghosh, A., Niazi, F. R., Basu, A. N., and Roychaudhuri, S. P. "The Role of Graminaceous Weeds in the Perpetuation of Rice Tungro Virus." *J. Indian Bot. Soc.* 52 (1973): 176.

34. Tarafder, P., and Mukhopadhyay, S. "Potential of Weeds to Spread Rice Tungro in West Bengal, India." *Int. Rice Res. Newslt.* 4 (1979): 11.

35. Tarafder, P., and Mukhopadhyay, S. "Further Studies on the Potential of Weeds to Spread Tungro in West Bengal, India." *Int. Rice Res. Newsl.* 5 (1980): 10.

36. Ling, K. C., and Palomer, M. K. "Studies on Rice Plants Infected with the Tungro Virus at Different Ages." *Philipp. Agriculturist* 50 (1966): 165.

37. Rao Prasada, R.D.V.J., and John, V. T. "Alternate Hosts of Rice Tungro Virus and Its Vectors." *Plant Dis. Rept.* 62 (1974): 955.

38. Anjaneyulu, A., and John, V. T. "Rice Stubbles and Self Sown Rice Seedlings, the Reservoir Hosts of the Tungro (Virus Disease) During Off Season." *Sci. Cult.* 41 (1975): 298.

39. Kondaiah, A., and John, V. T. "Recovery of Rice Tungro Virus from Dried Infected Rice Leaves." *Phytopathol. Z.* 90 (1977): 311.

40. Tarafder, P., and Mukhopadhyay, S. "Potential of Stubbles in Spreading Tungro in West Bengal, India." *Int. Rice Res. Newslt.* 4 (1979): 18.

41. Tarafder, P., and Mukhopadhyay, S. "Further Studies on the Potential of Rice Stubbles for Spreading Tungro in West Bengal, India." *Int. Rice Res. Newslt.* 5 (1980): 12.

42. Mukhopadhyay, S. "Management of Rice Tungro Virus Disease." *Rev. Trop. Plant Pathol.* 1 (1984): 181.

43. Mukhopadhyay, S. "Ecology of *Nephotettix* spp. and Its Relation with Rice Tungro Virus." Final Report, Department of Plant Pathology, Bidhan Chandra Krishi Viswavidyalaya, West Bengal, India, 1980.

44. Chakravarti, S. K., Nath, P. S., Chowdhury, A. K., and Mukhopadhyay S. "Studies on the Off Season Incidence of Rice Green Leafhoppers," in *Use of Traps for Pest/Vector Research and Control.* eds. S. Mukhopadhyay and M. R. Ghosh. (Department of Plant Pathology, Bidhan Chandra Krishi Viswavidyalaya, West Bengal, 1985), 87.

45. Mukhopadhyay, S., Chowdhury, A. K., and Chakravorti, S. K. "Epidemiology and Control of Rice Tungro Virus Disease in West Bengal," in *Integrated Management of Rice Hoppers and Hopper Borne Viruses,* eds. S. Mukhopadhyay and M. R. Ghosh. (Department of Plant Pathology, Bidhan Chandra Krishi Viswavidyalaya, West Bengal, 1987).

46. Mukhopadhyay, S. "Virus Diseases of Rice in India," in *Vistas in Plant Pathology,* eds. A. Verma and J. P. Verma (New Delhi: Malhotra Publishing House, 1986), 111.

47. Ling, K. C., "Experimental Epidemiology of Rice Tungro Disease II. Effect of Virus Source on Disease Incidence." *Philipp. Phytopathol.* 11 (1975): 21.

48. Mukhopadhyay, S., and Chowdhury, A. K. "Some Epidemiological Aspects of Tungro Virus Disease of Rice in West Bengal." *Int. Rice Comm. Newslt.* 19 (1973): 9.

49. Rao, G. M., and Anjaneyulu, A. "Carbofuran Prevents Rice Tungro Virus Infection." *Curr. Sci.* 48 (1979): 116.

50. Narayanasami, P. "Influence of Age of Rice Plants at the Time of Inoculation on the Recovery of Rice Tungro Virus by *Nephotettix impicticeps* (Ishihara)." *Phytopathol. Z.* 74 (1972): 109.

51. Rao, G. N., and Narayanasamy, P. "Effect of Plant Age on Inoculation and Rice Tungro Virus Development." *Int. Rice Res. Newslt.* 12 (1987): 20.

52. Baltazar, R. B., and Tangonan, N. G. "Disease Occurrence as Affected by Age of Transplanted Seedlings." *Int. Rice Res. Newslt.* 12 (1987); 21.

53. Tarafder, P., Ghosh, A. B., and Mukhopadhyay, S. "Effect of Transplanting Date and Transplanting Age on the Spread of Tungro." *Int. Rice. Res. Newslt.* 5 (1980): 12.

54. Tiongco, E. R., Hibino, H., and Ling, K. C. "The Use of Light Trap and Other Means in Rice Tungro Studies in the Philippines," in *Use of Traps for Pest/Vector Research and Control,* eds. S. Mukhopadhyay and M. R. Ghosh. (Department of Plant Pathology, Bidhan Chandra Krishi Viswavidyalaya, West Bengal, 1985).

55. Ling, K. C., Tiongco, E. R., and Flores, Z. M. "Epidemiological Studies on Rice Tungro," in *Plant Virus Epidemiology,* eds. R. T. Plumb and J. M. Thresh. (Oxford: Blackwell Scientific Publications, 1983), 249.

56. Ling, K. C. "Experimental Epidemiology of Rice Tungro Disease I. Effect of Some Factors of Vectors (*Nephotettix virescens*) on Disease Incidence." *Philipp. Phytopathol.* 10 (1975): 42.

57. Tinogco, E. R., Cabunagan, R. C., and Hibino, H. "Resistance of Five IR Varieties to Tungro." *Int. Rice Res. Newslt.* 8 (1983): 6.

58. Hibino, H., Tiongco, E. R., Cabunagan, R. C., and Flores, Z. M. "Resistance to Rice Tungro Associated Viruses in Rice under Experimental and Natural Conditions." *Phytopathology* 77 (1987): 871.

59. Hibino, H., Daquioag, R. D., Cabunagan, P. Q., and Dahal, G. "Resistance to Rice Tungro Spherical Virus in Rice." *Plant Dis.* 72 (1988): 843.

60. Mohnaty, S. K., Bhaktavatsalam, G., and Anjaneyulu, A. "Identification of Field Resistant Rice Cultivars for Tungro Disease." *Trop. Pest Manag.* 35 (1989): 48.

61. Rapusas, H. R., and Heinrich, E. A. "Plant Age and Levels of Resistance to Green Leafhopper, *Nephotettix virescens* (Distant), and Tungro Virus in Rice Varieties." *Crop Prot.* 1 (1982): 91.

62. Mukhopadhyay, S. *Epidemiology of Rice Tungro Virus and Its Vectors in West Bengal,* Plant Virus Research Centre, Bidhan Chandra Krishi Viswavidyalaya, West Bengal, 1986, 118.

63. Satapathy, M. K., and Anjaneyulu, A. "Use of Cypermethrin, a Synthetic Pyrethroid, in the Control of Rice Tungro Virus Disease and Its Vector." *Top. Pest Manag.* 30 (1984): 170.

64. Satapathy, M. K., and Anjaneyulu, A. "Prevention of Rice Tungro Virus Disease and Control of the Vector with Granular Insecticides." *Ann. Appl. Biol.* 108 (1986): 503.

65. Satapathy, M. K., and Anjaneyulu, A. "Management of Tungro Virus Disease by Application of Wettable Powder and Flowable Insecticides." *Trop. Pest Manag.* 35 (1989): 41.

66. Satapathy, M. K., and Anjaneyulu, A. "Effect of Root Zone Placement of Granular Insecticides for Tungro Prevention and Its Control." *Trop. Pest Manag.* 35 (1989): 51.

67. Jones, M., Dasgupta, I., Cliffe, J., Lee, G., Blakeborough, M., Davies, J. W., and Hull, R. "The Molecular Biology of the Rice Tungro Viruses—Possibilities for Resistance," Presented at *ICGEB Symposium,* New Delhi, Feb. 14–17, 1990.

9

WHITE TIP DISEASE
OF RICE

E. B. GERGON

Department of Plant Pathology, International Rice Research Institute
P.O. Box 933, Manila, Philippines

J. K. MISRA

Mycological Research Unit, Department of Botany
Sri J.N. Mahavidyalaya, Lucknow 226019, India

9-1 HISTORICAL BACKGROUND

This disease was first described by Kakuta[1] in 1915 in Kyushu, Japan as black grain disease of rice. Infection was due to a nematode causing ear blight of Italian millet. In the United States, white tip disease of rice was reported by several researchers but this was attributed to mineral deficiency. In 1941, an outbreak of the disease occurred in Japan which caused serious damage to rice crop. It was only in 1944 that nematodes were found within the pubescence of infected young rice and panicles and later inside the hulls of the grains.[2] These nematodes were identified by Yokoo[3] as *Aphelenchoides oryzae*. Cralley[4] discovered that the white tip disease described by the Americans was similar to the disease reported by the Japanese. In 1952, Allen[5] confirmed that the nematode causing white tip on rice was morphologically identical to the nematode attacking strawberry which had been described by Christie in 1942 as *A. besseyi*, the name that is accepted up to this time. Its common names include rice white tip nematode, strawberry bud nematode, summer crimp nematode, or summer dwarf nematode.[6]

Many aspects of the disease have been studied and published. Fortuner and Williams[7] extensively reviewed the literature on *A. besseyi* and Ou[8] presented a comprehensive account of the subject. This chapter attempts to provide an update of literature on this disease and its pathogen.

9-2 SYMPTOMS OF THE DISEASE

Infected plants may or may not exhibit any symptoms even if the seeds contain the nema-todes and show yield reduction.[9-10] Symptoms, if expressed by the diseased plant, usually vary depending on the environment, variety, or soil conditions. In susceptible varieties, the disease may appear on seedlings at 2–3 leaf stage as a chlorotic discoloration on the leaf base just below the collar.[11] On mature plants, whitening of the leaf tips, from 3 to 5 cm, becomes evident with lower parts greener than normal (Figure 9-1). The name 'white tip' comes from this symptom. Sometimes, chlorosis extends to the basal or middle parts of the leaves. Eventually, the affected areas turn dark brown or necrotic and frayed. The diseased plants are stunted, especially so when infection sets upon germination and may give rise to tillers at high nodes.[12] At booting stage, the upper leaves, particularly the flag leaf that is the most affected, conspicuously become short, markedly twisted at the apical portion, crinkled, or split longitudinally. The panicles that emerge either partially or completely from infected plant are prominently reduced in size and mature late. It bears whitish spikelets on the tip or throughout the panicle which are mostly sterile and reduced in weight and number. Fertile florets that mature produce twisted or distorted glumes and small deformed kernels. Black wedge-shaped spots may also appear on infested grains.[13]

9-3 DISTRIBUTION AND ECONOMIC IMPORTANCE

Since the discovery of *A. besseyi*, the nematode has been reported in nearly all rice-growing countries in addition to Japan and the United States. It is widespread in Africa, Asia, USSR, Latin America, and some countries in Oceania and Europe[14] (see distribu-tion map, Figure 9-2). In Havana, Cuba, the disease has been found to infest 61% of rice fields.[15]

Losses in crop production due to white tip vary with country, year, variety, and agronomic practices. Yoshii and Yamamoto[2] estimated 10% to 35% losses in yield from this disease. In Taiwan, 29% to 46% losses were attributed to white tip,[16] and about 60% in Bangladesh.[11] Up to 20% yield losses were reported in southern and eastern states of

Figure 9-1 The characteristic symptom of white tip disease on rice.

Figure 9-2   Worldwide distribution of *Aphelenchoides besseyi* on rice.

India[17] and about 60% in the State of Tamil Naidu (India).[18] In China, Zhang[19] reported 7% to 17% economic loss of rice due to *A. besseyi*. Atkins and Todd,[9] based on a 3-year study, estimated 17% and 7% losses respectively in susceptible and resistant varieties due to reduction in the number of grains and increase in the percentage of sterile spikelets per panicle. The nematode can also cause blackening of the grains when present in high number, thereby affecting the grain quality.[20]

9-4 THE PATHOGEN

> *Aphelenchoides besseyi* Christie
> (syn. *Aphelenchoides oryzae* Yokoo;
> *Asteroaphelenchoides besseyi* (Christie)
> Drozdovski)
>   Class: Nematoda
>   Order: Tylenchida
>   Family: Aphelenchoididae
>   Superfamily: Aphelenchoidea

The nematode is characterized by slender body, rounded lip region which is slightly offset and wider than the body (Figure 9-3), transverse vulva, elongated and oval spermatheca, short ovary, narrow and tiny post vulval uterine sac, tail bearing mucro with 3–4 processes, and males' tail not having dorsal processes at the proximal end.[21]

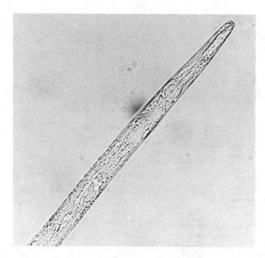

Figure 9-3   The upper portion of
Aphelenchoides besseyi.

Female and male species of this nematode occur in rice in equal proportion, although males outnumber females when food supply is exhausted.[22] Gokte and Mathur[23] also observed that males developed faster and appeared earlier than females during their postembryonic development. In spite of being higher in number, however, males may not be involved in reproduction. The female can reproduce parthenogenetically.[22] The length of the life cycle of *A. besseyi* varies depending on factors like medium, temperature, and atmospheric humidity. In a medium, the nematode develops from egg to egg in 6.5 to 7 days where egg stage takes 0.5 days, second larval stage 0.13 days, third 0.65 days, fourth 0.9 days, and female sexual maturation, 4.3 to 4.8 days.[22] In culture of *Fusarium solani*, it completed its life cycle in 24 ± 4 days at 16°C, 15 ± 2.9 days at 20°C, 9 ± 2 days at 23°C, 10 ± 2 days at 25°C, and 8 ± 2 days at 30°C. At 35°C, multiplication ceased.[24] Vuong[25] reported completion of life cycle in 3 to 6 days at 25–31.3°C and 9 to 24 days at 14.7–20.6°C with optimum temperature for development at 28°C and marginal temperatures at 13°C and 42°C. He observed that at least 70% atmospheric humidity is needed by the nematode for its normal development. Sudakova[26] reported a race developing at 13°C with optimum development at 21–23°C and life cycle completed in 10 days at 21°C and 8 days at 23°C. Sivakumar[27] recorded a range of 10 to 12 days for *A. besseyi* to complete its life cycle with 6 to 8 days devoted to development of J2 to adult at 29–32 C. Gokte and Mathur[23] noted that from egg to egg, the cycle takes 11 days. Differences in optimum temperature for the normal development of the nematode could be due to difference in the host used.[24]

A. besseyi can be multiplied in the laboratory using agar culture media containing fungal species of *Alternaria, Pyricularia, Cochliobolus, Nakataea, Colletotrichum,* and *Phytophthora*[28] as well as *Curvularia, Helminthosporium,* and *Fusarium.*[9] Rao[29] studied the pathogen using different monoxenic cultures and found that 2% oatmeal followed by 2% peptone supported population buildup of the nematode. He noted that it failed to multiply on *Pyricularia oryzae* but population was highest on *Fusarium moniliforme*, followed by *Alternaria padwickii, Helminthosporium oryzae,* and *Curvularia* sp. He obtained the highest male sex ratio of 5.6% in culture of *Curvularia* sp and only 2.8% in *F. moniliforme*.

A. *besseyi* is a specific parasite of rice but it has a wide host range. Some of the plants where it can survive besides strawberry and Italian millet are *Setariae viridis*,[2] *Panicum sanguinale*,[2] *Cyperus iria*,[2] *Setaria italica*,[2] *Cyperus polystachyus*,[25] and *Imperata cylindrica*.[25]

The nematodes can be located in the seed between the husks and caryopsis.[30] Both larvae and adults are capable of withstanding dessication. Rao[31] demonstrated that *A. besseyi* is coiled up inside the palea and on the surface of the lodicules which appeared dried up and papery as the seed matured.

9-5 DISEASE CYCLE

A. besseyi remains only in the seeds but not in other parts of the buds.[2] When the seed germinates, the nematodes revive and move out of the seed. They are attracted toward aqueous extracts of the germinating seeds of certain cultivars of rice.[33] The nematodes migrate vertically through the film of water clinging to the stem in the interior of the folded young leaf. They reach the growing point of the seedling in 1 to 5 days[23] where they assemble until tillering stage. As the plant grows, the nematodes continue moving toward the upper part of the larger rolled leaf completely enclosed in the sheath and transfer to the panicle when it reaches 5 to 6 cm long at 19–23 °C.[33,34] At booting stage, a greater number of the nematodes move toward the middle of the panicle than the terminal parts. Many stay on the surface of the glumes but some nematodes penetrate the florets before anthesis through a 30 μm opening circumscribed by the apiculi of the lemma and palea. Inside the florets, they aggregate around the ovary but not in other floral tissues and multiply rapidly in the inflorescence up to the second week of heading (soft dough stage of the grain). Nematode eggs that were hatched develop to adult and stay only between the inner surface of palea and lemma and the pericap. As the grain matures, they coil up and become quiescent in the seeds due to dehydration and subsequently anabiotic when grains are dried.[35-36]

9-6 EPIDEMIOLOGY

A. besseyi is seedborne and its primary means of spread is through infested seeds.[37-38] They can survive in the seed in quiescent state up to 23 months.[7,12] Yoshii and Yamamoto[2] obtained living nematodes from seeds that had been stored for 3 years, although their survival decreased from 62.9% for the newly harvested to 46.9% for stored rice grains. Sivakumar[27] detected nematodes from seeds stored for less than 25 months but not from seeds kept for more than 25 months. After 8 to 9 months of storage nematode population in the seeds was 83% of the original population and declined by 27.1% after 15 months. Zem and Monterio[39] reported survival of *A. besseyi* in infected seeds for as long as 8 years.

The nematodes could be present in discolored, deformed, or healthy-looking grains where the number of viable nematodes may vary.[40] The highest infestation obtained per seed was 121 in the case of deformed seeds.[40] The critical inoculum density for susceptible paddy rice was estimated to be 30 viable nematodes per 100 seeds.[34] Considering the number of nematodes that may infest a seed as found by Gergon and Mew,[40] one seed is likely enough to carry out the dissemination of the pathogen under favorable conditions. The nematodes can be reactivated from the seeds in 12 hr soaking in water but maximum recovery was achieved in 24 hr at $25 \pm 3°C$.[30]

A. besseyi is a parasite of warm regions. It can infect upland, irrigated lowland, and deepwater rice.[14] Although infection through seeds is the main form of disease dissemination, second growth of the crop[41] and irrigation water can be contaminated which may initiate infection especially when the prevailing temperature is between 20°C and 40°C. Sivakumar[42] demonstrated occurrence of diseased fields even when nematode-free seeds were used. He indicated that the nematodes came from rice stubbles which remained in the fields after harvest and survived by feeding on *Fusarium* or *Curvularia* present in the straw. The nematodes undergo diapause and get reactivated when rain comes and migrate to the panicle after flower initiation when relative humidity is over 80%.

Terry[43] found *A. besseyi* in cabbage and watermelon grown after rice, in rice regrowth, and fallow soil after harvesting the infested crop. McGawley et al.[44] however, could not recover nematodes from roots and soil but only from aboveground plant parts.

9-7 MANAGEMENT

9-7-1 Regulatory

A. besseyi plays an important role in the movement of rice germplasms because of its seedborne nature. In California, the disease does not occur but it was found in fields where infested seeds from southeastern states were planted.[45] To prevent its international spread, many countries have imposed strict quarantine regulations on seeds infected by this pathogen. The plant quarantine classification of rice diseases in West Africa has classified it Category A and given 0% tolerance level of infection for rice introduction into Nigeria.[46]

Prot,[47] however, based on data generated by the International Rice Research Institute (IRRI) in 1988, expressed that the potential risk of disseminating *A. besseyi* through seeds in the international exchange is only 0.48%, which does not justify the cost and inconvenience of quarantine. He added that the nematode already exists in most rice-growing countries and that more than 99% of seedlots are exchanged for scientific research purposes. He suggested the application of quarantine only on seeds intended for planting farmers' fields and the use of hot water treatment to all incoming seeds to prevent entry of possible different pathotypes.

The frequency of detecting *A. besseyi* from rice seedlots actually varies from year to year. In Seed Health Unit of IRRI where a lot of samples are examined for quarantine pests prior to release of seed to different consignees, *A. besseyi* was detected in less than 1% of the outgoing seedlots and from 5–33.8% of the total seedlots coming from other countries during 1985 to 1987.[48]

Since white tip produces distinct symptoms and causes yield losses, it is very important that importing countries take all necessary precautions to prevent the spread of seeds infected by the pathogen. For international exchange of seeds as germplasm, safeguard at entry, upon entry, and after entry of seeds may be used to reduce the chances of seeds to become the pathway for establishment of the pest or pathogen.[49]

9-7-2 Physical Control

Since *A. besseyi* is seedborne, the most appropriate method of its control is seed treatment by physical or chemical means. Hot water treatment for 15 min at 55–61°C after pre-soaking the seeds in water was found effective in controlling the disease with less seed injury.[38] Presoaking in water activates the nematodes from dormant state and subsequent heat treatment kills the nematodes.[50] Vuong[25] recommended hot water treatment at 55°C to 57°C for 15 min followed by a fungicidal dip.

The use of irradiation on rice seeds was also found effective in reducing the nematode population in infested seeds from 61% to 4% in pot experiment and in increasing plant height and weight of harvested rice.[51]

9-7-3 Chemical Control

The use of chemicals in the control of white tip disease is being widely advocated. In Japan, nematicides cartap (Padan), fenitrothion (Sumithion), or fenthion (Baycid) are used as seed disinfectants and Sumithion as spray for diseased heads.[52] Cho et al.[53] reported that fenitrothion 50EC, fenthion 50EC, and Dasuzin 34EC were effective as seed disinfectants without any phytotoxic effect even when mixed with fungicides Benoram 40WP, Proraz 25EC, and TCM 30EC. Treatment of the seeds before sowing with any of these chemicals followed by application of carbofuran 3G one day before transplanting controlled the disease in paddy fields. Sivakumar[54] found that soaking of seeds in 1% potassium or sodium chloride solution followed by sun drying for 6 hr at 40–41°C resulted in 95% to 97.6% seed disinfestation as compared to 87.3% by sun drying alone.

In the field, Prasad et al.[55] reported that spraying of infected plants 50 days after transplanting with isofenphos and carbosulfan at 1% a.i. was effective in reducing nematode population and increasing yield. Etaphos, a thiophosporic acid ester, at 2 kg a.i./1000 kg seeds was likewise effective against *A. besseyi* giving 29.6% yield increase over the control when tried in pot experiment. When used for soaking seeds for 18 hr using solution of 0.25%, 0.5%, and 1%, it gave yield improvement of up to 35.5%. Under field condition, its technical efficacy ranged from 65–92% when used as spray and 92–86% when used as 18-hr seed soaking.[56]

Efficacy of chemical treatment on the seeds can be improved by the use of ultrasound or vacuum infiltration dressing.[57] Nagy[57] found that exposure of seeds to ultrasound (21.5 khz) increased the permeability and diffusion of nematicidal solution of 0.5 to 1% oxamyl 25EC into the seeds.

9-7-4 Use of Resistant Variety

Resistance to *A. besseyi* has been found by some researchers although a wide difference in susceptibility exists between varieties. Popova and Dzyuba[58] described a method of screening for resistance to *A. besseyi* under greenhouse conditions using 500 nematodes per plant. The rate of symptom development at 4 to 5 leaf stage is first assessed, followed by reassessment of visually resistant plants for external response and rate of infestation before going to field trials of promising materials. Some of the varieties that have been found highly resistant to *A. besseyi* were Tongil, Early Tongil, Tongil-chal, and Shirogane.[59] Varieties with moderate resistance are Namyeongbyeo[60] and TPS2 from cross IR20 × Co40, which is a semidwarf variety.[61]

9-7-5 Cultural Control

Some of the cultural practices that were recommended to reduce the nematode population are direct sowing of seeds in water rather than sowing followed by flooding,[62] timing of planting in such a way that flowering occurs when there is less rain, or by early sowing, burying of plant debris, straws and weeds,[25] and crop rotation.[63] Ordinary hand transplanting culture can also reduce nematode buildup compared to mechanical transplanting in which seeds are sown densely.[64]

9-8 CONCLUSION

Many aspects of the disease and its pathogen have been studied in detail including its life cycle, biology, and control. However, *A. besseyi* is still being encountered frequently in rice seeds either for export or import despite quarantine regulations. The development of agriculture that calls for increased international exchange of plant germplasms also increases the risk of dispersal of this pathogen. Perhaps we need to improve certification schemes to produce healthy seed, which is the basic requirement of good disease management.

Information on the pathogen's dissemination, survival, and control in the field is still lacking. We also have to formulate seed treatment methods that will not have any adverse effect on the germination and development of the seeds. Correlation between the number of *A. besseyi* in the grain and the performance of the subsequent crop must be determined as well as the tolerance level of rice for seed health evaluation and quarantine regulation.

Further, evaluation of varieties for resistance to *A. besseyi* should be continuously explored and if possible, be integrated in the evaluation scheme for disease resistance.

9-10 REFERENCES

1. Kakuta, T. "Black Grain Disease of Rice." *J. Plant Prot. Tokyo* 2 (1913): 214. (In Japanese)

2. Yoshii, H., and Yamamoto, S. "A Rice Nematode Disease Senchu Shingare byo, I. Symptoms and Pathogenic Nematode. II. Hibernation of *Aphelenchoides oryzae*. III. Infection Course of the Present Disease. IV. Prevention of the Present Disease." *J. Fac. Agric. Kyushu Univ.* 9 (1950): 209, 223, 289, 293.

3. Yokoo, T. "*Aphelenchoides oryzae* n. sp., Parasitic Nematode of Rice." *Ann. Phytopath. Soc. Japan* 13 (1948): 40.

4. Cralley, E. M. "White tip of Rice." *Phytopathology* 39 (1949): 5 (Abstract).

5. Allen, M. W. "Taxonomic Status of the Bud and Leaf Nematodes Related to *Aphelenchoides fragariae* (Ritzema Bos, 1981)." *Proc. Helminth. Soc. Wash.* 19 (1952): 108.

6. Manalo, P. L., ed. "Proposed Common Names for Plant-Parasitic Nematodes." *PLANTI News* 7 (1988): 2.

7. Fortuner, R., and Orton Williams, K. J. "Review of the Literature on *Aphelenchoides besseyi* Christie, 1942, the Nematode Causing 'White Tip' Disease in Rice." *Helminthol. Abst.* Series B 44 (1975): 1.

8. Ou, S. H. *Rice Diseases*, 2nd ed. (Kew, Surrey, UK: Commonwealth Mycological Institute, 1985), 380.

9. Atkins, J. G., and Todd, E. H. "White Tip Disease of Rice, III. Field Tests and Varietal Resistance." *Phytopathology* 49 (1959): 189.

10. Ray, S., Das, S. N., and Catling, H. D. "Plant Parasitic Nematode Associated with Deepwater Rice in Orissa, India." *IRRN* 12 (1987); 20.

11. Rahman, M. "Occurrence of White Tip Disease in Deepwater Rice in Bangladesh." *IRRN* 7 (1982): 15.

12. Todd, E., and Atkins, J. "White Tip Disease of Rice, I. Symptoms, Laboratory Culture of Nematodes and Pathogenicity Tests." *Phytopathology* 48 (1958): 632.

13. Mishizawa, T. "Occurrence of Abnormal Rice Kernels with Black Wedge-Shaped Spots Associated with Infestation of White Tip Nematode, *Aphelenchoides besseyi*." *Jap. J. Nematol.* 61 (1976): 73 (English Summary).

14. Mew, T. W., Bridge, J., Hibino, H., Bonman, J. M., and Merca, S. D. "Rice Pathogens of Quarantine Importance," in *Proc. Int. Workshop on Rice Seed Health*, International Rice Research Institute, Philippines, 1988, 362.

15. Fernandez, M., and Ortega, J. "Distribution of Rice Parasitic Nematodes in Cuba. III. Province in Havana." *Cien. Agric.* 23 (1985): 25.

16. Hung, Y. "White Tip Disease of Rice in Taiwan." *Plant Prot. Bull., Taiwan* 1 (1959): 1 (In Chinese).

17. Prasad, J. S., Panwar, M. S., and Rao, Y. S. "Nematode Problems of Rice in India." *Trop. Pest Manage.* 33 (1987): 127.

18. Rajendran, G., Mutheekrishnan, T., and Balasubramanian, M. "Rice Varieties and the White Tip Nematode." *IRRN* 2 (1977): 3.

19. Zhang, Y. L. "The Occurrence and Control of *Aphelenchoides besseyi*." *Hubei Agric. Sci.* 1 (1987): 15 (In Chinese).

20. Park, S. D., Youn, J. T., Choi, Y. E., Choi, D. W. and Son, S. G. "On the Occurrence Status of the White Tip Nematode *Aphelenchoides besseyi* Christie and the Appearance of Blackened Rice in Kyeong Buk Province." *Res. Rep. Rural Dev. Adm., Plant Envi. Mycology and Farm Prod. Uti., Korea* 29 (1987): 290.

21. Franklin, M., and Siddique, M. K. *Aphelenchoides besseyi*, CIH Descriptions of Plant Parasitic Nematodes, *Set* 1, No. 4 (1972): 3.

22. Sudakova, I. M., and Stoyakov, A. V. "The Reproduction and Longevity of *Aphelenchoides besseyi*." *Zool. Zh.* 46 (1967): 1097 (English Summary).

23. Gokte, N., and Mathur, V. K. "Studies on Embryonic and Post Embryonic Development of *Aphelenchoides besseyi*." *Nematol. Medit.* 17 (1989): 57.

24. Huang, C. S., Huang, S. P., and Lin, L. H. "The Effect of Temperature on Development and Generation Periods of *Aphelenchoides besseyi*." *Nematologica* 18 (1972): 432.

25. Vuong, H. A. "The Occurrence in Madagascar of the Rice Nematodes, *Aphelenchoides besseyi* and *Ditylenchus angustus*" in Nematodes of Tropical Crops, ed. J. E. Peachey. *Tech. Commun. Commonw. Bureau Helminth* 40 (1969): 274.

26. Sudakova, M. I. "Effect of Temperature on the Life Cycles of *Aphelenchoides besseyi*," *Parazitologiya* 2 (1968): 71 (English Summary).

27. Sivakumar, C. V. "Postembryonic Development of *Aphelenchoides besseyi* in Vitro and Its Longevity in Stored Rice Seeds." *Indian J. Nematol.* 17 (1987): 147.

28. Iyatomi, K., and Nishizawa, T. "Artificial Culture of the Strawberry Nematode, *Aphelenchoides fragariae* and the 'White Tip' Nematode of Rice, *Aphelenchoides besseyi*." *Jap. J. Appl. Ent. Zool.* 19 (1954): 8 (In Japanese).

29. Rao, J. "Population Build-up of White Tip Nematode (*Aphelenchoides besseyi* Christie, 1942) in Different Monoxenic Cultures." *Oryza* 22 (1985): 45.

30. Huang, C. S., and Huang, S. P. "Dehydration and the Survival of Rice White Tip Nematode, *Aphelenchoides besseyi*." *Nematologica* 20 (1974): 9.

31. Rao, Y. "Position of the Nematodes in Seeds." Ann. Rep., Cuttack: Central Rice Res. Inst., 185, 1972.

32. Goto, K., and Fukatsu, R. "Studies on White Tip of Rice Plant Caused by *Aphelenchoides oryzae* Yokoo II. Number and Distribution of the Nematode of the Affected Plants." *Ann. Phytopathol. Soc. Japan* 16 (1952): 57.

33. Goto, K., and Fukatsu, R. "Studies on White Tip of Rice Plant, III. Analysis of Varietal Resistance and Its Nature." *B. Natl. Inst. Agric. Sci.* 6 (1956): 123.

34. Fukano, H. "Ecological Studies on White Tip Disease of Rice Plant Caused by *Aphelenchoides besseyi* Christie and Its Control." *Bul. Fuokuoka Agric. Expt. Sta.* 18 (1962): 1 (In Japanese).

35. Chiu, R. "Current Situation and Research of Rice Diseases in Taiwan," in Proc. Integrated Rice Pest Control in Southeast Asia, IRRI, Philippines, 1972.

36. Huang, C. S., and Huang, S. P. "Bionomics of White Tip Nematode, *Aphelenchoides besseyi* in Rice Florets and Developing Grains." *Bot. Bull. Acad. Sinica* (Taiwan) 13 (1972): 1.

37. Cralley, E. M. "Control of White Tip of Rice." *Arkansas Farm Res.* 1 (1952): 6.

38. Todd, E. H., and Atkins J. G. "White Tip Disease of Rice II. Seed Treatment Studies." *Phytopathology* 49 (1959): 184.

39. Zem, A. C., and Monterio, A. R. "Bahia: The White Tip Nematode Also Occurs in Rice Seeds." *Rev. Agric., Piracicaba, Brazil* 52 (1977): 81.

40. Gergon, E. B., and Mew, T. W. "Number of *A. besseyi* in Rice Seeds." *Rice Seed Health Newsl* 1 (1989): 5.

41. Vuong, H., and Rodriguez, H. "Lutte Contre les Nematodes du Riz a Madagascar (Resultats d' Experimentation 1968–1969)." *Agron. Trop.* 25 (1970): 52 (English Summary).

42. Sivakumar, C. V. "The Rice White Tip Nematodes in Kanyakumari District, Tamil Nadu, India." *India J. Nematol.* 17 (1987): 72.

43. Terry, E. R. "The Incidence of the Rice 'White Tip' Nematode in Sierra Leone. A Preliminary Study." *Njala Univ. Coll., Sierra Leone* 2 (1972): 1.

44. McGawley, E. C., Rush, M. C., and Hollis, J. P. "Occurrence of *Aphelenchoides besseyi* in Louisiana Rice Seed and Its Interaction with *Sclerotium oryzae* in Selected Cultivars." *J. Nematol.* 16 (1984): 65.

45. Havens, O. "Nematodes of Phytosanitary Significance to Importing Countries and States," in *Plant Pests of Phytosanitary and Regulatory Significance to Importing Countries and States,* ed. Q. L. Holdeman. (California: Department of Food and Agriculture suppl. 1986), 107.

46. Aluko, M. "Plant Quarantine Requirements for Rice in West Africa," in WARDA Seminar in Plant Protection for the Rice Crop, Monrovia, 1973.

47. Prot, J. C. "Is *Aphelenchoides besseyi* an Important Plant Quarantine Subject?" *Rice Seed Health Newsl.* 1 (1989): 1.

48. Mew, T. W., Gergon, E. B., and Merca, S. D. "Impact of Seedborne Pathogens on Germplasm Exchange," *Seed Sci. Technol.* 18, 1990, 441.

49. Kahn, R. P. "The Importance of Seed Health in International Seed Exchange," in *Proc. Int. Workshop on Rice Seed Health,* Los Banos: International Rice Research Institute, 1988, 362.

50. Caubel, G. "Epidemiology and Control of Seed-Bourne Nematodes." *Seed Sci. Technol.* 11 1983, 989.

51. Aleksandrova, I. A. "Efficacy of Irradiation of Rice Seeds Infected with *Aphelenchoides besseyi* Christie 1942 with 60Co Gamma Rays." *Byull. Vsesoyuznogo Inst. Gelmintol. K. I. Skyabina* 41 (1985): 5.

52. Nakamura, S. "Treatment of Rice Seeds in Japan," in *Seed Treatment*, 2nd ed., K. A. Jeffs. ed (Great Britain: Lavenham Press Ltd., 1986), 332.

53. Cho, S. S., Han, M. J. and Yang, J. S. "Chemical Control of Rice White Tip Nematode (*Aphelenchoides besseyi* C.) by Seed Disinfectant and in the Paddy Field." *Korean J. Plant Prot.* 26 (1987): 107.

54. Sivakumar, C. V. "Disinfestation of White Tip Nematode in Rice Seeds." *Indian J Nemathol.* 17 (1987): 148.

55. Prasad, J. S., Panwar, M. S. and Rao, Y. S. "Chemical Control of White-Tip Nematode, *Aphelenchoides besseyi.*" *Oryza* 24 (1987): 391.

56. Chumakova, E. I., Guschin, E. E., Doroshenko, N. D., and Popova M. B. "Etaphos—An Effective Preparation against *Aphelenchoides besseyi* on Rice." *Byull. Vsesoyuznogo Inst. Gelmintol. K. I. Skryabina* 41 (1985): 80.

57. Nagy, J. "Increased Efficacy of Seed-Dressing Agents by Ultrasound Exposure of Rice Nematodes." *Inter. Agrophysics* 3 (1987): 291.

58. Popova, M. B., and Dzyuba, V. A. "Methodology for Rapid Diagnosis of Resistance to Rice Leaf Nematode in Rice." *Byulleten Vsesoyuznogo Inst. Gelmintol. K. I. Skryabina* 45 (1986): 27.

59. Park, J. S. and Lee. J. O. "Varietal Resistance of Rice to the White-Tipped Nematode *Aphelenchoides besseyi.*" *IRRN* 1 (1976): 15.

60. Sohn, J. K., Lee, S. K., Koh, J. C., Kim, H. Y., Yang, S. J., Hwang, H. G., Hwang D. Y., and Chung, G. S. "A New Rice Variety with Multiple Resistance to Diseases and Insect Pests: Namyeongbyeo." *Res. Rep. Rural Dev. Adm. Crops, Korea* 29 (1987): 18.

61. Kalaimaini, S., Pillai, O. R., Manuel, W. W., and Subramania M. "TPS2—A New Variety for Kanyakumari." *IRRN* 12 (1987): 4.

62. Cralley, E. M. "A New Control Measure for White Tip." *Arkans. Fm. Res.* 5 (1956): 5.

63. Tikhonova, L. V. "*Aphelenchoides besseyi* Christie 1942 (Nematoda Aphelenchoididae) on Rice and Method of Control." *Zool. Zh.* 45 (1966): 1759 (English Summary).

64. Kobayashi Y., and Sugiyama, T. "Dissemination and Reproduction of White Tip Nematode, *Aphelenchoides besseyi* in Mechanical Transplanting Rice Culture." *Jap. J. Nematol.* 7 (1977): 74 (English Summary).

10

DOWNY MILDEWS OF MAIZE

S. S. BAINS AND H. S. DHALIWAL

Punjab Agricultural University, Regional Research Station,
Gurdaspur-143 521, Punjab, India

10-1 INTRODUCTION

Downy mildews of maize are caused by specialized biotrophic fungi of the Peronospora-ceae. They are a principal factor limiting maize production in Asia and certain other regions of the world. Some of them also parasitize species of *Sorghum, Saccharum, Pennisetum, Eleucine,* and *Setaria* and cause significant losses. Java downy mildew is the first downy mildew recorded on maize. Sugarcane downy mildew, sorghum downy mildew, graminicola downy mildew, crazy top downy mildew, philippine downy mildew, and spontaneum downy mildew were, subsequently, recorded between 1902 and 1921.[1] Brown stripe downy mildew was reported in 1967.[2] Two more fungal species were reported between 1980 and 1984 to cause maize downy mildews.[3-4] *P. sorghi* occurs in wider areas in India and other parts of the world (Figure 10-1) as compared with other species of *Peronosclerospora*. It has been reported from Argentina, China, Egypt, East Africa, India, Israel, Italy, Kenya, Malwi, Nigeria, Philippines, Rhodesia, South Africa, Sudan, Tanzania, Thailand, Uganda, and Zambia.

10-2 THE PATHOGENS

Eleven different fungal species belonging to three genera of Peronosporaceae are reported to cause downy mildew of maize. Eight of the species belong to the genus *Peronosclerospora*, two to *Sclerophthora*, and one to *Sclerospora*. The present binomials

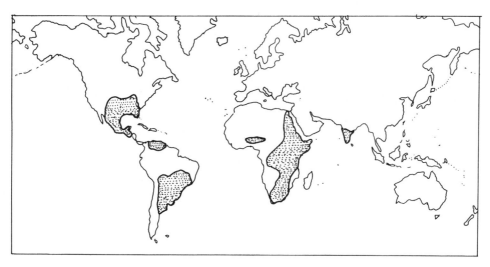

Figure 10-1   Distribution of *Peronosclerospora sorghi* in different parts of the world.

of different downy mildew pathogens, their synonyms, and the common names[5] of downy mildews caused by them are given in Table 10-1.

10-2-1 Genera: *Sclerospora, Peronosclerospora,* and *Sclerophthora*

De Bary (1881) was the first to include *Sclerospora* in the valid list of genera of the Peronosporaceae, though it was recognized as a subcategory with *Peronospora* as early as 1879 by Schroeter.[1] Ito in 1913, realizing differences in asexual spores of species within the genus, divided it into subgenera, *Sclerospora* and *Peronosclerospora*. Shirai and Hara[6] later raised the subgenus *Peronosclerospora* to generic level. Shaw[7] finalized this split and transferred conidia forming species in *Sclerospora* into *Peronosclerospora*. The sporangia forming species and the species where asexual spores were not known were retained in the genus *Sclerospora*. This classification has been followed in several recent publications.

  Sclerophthora macrospora was first described as *Sclerospora macrospora*. Later, it was found that sporangial phase of *S. macrospora* resembled that of *Phytophthora*. Thirumalachar et al.[8], realizing significant differences in the asexual and sexual phases of the species, subdivided it and erected a new genus, *Sclerophthora*. The separate identity of this genus has now been accepted. Within *Sclerophthora, S. macrospora* appears in aggregate species as oospores within it vary in size and structure. Information on germination of oospores of this species may decide its eventual taxonomical position.

10-2-2 Relationship among the Three Genera

An evolutionary trend is traceable within the genera *Sclerophthora, Sclerospora,* and *Peronosclerospora*. This, along with their possible relationship with other members of Peronosporaceae, is presented in Figure 10-2. *Sclerophthora macrospora*, with sporangia germinating through zoospores and having the least differentiated sporangiophores, ap-

TABLE 10-1: BINOMIALS AND SYNONYMS OF MAIZE DOWNY MILDEW PATHOGENS AND COMMON NAMES OF DOWNY MILDEWS CAUSED BY THEM

Present Binomial	Synonyms	Common name
Peronosclerospora maydis (Racib.) C. G. Shaw	*Peronospora maydis* Racib. *Sclerospora maydis* (Racib.) Butl. *Sclerospora javanica* (Palm)	Java downy mildew
Peronosclerospora miscanthi (Maiyake) C. G. Shaw	*Sclerospora miscanthi* Miyake	Leaf-splitting downy mildew
Peronosclerospora philippinensis (Weston) C. G. Shaw	*Sclerospora philippinensis* Weston *Sclerospora indica* Butl.	Philippine downy mildew
Peronosclerospora sorghi (Weston & Uppal) C. G. Shaw	*Sclerospora graminicola* (Sacc.) de By. *Sclerospora graminicola* var. *andropogonis-sorghi* Kulkarni *Sclerospora sorghi* Weston & Uppal *Sclerospora andropogonis-sorghi* (Kulkarni) Mundkur *Sclerospora-vulgaris* (Kulkarni) Mundkur	Sorghum downy mildew
Peronosclerospora spontanea (Weston) C. G. Shaw	*Sclerospora spontanea* Weston	Spontaneum downy mildew
Peronsclerospora sacchari (Miyake) Sharai and K. Hara	*Sclerospora sacchari* Miyake	Sugarcane downy mildew
Peronosclerospora heteropogoni Siradhana, Dange, Rathore, & Singh	*Sclerospora sorghi* Weston & Uppal	Rajasthan downy mildew.*
Peronosclerospora sp. Nov. Prabhu et al.	—	—
Sclerospora graminicola (Saac.) de By.	*Protomyces graminicola* Sacc. *Peronospora graminicola* Sacc. *Peronospora setariae* Pass. *Ustilago* (?) *urbani* Magnus	Graminicola downy mildew
Sclerophthora macrospora (Sacc.) Thirum. Shaw & Naras.	*Scloerospora macrospora* Sacc. *Phytophthora macrospora* (Sacc.) Ito & Tanaka *Scloerospora kriegeriana* Magnus *Sclerospora aryzae* Brizi *Nozenia macrospora* (Sacc.) Tsugi	Crazy top downy mildew
Sclerophthora rayssiae var. *zeae* (Payak & Renfro)	—	Brown stripe downy mildew (BSDM)

*Proposed one.

214

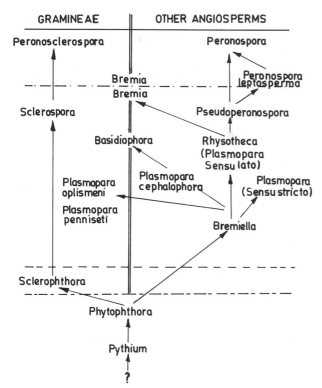

Figure 10-2   Phyllogeny of the Peronosporales (from Shaw) [7]

– · – · – Above: Germination always direct (germ tube; no operculum; conidia)

Below: Germination usually indirect (zoospores; operculum present; sporangia)

– – – – – Above: Obligate parasites

Below: Facultative saprophytes

– – – – – Above: Determinate conidiophores/sporangiophores

Below: Indeterminate sporangiophores

In the asexual state of *Bremia* an operculum is present, and thus a sporangium is produced. Although liberation of zoospores has been reported, most workers report germ tube germination.

One species of *Basidiophora* and *Bremia* each and two species of *Plasmopara* are reported on Gramineae.

pears the most primitive species among maize downy mildew pathogens. It is followed by relatively more evolved species, *Sclerospora graminicola*, having clearly differentiated sporangiophores and with sporangia germinating either through zoospores or directly through germ tubes. *Peronosclerospora* appears the most advanced genus as species within it develop clearly differentiated conidiophores and have conidia which germinate exclusively through germ tubes.[7] Within *Peronosclerospora*, *P. sacchari* is believed to have been evolved from *P. spontaneum* and *P. philippinensis* likewise is considered to be evolved from *P. miscanthi*.[9] The method of oospore germination, which remains to be investigated in detail, may provide further clues to the relationships among these genera.

10-2-3 Asexual Phase

Morphology.   The conidiophores or sporangiophores of species of *Sclerophthora*, *Sclerospora*, and *Peronosclerospora* emerge, as a rule, through stomata on the host surface. The sporangiophores of *Sclerophthora* species are least differentiated from the vegetative hyphae. They lack the extensive branching system characteristic of other members of Peronosporaceae and develop sporangia basipetally. The development of sporangia/conidia in *Peronosclerospora* and *Sclerospora* is typically basifugal. Sporangiophores of *S. graminicola* are distinguishable from those of *P. sorghi* (Figure 10-3) and, apparently, from those of *P. maydis, P. philippinensis, P. sacchari,* and *P. spontanea* also (Figure 10-4). In *S. graminicola* there exists a definite continuation of the main

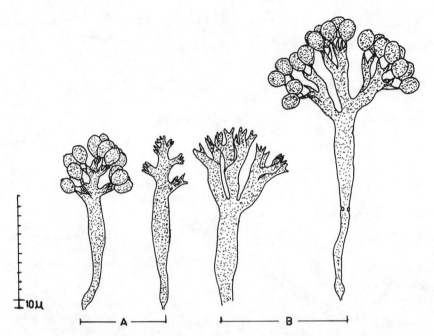

Figure 10-3 Comparative morphology of sporangiophores of *Sclerospora graminicola* (A) and conidiophores of *Peronosclerospora sorghi* (B) (from Weston and Uppal, *Phytopathology* 22, 1932: 573).

axis which branches abruptly and at irregular intervals so that conidia are placed at irregularly disposed distances. In *P. sorghi*, in contrast, branches are so borne that conidia developing on the tips of branchlets are in a hemispherical plane (Figures 10-3 and 10-4). The dimensions of sporangiophores/conidiophores of different downy mildew pathogens vary and are given in Table 10-2. It should be borne in mind that the morphology of conidiophores and conidia is influenced by the host cultivars and the prevailing environmental conditions.[10] Kimigafukuro[11] reported that the length and width of conidia of *P. sorghi* increased as the temperature increased from 14°C to 28°C. Length of *P. philippinensis* and *P. sacchari* conidia also increases with rise in temperature. Leu[12] found that the length of *P. sacchari* conidia was 41 to 45 μm at 22°C to 30°C, 36 μm at 18°C, and 29 to 30 μm at 10°C to 14°C. Rao et al.[13] revealed that the morphology of conidia and conidiophores of *P. sorghi* varied with the host genotype. Leu[12] observed significant differences in size of conidia of *P. sacchari* produced on different host cultivars. Various other factors that may affect spore morphology and size include the host species, time and place of spore collection, and the medium in which spores are mounted for measurements.

Factors affecting sporulation.   The process of sporangiophore/conidiophore initiation to the eventual production of mature sporangia/conidia takes several hours and is controlled by different environmental conditions. The role of important factors in this process is discussed as follows.

Temperature.   Pupipat[14] showed that *P. sorghi* did not develop conidia at 16°C or

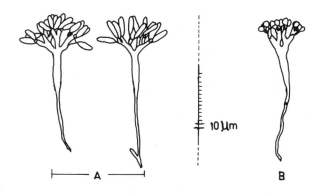

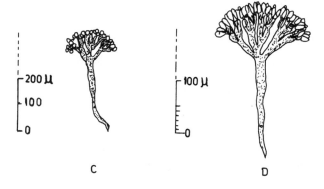

Figure 10-4 Conidiophores of *Peronosclerospora spontanea* (A), *P. sorghi* (B), *P. maydis* (C), and *P. philippinensis* (D) (A and B from Titatarn and Syamananda, *Plant Dis. Reptr.* 62, 1978:29; C and D from Semangoen[17]).

TABLE 10-2: MORPHOLOGICAL CHARACTERISTICS OF ASEXUAL SPORES OF DIFFERENT MAIZE DOWNY MILDEW PATHOGENS

Pathogen	Nature of spores	Shape of spores	Size of spores (μm)	Size of sporangiophores (μm)
P. sorghi	Conidia	Globose, ovoid-round	15–29 × 15–27	180–300
P. heteropogoni	Conidia	Globose, ovoid-round	17.7 × 16.2	87.2–142.8
Peronosclerospora spp.	Conidia	Round	15.6 – 18.7	–
P. maydis	Conidia	Round	28–45 × 16–22	150–200
P. sacchari	Conidia	Long, cylindrical-ovoid	24–35 × 11–23	190–280
P. miscanthi	Conidia	Long, cylindrical-ovoid	31–48 × 14–23	97–300
P. philippinensis	Conidia	Long, cylindrical-ovoid	14–44 × 11–27	150–400
P. spontanea	Conidia	Long, cylindrical-ovoid	25–65 × 11–21 (39–45 × 15–17)	350–500
S. graminicola	Sporangia	Globose, ovoid-round	19–21 × 12–21	100
S. macrospora	Sporangia	Elongated, apex, truncate or rounded poroid	60–100 × 43–64	—
S. rayssiae var. *zeae*	Sporangia	elongated, apex, truncate or rounded poroid	29–66.5 × 18–26	—

at 30 °C. Maximum conidia developed at 24 °C to 26 °C. Optimum temperature for sporulation of *P. sacchari* was 22 °C to 26 °C.[12] Sun[15] found that *P. sacchari* did not develop conidia at 13 °C and 31 °C. Bonde and Melching,[16] on the other hand, obtained heavy sporulation of *P. sacchari* at a temperature range of 15 °C to 23 °C. Conidia of *P. maydis* developed when the night temperature was ⟨24 °C.[17] Highest sporulation of *P. philippinensis* was obtained at 21 °C to 26 °C[18] or at 20 °C to 28 °C.[19] Minimum, maximum, and optimum temperatures for sporulations of *S. rayssiae* var. *zeae* were ⟨15 °C, ⟩35 °C, and 22 to 25 °C, respectively.[20]

Relative Humidity and Moisture.   Without exception, downy mildew pathogens require high RH for sporulation. *P. sacchari* and *P. sorghi/P. philippinensis* sporulate well at RH regimes of 86% and 90%, respectively.[14,18,21] The presence of free moisture was required for sporulation of *P. sorghi*, *P. maydis*, *S. rayssiae* var. *zeae*, and *P. philippinensis*.[17,20] However, the excessive moisture on the plant surface or rain during or immediately prior to sporulation delays, reduces, or completely inhibits sporulation.[11]

Light.   Light during the presporulation period enhances sporulation. It has been reported that a 5–6 hr presporulation period of light was required for sporulation of *P. sorghi*. Light during the period of sporulation, however, appears to adversely affect sporulation. Sporulation of *P. sacchari* was reduced with increasing light intensity during the sporulation period.

Photosynthesis.   Sporulation is closely related to photosynthesis of the host in several members of Peronosporaceae. This aspect of maize downy mildew pathogens has not been thoroughly investigated. Increase in sporulation due to presporulation exposure of infected plants to light can be expected to be through its effects on host photosynthesis.

Host, Host Nutrition.   Sporulation of species of *Peronosclerospora* is affected by the host species/cultivars and the physiological age of the tissues. *P. sorghi* (pathotype-3) did not sporulate on sorghum line, QL–3.[22] *P. sacchari* produced more conidia on younger than on older maize leaves.[23] The apical portions of diseased leaves likewise, supported more conidia than the basal parts of the same leaf. *S. macrospora* developed more sporangia on young, succulent leaves, and sporangial production decreased as the leaves increased in age.[24]

Gupta and Siradhana[25] reported that deficiency of nitrogen, phosphorus, and potassium did not affect sporulation of *P. sorghi* on maize.

Diurnal periodicity in sporulation.

Maize downy mildew pathogens, like other members of Peronosporaceae, exhibit diurnal periodicity in sporulation.[26] Conidial production in *Peronosclerospora* spp., for instance, initiates during or after midnight (24.00–02.00 hr). Conidia are released during early morning hours (0200–0600 hr) with optimum time between 0200–0400 hr. Singh and Renfro[27] trapped maximum sporangia of *S. rayssiae* var. *zeae* between 1200–1600 hr. The favorable conditions for sporulation that prevail during the night time apparently govern the periodicity of sporulation.[28] If these conditions are made available, the pathogens sporulate independent of the diurnal rhythm. The critical factors include the temperature, moisture periods, light, and time lapse between successive crops of spores.[29]

Discharge and dispersal of conidia.   Weston[30] reported the forcible discharge of conidia of *P. philippinensis*. Usually all conidia were released at the same time from a given conidiophore and propelled to an estimated distance of 1 to 2 mm. This distance allowed spores to be caught by air currents. Given that sporangia and conidia are short lived, their dissemination over long distances appears unlikely. Rajasah et al.[31] showed that conidia of *P. sorghi* reached 80 m from the source. Tantera[32] reported that most of *P. maydis* conidia remained within a 16 m radius from the source. Mikoshiba et al.[33] concluded that secondary infections from single disease cycle were limited to 42 m from the inoculum source. Sun[15] found that susceptible maize varieties were seldom infected when located more than half a mile from the inoculum source. Bains and Jhooty[34] observed that the spread of *P. philippinensis* was not equal in all directions. It was maximum in the downwind directions, and the resulting gradients of infection (Figure 10-5) persisted till harvest of the crop.

10-2-4 Sexual Phase (Oospores)

Oospore production.   Species of *Sclerophthora*, *Sclerospora*, and *Peronosclerospora*, with the exception of *P. maydis*,[17] are reported to develop oospores in maize tissues.[35-37] The oospore production depends on several factors including the host species or varieties, the pathogen and its isolate, and the prevailing weather conditions.[37] Because of these reasons, the formation of oospores by *Peronosclerospora* spp. probably has been found present by certain[38] and absent by other workers.[34,39,40] *P. sacchari* produced oospores more readily and more abundantly in sugarcane than in maize.[14] *S. rayssiae* var. *zeae* developed oospores in *D. sanguinalis* earlier in the season than in maize leaves.[41] *S. macrospora*, which formed abundant oospores in wheat, infected maize but formed no oospores.[42]

Spread of oospores.   Spread of oospores, locally or to geographically separated areas, may occur through the agency of wind, animals, insects, running and splash-

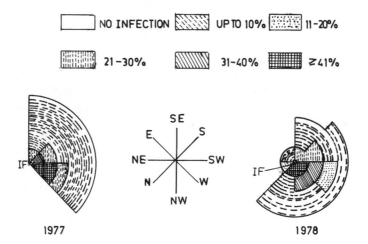

Figure 10-5 *Peronosclerospora philippinensis* infected plants of cv. MS-1 of maize around infection focus (IF).

ing water, and soil. Oospores adhered to seeds or those on leafbits in the seed may also be carried along to distant places. Rao et al.[13] found oospores of *P. sorghi* in pericarp, endosperm, and embryo of maize seeds. This phenomenon ensures the long-distance spread of *P. sorghi* along with the maize seed.

Germination of oospores. The mode of oospore germination in maize downy mildew pathogens varies with the organism or probably also with the conditions available. Oospores of *S. macrospora* germinate to develop a sporangiophore with a terminal sporangium.[43] Oospores of *Peronosclerospora* form coenocytic germ tubes[44] or zoospores[39] or spherical bodies.[29] Oospores of *Sclerospora graminicola* germinate directly by putting out a germ tube[35] or a germ sporangium.[45] Oospore germination in the soil depends on a variety of factors, including their release from the tissues, soil moisture, and soil temperature. External stimuli in the form of synthetic media are reported to have no effect on germination of oospores of *P. philippinensis, P. spontanea, P. sacchari,* and *P. miscanthi.* However, stimuli from plant roots in the soil are reported to have positive effects on oospore germination.[14,29,46] Oospores of *S. graminicola* exhibit no dormancy,[45] whereas there are conflicting reports regarding the dormancy in oospores of *P. sorghi.*[40]

10-2-5 Physiological Specialization

The published information, though not extensive, reveals the presence of physiological specialization in maize downy mildew pathogens. Two to three pathotypes have been reported in *P. sorghi*[4,48] and six in *P. philippinensis.*[49] Gowda et al.[50] found that *P. sorghi* from maize and sorghum was different pathologically. Bonde et al.[51] revealed that an isolate of *P. sorghi* from Thailand tolerated higher temperature, compared with those from India, the US, and Brazil. The wider range of temperatures required for germination of *P. sorghi* conidia probably also indicates the occurrence of ecological races in this pathogen. Circumstantial evidence reveals pathological variation in *S. rayssiae* var. *zeae* also.

10-2-6 Growth in Tissue Culture and Axenic Culture

A limited success has been achieved in culturing maize downy mildew pathogens in tissue or in axenic cultures. Tiwari and Arya[52] and Arya and Tiwari[53] successfully cultured *S. graminicola* in callus tissues of *Pennisetum typhoides* raised on a synthetic medium. Safeeulla[29] cultured *P. sorghi* in the sorghum callus supported on the modified White's basal mineral salts. *S. macrospora* was isolated from rice plants on to potato sucrose, watermelon sucrose, and oatmeal sucrose agar media.[54–55] *S. graminicola* was cultured axenically on the medium used for the growth of callus tissues.[56] Efforts to culture *S. rayssiae* var. *zeae* on sucrose potato, sucrose oatmeal, and sucrose watermelon media,[35] and *P. sorghi* on media used for growing sorghum callus did not succeed.[29,55]

10-2-7 Host Range

All known hosts of *Peronosclerospora* spp. and *Sclerospora graminicola* belong to Andropogoneae and Maydeae. Hosts of *S. macrospora* are found in assemblages belonging to 46 species in 31 genera of Gramineae, in addition to *Cyperus*.[2] The names of host plants are as follows:

Agropyron repens, A. tsukushiense var. *transsiensiens, A. smithii, A. cristatum, A. trachycaulun, Agrostis clavata* var. *nukabo, A. hiemalis, A. palustris, A. stolonifera, A. alba, Alopecurus agrestis, Alopecurus* sp., *A. aequalis, A. fulvus, Andropogon nardus* var. *georingii, Anthraxon hispidus, A. ciliaris, Arrhenatherum elatius, Arundinella hirta, A. anomala, Avena fatua, A. sativa, A. nigra, Axonopus compressus, Beckmannia syzigachne, Brachiaria mutica, B. reptans, Bromus japonicus, B. commutatus, B. inermis, Cymbopogon tortilis* var. *georingii, Crypsis aculeata, C. areopecuroides, C. schonoides, Dactylis glomerata, Dactyloctenium aegyptium, Digitaria sanguinalis, Digitaria marginata* var. *fimbriata, D. ischaemum, D. violascens, D. adscendens, Dinebra retroflexa, Echinochloa crus-galli, E. crusgalli* var. *hispidula, E. colonum, Eleusine coracana, E. indica, Elymus macounii, Elytrophorus spicatus, Eragrostis cilianensis, E. pectinacea, E. ferruginea, E. niwahokori, E. major, E. multicaulis, E. nutans, Euchlaena mexicana, E. luxurians, Festuca elatior, F. arundanacea, Glyceria martina, G. acutiflora, G. festucaeformis, Hemarthria compressa, Holcus lanatus, H. sorghum* var. *japonicus, Hordeum jubatum, H. sativum* var. *vulgare, H. sativum* var. *hexastichon, Imperata cylindrica* var. *koenigii, I. arundianacea, Iseilema laxum, Leersia japonica, Lolium perenne, L. multiforum, L. temulentum, Microstegium vimineum, M. sinensis, Muhlenbergia asperifolia, Miscanthus floridulus, Oplismenus undulatifolius* var. *japonicus, O. burmanni, Oryza sativa, Panicum antidotale, P. bisulcatum, P. violascens, P. capillare, P. sanguinale,* var. *ciliare, P. barbides, P. crus–galli* var. *frumentaceum, P. maximum, P. crus-galli* var. *submuticum, P. virgatum, P. indicum* var. *oryzetorum, P. miliaceum, P. acroanthum, Pennisetum alopecuroides, P. pedicillatum, P. typhoides, P. purpureum, Paspalum thunbergii, Phalaris arundinacea, P. canariensis, P. coerulescens, P. tuberosa, Phleum pratense, Phragmites communis, P. coerulescens, P. longivalvis, Poa pratensis, P. acroleuca, P. annua, P. nipponica, Pollinia imberbis* var. *genuina, Polypogon interruptus, Rottboellia compressa, Sacciolepsis indica* var. *oryzetorum, S. interrupta, Schedonnardus paniculatus, Saccharum officinarum, Secale cereale, Setaria lutescens, S. viridis, S. glauca, S. pumilla, S. viridis* var. *purpurascens, S. verticillata, S. tomentosa, Sorghum vulgare, S. vulgare* var. *sudanense, S. halepense, S. verticilliforum, Sporobolus neglectus, S. elongatus, Stenotaphrum secundatum, Syntherisma sanguinalis, Triticale, Triticum aestivum* subsp. *vulgare, T. durum, T. vulgare* and *Zea mays, Zizania latifolium, Zea mays* × *E. mexicana* hybrids.

S. graminicola parasitize: *Agrostisalba, Chaetochloa magna, C. viridis, Echinochloa crus-galli, E. crus-galli* var. *frumentacea, Eucheaena luxuriana, E. mexicana, Panicum miliaceum, P. viridis, Pennisetum leonis, P. spicatum, P. glaucum, P. typhoides, Saccharum officinarum, Setaria glauca, S. italica, S. lutescens, S. magna, S. verticillata, S. viridis, Sorghum halepense, S. sudanense, S. verticilliflora, S. vulgare,* and *Zea mays.*

Hosts other than maize of different downy mildew pathogens (excluding *S. macrospora* and *S. graminicola*) are as follows:

S. rayssiae var. *zeae: Digitaria sanguinalis, D. bicornis, Z. mays* var. *luxuriana, Z. perennis, Z. mays* ssp. *mexicana (Euchlaena mexicana)* and *Z. mays* × *Z. mays* spp. *mexicana.; P. maydis: E. mexicana,* species of *Tripsacum* and *Pennisetum* and *E. mexicana* × *Z. mays* hybrids, *Sorghum plumosum; P. miscanthi: Sorghum phemosum, Mis-*

canthus spp., *Saccharum* spp.; *P. philippinensis: Sorghum* spp., *Avena sativa, E. mexicana, Saccharum officinarum, S. spontaneum, Sorghum* spp.; *P. sacchari: Saccharum* spp., *E. mexicana, Sorghum* spp., *T. lactyloides; P. sorghi: S. bicolor, S. controversum, S. hewisonii, S. niloticum, S. plumosum, S. propinguum, S. pugionifolium, S. sudanense, S. versiolour, S. mexicana, S. vulgare, S. arundinaceum, S. caffrosum, S. halepens, S. verticilliflorum, S. spontaneae, Saccharum* spp., *E. mexicana, M. japonicus, Sorghum* spp.; *P. heteropogoni: Heteropogon contortus.*

10-3 SYMPTOMS

S. rayssiae var. *zeae* develops local infections (Figure 10-6) in the form of chlorotic to yellow streaks of varying lengths. They turn purplish with age. Severe infections result into blighting of the foliage. Malformation of plants and shredding of leaves, as is found in certain other downy mildews, does not occur. The infected tissues may support abundant oospores. Whitish "down" is visible but not as conspicuously as is the case with *Peronosclerospora* spp.

 P. philippinensis principally causes systemic infections. Long, light yellow stripes (Figure 10-7a,b) become conspicuous first in the basal regions of the young and newly developed leaves. Stripes grow in length and in severe cases the entire leaf becomes yellow (Figure 10-7b). Systemically infected plants are severely stunted, especially when infection develops in early growth stages of seedling development. Tassels of plants may become variously malformed. Infected plant parts support a conspicuous white growth of conidiophores and conidia (Figure 10-7(a)). In resistant plants yellowish green, discontinuous stripes may develop in place of typical yellow ones. Such stripes are either free of sporulation or they support traces of sporulation.

 S. macrospora induces varied types of symptoms. This variation in symptoms is considered to result from the time of infection in relation to developmental stage of the

Figure 10-6 Symptom of BSDM caused by *S. rayssiae* var. *zeae*.

TABLE 10-3:   OPTIMUM, MINIMUM, AND MAXIMUM TEMPERATURES FOR THE GERMINATION OF SPORANGIA/CONIDIA OF DOWNY MILDEW PATHOGENS OF MAIZE

Pathogen	Temperature (°C)*		
	Optimum	Minimum	Maximum
P. philippinensis	19–20	16	28
	—	12	32
P. sacchari	25	10	34
P. sorghi	21–25	10	32
	10–19	10	27
S. graminicola	16–22	4	32
S. rayssiae var. *zeae*	22–25	—	—

*The figures with respect to different pathogens from *P. philippinensis* to *S. rayssiae* var. *zeae* in that order are from Exconde,[67] Bonde and Melching,[16] Sun,[15] Safeeulla,[29] Bonde et al (*Phytopathology* 68, 1978:219), Suryanarayana (*Bull. Indian Phytopathol. Soc.* 3, 1966:72), and Singh et al.[71]

Figure 10-7 Symptom of downy mildew caused by *S. phillipinensis* (a) downy growth (b) chlorotic stripes.

host and, perhaps, the extent of colonization of the young plants by *S. macrospora*.[36] The earliest symptoms include excessive tillering and rolling and twisting of the upper leaves. The failure of the upper leaves to unfurl may result in buckling, bending, and distortion of the plants. Leaves of severely infected plants are narrow, straplike, and leathery in texture. Stunting of infected plants is a common symptom but the most pronounced symptom is the replacement of normal tassels by leaves. The name "crazy top" (Figure 10-8) of this down mildew is given due to these symptoms. Other symptoms include the formation of green islands, the chlorotic stripes, and elongation of ear shanks.[36]

Yellowing develops from the basal regions of the leaves toward their tips in plants systemically infected by *P. sorghi*. The infected leaves support a thick white 'down'. Infected plants produce smaller cobs and malformed tassels.

P. sacchari, like *P. philippinensis*, develops chlorotic stripes supporting conspicuous white 'down' on maize leaves. The stripes due to *P. sacchari* infection are, however, reddish yellow in color. Infected maize plants exhibit varied types of distortions and malformations. Compared to healthy plants, the infected ones support small, poorly filled but more ears.

10-4 EPIDEMIOLOGY

10-4-1 Overwintering and Next Season Infections

Oospores.   With the exception of *P. maydis*, all other maize downy mildew pathogens produce oospores. The role of oospores in initiating fresh infections (Fig.10-9) in nature and under artificial inoculation conditions has been established beyond doubts.[57] Oospores seem to be the sole source of primary inoculum of *S. rayssiae* var. *zeae* in Punjab and, probably, elsewhere also.

Collateral and overwintering hosts.   *Sorghum halepense* and *S. bicolor* support oospores and conidia of *P. sorghi* in North America where they serve as a primary source of inoculum for infection of maize and sorghum. Kenneth and Klein[58] reported that

Figure 10-8 "Crazy top" symptom caused by *S. macrospora*.

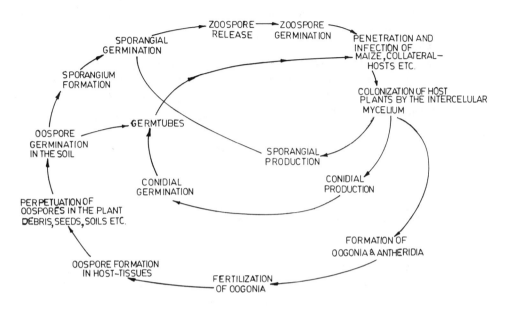

Figure 10-9 Disease cycle of downy mildew of maize.

P. sorghi overwinter in rhizomes of Johnson grass in Israel. Perennial wild sorghum that commonly grows along water channels serves to provide *P. sorghi* inoculum in Vanzuella. *H. contortus*, a wild host, supporting conidia and oospores of *P. sorghi* and *P. heteropogoni*, provides inocula of these pathogens in Rajasthan, India.[3,59] *P. philippinensis* parasitizes *S. spontaneum* in North India[34] where it develops asexual spores, most actively during June to November. In winters, conidia do not develop, but mycelium of the pathogen survives in the infected plants. Thus, in the absence of oospores and other overwintering hosts, the primary inoculum of *P. philippinensis* in this region originates from *S. spontaneum*.[34] The continuity of cultivation of maize in different months of the year serves to maintain inoculum of *P. philippinensis* across crop seasons in the Philippines.[60] Sugarcane is the only alternative host of *P. sacchari* in Taiwan. Spread of *P. sacchari* from sugarcane to maize in its proximity has been shown by Sun.[15] *P. maydis* is considered to perpetuate on wild grasses in Java. Cultivation of successive crops of maize around the year also help in perpetuation of this pathogen.[17] *S. rayssiae* var. *zeae* perpetuates in its oospore phase. It is known to develop oospores in leaves of maize[2] and *D. sanquinalis*,[41] a weed host in India. *S. macrospora* parasitizes a wide range of plants and develops oospores which help in perpetuation of this pathogen. The authors successfully inoculated maize and oats with *S. macrospora* oospores from wheat.

Transmission through seeds. The presence of mycelia of *P. maydis*, *P. philippinensis*, *P. sacchari*, *S. rayssiae* var. *zeae*, *P. sorghi*, and *S. macrospora* in maize seeds has been reported[17,34,61-64] and downy mildew supporting plants from such seeds have been obtained. Transmission of these pathogens through infected seeds has been found to depend principally on moisture content of the seeds.[63,65] The critical moisture level at which no downy mildew transmission occurs depends on the cultivar–pathogen system. Summartaya et al.[66] found that the seed moisture level related to no-transmission of downy mildew varied from 10.0% to 18.5% in different varieties. Survival of downy mildew pathogens between seasons through this source appears less likely as dry seeds are stored and such seeds fail to develop downy mildew supporting plants. However, *P. sorghi* where oospores are formed within the maize seed[13] appears an exception to this. Seed transmission of *S. macrospora* does not appear important epidemiologically as the small ears from infected maize plants are usually discarded during processing.[36]

10-4-2 Onset and Progress of Downy Mildew Epiphytotics

Onset of downy mildew epidemics in nature depends on inoculum availability and the prevailing favorable weather factors. In *Peronosclerospora* spp., conidia are dispersed soon after their production. They establish infections before the unfavorable weather commences in the daytime to follow. The brevity of infection periods enables these pathogens to get benefit of even very short spells of favorable conditions. In case of *P. philippinensis* the presence of inoculum in abundance, temperature, and RH appear as the more important factors in its overall progress in nature.[67] The incidence of systemically infected plants due to *P. philippinensis* was positively correlated with rainfall. Solar radiations did not relate with the progress of this downy mildew.[67] In case of *P. sacchari*, the infection rates were high when the night temperature was 12°C to 28°C.[68] Frederiksen[69] stated that *P. sorghi* development in the United States in 1967 was favored partly by cool, moisture-saturated air which prevailed for 5 days, i. e., during and shortly after the emergence of plants. Govindu et al.[38] reported that moist, cool, and cloudy weather with intermittent rainfall favored *P. sorghi* development on maize. Payak et al.[35] observed that severity of brown stripe downy mildew depended on rainfall. The disease intensity was low where the annual rainfall varied from 40–60 cm. It was moderate in areas with 60 to 100 cm rainfall and high in regions with 100–200 cm of rains.

Isolated reports are available on the role of host nutrition in downy mildew development on maize. Goor[70] reported that an overdose of nitrogen increased susceptibility of maize plants to *P. maydis*, whereas potassium had an opposite effect on this downy mildew. Singh et al.[71] reported that zinc deficiency predisposed maize plants to increased susceptibility to *P. sacchari* and *S. rayssiae* var. *zeae*. Application of nitrogen, phosphorus, potash (NPK) mixture with foliar application of zinc, iron, boron, copper, and molybdenum reduced incidence of these pathogens. Gupta and Siradhana[72] reported that there was less disease in nitrogen-deficient plants, but absence of phosphorus increased disease resistance. There was more disease in potash-deficient plants.

Soil moisture also affects downy mildew infection levels. Balasubramanian[73] revealed that the sorghum downy mildew was suppressed considerably by an available soil moisture of 76% to 79% and favored by that of 44% to 47%.

Proximity of maize in its susceptible phase to inoculum source ensures early and greater infection, as has been found in cases of *P. maydis*, *P. philippinensis*, *P. sacchari*, and *S. rayssiae* var. *zeae*.[17,34,41]

10-5 CONTROL MEASURES

10-5-1 Control Through Cultural Practices

Crop rotation.   Pratt[44] showed that oospores in the soil were induced to germinate if the infected soil supported non-host crops like wheat, oats, cotton, and soybean. Tuleen *et al.*[74] reported that *P. sorghi* incidence decreased if oospore-infested soil was sown with nonhost plants, 15 days prior to planting of maize. These investigations suggest that rotation might greatly reduce *P. sorghi* and conceivably other downy mildew pathogens also, in fields with high levels of oosporic inoculum.

Deep ploughing.   Deep ploughing of fields with moldboard plough significantly reduced *P. sorghi* incidence in susceptible sorghum.[74] Frederiksen et al.[75] found a fourfold decrease due to deep ploughing compared with shallow ploughing of fields where the downy mildew was prevalent during the previous year. The adoption of this practice by the farmers will, however, depend on its suitability and economical feasibility, compared with the existing practice of cultivation.

Higher seed rates.   The use of seeds at rates more than normal is being practiced in Texas to neutralize losses due to *P. sorghii*.[48] Increasing plant populations 1.5 times the normal appears to neutralize losses from 20% to 30% infected plants.

Date of sowing.   Adjustments in sowing times often avoid severe infection from both primary and secondary inoculum. For instance, early planted maize escapes infection in several areas as plants develop resistance by the time significant quantities of asexual spores are produced in the collateral hosts. *P. philippinensis* in Punjab[34] and *P. sacchari* in Utter Pradesh[76] are less severe on early sown maize, compared with the late sown one. *P. maydis* was more severe on late planted early season maize in Indonesia.[17] *P. philippinensis* was most prevalent in the Philippines on crops sown in July, November, or December.[67]

10-5-2 Biological Control

Oospores of certain downy mildew pathogens of maize are parasitized by fungi and bacteria.[37,44,77] Likewise, plants supporting systemic infections of downy mildew develop secondary infections of other organisms[67,78] which partially or completely inhibit asexual spore formation on such plants.[78] *P. philippinensis* infected plants are commonly parasitized by *Helminthosporium maydis* and such plants are killed, thus eliminating the source of asexual spores from the field much earlier than the plants supporting *P. philippinensis* only. Doggett[79] suggested that secondary invaders like *H. turcicum* and *Colletotrichum* spp. inhibit oospore development by *P. sorghi*. The use of organisms which themselves are not a threat to maize but have a promise for the control of maize downy mildew pathogens can be used. To date, however, none appear to have been used in actual practice.

10-5-3 Control with Chemicals

A wide range of nonsystemic fungicides has been evaluated against maize downy mildew pathogens in different parts of the world.[80-84] Dithane M-45, Dithane M-22, captan, ziram, Blitox, maneb, bleaching powder, and Bordeaux mixture are among those found effective against different downy mildew pathogens. The level of control has varied with the fungicide used, however. The other factors that affect efficacy of these fungicides include the dosages applied, the frequency of application of fungicides, the inoculum potential of the pathogen, host cultivars, the mode of application of fungicides, i. e., whether used for seed treatment or applied to the soil or sprayed on the foliage, and the time of initiation of sprays.

Of the systemic fungicides evaluated and found effective against maize downy mildew, pathogens include Demosan (chloroneb), Plantvax (oxycarboxin), and metalaxyl. Demosan applied to the seeds or in the soil significantly reduced incidence of Philippine downy mildew.[85-86] Sprays of Plantvax effectively controlled *S. rayssiae* var. *zeae*.[81] The treatment of seeds with metalaxyl have resulted in excellent downy mildew control at many localities. Seed treatment with this fungicide is being commercially used in the Philippines and Thailand.[37]

10-5-4 Host Resistance

The control of maize downy mildews through the cultivation of resistant genotypes is the most effective and economical. However, the evolution of resistant cultivars is a long-term process and it depends on the availability of sources of resistance and their proper utilization in breeding for downy mildew resistance programs.

Sources with complete resistance (immunity) to maize downy mildew pathogens are unknown and the resistance so far available is of relative degree only. Such sources, i. e., with relative degree of resistance, against *Peronosclerospora* species have been found in local cultivars in the Philippines, Indonesia, Taiwan, and Vietnam. Varieties rated as resistant to *P. philippinensis* in the Philippines include Aroman White Flint (WF), Kabacan WF, Bukidnon WF, Cebu WF, College WF, Bicol WF, Tiniguib WF, A 206 DMR, and Ph. 9 DMR. In Thailand downy mildew resistance has been found in Sudan lines 1 DMR through 12. In Taiwan, lines TW 79, A 117, Ph 7, and TW 25 were identified as sources of resistance against this pathogen. Local varieties with downy mildew resistance have been identified in Indonesia. They include Medok, Putinhusa, Impa, Genjah Koclock, Genbah Warangen, Penjabins, Kresek.[87-88] Cultivars and lines resistant to *P. sorghi* in India include Phil. DMR 1, Phil DMR 2, Phil. DMR 3, Phil. DMR 4, Phil. DMR 5, Phil. DMR 8, Phil. DMR 9, Bogosyn-2, MDR 1, MDR 2, Narino 330, A 297, A 545, C 166, H 35, K 166, K 175, P 14, Tx 601, W 23. Lines resistant to brown stripe downy mildew in India include EHVL-45, VLD 68, VLD 90, VL 54A, several EH lines, Ganga-5, certain Hungarian maize hybrids and some materials from Mexico.[82] Resistance to *P. sacchari* in India has been found in populations including College WF × Texpeno, MIT × Cuba Gr I, MIT × Flint, Comp. Amarillo, Chain DMR Syn., Bogor Syn. 1, Bogor Syn. 2, Tainan DMR Comp. 2, Phil. DMR 1, Phil. DMR 2, Phil. DMR 3, Phil. DMR 4, and Phil. DMR 6. Several more cultivars and lines resistant to different downy mildew pathogens are known and are listed by different scientists.[68,87,89,90-94]

Resistance against *P. philippinens* is inherited as a quantitative character polygenically controlled by partially dominant[95] or dominant genes.[96-97] Similar observations have been recorded in the case of *P. sorghi* also.[98-99] Resistance against *P. sacchari* has been found to be controlled by a single pair of genes[68] located on the short arm of chromosome II. Resistance to *P. maydis* was reported as polygenic and it was inherited in an additive manner though dominant effects were also detected.[100] Control of resistance against brown stripe downy mildew was polygenic.[101-102]

10-5-5 Integrated Control

The simultaneous application of more than one method of downy mildew control can provide greater benefits than their use individually. For instance, the treatment of seeds of a resistant cultivar with a fungicide like metalaxyl will work against the pathogen with the added advantage as the fungicide sensitive strains will be checked by the chemical and tolerant types by the host resistance. These measures, when further combined with other methods of disease control, i. e., those aimed to reducing the primary inoculum and limiting the spread of pathogens, will provide more benefits. However, the adoption of these measures by the farmer will depend on their effectiveness and economic feasibility of their use. *P. sacchari* was brought under control in Taiwan by using a combination of different measures[68] which include an islandwide eradication campaign of the infected plants, prohibition of planting of susceptible maize and sugarcane varieties in severely infected areas, and substitution of susceptible sugarcane varieties with tolerant ones.

10-6 CONCLUDING COMMENTS

With the acceptance of *Sclerophthora* as a genus of family *Peronosporaceae,* all downy mildew pathogens of maize are now classified under it. The separate taxonomic position of the three genera, i. e., *Sclerospora, Sclerophthora,* and *Peronosclerospora,* has been accepted. The need exists, however, to clarify taxonomic confusions that still exist at specific levels in certain cases.

A comparatively better understanding of the biology and control measures of certain downy mildew pathogens has been achieved. The application of this information in practice has made it feasible to contain maize downy mildews in certain areas of the world. However, experience has shown that maize downy mildew pathogens have a considerable potential for developing new pathotypes. Efforts are needed to be intensified to determine the occurrence and the frequency of appearance of physiologically specialized types within these pathogens. The present state of knowledge on this aspect of maize downy mildew pathogens is miserably inadequate.

10-7 REFERENCES

1. Shaw, C. G. "The Taxonomy of Graminicolous, Downy Mildews, with Emphasis on Those Attacking Maize." *Trop. Agric. Res.* 8 (1975): 47.

2. Payak, M. M., and Renfro, B. L. "A New Downy Mildew Disease of Maize." *Phytopathology* 57 (1967): 394.

3. Siradhana, B. S., Dange, R. S., Rathore, R. S., and Singh, S. D. "A New Downy Mildew on Maize in Rajasthan, India." *Curr. Sci.* 49 (1980): 316.

4. Prabhu, M. S. C., Safeeulla, K. M., and Shetty, H. S. "A New Downy Mildew of *Heteropogon contortus*—Threat to Maize Crop." *Curr. Sci.* 53 (1984): 1046.

5. Renfro, B. L. "Introduction and Decision Taken." *Indian Phytopathol.* 23 (1970): 177.

6. Shirai, M., and Hara, K., *A List of Japanese Fungi Hitherto Known.* Japan: Shizuoka, 1972, p. 448.

7. Shaw, C. G. "Peronosclerospora Species and Other Downy Mildews of the Gramineae." *Mycologia* 70 (1978): 594.

8. Thirumalachar, M. J., Shaw, C. G., and Narasimhan, M. J. "The Sporangial Phase of the Downy Mildew on *Eleucine coracane* with a Discussion of the Identity of *Sclerospora macrospora* Sacc." *Bull. Torrey Bot. Club* 80 (1953): 299.

9. Frederiksen, R. A., and Renfro, B. L. "Global Status of Maize Downy Mildew." *Ann. Rev. Phytopathol.* 15 (1977): 249.

10. Schmitt, C. G., Woods, J. M., Shaw, C. G. and Stansbery, E. "Comparison of some morphological characters of several corn downy mildew incitants." *Plant Dis. Rept.* 63 (1979): 621.

11. Kimigafukoro, T. "Effect of Temperature on Conidial Size of *Sclerospora maydis*, *S. philippinensis* and *S. sorghii.*" *JARQ* 13 (1979): 76.

12. Leu, L. S. "Effects of Temperature on Conidial Size and Sporulation of *Sclerospora sacchari.*" *Plant Prot. Bull. Taiwan* 15 (1973): 106.

13. Rao, N. M., Prakash, H. S., and Shetty, H. S. "Relationship of Cultivars with Sporulation and Morphology of Asexual Propagules of *Peronosclerospora sorghi* on Maize." *Int. J. Trop. Pl. Dis.* 2 (1984): 175.

14. Pupipat, U. "Corn Downy Mildew Research at Kasetsart University." *Kasetsart J.* 10 (1976): 106.

15. Sun, M. H. "Sugarcane Downy Mildew of Maize." *Indian Phytopathol.* 23 (1970): 262.

16. Bonde, M. D., and Melching, J. S. "Effects of Dew Period, Temperature on Sporulation, Germination of Conidia and Systemic Infection of Maize by *Peronosclerospora sacchari.*" *Phytopathology* 69 (1979): 1084.

17. Semangoen, H. "Studies on Downy Mildew of Maize in Indonesia, with Special Reference to Perennation of the Fungus." *Indian Phytopathol.* 23 (1970): 307.

18. Esconde, O. R. "Philippine Corn Downy Mildew. Assessment of Present Knowledge and Future Research Needs." *Kesetsart J.* 10 (1976): 94.

19. Shah, S. M. "Downy Mildew of Maize in Nepal." *Kesetsart J.* 10 (1976): 137.

20. Singh, J. P., Renfro, B. L., and Payak, M. M. "Studies on the Epidemiology and Control of Brown Stripe Downy Mildew of Maize (*Sclerophthora rayssiae* var. *zeae*). *Indian Phytopathol.* 23 (1970): 194.

21. Dalmacio, S. C., and Raymundo, A. D. "Spore Density of *Sclerospora philippinensis* in Relation to Field Temperature, Relative Humidity and Downy Mildew." *Philip. Phytopathol.* 8 (1972): 72.

22. Craig, J., and Frederiksen, R. A. "Differential Sporulation of Pathotypes of *Peronosclerospora sorghi* on Inoculated Sorghum." *Plant Dis.* 67 (1983): 278.

23. Kimigafukoro, T., and Leu, L. "Sporulation of *Sclerospora sacchari* on Corn." *Plant Prot. Bull.* 14 (1972): 153.

24. Sun, M. H., and Ullstrup, A. J. "Production and Germination of Sporangia of *Sclerophthora macrospora* from Corn." *Phytopathology* 60 (1970): 1316.

25. Gupta, A. K., and Siradhana, B. S. "Effect of Nutrition on the Incidence and Sporulation of *Sclerospora sorghi* of Maize." *Indian Phytopathol.* 30 (1977): 424.

26. Kajiwara, T. "Some Experiments on Downy Mildew of Maize." *Trop. Agric. Res.* 8 (1975): 121.

27. Singh, J. P., and Renfro, B. L. "Studies on Spore Dispersal in *Sclerophthora rayssiae* var. *zeae*." *Indian Phytopathol* 24 (1971): 457.

28. Safeeulla, K. M., and Thirumalachar, M. J. "Periodicity Factor in the Production of Asexual Phase in *Sclerospora graminicola* and *Sclerospora sorghi* and the Effect of Moisture and Temperature on the Morphology of Sporangiophores." *Phytopathol. Z.* 26 (1956): 41.

29. Safeeulla, K. M. *Biology and Control of the Downy Mildews of Pearl Millet, Sorghum and Finger Millet.* Mysore: Mysore University 1976, p.304.

30. Weston, W. H., Jr. "Another Conidial *Sclerospora* of Philippine Maize." *J. Agric. Res.* 20 (1923): 669.

31. Rajasah, A. H., Shenoi, M. M., and Ramalingam, A. "Epidemiology of Sorghum Downy Mildew. II. Dispersal and Deposition of Inoculum." *Kavaka* 7 (1979): 63.

32. Tantera, D. M. "Cultural Practices to Decrease Losses due to Corn Downy Mildew Disease." *Trop. Agric. Res.* 8 (1975): 165.

33. Mikoshiba, H. Mas Sudajadi, and Soardiaxto, A. "Dispersion of Conidia of *Sclerospora maydis* in Outbreak of Maize Downy Mildew Disease in Indonesia." *Jap. Agric. Res. Quart.* 11 (1977): 186.

34. Bains, S. S., and Jhooty, J. S. "Distribution, Spread and Perpetuation of *Peronosclerospora philippinensis* in Punjab." *Indian Phytopathol.* 35 (1982): 566.

35. Payak, M. M., Renfro, B. L., and Lal, S. "Downy Mildew Diseases Incited by *Sclerophthora.*" *Indian Phytopathol.* 23 (1970): 183.

36. Ullstrup, A. J. "Crazy Top of Maize." *Indian Phytopathol.* 23 (1970): 250.

37. Williams, R. J. "Downy Mildews of Tropical Cereals." *Adv. Plant Pathology* 2 (1984): 1.

38. Govindu, H. C., Kulkarni, B. G. P., and Ranganathiah, K. G. "Present Status of Downy Mildew Diseases of Sorghum, Millets and Maize in Mysore." *Indian Phytopathol.* 23 (1970): 378.

39. Frederiksen, R. A., Bockholt, A. J., Rosenow, D. T., and Reyes, L. "Problems and Progress of Sorghum Downy Mildew in United States." *Indian Phytopathol.* 23 (1970): 321.

40. Kenneth, R. "Downy Mildews of Gramineae in Israel." *Indian Phytopathol.* 23 (1970): 371.

41. Bains, S. S., Jhooty, J. S., Sokhi, S. S., and Rewal, H. S. "Role of *Digitaria sanquinalis* in Outbreak of Downy Mildew of Maize." *Plant Dis. Reptr.* 62 (1978): 143.

42. Bains, S. S., and Dhaliwal, H. S. "Infectivity of *Sclerophthora macrospora* from Wheat to Maize." *Indian J. Pl. Path.* 4, 1986: 27.

43. Semeniuk, G., and Mankin, C. J. "Occurrence and Development of *Sclerophthora macrospora* on Cereals and Grasses in South Dakota." *Phytopathology* 54 (1964): 409.

44. Pratt, R. G. "Germination of Oospores of *Sclerospora sorghi* in the Presence of Growing Roots of Host and Non-host Plants." *Phytopathology* 68 (1978): 1606.

45. Pandey, A. "Germination of Oospores in *Sclerospora graminicola.*" *Mycologia* 64 (1972): 426.

46. Lucas, J. A. and Sherriff, C. "Pathogenesis and Host Specificity in Downy Mildew Fungi." in *Experimental and Conceptual Plant Pathology. Vol II-Pathogenesis and Host-Parasite Specificity in Plant Diseases,* eds. R. S. Singh, U. S. Singh, W. M. Hess, and D. J. Weber. New York: Gordon and Breach Scientific Publishers. 1988, 321.

47. Craig, J., and Frederiksen, R. A. "Pathotypes of *Peronosclerospora sorghi.*" *Plant Dis.* 64 (1980): 778.

48. Bonde, M. R. "Epidemiology of Downy Mildew Diseases of Maize, Sorghum and Pearl Millet. *Trop. Pest Manag.* 28, 1982: 49.

49. Titatarn, S., and Exconde, O. R., "Comparative Virulence and Gross Morphology of Isolates of *Sclerospora philippinensis,* Weston on Corn." *Philip. Agr.* 58 (1974): 90.

50. Gowda, K. T. P., Rajasekharaiah, S., Jaya Ramagowda, B., and Naidu, B. S. "Occurrence of Sorghum Downy Mildew on Maize and Its Epidemiology." *Curr. Res.* 16 (1987): 42.

51. Bonde, M. R., Peterson, G. L., and Duck, N. C. "Effects of Temperature on Sporulation, Conidial Germination and Infection of Maize by *Peronosclerospora sorghi* from Different Geographical Areas." *Phytopathology* 75 (1985): 122.

52. Tiwari, M. M., and Arya, H. C. "Growth of Normal and Diseased *Pennisetum typhoides* Tissues Infected with *Sclerospora graminicola* in Tissue Culture." *Indian Phytopathol.* 20 (1967): 356.

53. Arya, H. C., and Tiwari, M. M. "Growth of *Sclerospora graminicola* on Callus Tissues of *Pennisetum typhoides* and in Culture." *Indian Phytopathol.* 22 (1969): 446.

54. Katsura, K. "Some Information on the Downy Mildew of the Rice Plant." *Ann. Phytopath. Soc. Japan* 16 (1952): 170.

55. Tokura, R. "Axenic or Artificial Culture of the Downy Mildew Fungi of Gramineous Plants." *Trop. Agric. Res.* 8 (1975): 57.

56. Tiwari, M. M., and Arya, H. C. "*Sclerospora graminicola*—Axenic Culture." *Science* 163 (1969): 291.

57. Payak, M. M. "Epidemiology of Maize Downy Mildew with Special Reference to Those Occurring in Asia." *Trop. Agric. Res.* 8 (1975): 81.

58. Kenneth, R., and Klein, Z. "Epidemiological Studies of Sorghum Downy Mildew (*Sclerospora sorghi*) Weston and Uppal on Sorghums and Corn in Israel." *Israel J. Agric. Res.* 20 (1970): 183.

59. Dange, S. R. S., Jain K., Siradhana, B. S., and Rathore, R. S. "Perpetuation of Sorghum Downy Mildew (*Sclerospora sorghi*) of Maize on *Heteropogon contotus* in Rajasthan, India." *Plant Dis. Rept.* 58 (1974): 285.

60. Exconde, O. R., Adversario, J. Q., and Advincula, B. A. "Incidence of Corn Downy Mildew in Relation to Planting Dates and Meteorological Factors." *Philip Agr.* 52 (1968): 189.

61. Ullstrup, A. J. "Observations on Crazy Top of Corn." *Phytopathology* 42 (1952): 675.

62. Singh, R. S., Chaube, H. S., Khanna, R. N., and Joshi, M. M. "Internally Seed-borne Nature of Two Downy Mildews on Corn." *Plant Dis. Rept.* 51 (1967): 1010.

63. Jones, B. L., Leeper, J. C., and Frederiksen, R. A. "*Sclerospora sorghi* in Corn: Its Location in Carpellate Flowers and Mature Seeds." *Phytopathology* 62 (1972): 817.

64. Advinula, B. A., and Exconde, O. R. "Seed Transmission of *Sclerospora philippinensis* Weston in Maize." *Philip. Agr.* 59 (1975): 214.

65. Rathore, R. S., Siradhana, B. S., and Mathur, K. "Transmission and Location of *Peronosclerospora heteropogoni* in Maize Seeds." *Seed Sci. Technol.* 15 (1987): 101.

66. Sommartaya, T., Pupipat, U., Intrama, S., and Renfro, B. L. "Seed Transmission of *Sclerospora sorghi* (Weston and Uppal) the Downy Mildew of Corn in Thailand." *Kasetsart. J.* 9 (1975): 12.

67. Exconde, O. R. "Downy Mildew of Maize in Southeast Asia. Present Situation and Future Outlook." *Indian Phytopathol.* 23 (1970): 389.

68. Chang, S. C. "A Review of Studies on Downy Mildew in Taiwan." *Indian Phytopathol.* 23 (1970): 270.

69. Frederiksen, R. A. "Sorghum Downy Mildew in the United States: Overview and Outlook." *Plant Dis.* 64 (1980): 903.

70. Goor, G.A.W. van de. "Agronomical Research on Maize in Indonesia." *Landbouw* 24 (1953): 393.

71. Singh, R. S., Chaube, H. S., Singh, N., Asnani, V. L., and Singh, R. Observations on the Effect of Host-Nutrition and Seed, Soil and Foliar Treatments on the Incidence of Downy Mildews I. A Preliminary Report." *Indian Phytopathol.* 23 (1970): 209.

72. Gupta, A. K., and Siradhana, B. S. "Effect of Nutrition on the Incidence and Sporulation of *Sclerospora sorghii* of maize." *Indian Phytopathol.* 30 (1977): 424.

73. Balasubramanian, K. A. "Role of Date of Sowing Soil Moisture Temperature and pH in the Incidence of Downy Mildew of Sorghum." *Plant Soil* 41 (1974): 232.

74. Tuleen, D. M., Frederiksen, R., and Vudhivanich, P. "Cultural Practices and the Incidence of Sorghum Downy Mildew in Grain Sorghum." *Phytopathology* 70 (1980): 905.

75. Frederiksen, R. A., McCombs, D. B., Tuleen, D., and Reyes, L. "Cultural Control of Downy Mildew. I. The Effect of Deep Ploughing on the Distribution of Soil-borne Inoculum and the Incidence of Downy Mildew." *Texas Agr. Expt. Stn. Prog. Rept.* 2950, 1971: 7.

76. Lal, S., and Saxena, S. C. "Date of Planting in Relation to Sugarcane Downy Mildew of Maize." *Indian J. Mycol. Pl. Path.* 11 (1981): 195.

77. Rao, N. N., and Pavgi, M. S. "A Mycoparasite on *Sclerospora graminicola.*" *Can. J. Bot.* 54 (1976): 210.

78. Meenakshi, M. S., and Ramalingam, A. "Newsletter, Intl. Working Group on Graminaceous Downy Mildews" 3 (1981): 5.

79. Doggett, H. "Downy Mildew in East Africa." *Indian Phytopathol.* 23 (1970): 350.

80. Nene, Y. L., and Saxena, S. C. "Studies on the Fungicidal Control of Downy Mildew Maize Caused by *Sclerophthora rayssiae* var. *zeae.*" *Indian Phytopathol.* 23 (1970): 216.

81. Singh, J. P., Renfro, B. L., and Payak, M. M. "Studies on the Epidemiology and Control of Brown Stripe Downy Mildew of Maize (*Sclerophthora rayssiae* var. *zeae*)." *Indian Phytopathol.* 23 (1970): 194.

82. Lal, S. "Brown Stripe and Sugarcane Downy Mildews of Maize: Germplasm Evaluation, Resistance Breeding and Chemical Control." *Trop. Agric. Res.* 8 (1975): 235.

83. Exconde, O. R. "Chemical Control of Maize Downy Mildew." *Trop. Agric. Res.* 8 (1975): 157.

84. Sharma, S. C., Khehra, A. S., Bains, S. S., and Malhi, M. "Efficacy of Fungitoxicant Sprays and Seed Treatment against Philippine Downy Mildew of Maize." *Indian Phytopathol.* 34 (1981): 498.

85. Schultz, O. E. "Evaluation of Chloroneb for Control of Philippine Downy Mildew of Corn (*Sclerospora philippinensis*)." *Phytoparasitology* 62 (1972): 500.

86. Schultz, O. E., and Dalmacio, S. C. "Investigations of Chemical Control of Philippine Downy Mildew." *Proc. 2nd Annu. Cong. Com. Sorghum. Soybean, Mango and Peanut. Univ. Philip.,* Los Banos, *Laguna,* 1971, 41.

87. Leon, C. De. "Selection for Disease Resistance in CIMMYT's Maize Programme." *Kesetsart J.* 10 (1976): 168.

88. Leu, L. S. "Effects of Temperature on Conidial Size and Sporulation of *Sclerospora sacchari.*" *Plant Prot. Bull., Taiwan,* 15 (1973): 106.

89. Frederiksen, R. A., Miller, F. R., and Bockholt, A. J. "Reaction of Corn and Sorghum Cultivars to *Sclerospora sorghi.*" *Phytopathology* 55 (1965): 1058.

90. Frederiksen, R. A., Bockholt, A. J., Rayes, L., and Ullstrup, A. J. "Reaction of Selected Mid-western Corn Inbred Lines to *Sclerospora sorghi.*" *Plant Dis. Reptr.* 55 (1971): 202.

91. Craig, J., Bockholt, A. J., Frederiksen, R. A., and Zurber, M. S. "Reaction of Important Corn Inbred Lines to *P. sorghi.*" *Plant Dis. Reptr.* 61 (1979): 563.

92. Sharma, R. C., Payak, M. M., Mukherjee, B. K., and Lilaramani, J. "Multiline Disease Resistance in Maize." *Kasetsart J.* 10 (1976): 135.

93. Schmitt, C. G., Scott, G. E., and Freytag, R. E. "Response of Maize Diallel Cross to *Sclerospora sorghi* Caused by Sorghum Downy Mildew." *Plant Dis. Reptr.* 61 (1977): 607.

94. Yamada, M., and Aday, B. A. "Usefulness of Local Varieties for Developing Resistant Varieties to Philippine Downy Mildew Disease." *Maize Genet. Crop News* 57 (1977): 68.

95. Gomez, A. A., Aquilizan, F. A., Pyason, R. M., and Calub, A. G. "Preliminary Studies on the Inheritance of the Reaction of Corn to Downy Mildew Disease." *Philip. Agr.* 47 (1963): 113.

96. Mochizuki, N. "Inheritance of Host-Resistance to Downy Mildew Disease of Maize." *Trop. Agr. Res.* 8 (1975): 179.

97. Mochizuki, N., Carangal, V. R., and Aday, B. A. "Diallel Analysis of Host-Resistance to Philippine Downy Mildew of Maize Caused by *Sclerospora philippinensis.*" *Jap. Agr. Res. Quart.* 8 (1974): 185.

98. Frederiksen, R. A., and Ullstrup, A. J. "Sorghum Downy Mildew in the United States." *Trop. Agric. Res.* 8 (1975): 39.

99. Singburaddom, N., and Renfro, B. L. "Heritability of Resistance in Maize to Sorghum Downy Mildew (*Peronosclerospora sorghi*)." *Crop Protect.* 1 (1982): 323.

100. Hakim, R., and Maesum, D. "Segregating Behaviour of *Sclerospora maydis* Resistance of Corn." Proc. South East Asia Regional Symp. *Pl. Dis. Trop.* Joggakarta (Indonesia), 1972.

101. Asnani, V. L., and Bhusan, B. "Inheritance Study on the Brown Stripe Downy Mildew of Maize." *Indian Phytopathol.* 23 (1970): 220.

102. Handoo, M. B., Renfro, B. L. and Payak, M. M. "On the Inheritance of Resistance to *Sclerophthora rayssiae* var. *zeae* in Maize." *Indian Phytopathol.* 23 (1970): 231.

11

MAIZE DWARF MOSAIC, MAIZE CHLOROTIC DWARF, AND MAIZE STREAK

J. K. KNOKE

Research Entomologist, Agricultural Research Service (ARS)
U.S. Department of Agriculture (USDA), Department of Entomology

R. E. GINGERY

Research Chemist, ARS, USDA and Adjunct Professor
Department of Plant Pathology

R. LOUIE

Research Plant Pathologist, ARS, USDA and Adjunct Professor
Department of Plant Pathology, The Ohio State Universtiy
Ohio Agricultural Research and Development Center, Wooster 44691

11-1 MAIZE DWARF MOSAIC

11-1-1 Introduction and History

A mosaic disease of maize (*Zea mays* L.) was first reported in 1920 in Georgia and Louisiana in fields adjacent to sugarcane (*Saccharum officinarum* L.).[1] A similar mosaic disease in this sugarcane had been previously reported.[2] Both diseases were caused by the sugarcane mosaic virus (SCMV) but damage in maize remained minimal and localized.[3-4] Some 40 years later, in the early to middle 1960s, epiphytotics of a mosaic disease of maize were reported in several areas of the United States, the midwest and southern states in the east and California in the west.[4] In Ohio, the new disease and associated pathogen were named maize dwarf mosaic (MDM) and maize dwarf mosaic virus (MDMV), respectively.[5-7] Maize dwarf mosaic is now known to be prevalent in many locations around the world and should be considered synonymous with abaca mosaic,[8] California corn mosaic,[9] corn mosaic in Hawaii,[10] European maize mosaic,[11] maize mosaic in India,[12] sugarcane mosaic,[3] sorghum concentric ringspot,[13-14] sorghum red stripe,[15] and Transvaal grass mosaic.[13]

11-1-2 Geographical Distribution and Economic Importance

MDM is the most widely distributed and important virus disease of maize.[16] It has been reported from 49 countries where about 95% of the world's 450 million metric tons of maize are produced annually on 128 million hectares.[4,17-20] The disease causes significant losses in the United States,[4] China,[21] Yugoslavia,[11] France,[22] India,[23] South Africa,[24] and Italy.[25] Additional major maize-producing countries where MDM is present but losses are more limited include Brazil,[26] Romania,[27] Argentina,[28] Mexico,[29] and Canada.[30] In the United States, MDM has been reported from about 40 states.[4,31] These states produce about 98% of the 195 million metric tons of maize for grain produced annually in the United States. MDM has not been reported in Colorado, Oregon, or Washington, states that produce most of the remaining 2%. About 94% of the commercial sweet corn for fresh market and 75% of the sweet corn grown for processing are also produced in states reporting MDM. Although this virus disease has been detected in most areas of the United States, the greatest losses to MDM occur in areas from the southern edge of the corn belt southward, where johnsongrass [*Sorghum halepense* (L.) Pers.], the major overwintering host for the pathogen, is prevalent. In Ohio, for example, the disease is common each year in riverbottom areas in the southern half of the state. Outbreaks of MDM in the main corn belt area of Ohio and other states are relatively rare and occur primarily in late-planted sweet corn or dent corn grown for silage. For the United States as a whole, over 80% of the maize grown for grain is produced north of the areas of prevalent johnsongrass. Satisfactory estimates of economic loss to MDM are unavailable.[4] For the United States, even for individual states, such estimates would require the accumulation of large amounts of information about hybrid seed usage, the resistance or tolerance of the hybrids to the disease, disease incidence in maize, overwintering hosts of virus, and the effect of meteorological conditions on resistance, inoculum pressure, or plant growth.[32] Information on hybrid seed usage or knowledge of the genotypic makeup of the most-used hybrids is largely unavailable in the United States since most commercial plantings are seeded with proprietary hybrids made from private inbreds.[33] Some information is available on MDM losses in individual genotypes in local areas.[4] Valid yield loss data may be obtained from paired-row trials comparing grain yields from inoculated plants with yields from noninoculated plants.[4] If the test line used is highly susceptible, and the trial is conducted in an area where cross-row contamination from this or other virus diseases is minimal, then yields from diseased plants can be compared with yields from healthy plants of the same genotype. Studies of this type have indicated that maximum yield losses in dent corn are in the 20–40% range.[4,34-38] Losses in fresh market and processing sweet corn are often greater because ears with poor seed set are unmarketable.[31,39] Most yield-loss estimates are taken from small test areas and consequently are of little use in determining losses over larger geographical regions.

11-1-3 Symptoms

Disease symptoms in plants are the visible manifestations of an abnormality. With infectious diseases, symptoms are the result of the interaction of the pathogen, the host genotype, and the environment. Maize dwarf mosaic symptoms vary with the host plant species, host plant genotype, virus strain, method of inoculation, age and physiological condition of host when inoculated, presence of other pathogens in host, and environmental conditions at inoculation or during incubation.[40-44]

Initial symptoms of MDM in susceptible maize seedlings typically appear 5–7 days after inoculation and include the development of pale chlorotic spots and short streaks at the base of leaves emerging after inoculation. As the leaves mature, these spots and streaks enlarge and coalesce to produce patterns of alternating light and dark green areas in the form of mosaic (Figures 11-1, 11-2), rings, or flecks.[4,40,42] In most susceptible maize genotypes, the diseased plants do not appear dwarfed, although plant height may be slightly reduced due to proportionally shortened internodes.[4,42] Sweet corn lines may have more severe mosaic symptoms, substantial height reduction, and reduced kernel set, but an apparent late-season recovery often limits visual detection of mosaic to leaf sheaths and flag leaves on ears.[31,39] In more resistant maize, symptoms may be delayed for several weeks after inoculation and may be typical mosaics or, more often, narrow bands of chlorotic streaks extending the length of the leaves.

Typical MDM symptoms in sorghum are often more severe than in maize and include mosaics, leaf reddening and red striping, severe stunting, delayed flowering, significant yield reduction, and plant necrosis.[3,21,45] Symptoms in sugarcane are generally less severe than in maize and are mainly mild mosaics and striping.[2,3]

Physiological changes in maize infected with MDMV and the environmental influences on disease and symptom development have been reviewed by Gudauskas and Ford[46] and Ford and Milbrath.[47] Plants with MDM have reduced photosynthesis associated with a reduction in chlorophyl content caused by a decrease in number and size of chloroplasts.[48] The respiration rate in plants with MDM increases about 30% at the time symptoms appeared.[49,50] Other disease associated physiological changes are reduced carbohydrate levels,[51] altered enzymatic activity[52,53] and nitrogen metabolism,[53-57] and reduced transpiration caused by restriction of stomatal openings.[58]

Moisture stress and host nutrition appeared to have only limited effects on MDM symptom development,[59-60] although symptoms are inhibited if MDMV-infected plants are grown in 10-15 times normal levels of iron and magnesium.[61] Plant age and temperature,

Figure 11-1 MDM symptoms on maize. Field view showing stunting and chlorosis of plants. (Courtesy of CIMMYT)

Figure 11-2 MDMV symptoms on maize.
Typical mosaic pattern on young leaves.
(Courtesy of CIMMYT)

on the other hand, markedly affect symptom development. In older plants, symptoms are often delayed with mosaics appearing on only the younger unfurling leaves, while plant height is reduced less than in younger plants.[40,62] The development of systemic symptoms is temperature dependent; mosaic patterns generally become brighter with temperature increases up to 30°C[63-65] while some varieties of MDMV-infected maize and sorghum become reddened at lower temperatures.[66]

Electron microscopy visualizes virus-induced inclusions that are found in cells infected with MDMV.[67] Pinwheel and tube type cylindrical inclusions and laminated aggregates occur in leaf cells while membrane-bound inclusion bodies occur between the cell wall and plasmalemma.[67-70]

11-1-4 Causal Organism

MDMV is synonymous with or closely related to SCMV.[3-4] These pathogens belong to the potyvirus group of plant viruses characterized by flexuous rod-shaped particles with helical symmetry. These viruses are typically transmitted nonpersistently by aphids, sometimes through seeds, and artificially by mechanical means.[3,71] Several strains of MDMV have been identified, primarily on the basis of host range or symptoms. These include strains A, B, C, D, E, F, G, and O.[43,72-74] Additional designated isolates or strains include MDMV-Ap, -A(1), and -A(ll),[75-78] MDMV-KS1,[79] MDMV-V,[80] and MDMV-YU

(=MMV-YU).[11,81] Many of these strains or isolates may be similar or identical to each other or to one or more of the 13 strains of SCMV.[3,4,9,79,81-84]

The earliest identified and most extensively studied strains of MDMV are MDMV-A and MDMV-B.[4,72] MDMV-A infects johnsongrass and is the most prevalent strain in maize where johnsongrsss is abundant. MDMV-B does not infect johnsongrass and is more common in maize in areas north of the range of johnsongrass.[4,42,85,86]

Viruses must be purified before detailed studies can be completed on their physical and chemical properties, serological relationships, and classification. Procedures for the purification of MDMV and SCMV have been summarized and compared by Tolin and Ford.[84] The use of various buffers for extraction of virus from plant tissues followed by centrifugation and/or the use of polyethylene glycol for virus concentration have produced purified virus suitable for characterization and the production of antisera. Selected procedures have been undertaken to purify MDMV-A,[87-94] MDMV-B,[95-96] single strains of SCMV,[14,97-99] multiple strains of MDMV,[100-103] multiple strains of SCMV,[82,103-105] or to compare one or more strains of MDMV with one or more strains of SCMV.[9,82,103-104] Final virus preparations, with yields as high as 30 mg of virus per gram of leaf tissue, frequently retain their infectivity, but these yields are often accompanied by a lesser purity.

MDMV is readily transmitted by mechanical methods, is easily purified, and is the most fully characterized maize virus.[44,84,106-108] Chemical and physical properties of MDMV have been reviewed by Gingery.[108] Measurements of MDMV particle stability in plant sap suggest an average thermal inactivation point of 56°C and a dilution end point between 10^{-2} and 10^{-5}.[3,6,9,11,12,21,23,74,88,89,109-116] Longevity in vitro values are 1–3 days at room temperature and 3–7 days at 0–7°C;[3,11,12,21,23,74,89,109,111,112,115-117] infectivity is retained between pH 4 and 10.[23,113] Virus particles have a sedementation coefficient of 146–174S, a maximum and minimum ultraviolet absorption at 260–262 nm and 244–247 nm, respectively, an A 260/280 ratio of 1.13–1.27, and a buoyant density in cesium chloride of 1.254-1.325 g/ml.[3,9,87,88,90,91,96,100,102,113,118] Particles are flexuous rods 700–755 nm by 12–16 nm (Figure 11-3) containing one molecule of single-stranded RNA of molecular weight $2.7–2.9 \times 10^6$.[9,87,90,92,95,96,112,115,119,120]

11-1-5 Host Range

Information on the host range of MDMV has most recently been reviewed by Rosenkranz.[85,121,122] MDMV has a wide host range within the Gramineae. Important susceptible cultivated species include maize, sorghum [*Sorghum bicolor* (L.) Moench], rice (*Oryza sativa* L.) sugarcane, sudangrass [*Sorghum sudanense* (Piper) Stapf], and johnsongrass. Several major grain and grass species that are immune to MDMV include wheat (*Triticum aestivum*L.), rye (*Secale cereale* L.), oats (*Avena sativa* L.), barley (*Hordeum vulgare* L.), timothy (*Phleum pratense* L.), orchardgrass (*Dactylis glomerata* L.), Kentucky bluegrass (*Poa pratensis* L.) and ryegrass (*Lolium perenne* L.) More species are susceptible to MDMV-A than to MDMV-B. The perennial johnsongrass, susceptible to MDMV-A but not to MDMV-B, has been reported as the major overwintering host for MDMV-A in Argentina,[28] Australia,[123] France,[22] Italy,[25] Morocco,[124] Peru,[125] the United States,[4] and Yugoslavia.[11,25] The overwintering host for MDMV-B remains undetected. Of about 525 Gramineae species tested, 72% were susceptible and 43% were not susceptible to MDMV-A or MDMV-B.[11,21,85,121,122,126] Where direct comparisons between MDMV-A and

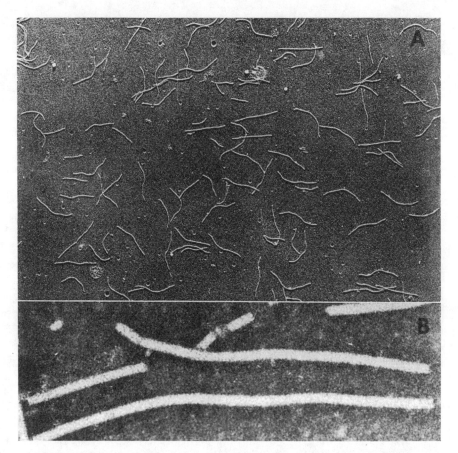

Figure 11-3 Electron micrographs of palladium-shadowed (A) and phosphotungstic acid-stained (B) MDMV particles at 20,000 X and 200,000 X, respectively (Bancroft et al.[87]).

MDMV-B were made, 298 and 247 species were susceptible to strain A and B, respectively. Of about 400 species tested for their reaction to both strains, 83% reacted similarly, 15% reacted differently, and 4% gave inconclusive reactions.[85,122] When species reacted differently, 88% were susceptible to MDMV-A but immune to MDMV-B, with only 7 species immune to MDMV-A but susceptible to MDMV-B.

About 43% of the 208 Gramineae genera found in the United States have been tested for reaction to MDMV.[85,122,127] With about 65% of the tested genera containing susceptible species, and 72% of the tested species reacting positively to MDMV, it is possible that about 390 grass genera containing upwards of 7500 grass species may be susceptible to MDMV worldwide.[85,122,128]

Among grass subfamilies a greater proportion of host species susceptible to MDMV were found within the Panicoideae (78%), Eragrostoideae (76%), and Oryzoideae (73%) than within the Bambusoideae (67%), Arundinoideae (54%), or Festicoideae (41%).[85,122] More annual than perennial grasses are susceptible to MDMV and

SCMV.[122] Direct comparisons between susceptibility to MDMV strains and SCMV strains suggest that more grass species are susceptible to MDMV than to SCMV.[121,122,129-133]

11-1-6 Vectors

Small, soft-bodied homopteron insects of the family Aphididae are responsible for local and distant transmission of several MDMV strains in nature. From the initial pioneering studies on the aphid vectors of the related SCMV by Brandes[134,135] and Kunkel[120] around 1920, through the demonstration by Stoner et al.[136] in 1964 that the corn leaf aphid [*Rhopalosiphum maidis* (Fitch)] is a vector of the virus that causes severe disease in maize in southern Ohio, at least 25 aphid species have been shown to be vectors of MDMV.[32,137-139] Information relating the insect transmission of MDMV has been reviewed.[32,137,138]

The aphid–MDMV association is nonpersistent. Under environmental conditions suitable for normal development of vector and host plants, aphids may acquire and transmit MDMV following insertion of their stylets first into virus-infected and then into virus-susceptible host tissue for a few seconds to minutes.[140-144] The virus is associated with the mouthparts and foregut of the aphid and is lost when the insect molts. Individuals remain inoculative for a few minutes[140,142,143,145] to a few hours[144,146] under natural conditions while the immobilization of aphids with nitrogen or argon gas, or cold temperatures may prolong virus persistence for possibly 1–3 days.[78,147-149]

Laboratory studies show that some species of aphids are better vectors of MDMV than others.[32,43,87,138,140,150-152] The most efficient vectors, averaging about 15% transmission in single aphid trials, were *Schizaphis graminum* (Rondani), *Aphis maidiradicis* Forbes, *Aphis craccivora* Koch, *Aphis fabae* Scopoli, and *Hyalopterus atriplicis* (L.). Biotypes of *S. graminum* also differ in their ability to transmit MDMV.[153] Vector species transmitting at about 10% in laboratory trials include *Acyrthosiphon pisum* (Harris), *Myzus persicae* (Sulzer), *Aphis gossypii* Glover, *Therioaphis maculata* (Buckton), and *Macrosiphum avenae* (F.). Of the above efficient vectors, only *A. maidiradicis, A. fabae,* and *M. persicae* have maize among their listed host plants.[154-155] Relatively inefficient vectors (<5% transmission), but species still capable of causing epiphytotics include *Hyadaphis erysimi* (Kaltenbach), *Rhopalomyzus poae* (Gillette), *R. padi* (L.), *Dactynotus ambrosiae* (Thomas), *Macrosiphum euphorbiae* (Thomas), and *R. maidis*. Both *R. maidis* and *R. padi* frequently develop high populations on maize.

To be considered as a significant natural vector of MDMV, aphids must be present in field areas in relatively high numbers and they must attempt to probe or feed on maize and alternate host plants of MDMV.[138] It is not necessary, however, for the aphid to colonize the maize plant to be a good natural vector; if maize is a nonhost, the aphid may probe and relocate more frequently than if maize were a host. Species considered as important natural vectors in designated countries include *R. maidis* in Australia,[156] China,[157] France,[158] Iraq,[159] Mexico,[29] and the United States;[138,160,161] *R. padi* in China,[157] France,[158] Hungary,[162-163] Iran,[164] and Spain;[165] *M. persicae* in China,[157] France,[158] India,[166] and northern United States;[138,160] *A. gossypii* in China[157] and India;[166] *M. avenae* in Spain and the United States;[160] *S. graminum* in China,[157] France[158] and southern United States;[167] and *M. euphorbiae* in northern United States.[160]

The natural inoculation of maize by aphid vectors may result in minimal to 100% infection in individual field areas. For experimental purposes, low or inconsistent infection levels are inadequate for evaluation of host plants for tolerance or resistance to MDMV. Further, the results of field studies conducted in viral epiphytotic areas are often compromised by the presence of other coexisting viral pathogens. These problems have been partially overcome by taking advantage of the opportunities to artificially transmit MDMV. Several suitable artificial inoculation methods have been developed and utilized to provide uniformly high inoculum pressure for standardized field trials. These methods have progressed from the inexpensive but slow leaf-rub techniques,[44] through semi-mobile to tractor-mounted, compressor-activated spray guns[168-170] or airbrushes,[171] to solid stream, recirculating inoculators[107] enabling one person to uniformly inoculate large numbers of seedling plants at a ground speed of ca. 5 Km/hr.

11-1-7 Epidemiology

Epidemiological studies attempt to relate the available information on vectors, pathogens, host plants, and environmental conditions to disease incidence in the field. For many virus diseases, the initial levels of virus inoculum and time of disease onset are more important for epiphytotics than are weather or other factors influencing subsequent spread of the pathogen.[172] Comprehensive reviews on maize virus disease epiphytotics[32,152,173-175] and on the factors that contribute to epiphytotics[4,47,85,137] have been published.

Maize dwarf mosaic epiphytotics are found in nature whenever adequate numbers of aphid vectors, MDMV-infected source plants, and susceptible host plants are present together in a suitable environment to permit repeated insect transfers of MDMV.[32] The abundance of MDMV-A infected johnsongrass appears to be the major contributing factor to the annual reappearance of MDM epiphytotics in such diverse areas as Argentina,[28] Australia,[123] France,[22] Hungary,[176] Italy,[25] Morocco,[124] Peru,[125] Yugoslavia,[11,25] and from the southern edge of the corn belt southward in the United States.[4,32,177] The source of virus for MDMV-A epiphytotics occurring infrequently in areas beyond the normal range of johnsongrass or for MDMV-B caused epiphytotics remains open to speculation. Circumstantial evidence on the prolonged retention of MDMV by aphids, combined with suitable weather fronts and wind currents, plus the absence of any known local source, has prompted the suggestion that the long-distance transport of virus by aphid vectors may be responsible for infrequent MDM outbreaks in sweet corn in Minnesota and Wisconsin[78,149] and possibly northern Illinois.[139] Other plausible explanations for outbreaks of MDM in non-johnsongrass areas include seed transmission of MDMV in maize or other annual grass hosts[4,21,32,178-182] and the overwintering of the virus in other perennial or winter annual grasses present in more northern areas.[4,85,121,122,133,183,184] The annual epiphytotics developing in late-planted sweet corn in northern Ohio[39] are caused largely by MDMV-B[42] and probably originate from latent infections in winter annuals or perennials.[185] In this area, long-distance transport by aphids appears less likely because virus host plants growing in source areas in southern Ohio and other southern U.S. areas are mostly infected with MDMV-A.

The exact requirements as to number, species, and proximity of vectors and host plants, and the precise environmental conditions suitable for the development of MDM epiphytotics, have yet to be determined.[32] Even though accurate models for predicting MDM disease outbreaks are yet unavailable, some progress has been made in identifying critical factors and relating them through statistical models to disease outbreaks.[161,186] Vectors from different geographical areas or from different times in the same areas have been identified, quantified, and related to disease incidence.[21,32,42,138,150,160-162,187,188] MDMV infections in maize were shown to be related to distance from infected maize[188] or johnson-grass,[189] but no relationship between direction of virus spread and prevailing wind was detected. Attempts at relating MDM seasonal intensity with winter temperatures and spring rains indicated that higher early-season disease incidence in Ohio follows warmer winters and drier springs.[186] Studies of the spatial distribution of MDMV-infected plants indicate that ordinary runs is the best test for determining randomness of diseased plants in the field.[190] With MDM epiphytotics, the initial random distribution of infected plants is usually followed by a clustered pattern[191] because aphid vectors are more likely to inoculate adjacent than more distant suscepts.[192]

11-1-8 Management

Control of MDM in dent corn,[193] sweet corn,[31,39] and grain sorghum[41] has largely been accomplished by the use of resistant or tolerant hybrids. Other attempts to reduce MDM incidence by limiting aphid vector populations with pesticides[41,152,194-197] or protecting the host plant with stylet oils[198] have been relatively unsuccessful. Reviews related to virus disease control by host plant resistance[36,41,193,199-203] and by alternate methods[41,204-206] have been published.

Although good progress has been made in breeding for resistance to MDMV in maize, the best sources of resistance often lack genes for high yield, standability, and early maturity.[199] Consequently, yield potential is often sacrificed when resistant or tolerant hybrids are grown, producing an indirect loss due to this virus. Less definitive progress has been made at determining the genetic mechanism of disease resistance. Different methods of mechanically inoculating crosses between selected resistant and susceptible inbreds with MDMV-A and/or MDMV-B have led to the conclusions that one dominant gene conditions resistance[23] in Oh7B[207] and Va35,[208] two genes control resistance in GA209, Mp339, Mp412, Mp71:222, AR254, and in B68 x sweet corn crosses,[36,208-210] three genes control resistance in T240 and GA203, while one to five genes contribute to resistance in Pa405.[199,209,211] Genes for resistance to MDMV strains appear to be located on both arms of chromosome 6.[199,212,213] Pa405 appears to have the highest level of resistance to the various MDMV strains.[44,199,209,211] Other good sources of resistance are inbreds B68 and Oh1EP.[211] These and other inbreds have enabled several hybrid seed companies to develop commercial hybrids with adequate MDMV resistance.[214]

Field trials to evaluate inbreds and/or hybrids for resistance or tolerance to MDMV have been routinely conducted in several states by exposing seedlings to natural or mechanical inoculations.[201] Trials often began in the mid-1960s and are continuing to date, with plants in individual rows evaluated for the presence or absence of viral disease symptoms and often rated for disease severity. Multiyear trials have been reported from Arkansas,[215-218] Georgia,[219-220] Kentucky,[221-222] Missouri,[223-226] Ohio,[227-230] Tennessee,[231-236] and Virginia.[237-238] Less extensive, short-term resistance trials have been conducted in Australia,[123,239] China,[240] India,[37] and Korea,[241] while multiple location and multiple disease evaluations on public inbreds in five maturity groups[242] and in inbreds adapted to the tropics[243] have been reported.

Other control strategies that may reduce MDM incidence include early planting,[21,244,245] eradication of major overwintering hosts such as johnsongrass,[204-205] and the use of vector-resistant maize plants.[246-249] Although disease control through the reduction of johnsongrass is theoretically possible, the eradication in continuous maize is difficult and costly and may require the alternate planting of a nonmaize crop to reduce the johnson-grass population.[204]

11-1-9 Conclusion

Maize dwarf mosaic is the most widespread and important virus disease of maize. Recurrent annual epiphytotics in many maize-producing areas may be attributed to an abundance of alternate host plants, large numbers of aphid vectors, easily transmitted virus strains, and generally susceptible maize cultivars coupled with the frequent use of unsound cultural practices such as late planting, following maize with maize, or using higher producing but highly susceptible hybrids when resistant or tolerant cultivars would be more appropriate. Major gaps in knowledge of maize dwarf mosaic and its viral pathogen relate to (a) identity of overseasoning hosts for MDMV-B, (b) seasonal and geographic distribution of the various MDMV strains, (c) methods for detection, identification, and location of genes for resistance to the different strains, (d) methods for the incorporation of the resistant genes into commercially suitable cultivars, (e) understanding the cellular and molecular basis for resistance to the virus, and (f) the completion, testing, and use of statistical models for predicting disease epiphytotics.

11-2 MAIZE CHLOROTIC DWARF

11-2-1 Introduction and History

In 1969, Rosenkranz[250] described a stunting disease of maize in southern Ohio with symptoms similar to those of the corn stunt disease. He named the new pathogen the corn stunt agent-Ohio strain (CSA-OH) and showed that it is transmitted by the blackfaced leafhopper, *Graminella nigrifrons* (Forbes). Shortly thereafter, a 30-nm isometric viruslike particle was associated with this disease[251-253] and the pathogen was renamed maize chlorotic dwarf virus (MCDV) to avoid potential confusion between it and the corn stunt spiroplasma.[254] Gordon and Nault[101] found MCDV associated with 76% of stunted maize plants from 16 states.

Properties distinguishing MCDV from most other viruses of similar morphology are a rapid sedimentation rate and semipersistent relationship with its leafhopper vectors, i.e., the insects can transmit the virus for only a few days following virus acquisition.[254] The only other virus sharing these properties is the rice tungro spherical virus.[255] All other known virus-leafhopper relationships are of the persistent type in which the virus is transmitted for weeks or the life of the vector.

11-2-2 Geographical Distribution and Economic Importance

MCDV is found throughout the southeastern United States, from the Gulf of Mexico on the south to states bordering the Ohio River plus Pennsylvania on the north, and from the Atlantic coast to eastern Texas.[4,101,256,257] This is essentially the area of overlap of the distributions of the overwintering host, johnsongrass and the vector, *G. nigrifrons*. There is one report of MCDV in Mexico[4] and it probably occurs in other countries south of the United States where both johnsongrass and *Graminella* species occur.

Although little actual loss data are available, maize chlorotic dwarf (MCD) is probably the second most damaging U.S. maize virus disease, the other important virus disease being MDM.[4] From experimental studies, losses caused by MCDV can be quite high[4] depending on the age of the plant at the time of infection and the susceptibility of the maize genotype. Main et al.[258] reported a 5% loss in North Carolina, and Scott et al.[259] noted that younger plants are more severely affected than older ones.

11-2-3 Symptoms and Host Range

MCDV causes plant stunting, leaf discoloration (reddening and yellowing),[101,250-252] and a diagnostic chlorosis (Figure 11-4) of the smallest leaf veins variously referred to as chlorotic striping of tertiary veins,[101] veinbanding,[260] or veinclearing.[261] Leaf discoloration and plant stunting are correlated with MCDV infection, but not diagnostic.[101] In most maize genotypes, a chlorotic mottle at the base of the whorl 5–8 days after inoculation is the first symptom of MCDV infection in the greenhouse.[261] Veinclearing follows, but sometimes not on the youngest leaves of tolerant genotypes. Other symptoms, reported mainly for field-infected plants, are "miniature" plants in which leaves and internodes are proportionately shortened,[250] necrosis at the base of the stalk resulting in early death of the plant, high incidence of tassel seed,[262] leaves with dull and rough upper surfaces that are less easily torn from leaf sheaths then are healthy leaves,[260] and chlorosis and tearing of leaf margins in severely infected plants.[262] It is puzzling that only veinclearing is often observed in maize plants infected with MCDV in the greenhouse. Perhaps selecting for veinclearing during serial transfers attenuates the virus.[263] or perhaps a second particle is sometimes associated with MCDV in the field, but is lost in greenhouse transfers.

MCDV can be serologically diagnosed by agar gel double-diffusion,[264] immune density gradient centrifugation,[101] immunofluorescence,[265] enzyme-linked immunosorbent assay (ELISA),[266] and immuno-specific electron microscopy.[267] Antisera to MCDV have been prepared by Gordon and Nault[101] and Reeves et al.[266]

MCDV infects only gramineous species[268] Susceptible cultivated species besides maize include sorghum, sudangrass, proso millet (*Panicum miliaceum* L.), pearl millet [*Pennisetum glaucum* (L.) R. Br.], and wheat. Wheat and sorghum are symptomless and

Figure 11-4 MCDV symptoms on maize.
(Courtesy of CIMMYT)

symptoms on the others are mild. Johnsongrass is the only perennial grass known to be susceptible to MCDV. Susceptible annual grasses include crabgrass [*Digitaria sanguinalis* (L.) Scop.], *Coix lacryma-jobi* L., *Echinochloa crusgalli* (L.) Beauv., *Eleusine indica* (L.) Gaertn., *Panicum capillare* L., the foxtails *Seteria faberi* Herrm., *S. pumila* (Poiret) Roemer and Schultes, *S. magna* Griseb., and *S. viridis* (L.) Beauv., and the teosinte, *Zea luxurians* (Durieu and Ascherson) Bird. Forty-two other gramineous species are not susceptible.[268]

11-2-4 Causal Agent

In 1981, MCDV was designated as the type member of the Maize Chlorotic Dwarf Virus Group.[269] MCDV has an isometric particle about 30 nm in diameter [251-253] with a sedimentation rate of $183 \pm 6S$[270] and a buoyant density in cesium chloride of 1.507 g/ml (Figure 11-5).[271] The single-stranded RNA viral genome has a molecular weight of about 3.2×10^6 daltons and comprises about 36% of the virion by weight.[271] The molar percentages of nucleotides are 24 Gp, 30 Ap, 17 Cp, and 29 Up. The calculated molecular weight of the virion, based on the above properties, is about 8.8×10^6 daltons, which is high relative to other virus particles of similar size and may reflect a compact structure for MCDV. Such a structure may explain MCDV's relatively high buoyant density and rapid sedimentation rate.

There are two or three structural proteins ranging in size from 18,000 to 30,000

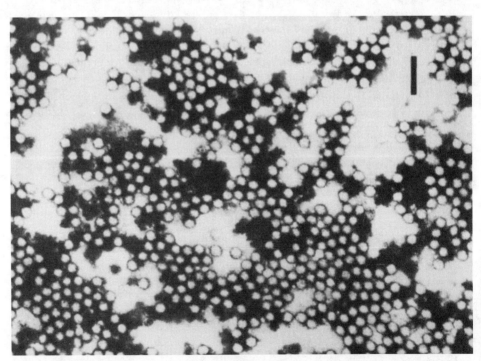

Figure 11-5 Purified MCDV. Potassium phosphotungstate negative stain. The bar represents 100 nm (Gingery[271]).

daltons as determined by degradation of the virus by sodium dodecyl sulfate and electro-phoresis on polyacrylamide gels.[263,272]

11-2-5 Transmission and Epidemiology

MCDV has been transmitted only by leafhoppers and only from plant to plant. Both adult and nymphal *G. nigrifrons* males and females transmit MCDV, but females do not trans-mit the virus transovarially[254,273] and nymphs lose their inoculativity following a molt.[254] The lesser lawn leafhopper, *Graminella sonora* (= *Deltocephalus sonorus*) (Ball), and the gray lawn leafhopper, *Exitianus exitiosus* (Uhler), transmit MCDV, but less effi-ciently than does *G. nigrifrons*. In one study, 41.4% of *G. nigrifrons* transmitted MCDV compared to only 7.1% of *G. sonora*.[137] In another, the rates of transmission by *G. nigri-frons, G. sonora,* and *E. exitiosus* were about 33%, 6%, and 8%, respectively. In a comprehensive study of phylogenetic relatedness of leafhoppers and their rate of MCDV transmission, Nault and Madden[274] reported the following vectors and comparative trans-mission rates: *Ambysellus grex* (Oman) (24.8%); *Endria inimica* (Say) (1.5%); *E. ex-itiosus*(12.6%); *G. nigrifrons* (35.9%); *G. sonora* (10.5%); *Macrosteles severini* (Hamil-ton) (1.9%); *Planacephalus flavicostus* (Stal) (12.9%); and *Stirellus bicolor* (Van Duzee) (13.7%). Apparently, most individual leafhoppers can transmit because at least 86% of *G. nigrifrons* individuals eventually transmitted MCDV during a 6-week experiment in which they were given a 48-hr acquisition-access period on infected maize at the begin-ning of each week.[138] Known nonvectors include *Baldulus tripsaci* (Kramer and Whit-comb), nine *Dalbulus* species, *Eucelidius variegatus* (Kirschbaum), *G. fitchii* (Van Du-zee), *G. oquaka* Delong, *Macrosteles fascifrons* (Stal), *Ollarianus strictus* (Ball), and *Psammotettix lividellus* (Zetterstedt).[261,274-275]

G. nigrifrons can acquire MCDV from diseased plants and inoculate healthy plants, both in as little as 15 min with no intervening latent period in the vector.[254,273] However, transmission rates increase with longer acquisition-access periods.[276] G. nigrifrons retains MCDV for 2-4 days after acquistion in what has been described as a semipersistent relationship.[254] The retention time ranges from 2 days at 30°C to 4 days at 15°C, 20°C, and 25°C.[273,276] Early reports of persistent transmission of MCDV, in which inoculativity persisted for weeks,[250] were later attributed to reacquisition of the virus by the leafhoppers during the week-long intervals that they were held on healthy plants.[261,273]

No transstadial passage, no latent period, and short retention times prompted Nault and Bradfute[261] to speculate that MCDV may adsorb to a site or sites in the foreguts of vectors. Expanding this idea, Harris[277] proposed an ingestion-egestion hypothesis in which virus is ingested during feeding, attaches to the foregut, and is egested during later probes. This hypothesis is in accord with observations of MCDV-like particles attached to the foregut of insects that had fed on MCDV-infected maize, but not those that had fed on healthy controls.[278,279] Unfortunately, these particles were not shown to be infective, trans-missible, or even to be MCDV. Also, no mechanism of attachment and detachment of the virus was hypothesized. Ammar et al.[280] observed MCDV-like particles embedded in a matrix material associated with the lining of the foregut in viruliferous *G. nigrifrons.*

Recently, Hunt et al.[281] obtained evidence for a "helper" component involved in MCDV transmission. A helper component is a substance other than the virus particle itself that is required for insect transmission. The helper component in potyvirus transmission may facilitate binding of viral particles to aphid mouth parts,[282] but no role for the putative MCDV helper component has been determined.

Johnsongrass, a perennial weedgrass, appears to be the main and perhaps only overwintering host for MCDV.[268] MCDV was found in 4 of 27 field-collected johnson-grass plants,[253,261,266] and MCD epiphytotics are limited to areas where johnsongrass is abundant.[101] All other known plant hosts of MCDV are annuals[268] and are unlikely to function as overwintering hosts.

G. nigrifrons is probably involved in most of the natural spread of MCDV because MCD incidence is highly correlated with the numbers of *G. nigrifrons*.[196,283] However, it is unclear whether *G. nigrifrons* overwinters in the northern areas of MCD occurrence or migrates into these regions each year.[284-285] In either case, large numbers of *G. nigrifrons* build up in these areas each year on small grains and grasses, especially ryegrass, barn-yardgrass, crabgrass, and bermudagrass (*Cynadon dactylon* (L.) Pers.).[138]

The probability of infection by MCDV appears to vary throughout the growing season. In Ohio, disease potential, as measured by the percentage of maize seedling trap plants becoming infected after a week's exposure to field conditions, is first detected in early June, peaks in mid-July, and then sharply declines.[260] In other studies, variations in *G. nigrifrons* populations have been noted, but were not correlated with disease potential. Peak populations of *G. nigrifrons* were observed in late May and early June in Alabama[286] and Kentucky,[287] in late July, with lesser peaks in late May–early June and early Septem-ber, in one study in Ohio,[32] and in June and late July–early August in another.[138]

G. nigrifrons prefers grasses and small grains to maize as feeding hosts,[138] a trait which Stevens et al.[286] felt might result in its being more mobile in maize fields and thus more efficient at spreading MCDV than if it were well-adapted to maize and moved about less. However, the rate of movement of *G. nigrifrons*, measured by following the movement of rubidium-tagged insects released from a point source, was quite slow, only 15 m in 4 days.[288] Knoke[289] also estimated slow movement, about 6 m per month, based on the rate of disease spread from a point source. *G. nigrifrons* appears to be attracted to maize in the early whorl (4-leaf) stage, especially in no-tillage fields that have a yellow background. The numbers of *G. nigrifrons* on maize drop dramatically after this stage.[287,290]

11-2-6 Management

Control of MCD is attempted most often through the use of tolerant maize genotypes. Although no immune lines were found among 104 dent corn inbreds and 158 commercial and experimental sweet corns, some were more tolerant than others.[39,291]

Experimental reduction of MCD incidence follows application of the systemic in-secticide carbofuran.[292-294] Kuhn et al.[196] applied carbofuran in the furrow at the time of planting and reduced MCD incidences about 75% and 50% and increased yields 37% and 125% for a moderately susceptible and a highly susceptible hybrid, respectively. The *G. nigrifrons* populations was reduced 65–74% in these plots. Other studies have confirmed the efficacy of carbofuran for reducing disease incidence in the field[295] and greenhouse.[195]

Other recommended controls include spraying with oils that interfere with leafhop-per probing,[279,296] planting early so that only older, more tolerant plants are exposed to high *G. nigrifrons* populations,[47,204,244,294] and eradicating johnsongrass, the overwintering

MCDV host, and other grasses that could support large numbers of leafhoppers.[196,204] Several workers have favored a combined approach using resistant maize, johnsongrass eradication, and systemic insecticide.[204,283,297,298] However, high cost often precludes such comprehensive control efforts, and the single most effective control strategy remains the planting of tolerant varieties.[283,298-300]

The genetics of MCDV resistance is poorly understood, in large part because MCDV is not mechanically transmissible and so genetic studies are subject to variability introduced by the insect vector. For example, it is difficult to control disease pressure from experiment to experiment and impossible to separate resistance to the virus from resistance to the vector. Other complicating factors include: (a) the confounding effect of pathogens other than MCDV in field studies, particularly MDMV; (b) masking or altering of symptom expression (and therefore disease rating) caused by hybrid vigor; and (c) variable symptom expression resulting from plants infected at different ages.[301] Moreover, because most current breeding efforts are aimed at tolerance, not immunity, even "resistant" plants are often infected and symptomatic, making genetic studies difficult to quantify and interpret.[193] This may explain the conflicting views of disease resistance. For example, Naidu and Josephson[302] reported as many as four genes for resistance, but Scott and Rosenkranz[303] found only three, on chromosomes 1, 3, and 4. In another study, Grogan and Rosenkranz[304] found no evidence for dominant resistance genes and concluded that resistance results from an additive effect of several genes.

11-2-7 Cytopathology

Two unique inclusions are found in MCDV-infected leaves, primarily in vascular parenchyma and phloem cells, but occasionally in mesophyll cells. One is a quasi-spherical, electron-dense granular inclusion that contains embedded 31-nm MCDV-like particles; the other is an elongate fibrous inclusion.[67,251-253,280,305] Isometric viruslike particles are also found individually and aggregated in the cytoplasm and central vacuole of phloem cells.[252,306] Harris and Childress[306] also reported laminate inclusions, numerous vesicles, and fibrillar material of unknown composition in MCDV-infected cells.

11-2-8 Conclusion

MCDV is clearly one of the most important viruses of maize in the United States. Control of the disease, although acceptable in most instances as a result of planting tolerant hybrids, is unsatisfying in that the genetics of resistance or tolerance is not clear and even the most tolerant lines are susceptible to infection.

There are several interesting aspects of MCDV that make it attractive for future work, including: (a) the possibility that the capsid contains three proteins, unprecedented in plant viruses of its size; (b) the existence of a helper component, the first to be described for a leafhopper-borne virus; (c) the variability in symptomatology, especially between field- and greenhouse-infected plants that suggests other agents or strains may be involved in the disease in the field; and (d) the absence of knowledge about the replication and translation strategies of this type of virus.

11-3 MAIZE STREAK

11-3-1 Introduction and History

Fuller[307] often is credited for the first description of maize streak (MS) disease, which he called "mealie variegation." He attributed nutrient deficiencies in the soil for the cause of the disease. Storey,[308] however, was first to use the term "streak" for a disease in sugarcane that he showed was distinct from sugarcane mosaic. Later he also applied the term to maize infected with "mealie variegation."

Maize streak virus (MSV) is the type virus for the geminivirus group; it was the first geminate particle virus discovered and has a ssDNA genome. The phenomenon of biotypes of active and inactive leafhopper vectors also was first found in transmission studies of MSV.

Some synonyms for related strains of MSV found in other crops include: Bajra streak,[309] maize streak virus A,[310] and sugarcane streak virus.[308,311]

Bock[312,313] and Damsteegt[314] have recently reviewed both the virus and the disease. Recent reviews of the virus include those by Francki, Milne, and Hattar,[315] Goodman,[316-317] Harrison,[318] and Howarth and Goodman.[319] General reviews on MS by country describe the disease in Egypt,[320] India,[321] Ivory Coast,[322] Mauritius,[126] Nigeria,[323] Zimbabwe (-Rhodesia),[324] and South Africa.[325]

11-3-2 Geographical Distribution and Economic Importance

The distribution of MS includes countries of Africa, southern Asia, and islands in the eastern hemisphere; it is not found in the western hemisphere[326] with the exception of one observation of MS in Panama.[327] In Africa, it has been reported from Angola,[328] Dahomey,[329] East Africa,[330] Egypt,[331] Ivory Coast,[322] Nigeria,[323,332-333] Ghana,[334-335] Malawi,[336] Mozambique,[337] Zimbabwe (=Rhodesia),[324,338-339] and South Africa.[24,307,310,325,340-341] In the southern Asia and Indian ocean area, MS has been found in India,[309,342-344] Burma,[345] Mauritius,[346-348] and Vanauatu.[349]

Serological confirmation of the virus, evidence of transmission by *Cicadulina* vectors, and development of diagnostic and typical disease symptoms are definitive evidence for the actual distribution of MSV. Soto et al.[350] further advocated extensive field observations combined with conclusive studies on vectors, virus transmission, and serological tests before reaching a positive conclusion about the geographical distribution of MSV. Minimally, MSV identification should be based on transmission with known *Cicadulina* vectors and expression of typical disease symptoms rather than diagnoses based on symptoms alone. The distribution of MSV indicated in the reviews by Edgerton[326] and Bock[312] is based primarily on transmission studies and symptoms.

Leafhopper transmission alone is insufficient for virus identification because *Cicadulina* species can transmit other viruses or cause viruslike symptoms. Furthermore, different species of *Cicadulina* are MSV vectors in different areas. For example, *C. mbila* is the most important vector of MSV in South Africa and East Africa,[351,352] but *C. bipunctella* and *C. triangula* are most important in Egypt[353] and Nigeria,[354] respectively. *C. triangula*, however, also transmits maize mottle chlorotic stunt virus in Nigeria[355] and *C. chinai* is a vector of maize fine stripe and maize yellow stripe viruses.[353] Eastern wheat striate virus in India is transmitted by *C. mbila*.[356] Also, initially, maize wallaby ear was thought to be transmitted by *C. bipunctata bimaculata* (Evans) and *C. bipunctata* (Melichar).[357] More recently, however, the viral nature of maize wallaby ear has become suspect and injury from leafhopper feeding is now believed to be the cause of that disease.[123] Similarly, swellings on maize leaf veins are part of an enanismo disease syndrome associated with by *C. pastusa*.[358] In the Philippines,[359] feeding of *C. bipunctella* causes galls to form on maize and rice. Galls are also caused by the feedings of *C. mbila* and *C. bipunctata bipunctata* on maize in East Africa, *C. storeyi* in oats in Rhodesia, and *C. mbila* on oats in Kenya.[355]

The reported distribution of the leafhopper vectors is greater than the recorded distribution of MS.[358] The occurrence of *Cicadulina* spp. in Australia and South America suggests that MS also could occur on maize in these places, provided that MSV-infected source plants were also present.

Plants susceptible to MSV have been reported in the subfamilies Panicoideae, Andropogonioideae, Chloridoideae (=Eragrostoideae), Pooideae (=Festucoideae), and Oryzoideae. None has been reported in the subfamily Bambusoideae. Early reports of MSV hosts were based on the similarity in symptoms in the new host to those on maize. However, sometimes other viruses cause similar symptoms. Etienne[360] found that symptoms of MS were similar to those of maize stripe virus transmitted by *Peregrinus maidis* (Ashm). Later, MSV infection was confirmed by leafhopper transmission.[347,361] Other viral diseases sometimes confused with maize streak are maize line and maize mosaic,[312] but they differ from maize streak in particle morphology and in leafhopper vectors. More recently, Thottappilly[362] and Ammar,[363] among others, have been able to confirm their diagnoses of MS based on symptoms and leafhopper transmissions by using MSV antisera developed by Bock.[364] Damsteegt[365-366] determined the MSV host range by transmission of a known MSV isolate with *C. mbila*. Although this approach does not necessarily implicate the sources of MSV in nature, it does provide knowledge on potential candidates. Damsteegt[365] found 31 susceptible of 62 perennial and annual grasses in his tests. Interestingly, the number of susceptible species among the Festucoids, Chloridoids, Panicoids, and Andropogonoids were 16/35, 3/6, 7/12, and 5/9, respectively. At least from these tests, it appears that the proportion of susceptible species is equally divided among the four subfamilies. No symptomless hosts were found by backassays.

Ricaud[367] listed *Coix lachryma-jobi* L. and *Brachiaria reptans* (L.) Gardner and Hubb. as natural hosts; he also listed *Digitaria horizontalis* Willd., *Panicum maximum* Jacq., and *Paspalum conjugatum* Berg. as natural hosts, but the viruses from these hosts were not the same as the one that infects maize.

Maize streak is economically important in maize in South Africa[24,325] and Nigeria;[332] in sugarcane in South Africa,[308] Egypt,[320,353] India,[344] Burma;[345] in finger millet [*Eleusine coracana* (L.) Gaertner] in India;[368] in wheat in South Africa[341,361] and in India;[343] in pearl millet in India;[342] in barley, oats, and rye in South Africa;[361] in rice in Nigeria;[323] in Sorghum in Malawi;[336] and in napier grass (*Pennisetum purpureum* Schumacher).[335]

In the First Eastern, Central, and Southern Africa Regional Maize Workshop in 1985, 11 of 19 African countries considered MS a significant constraint on maize production.[369] However, in other countries, other diseases in maize or the occurrence of MSV in other crops were more important. In southern Nigeria,[370] the losses caused by *Puccinia polysora*, *Helminthosporium maydis*, *H. turcicum*, and stem borers exceeded those caused by MSV. Sugarcane streak in Egypt[353] and MS in wheat[371] and pearl millet[372] in India also were more important diseases than MS in maize.

Yield losses in maize to MSV are dependent on the stage of plant growth at the time of infection. Yield losses are greatest with young plants. Van Rensburg[373] reported no yield, half yield, and nearly full yield from plants infected at 1, 3, and 8 weeks after planting. At 8 weeks after planting during an epidemic, almost 100% of the plants were infected. Most losses occurred during early infection and were related to a lower plant stand, whereas the losses from late infections resulted from fewer harvestable ears associated with poorly filled ears and poor husk cover, shorter plants, narrower stem diameter, smaller leaf size, and smaller tassels.[374] Mzira[375] found a closer relationship between yield loss and time of infection than between yield loss and amount of disease incidence. Incidence of MSV in wheat was less in early plantings and seeding rates of 200 kg/ha gave the most yield in the presence of streak.[376]

The reported yield loss also varies with the method of determination. Guthrie[377] found a 33% and 61% loss in grain weight, respectively, in naturally infected plants compared to inoculated plants that were transplanted from the greenhouse to the field. In field plants inoculated at different stages of growth, he confirmed Van Rensburg's findings[373] that loss was related to age of plant at the time of infection and was greatest when infections occurred in young seedlings. He also noted that cob weight correlated closely with grain weight and suggested the use of cob weights as a measurement of yield loss. Additional information, such as interactions of MSV strains, predisposition to stress or other maize diseases, and influence of maize genotypes on yield loss are still needed.

11-3-3 Symptoms

The primary MS symptom on maize consists of leaf discolorations. Storey's[308,352] original descriptions remain essentially complete. The first symptoms of infection are circular, colorless spots 0.5 to 2 mm in diameter separated one from another by 2 to 5 cm on the lowest exposed portion of the youngest leaves (Figure 11-6). Later, the frequency of the elongating spots on the developing leaf increases. Finally, the spots become general over the whorl of that leaf and subsequently formed leaves. There are always some fully green leaves at the base of the plant, which suggests infection occurred after development and exposure of these leaves.

Storey[352] was very certain about the differences in symptoms between the mosaic and streak diseases of sugarcane and maize and considered them diagnostic. In SCMV infections in maize, the patterns are predominantly dark green areas against a background of pale green. The edges of the mosaic areas are irregular and diffused, frequently elongating in the directions of the leaf axis. In streak virus infections, the patterns consist of distinct light areas against a dark background (Fig. 11-6). The chlorotic areas are elongated, and narrow in width. The chlorotic areas may become transparent and, early in development, are restricted to particular interveinal areas. Streak symptoms on other gramineous hosts are similar to those observed on maize.

Damsteegt reported[314] morphological teratology including leaf margin splitting, leaf tip twisting and necrosis, and shoot stunting. Fajemisin et al.,[374] reported stunting, narrower stem diameter, and reduced leaf size and tassels.

The predominant microscopic indications of MSV infection in leaf tissue are the

Figure 11-6 MS symptoms on maize. Chlorosis with broken yellow streaks along the veins, contrasting with dark green color of normal foliage. (Courtesy of CIMMYT)

crystalline nuclear inclusions observed in ultrathin sections.[364] These inclusions contain particles 13–20 nm in diameter which are similar in size to the unpaired particles observed in purified preparations.

11-3-4 Causal Organism

MSV was first postulated to be a small RNA virus.[378] The evidence for its viral nature was the small polyhedral viruslike particles observed in the nuclei of diseased maize leaves in ultrathin sections from Nigeria.[379] Viruslike particles of 18–20 nm in diameter also were observed in the nuclei, but not on the cytoplasm of ultrathin sections from diseased leaves from Uganda.[380] Sylvester et al.,[380] however, were not able to confirm positively MSV infection in these samples. Bock et al.[364] purified MSV and reported that purified preparations contained particles 20nm in diameter. Most particles occurred in pairs and these paired particles measured 30 × 20 nm (Figure 11-7). The single and paired particles had sedimentation coefficients of 54 and 76 S, respectively. In earlier work,[364,381] MSV was thought to contain single-stranded RNA. Later work[382] showed that MSV is a plant virus with a circular single-stranded DNA genome and the name geminivirus was proposed for viruses with paired particles. MSV is the type member of this virus group.[269] Although two bands of nucleic acid were observed in polyacrylamide gel electrophoresis,[381] it now is considered unlikely that there is more than one component to the MSV genome.[382,383] The protein subunit molecular weight is estimated at 28,000 daltons.[381] The molecular weight of the single-strand circular DNA is estimated at 0.71×10^6.[384]

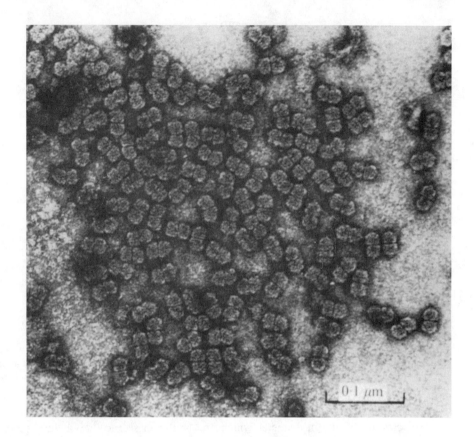

Figure 11-7 Paired and single particles of MSV (maize isolate) stained with uranyl acetate, chloroform preparation (Bock et al.[364])

Collections of the viruses causing streak symptoms have been considered as forms or isolates of MSV.[310,385] The form separation and relationships are based on host suscepti- bility and reactions. Cross-protection tests do not establish any relationships between the streak virus from cane and the one from maize; cross-protection also does not occur in leafhoppers.[385] Infection with the B-form of MSV, however, does provide protection against the A-form. Serological studies[364] among isolates from maize, sugarcane, and guinea grass (*Panicum maximum* Jacq.) showed that these plants were all infected with MSV and that the sugarcane isolate is closely related and the guinea grass isolate more distantly related to the MSV from maize.

The notion of host adaptation[364] or host-adapted strains,[386] i.e., each MSV isolate becomes specialized on its own hosts,[310,385] was put forth to account for variations in host ranges among the various MSV isolates. Rather than a response conditioned by the host plant, each virus was considered a separate entity that caused different symptoms. Markham et al.,[387] however, suggested that virus acquisition and transmission by vectors played a major role in host range determinations. They demonstrated that various strains could be efficiently transmitted to different hosts by *C. mbila, C. triangula,* and *C. chinai* when the insects were injected with virus. In this system, the virus host plant and the geographical origin of the virus were not critical for efficient transmission.

Recent studies on MSV have concentrated at the molecular level. One intent is to develop the virus as a vector for plant genetic engineering.[319] Sequencing of MSV[388-389] as well as the comparison of the nucleotide sequence for homologies among different isolates[390-391] and among different geminiviruses[389,391-393,395] have been accomplished. Studies to map RNA transcripts on the MSV genome and locate possible control regions that are involved in gene expressions[394] also have been made. Grimsley et al.[410] have shown that tandemly repeated copies of the MSV genome in TDNA from *Agrobacterium* have the potential to replicate and produced symptoms independently of the other viral or insect components.

As determined by feeding treated crude sap through membranes, the thermal-inactivation-point of MSV is not less than 50°C, the dilution-end-point not less than 1/1000, and longevity-in-vitro at room temperature is not less than 24 hr and can be as long as 4 days.[312] Infectivity of either partially purified preparations or of infected leaves is maintained for more than 6.5 yr at -125°C to -180°C.[396] MSV also remains infectious after five cycles of freezing and thawing during a 5-month period when stored at -20°C.[397]

11-3-5 Vectors

The species of *Cicadulina* reported as vectors of MSV include: *C. mbila* (Naude) (=*Balclutha mbila* Naude),[398] *C. storeyi* China (=*C. nicholsi*),[399,400] *C. latens* Fennah,[401] *C. parazeae* Ghauri,[351] *C. triangula* Ruppel,[350] *C. bipunctella zeae* China (=*C. zeae* China),[363,399] *C. arachidis* China,[402] *C. similis* China,[402] and *C. ghauri*.[402]

Taxonomic reviews on the *Cicadulina* are reported by Nielson[403-404] and Ruppel.[358] Details on the leafhopper biology are reported for *C. mbila,*[405-408] *C. triangula,*[408] and *C. bipunctella zeae*.[409]

Under experimental conditions, MSV is inefficiently transmitted by *D. maidis* to maize when injected with virus.[397] MSV is also transmitted (50–88%) to maize by *Agrobacterium* with tandemly repeated copies of MSV genome carried in TDNA.[410] One mechanical transmission of MSV by electro-endomosis was reported by Polson.[411]

Storey's detailed studies[398,412-417] on the relationship between MSV and *C. mbila* are classics. He reported that the virus acquisition access period (AAP) required by *C. mbila* is generally less than an hour and could be as short as 5 sec. The short AAP suggests that virus is acquired from mesophyll as well as from phloem cells. Recently Boulton *et al.*[418] reported high but variable levels of MSV in *C. mbila* given a 7-day AAP when assayed with spot hybridization techniques. MSV is detected in 55% or 100% of the leafhoppers after a 3- or 6-hour AAP, respectively. Storey reported a minimum latent period of 6 hr after AAP, but it was usually 12 to 48 hr. A minimal inoculation access period (IAP) of 5 min is required for virus transmission. IAPs of 15, 30, and 75 secs, however, are generally insufficient. Leafhoppers in all stages of growth can acquire and transmit MSV and females are better vectors than males (86% vs. 26%). Although leafhoppers retain the ability to transmit MSV after a molt, the virus is not transovarially transmitted. The capability of leafhopper acquisition and transmission of MSV is temperature dependent; 30°C appears optimal for these activities. Some biotypes do not transmit MSV (inactive race), but can be made active by injecting virus into their hemoecel. This trait was determined to be a dominant, sex-linked factor. Multiplication of MSV in *C. mbila* has not been demonstrated, but the ability to transmit MSV remains with the insect throughout its life. MSV can be recovered from the leafhopper's salivary glands and various body fluids. This virus–vector relationship is classified as circulative but not propagative.

11-3-6 Epidemiology

Gorter[361] and later Rose[335] published comprehensive reviews on the epidemiology of maize streak.

The spread of MSV is critically dependent on the various *Cicadulina* species. The leafhoppers themselves are not an important pest on the crops.[419] However, they constitute a serious potential for disease outbreaks because there are no less than nine vector species whose distribution coincides or extends beyond the crop and virus source distributions. Autry[386] suggested that three major factors influence the probability of disease outbreaks: the maize genotype, scrubland proximity, and cropping sequences. The ability of leafhoppers to produce many generations, from four to five for *C. mbila, C. parazea,* and *C. storeyi*[351] to eight or nine for *C. bipunctella*[409] and *C. chinai,*[331] provides ample opportunities to accommodate climatic and plant growth variations and assures the transmission and dispersal of the virus to new hosts during a growing season.

The patterns of virus spread, however, are dependent the sexual composition of the vector population as well as their physiological state arrival within a maize crop. Gravid females are less likely to fly; feeding on young wheat seedlings also inhibits flight. Rose[420] found that males and nongravid females from drying, mature wheat stems are most likely to fly. Short-bodied forms also are more likely to be long-distance fliers than are long-bodied forms. Females fly higher than males and *C. mbila* fly higher than *C. storyei.*[421] A steep infection gradient within 10 m of a field's edge is due largely to short-distance fliers and a more gradual gradient decline is due largely to long-distance fliers.[422]

The rate of spread was found to be correlated with leafhopper population development and with the efficiency of vector transmission. Disease increase is arithmetic when vector populations are low and becomes exponential when populations are high.[423] *C. mbila* is the most efficient vector[397,412] and *C. arachidis, C. similis,* and *C. gharurii* are relatively inefficient.[402] The environment may be the overriding factor in disease development when the availability of leafhopper vectors and susceptible plants is not limiting. Disease development is more likely when rainfall distribution is uniform throughout a growing season than when rainfall is sporadic.[172,361]

Disease development also may be influenced by the climatic zones where the crop is produced. MSV is widespread in the warmer parts of South Africa, but is seldom found at altitudes about 1,200 m.[424] The growing of maize, instead of the traditional pearl millet and sorghum in the drier savanna zones of northern Nigeria, where graminaceous grass weeds support leafhopper development, may be responsible for the high streak incidence.[425] Changes in crop production practices, e.g., irrigation, have caused MSV to cross altitudinal and vegetation stratification zones that once limited its spread. Irrigated maize or maize planted adjacent to irrigated lands may be subjected to high disease incidence even when planted after major leafhopper flights.[335]

11-3-7 Management

Cultural practices employed for disease control usually include eradication and avoidance measures. Destroying volunteer maize and grass weeds to reduce numbers of overwintering leafhoppers, locating nonirrigated maize away from crops planted in irrigated lands, planting maize after the cereal crop harvest, and planting garden maize after the main crop are recommended cultural procedures.[424,426] A 10-m fallow area appears suitable as a barrier to virus spread.[361] Incidence of maize streak is decreased by increased levels of phosphorus and decreased levels of exchangeable soil acidity. Disease incidence is increased by increased levels of limestone ammonium nitrate.[427]

Contact insecticides (e.g., DDT, carbaryl),[335] systemic insecticides (e.g., dimethoate, methyl demeton),[335,426] and granular systemic insecticides (e.g., disulfonton, phorate, aldicarb, carbofuran)[428,429] afford some level of disease control. However, their use for field-grown maize is economically hard to justify.[335]

Maize with resistance to MSV has been reported from South Africa,[430–432] East Africa,[433–435] Reunion,[436] and West Africa.[374,437–438] The genetic basis for disease resistance has been considered non-Mendelian,[431] due to a single gene whose action is modified by the inbred genetic background,[433] or simply inherited.[350] Many of the discrepancies in disease resistance work are related to disease escapes when screening for resistance by relying on natural infections under field conditions. Progress in screening for disease at IITA was greatly advanced by the development of mass rearing and handling techniques for insect vectors.[350,439]

11-4 CONCLUSION

Maize streak is a unique and important virus disease of maize occurring primarily on the African continent. Major new discoveries related to MSV and its virus–vector associations include the presence of viral particles in doublets in purified virus preparations, a circular ssDNA genome, and the recognition of genetically conditioned "active" and "nonactive" vectors. Many challenges for research still exist. Information is needed on the interaction of the various MSV strains, predisposition to MSV infection by stress and other maize diseases, and the influence of maize genotypes on yield loss. Additional research needed at the cellular and molecular levels includes the determination of the priming mechanism(s) for the replication of the circular ssDNA, determination of the genetic or structural basis for MSV movement through the cytoplasm to establish infection in nuclei, determination of the mechanism of viral transcription and DNA replication, and determination of the molecular basis for pathogenicity and virus specificity.[319]

11-5 REFERENCES

1. Brandes, E. W. "Mosaic Disease of Corn." *J. Agric. Res.* 19 (1920): 517.

2. Brandes, E. W. "The Mosaic Disease of Sugar Cane and Other Grasses." *U.S. Dep. Agric. Bull.* 829 (1919): 26.

3. Pirone, T. P. "Sugarcane Mosaic Virus." *CMI/AAB Descriptions of Plant Viruses* No. 88, 1972.

4. Gordon, D. T., Bradfute, O. E., Gingery, R. E., Knoke, J. K., Louie, R., Nault, L. R., and Scott, G. E. "Introduction: History, Geographical Distribution, Pathogen Characteristics, and Economic Importance," in *Virus and Viruslike Diseases of Maize in the United States*, eds. D. T. Gordon, J. K. Knoke, and G. E. Scott. (Southern Coop. Ser. Bull., 247, Ohio Agric. Res. Dev. Cent., Wooster, 1981), 1.

5. Janson, B. F., Williams, L. E., Findley, W. R., Dollinger, E. J., and Ellett, C. W. "Maize Dwarf Mosaic: New Corn Virus Disease in Ohio." *Ohio Agric. Exp. Stn. Res. Circ.* 137 (1965): 16.

6. Williams, L. E., and Alexander, L. J. "Maize Dwarf Mosaic, a New Corn Disease." *Phytopathology* 55 (1965): 802.

7. Williams, L. E., Alexander, L. J., Findley, W. R., Dollinger, E. J., Rings, R. W., and Treece, R. E. "Maize Dwarf Mosaic." *Ohio Agric. Exp. Stn. Ohio Rep.* 49 (1964): 88.

8. Gordon, D. T., and Williams, L. E. "The Relationship of a Maize Virus Isolate from Ohio to Sugarcane Mosaic Virus Strains and the B Strain of Maize Dwarf Mosaic Virus." *Phytopathology* 60 (1970): 1293.

9. Shepherd, R. J. "Properties of a Mosaic Virus of Corn and Johnson Grass and its Relation to the Sugarcane Mosaic Virus." *Phytopathology* 55 (1965): 1250.

10. Kunkel, L. O. "The Corn Mosaic of Hawaii Distinct from Sugar Cane Mosaic." *Phytopathology* 17 (1927): 41 (Abstract).

11. Tosic, M. "Investigations of Maize Mosaic in Yugoslavia," in *Proc. Int. Maize Virus Dis. Colloq. and Workshop*, eds. D. T. Gordon, J. K. Knoke, L. R. Nault, and R. M. Ritter. (Ohio State Univ., Ohio Agric. Res. Dev. Cent., Wooster, 1983), 117.

12. Chona, B. L., and Seth, M. L. "A Mosaic Disease of Maize (*Zea mays* L.) in India." *Indian J. Agric. Sci.* 30 (1960): 25.

13. Gorter, G.J.M.A., and Klesser, P. J. "Sorghum Concentric Ring Blotch, a Newly Observed Virus Disease." *S. Afr. J. Agric. Sci.* 7 (1964): 329.

14. Von Wechmar, M. B., and Hahn, J. S. "Virus Diseases of Cereals in South Africa. II. Identification of Two Elongated Plant Viruses as Strains of Sugar Cane Mosaic Virus." *S. Afr. J. Agric. Sci.* 10 (1967): 241.

15. Dijkstra, J., and Grancini, P. "Serological and Electron Microscopical Investigations of the Relationship between Sorghum Red Stripe Virus and Sugar Cane Mosaic Virus." *Tidjdschr. Plantenziekten* 66 (1960): 295.

16. Gordon, D. T., Knoke, J. K., Nault, L. R., and Ritter, R. M. "Introduction; Interpretive Summary of Proceedings," in *Proc. Int. Maize Virus Dis. Colloq. and Workshop*, eds. D. T. Gordon, J. K. Knoke, L. R. Nault, and R. M. Ritter. (Ohio State Univ., Ohio Agric. Res. Dev. Cent., Wooster, 1983), v.

17. Anonymous. *Agricultural Statistics 1986*. U.S. Department of Agriculture, U.S. Government Printing Office, Washington, 1986, p.551.

18. Williams, L. E., Gordon, D. T., and Nault, L. R., eds. *Proc. Int. Maize Virus Disease Colloq. and Workshop*. Ohio Agric. Res. Dev. Cent., Wooster, 1976, p.145.

19. Gordon, D. T., Knoke, J. K., Nault, L. R. and Ritter, R. M., eds. *Proc. Int. Maize Virus Dis. Colloq. and Workshop*. Ohio State Univ., Ohio Agric. Res. Dev. Cent., Wooster, 1983, p.261.

20. Anonymous. *1986 CIMMYT World Maize Facts and Trends. The Economics of Commercial Maize Seed Production in Developing Countries.* Mexico, D. F.: CIMMYT, 1987, p.50.

21. Zhu, F. C., Chen, Y. T., Zhang, H. F., and Tsai, J. H. "Identification, Transmission, Host Range, and Epidemiology of Maize Dwarf Mosaic Virus in Northwestern China," in *Proc. Int. Maize Virus Dis. Colloq. and Workshop,* eds. D. T. Gordon, J. K. Knoke, L. R. Nault, and R. M. Ritter. (Ohio State Univ., Ohio Agric. Res. Dev. Cent., Wooster, 1983), 194.

22. Signoret, P. A. "Maize Virus Diseases in France," in *Proc. Int. Maize Virus Dis. Colloq. and Workshop,* eds. D. T. Gordon, J. K. Knoke, L. R. Nault, and R. M. Ritter (Ohio State Univ., Ohio Agric. Res. Dev. Cent., Wooster, 1983), 113.

23. Sharma, R. C., and Payak, M. M. "An Overview of Virus and Viruslike Diseases of Maize in India," in *Proc. Int. Maize Virus Dis. Colloq. and Workshop,* eds. D. T. Gordon, J. K. Knoke, L. R. Nault, and R. M. Ritter (Ohio State Univ., Ohio Agric. Res. Dev. Cent., Wooster, 1983), 186.

24. Von Wechmar, M. B. "Viruses Affecting Maize in South Africa," in *Proc. Int. Maize Virus Dis. Colloq. and Workshop,* eds. D. T. Gordon, J. K. Knoke, L. R. Nault, and R. M. Ritter. (Ohio State Univ., Ohio Agric. Res. Dev. Cent., Wooster, 1983), 161.

25. Conti, M. "Maize Viruses and Virus Diseases in Italy and Other Mediterranean Countries," in *Proc. Int. Maize Virus Dis. Colloq. and Workshop,* eds. D. T. Gordon, J. K. Knoke, L. R. Nault, and R. M. Ritter. (Ohio State Univ., Ohio Agric. Res. Dev. Cent., Wooster, 1983), 103.

26. Kitajima, E. W., and Costa, A. S. "Diseases of Maize Caused by Viruses and Mycoplasmalike Organisms in Brazil," in *Proc. Int. Maize Virus Dis. Colloq. and Workshop,* eds. D. T. Gordon, J. K. Knoke, L. R. Nault, and R. M. Ritter. (Ohio State Univ., Ohio Agric. Res. Dev. Cent., Wooster, 1983), 100.

27. Pop, I., and Tusa, C. "Influence of Maize Mosaic on the Growth and Yield of Some Maize Hybrids," in *Proc. Int. Conf. Plant Viruses.* The Netherlands: Wageningen, 1966, p.170.

28. Teyssandier, E. E., Nome, S. F., and Dal Bo, E. "Maize Virus Diseases in Argentina," in *Proc. Int. Maize Virus Dis. Colloq. and Workshop,* eds. D. T. Gordon, J. K. Knoke, L. R. Nault, and R. M. Ritter. (Ohio State Univ., Ohio Agric. Res. Dev. Cent., Wooster, 1983), 93.

29. Lin, K. "The Identification of Maize Dwarf Mosaic Virus Strain A in Mexico City." *Scientia Agric. Sin.* 6, 1982: 15.

30. Wall, R. E., Paliwal, Y. C., Slykhuis, J. T., and Mortimore, C. G. "Corn Virus Diseases in Ontario in 1965," in *Corn (Maize) Viruses in the Continental United States and Canada,* ed. W. N. Stoner. (U.S. Dep. Agric. ARS 33–118, 1968), 80.

31. Boothroyd, C. W. "Virus Diseases of Sweet Corn," in *Virus and Viruslike Diseases of Maize in the United States,* eds. D. T. Gordon, J. K. Knoke, and G. E. Scott. (Southern Coop. Ser. Bull., 247, Ohio Agric. Res. Dev. Cent., Wooster, 1981), 103.

32. Knoke, J. K., and Louie, R. "Epiphytology of Maize Virus Diseases," in *Virus and Viruslike Diseases of Maize in the United States,* eds. D. T. Gordon, J. K. Knoke, and G. E. Scott. (Southern Coop. Ser. Bull., 247, Ohio Agric. Res. Dev. Cent., Wooster, 1981), 92.

33. Darrah, L. L., and Zuber, M. S. "1985 United States Farm Maize Germplasm Base and Commercial Breeding Strategies." *Crop Sci.* 26 (1986): 1109.

34. Genter, C. F., Roane, C. W., and Tolin, S. A. "Effects of Maize Dwarf Mosaic Virus on Mechanically Inoculated Maize." *Crop Sci.* 13 (1973): 531.

35. Rosenkranz, E., and Scott, G. E. "Effect of Plant Age at Time of Inoculation with Maize Dwarf Mosaic Virus on Disease Development and Yield in Corn." *Phytopathology* 68 (1978): 1688.

36. Scott, G. E., Findley, W. R., and Dollinger, E. J. "Genetics of Resistance in Corn," in *Virus and Viruslike Diseases of Maize in the United States,* eds. D. T. Gordon, J. K. Knoke, and G. E. Scott. (Southern Coop. Ser. Bull., 247, Ohio Agric. Res. Dev. Cent., Wooster, 1981), 141.

37. Raychaudhuri, S. P., Seth, M. L., Renfro, B. L., and Varma, A. "Principal Maize Virus Diseases in India," in *Proc. Int. Maize Virus Dis. Colloq. and Workshop,* eds. L. E. Williams, D. T. Gordon, and L. R. Nault. (Ohio Agric. Res. Dev. Cent., Wooster, 1977), 69.

38. Uyemoto, J. K., Claflin, L. E., Wilson, D. L., and Raney, R. J. "Maize Chlorotic Mottle and Maize Dwarf Mosaic Viruses: Effect of Single and Double Inoculations on Symptomatology and Yield." *Plant Dis.* 65 (1981): 39.

39. Knoke, J. K., Anderson, R. J., Findley, W. R., Louie, R., Abt, J. J., and Gordon, D. T. "The Reaction of Sweet Corn Hybrids to Maize Dwarf Mosaic Strains and Maize Chlorotic Dwarf Virus." Ohio Agric. Res. Dev. Cent. Res. Bull. 1135 (1981): 22.

40. Louie, R., and Knoke, J. K. "Symptoms and Disease Diagnosis," in *Virus and Viruslike Diseases of Maize in the United States,* eds. D. T. Gordon, J. K. Knoke, and G. E. Scott. (Southern Coop. Ser. Bull., 247, Ohio Agric. Res. Dev. Cent., Wooster, 1981), 13.

41. Toler, R. W. "Maize Dwarf Mosaic, the Most Important Virus Disease of Sorghum." *Plant Dis.* 69 (1985): 1011.

42. Knoke, J. K., Louie, R., Anderson, R. J., and Gordon, D. T. "Distribution of Maize Dwarf Mosaic and Aphid Vectors in Ohio." *Phytopathology* 64 (1974): 639.

43. Louie, R., and Knoke, J. K. "Strains of Maize Dwarf Mosaic Virus." *Plant Dis. Reptr.* 59 (1975): 518.

44. Louie, R. "Effects of Genotype and Inoculation Protocols on Resistance Evaluation of Maize to Maize Dwarf Mosaic Virus Strains." *Phytopathology* 76 (1986): 769.

45. Yossen, V., Dal Bo, E., Nome, S. F., and Teyssandier, E. "Frecuencia del Virus del Mosaico Enanizante del Maiz (MDMV) en la Republica Argentina." *Rev. Invest. Agropecu.* 18 (1983): 225.

46. Gudauskas, R. T., and Ford, R. E. "Physiology of Disease," in *Virus and Viruslike Diseases of Maize in the United States,* eds. D. T. Gordon, J. K. Knoke, and G. E. Scott. (Southern Coop. Ser. Bull., 247, Ohio Agric. Res. Dev. Cent., Wooster, 1981), 85.

47. Ford, R. E., and Milbrath, G. M. "Environmental Factors Influencing Disease Development: Corn Diseases Caused by Viruses and Spiroplasma," in *Virus and Viruslike Diseases of Maize in the United States,* eds. D. T. Gordon, J. K. Knoke, and G. E. Scott. (Southern Coop. Ser. Bull., 247, Ohio Agric. Res. Dev. Cent., Wooster, 1981), 88.

48. Tu, J. C., Ford, R. E., and Krass, C. J. "Comparisons of Chloroplasts and Photosynthetic Rates of Plants Infected and not Infected by Maize Dwarf Mosaic Virus." *Phytopathology* 58 (1968): 285.

49. Tu, J. C., and Ford, R. E. "Effect of Maize Dwarf Mosaic Virus Infection on Respiration and Photosynthesis of Corn." *Phytopathology* 58 (1968): 282.

50. Gates, D. W., and Gudauskas, R. T. "Photosynthesis, Respiration, and Evidence of a Metabolic-Inhibitor in Corn Infected with Maize Dwarf Mosaic Virus." *Phytopathology* 59 (1969): 575.

51. Moline, H. E., and Ford, R. E. "Sugarcane Mosaic Virus Infection of Seedling Roots of *Zea mays* and *Sorghum halepense.*" *Physiol. Plant Pathol.* 4 (1974): 197.

52. Beniwal, S. P., Gudauskas, R. T., and Truelove, B. "Polyphenol Oxidase Activity in Leaves of Corn Seedlings Infected with Maize Dwarf Mosaic Virus." *Phytopathology* 60 (1970): 1284 (Abstract).

53. Millikan, D. F., and Mann, D. R. "Influence of the Maize Dwarf Mosaic Virus and Simazine upon the Buffer Soluble Proteins and Catalase Activity of a Virus-Susceptible and a Virus-Tolerant Variety of *Zea mays.*" *Physiol. Plant.* 22 (1969): 1139.

54. Ford, R. E., and Tu, J. C. "Free Amino Acid Contents in Corn Infected with Maize Dwarf Mosaic Virus and Sugarcane Mosaic Virus." *Phytopathology* 59 (1969): 179.

55. Tu, J. C., and Ford, R. E. "Maize Dwarf Mosaic Virus Infection in Susceptible and Resistant Corn: Virus Multiplication, Free Amino Acid Concentrations, and Symptom Severity." *Phytopathology* 60 (1970): 1605.

56. Tu, J. C., and Ford, R. E. "Maize Dwarf Mosaic Virus Predisposes Corn to Root Rot Infection." *Phytopathology* 61 (1971): 800.

57. McCord, R. W., and Gudauskas, R. T. "Nitrogen Metabolism in Corn Infected by Maize Dwarf Mosaic Virus." *Phytopathology* 59 (1969): 116.

58. Lindsey, D. W., and Gudauskas, R. T. "Effects of Maize Dwarf Mosaic Virus on Water Relations of Corn." *Phytopathology* 65 (1975): 434.

59. Kuhn, C. W., and Jellum, M. D. "Evaluations for Resistance to Corn Stunt and Maize Dwarf Mosaic Virus Diseases in Corn." *Ga. Agric. Exp. Stn. Res. Bull.* 82 (1970): 37.

60. Tu, J. C., and Ford, R. E. "Influence of Host Nutrition on Susceptibility of, Multiplication in, and Symptom Expression by Corn to Infection by Maize Dwarf Mosaic Virus." *Phytopathology* 58 (1968): 1343.

61. Farrell, J. R. "Some Nutritional Effects on Maize Dwarf Mosaic Virus Infected Corn Growth and Chlorophyll Concentration." *J. Tenn. Acad. Sci.* 46 (1971): 149 (Abstract).

62. Batte, R. D., Toler, R. W., and Bockholt, A. J. "Effects of Time of Inoculation with Maize Dwarf Mosaic Virus Strain A on the Agronomic Characteristics of Grain Sorghum Hybrids." *Phytopathology* 60 (1970): 581 (Abstract).

63. Liu, L. J. "Effects of Temperatures on Symptom Expression by Sugarcane Infected with Different Strains of Mosaic Virus." *J. Agric. Univ. P. R.* 54 (1970): 128.

64. Scott, G. E., and Rosenkranz, E. E. "Frequency of Progenies Resistant to Corn Stunt and Maize Dwarf Mosaic in Maize Populations." *Crop Sci.* 15 (1975): 233.

65. Tu, J. C., and Ford, R. E. "Effect of Temperature on Maize Dwarf Mosaic Virus Infection, Incubation, and Multiplication in Corn." *Phytopathology* 59 (1969): 699.

66. Snow, J. P. "The Effects of Maize Dwarf Mosaic Virus (MDMV) Infection, Variety, Temperature, and Light on the Ultrastructure and Red Pigment Expression of *Sorghum bicolor* (L.) Moench." (Ph.D. thesis, Texas A & M University, 1970).

67. Bradfute, O. E., and Robertson, D. C. "Electron Microscopy of Viruses and Virus-Infected Cells of Maize," in *Virus and Viruslike Diseases of Maize in the United States*, eds. D. T. Gordon, J. K. Knoke, and G. E. Scott. (Southern Coop. Ser. Bull., 247, Ohio Agric. Res. Dev. Cent., Wooster, 1981), 25.

68. Edwardson, J. R. "Some Properties of the Potato Virus y-Group." *Fla. Agric. Exp. Stn. Monogr.* 4 (1974): 398.

69. Krass, C. J., and Ford, R. E. "Ultrastructure of Corn Systemically Infected with Maize Dwarf Mosaic Virus." *Phytopathology* 59 (1969): 431.

70. Langenberg, W. G., and Schroeder, H. F. "Electron Microscopy of Unstable Inclusions in Maize by Maize Dwarf Mosaic Virus." *Phytopathology* 63 (1973): 1066.

71. Francki, R. I. B., Milne, R. G., and Hatta, T. "Potyvirus Group," in *Atlas of Plant Viruses,* Vol. II. (Boca Raton, Fla: CRC Press, 1985), 183.

72. MacKenzie, D. R., Wernham, C. C., and Ford, R. E. "Differences in Maize Dwarf Mosaic Virus Isolates of the Northeastern United States." *Plant Dis. Reptr.* 50 (1966): 814.

73. Shi, Y., Zhang, Q., Wang, F., Xi, Z., and Xu, S. "Identification of Strains of Maize Dwarf Mosaic Virus." *Acta Phytopathol. Sin.* 16 (1986): 99.

74. McDaniel, L. L., and Gordon, D. T. "Identification of a New Strain of Maize Dwarf Mosaic Virus." *Plant Dis.* 69 (1985): 602.

75. Berger, P. H., Toler, R. W., and Harris, K. F. "Maize Dwarf Mosaic Virus Transmission by Greenbug Biotypes." *Plant Dis.* 67 (1983): 496.

76. Berger, P. H. "The Retention of Infectivity of Maize Dwarf Mosaic Virus in Aphids (Implications for Long-Distance Dispersal)." (M.S. thesis, University of Minnesota, St. Paul, 1980).

77. Hill, E. K., Hill, J. H., and Durand, D. P. "Production of Monoclonal Antibodies to Viruses in the Potyvirus Group: Use in Radioimmunoassay." *J. Gen. Virol.* 65 (1984): 525.

78. Berger, P. H., Zeyen, R. J., and Groth, J. V. "Aphid Retention of Maize Dwarf Mosaic Virus (Potyvirus): Epidemiological Implications." *Ann. Appl. Biol.* 111 (1987): 337.

79. Jarjees, M. M., and Uyemoto, J. K. "Serological Relatedness of Strains of Maize Dwarf Mosaic and Sugarcane Mosaic Viruses as Determined by Microprecipitin and Enzyme-Linked Immunosorbent Assays." *Ann. Appl. Biol.* 104 (1984): 497.

80. Garrido, M. J., and Trujillo, G. "Rango de Huespedes del Virus del Mosaico Enano del Maiz (MDMV)," (Abstract), in *VIII Congr. Venezolano Botanica,* Merida, 1985), 115.

81. Tosic, M., and Ford, R. E. "Sorghum Cultivars Differentiating Sugarcane Mosaic and Maize Dwarf Mosaic Virus Strains," in *Proc. Int. Maize Virus Dis. Colloq. and Workshop*, eds. D. T. Gordon, J. K. Knoke, L. R. Nault, and R. M. Ritter. (Ohio State Univ., Ohio Agric. Res. Dev. Cent., Wooster, 1983), 229.

82. Snazelle, T. E., Bancroft, J. B., and Ullstrup, A. J. "Purification and Serology of Maize Dwarf Mosaic and Sugarcane Mosaic Viruses." *Phytopathology* 61 (1971): 1059.

83. Langham, M. A. C., and Toler, R. W. "An Analysis of Maize Dwarf Mosaic Virus Strains by Dot Immunobinding Assay." *Phytopathology* 76 (1986): 1132 (Abstract).

84. Tolin, S. A., and Ford, R. E. "Virus Purification," in *Virus and Viruslike Diseases of Maize in the United States,* eds. D. T. Gordon, J. K. Knoke, and G. E. Scott. (Southern Coop. Ser. Bull., 247, Ohio Agric. Res. Dev. Cent., Wooster, 1981), 33.

85. Rosenkranz, E. "Host Range of Maize Dwarf Mosaic Virus," in *Virus and Viruslike Diseases of Maize in the United States,* eds. D. T. Gordon, J. K. Knoke, and G. E. Scott. (Southern Coop. Ser. Bull., 247, Ohio Agric. Res. Dev. Cent., Wooster, 1981), 152.

86. Ross, M. A. "Johnsongrass: Two Decades of Progress in Control." *Crops Soils Mag.* 39 (1986): 12.

87. Bancroft, J. B., Ullstrup, A. J., Messieha, M., Bracker, C. E., and Snazelle, T. E. "Some Biological and Physical Properties of a Midwestern Isolate of Maize Dwarf Mosaic Virus." *Phytopathology* 56 (1966): 474.

88. Jones, R. K., and Tolin, S. A. "Factors Affecting Purification of Maize Dwarf Mosaic Virus from Corn." *Phytopathology* 62 (1972): 812.

89. Sehgal, O. P. "Host Range, Properties and Partial Purification of a Missouri Isolate of Maize Dwarf Mosaic Virus." *Plant Dis. Reptr.* 50 (1966): 862.

90. Sehgal, O. P. "Purification, Properties and Structure of Maize Dwarf Mosaic Virus." *Phytopathol. Z.* 62 (1968): 232.

91. Sehgal, O. P., and Jean, J. "Purification of Maize Dwarf Mosaic Virus by Equilibrium Centrifugation in Cesium Chloride." *Phytopathology* 60 (1970): 189.

92. Von Baumgarten, G., and Ford, R. E. "Purification and Partial Characterization of Maize Dwarf Mosaic Virus Strain A." *Phytopathology* 71 (1981): 36.

93. Kerlan, C., Lapierre, H., and Moreau, J. P. "Observations Sur l'Apparition du Virus de la Mosaique Nanisante du Mais (Maize dwarf mosaic) dans le Nord de la France." *Ann. Phytopathol.* 6 (1974): 455.

94. Damirdagh, I. S., and Shepard, R. J. "Purification of the Tobacco Etch and other Viruses of the Potato Y Group." *Phytopathology* 60 (1970): 132.

95. Pring, D. R., and Langenberg, W. G. "Preparation and Properties of Maize Dwarf Mosaic Virus Ribonucleic Acid." *Phytopathology* 62 (1972): 253.

96. Hill, J. H., Ford, R. E., and Benner, H. I. "Purification and Partial Characterization of Maize Dwarf Mosaic Virus Strain B (Sugarcane Mosaic Virus)." *J. Gen. Virol.* 20 (1973): 327.

97. Pirone, T. P., and Anzalone, L., Jr. "Purification and Electron Microscopy of Sugarcane Mosaic Virus." *Phytopathology* 56 (1966): 371.

98. Baudin, P. "Etude d'une Souche du Virus de la Mosaique de la Canne a Sucre." *Agron. Trop.* 32 (1977): 66.

99. Baudin, P., Larbaight, G., and Dodin, A. "Purification Partielle de la Mosaique da la Canne a Sucre par Filtration sur Gel." *Ann. Inst. Pasteur* (Paris) 115 (1968): 288.

100. Gordon, D. T., and Gingery, R. E. "Purification of Maize Dwarf Mosaic Virus by Continuous-Flow Centrifugation." *Phytopathology* 63 (1973): 1386.

101. Gordon, D. T., and Nault, L. R. "Involvement of Maize Chlorotic Dwarf Virus and Other Agents in Stunting Diseases of *Zea mays* in the United States." *Phytopathology* 67 (1977): 27.

102. Langenberg, W. G. "Serology, Physical Properties, and Purification of Unaggregated Infectious Maize Dwarf Mosaic Virus." *Phytopathology* 63 (1973): 149.

103. Tosic, M., Ford, R. E., Moline, H. E., and Mayhew, D. E. "Comparison of Techniques for Purification of Maize Dwarf and Sugarcane Mosaic Viruses." *Phytopathology* 64 (1974): 439.

104. Gillaspie, A. G., Jr. "Sugarcane Mosaic Virus: Purification." *Int. Soc. Sugar Cane Technol.* 14 (1972): 961.

105. Bond, W. P., and Pirone, T. P. "Purification and Properties of Sugarcane Mosaic Virus Strains." *Phytopathol. Z.* 71 (1971): 56.

106. Louie, R., Knoke, J. K., and Findley, W. R. "Methodology and Evaluation of Resistance in Maize to Maize Chlorotic Dwarf and Maize Dwarf Mosaic Viruses," in *Proc. 3rd Int. Cong. of Plant Pathology*, ed. W. Laux. (Deutsche Phytomed. Ges., Paul Parey, Berlin, 1978), 30.

107. Louie, R., Knoke, J. K., and Reichard, D. L. "Transmission of Maize Dwarf Mosaic Virus with Solid-Stream Inoculum." *Plant Dis.* 67 (1983): 1328.

108. Gingery, R. E. "Chemical and Physical Properties of Maize Viruses," in *Virus and Viruslike Diseases of Maize in the United States,"* eds. D. T. Gordon, J. K. Knoke, and G. E. Scott. (Southern Coop. Ser. Bull., 247, Ohio Agric. Res. Dev. Cent., Wooster, 1981), 38.

109. Bergquist, R. R., and Ishii, M. "Maize Dwarf Mosaic Virus in Hawaii." *Plant Dis. Reptr.* 58 (1974): 495.

110. Dale, J. L. "Additional Data on Corn Virus in Arkansas." *Plant Dis. Rep.* 49 (1965): 202.

111. Gudauskas, R. T., and Gates, D. W. "Distribution and Some Properties of a Sap-Transmissible Virus Isolated from Corn in Alabama," in *Corn (Maize) Viruses in the Continental United States and Canada.* ed. W. N. Stoner. (U.S. Dept. of Agriculture ARS 33-118, 1968), 11.

112. Klein, M., Harpaz, I., Greenberger, A., and Sela, I. "A Mosaic Virus Disease of Maize and Sorghum in Israel." *Plant Dis. Reptr.* 57 (1973): 125.

113. Tosic, M., and Ford, R. E. "Physical and Serological Properties of Maize Dwarf Mosaic and Sugarcane Mosaic Viruses." *Phytopathology* 64 (1974): 312.

114. Zummo, N., and Gordon, D. T. "Comparative Study of Five Mosaic Virus Isolates Infecting Corn, Johnson Grass, and Sorghum in the United States." *Phytopathology* 61 (1971): 389.

115. Signoret, P. A. "Studies of a Mechanically Transmissible Virus Isolated from Sorghum in Southern France." *Plant Dis. Reptr.* 55 (1970): 1090.

116. Lawas, O. M., and Fernandez, W. L. "A Study of the Transmission of the Corn Mosaic and of Some of the Physical Properties of its Virus." *Philipp. Agric.* 32 (1949): 231.

117. Rosenkranz, E. E. "Present Status of MDM, Corn Stunt and Other Virus and Viruslike Diseases of Corn and Sorghum," in *Proc. 27th Annu. Corn Sorghum Res. Conf.* (American Seed Trade Assoc., Washington, D.C., 1972), 65.

118. Gingery, R. E. "Influence of pH and Divalent Anions on the Buoyant Density of Maize Dwarf Mosaic Virus in CsC1." *J. Gen. Virol.* 31 (1976): 257.

119. Hill, J. H., and Benner, H. I. "Properties of Potyvirus RNAs: Turnip Mosaic, Tobacco Etch, and Maize Dwarf Mosaic Viruses." *Virology* 75 (1976): 419.

120. Kunkel, L. O. "Insect Transmission of Yellow Stripe Disease." *Hawaii. Plant. Rec.* 26 (1922): 58.

121. Rosenkranz, E. "Susceptibility of Representative Native Mississippi Grasses in Six Subfamilies to Maize Dwarf Mosaic Virus Strains A and B and Sugarcane Mosaic Virus Strain B." *Phytopathology* 73 (1983): 1314.

122. Rosenkranz, E. "New Hosts and Taxonomic Analysis of the Mississippi Native Species Tested for Reaction to Maize Dwarf Mosaic and Sugarcane Mosaic Viruses." *Phytopathology* 77 (1987): 598.

123. Greber, R. S. "Characteristics of Viruses Affecting Maize in Australia," in *Proc. Int. Maize Virus Dis. Colloq. and Workshop,* eds. D. T. Gordon, J. K. Knoke, L. R. Nault, and R. M. Ritter. (Ohio State Univ., Ohio Agric. Res. Dev. Cent., Wooster, 1983), 206.

124. Lockhart, B. E. L., and Elyamani, M. "Virus and Viruslike Diseases of Maize in Morocco," in *Proc. Int. Maize Virus Dis. Colloq. and Workshop,* eds. D. T. Gordon, J. K. Knoke, L. R. Nault, and R. M. Ritter. (Ohio State Univ., Ohio Agric. Res. Dev. Cent., Wooster, 1983), 127.

125. Castillo, J. "Present Knowledge of Virus and Mollicute Diseases of Maize in Peru," in *Proc. Int. Maize Dis. Colloq. and Workshop,* eds. D. T. Gordon, J. K. Knoke, L. R. Nault, and R. M. Ritter. (Ohio State Univ., Ohio Agric. Res. Dev. Cent., Wooster, 1983), 87.

126. Autrey, L. J. C. "Maize Mosaic Virus and Other Maize Virus Diseases in the Islands of the Western Indian Ocean," in *Proc. Int. Maize Dis. Colloq. and Workshop,* eds. D. T. Gordon, J. K. Knoke, L. R. Nault, and R. M. Ritter. (Ohio State Univ., Ohio Agric. Res. Dev. Cent., Wooster, 1983), 167.

127. Anonymous. *National List of Scientific Plant Names,* Vol. 1. United States Department of Agriculture, Soil Conservation Service, SCS-TP-159, 1982, 416.

128. Hitchcock, A. S. *Manual of the Grasses of the United States.* Washington, D.C.: U.S. Department of Agriculture, Misc. Pub. No. 200, U.S. Government Printing Office, 1950, 1051.

129. Watson, L., and Gibbs, A. J. "Taxonomic Patterns in the Host Ranges of Viruses Among Grasses, and Suggestions on Generic Sampling for Host-Range Studies." *Ann. Appl. Biol.* 77 (1974): 23.

130. Tosic, M., and Ford, R. E. "Grasses Differentiating Sugarcane Mosaic and Maize Dwarf Mosaic Viruses." *Phytopathology* 62 (1972): 1466.

131. Ford, R. E., and Tosic, M. "New Hosts of Maize Dwarf Mosaic Virus and Sugarcane Mosaic Virus and a Comparative Host Range Study of Viruses Infecting Corn." *Phytopathol. Z.* 75 (1972): 315.

132. Rosenkranz, E. "Grasses Native or Adventive to the United States as New Hosts of Maize Dwarf Mosaic and Sugarcane Mosaic Viruses." *Phytopathology* 68 (1978): 175.

133. Rosenkranz, E. "Taxonomic Distribution of Native Mississippi Grass Species Susceptible to Maize Dwarf Mosaic and Sugarcane Mosaic Viruses." *Phytopathology* 70 (1980): 1056.

134. Brandes, E. W. "Mechanics of Inoculation with Sugar-Cane Mosaic by Insect Vectors." *J. Agric. Res.* 23 (1923): 279.

135. Brandes, E. W. "Artificial and Insect Transmission of Sugar-Cane Mosaic." *J. Agric. Res.* 19 (1920): 131.

136. Stoner, W. N., Williams, L. E., and Alexander, L. J. "Transmission by the Corn Leaf Aphid, *Rhopalosiphum maidis* (Fitch) of a Virus Infecting Corn in Ohio." Ohio Agric. Exp. Stn. Res. Circ. 136 (1964): 4.

137. Nault, L. R., and Knoke, J. K. "Maize Vectors," in Virus and Viruslike Diseases of Maize in the United States, eds. D.T. Gordon, J.K. Knoke, and G.E. Scott. (Southern Coop. Ser. Bull., 247, Ohio Agric. Res. Dev. Cent., Wooster, 1981), 77.

138. Knoke, J. K., Anderson, R. J., Louie, R., Madden, L. V., and Findley, W. R. "Insect Vectors of Maize Dwarf Mosaic Virus and Maize Chlorotic Dwarf Virus," in *Proc. Int. Maize Virus Dis. Colloq. and Workshop,* eds. D. T. Gordon, J. K. Knoke, L. R. Nault, and R. M. Ritter. (Ohio State Univ., Ohio Agric. Res. Dev. Cent., Wooster, 1983), 130.

139. Rest, E. B. "The Epidemiology of Maize Dwarf Mosaic in Northern Illinois." (M.S. thesis, University of Illinois, Urbana-Champaign, 1981).

140. Messieha, M. "Aphid Transmission of Maize Dwarf Mosaic Virus." *Phytopathology* 57 (1967): 956.

141. Nault, L. R., and Bradley, R. H. E. "Acquisition of Maize Dwarf Mosaic Virus by the Greenbug, *Schizaphis graminum.''* Ann. Entomol. Soc. Am., 62 (1969): 403.

142. Nault, L. R., Harlan, H. J., and Findley, W. R. "Comparative Susceptibility of Corn to Aphid and Mechanical Inoculation of Maize Dwarf Mosaic Virus." *J. Econ. Entomol.* 64 (1971): 21.

143. Tu, J. C., and Ford, R. E. "Factors Affecting Aphid Transmission of Maize Dwarf Mosaic Virus to Corn." *Phytopathology* 61 (1971): 1516.

144. Thongmeearkom, P., Ford, R. E., and Jedlinski, H. "Aphid Transmission of Maize Dwarf Mosaic Virus Strains." *Phytopathology* 66 (1976): 332.

145. Shaunak, K. K., and Pitre, H. N. "Comparative Transmission of Maize Dwarf Mosaic Virus and Sugarcane Mosaic Virus by the Green Peach Aphid, *Myzus persicae.''* Plant Dis. Reptr. 54 (1970): 876.

146. Tosic, M., and Sutic, D. "Medusobni Odnos Virusa Mozaika Kukuruza i Nekih Njegovih Vektora," in *Prvi Kongres Mikrobiol. Jugosl.* 1969 (1969): 685.

147. Berger, P. H., and Zeyen, R. J. "Aphid Retention of MDMV (Long-Distance Transport Hypothesis)," in *Sweet Corn Disease Report 1979,* eds. R. J. Zeyen, J. V. Groth, and D. W. Davis. (University of Minnesota, St. Paul, 1980), 10.

148. Berger, P. H. "The Retention of Maize Dwarf Mosaic Virus by the Greenbug, *Schizaphis graminum* Rondani and Its Implications for Transmission Mechanisms." (Ph.D. thesis, Texas A & M University, College Station, 1983).

149. Zeyen, R. J., Stromberg, E. L., and Kuehnast, E. L. "Long-Range Aphid Transport Hypothesis for Maize Dwarf Mosaic Virus: History and Distribution in Minnesota, USA." *Ann. Appl. Boil.* 111 (1987): 325.

150. Shaunak, K. K., and Pitre, H. N. "Seasonal Alate Aphid Collections in Yellow Pan Traps in Northeastern Mississippi: Possible Relationship to Maize Dwarf Mosaic Disease." *J. Econ. Entomol.* 64 (1971): 1105.

151. Shaunak, K. K., and Pitre, H. N. "Comparative Transmission of Maize Dwarf Mosaic Virus by *Aphis fabae, Aphis gossypii,* and *Schizaphis graminum.*" *Plant Dis. Reptr.* 57 (1973): 533.

152. Knoke, J. K., Anderson, R. J., and Louie, R. "Virus Disease Epiphytology: Developing Field Tests for Disease Resistance in Maize," in *Proc. Int. Maize Virus Dis. Colloq. and Workshop*, eds. L. E. Williams, D. T. Gordon, and L. R. Nault. (Ohio Agric. Res. Dev. Cent., Wooster, 1977), 116.

153. Anonymous. "Corn Breeding and Cultivation," in *Annu. Res. Rep. 1980* (Off. Rural Dev., Minist. Agric. Fish., Repub. Korea, 1981), 120.

154. Palmer, M. A. *Aphids of the Rocky Mountain Region*, Vol. 5. (Denver, Colo.: The Thomas Say Foundation, 1952), p. 452.

155. Quist, J. A. *A revised list of the aphids of the Rocky Mountain Region.* Colo. State Univ. Exp. Stn. Bull., 567S, 1978, 77.

156. Persley, D. M. "Maize Dwarf Mosaic Disease." *Queensl. Agric. J.* 102 (1976): 531.

157. Yang, S., Yang, L., Zhang, Q., Niu, R., Shi, Y., and Wang, F. "On the Occurrence and Dynamics of Aphid Vector in Relation to Epiphytotics of the Maize Dwarf Mosaic Disease in Shanxi." *Acta Phytopathol. Sin.* 12 (1985): 113.

158. Signoret, P., and Alliot, B. "Mais. Les Maladies a Virus." *Phytoma Def. Cult.* 381 (1986): 21.

159. Shawkat, A. L. B., Fegla, G. I., and Yuhya, S. "Maize Dwarf Mosaic in Iraq and Evaluation of Some Corn Cultivars for Resistance." *Iraqi J. Agric. Sci.* 1 (1983): 71.

160. Straub, R. W. "Occurrence of Four Aphid Vectors of Maize Dwarf Mosaic Virus in Southeastern New York." *J. Econ. Entomol.* 75 (1982): 156.

161. Madden, L. V., Knoke, J. K., and Louie, R. "The Statistical Relationship Between Aphid Trap Catches and Maize Dwarf Mosaic Virus Inoculation Pressure," in *Plant Virus Epidemiology*, eds. R. T. Plumb and J. M. Thresh. (Oxford, UK: Blackwell Scientific Publications, 1983), p. 159.

162. Milinko, I., Rakk, Z. V., and Kovacs, G. "Population Dynamics of Aphids, Vectors of Maize Dwarf Mosaic Virus and Aphid Resistance of Some Maize Hybrids." *Acta Phytopathol. Acad. Sci. Hung.* 18 (1983): 201.

163. Milinko, I., Vidos, Z., and Kovacs, G. "Investigations on Population Dynamics of Maize Dwarf Mosaic Virus Vector Aphids and Vector Resistance of Corn Hybrids." *Proc. Int. Conf. Integr. Plant Prot.* 1 (1983): 63.

164. Izadpanah, K. "Difference in the Etiology of Maize Mosaic in Shiraz and Karaj.." *Iran. J. Plant Pathol.* 18 (1982): 3.

165. Albajes, R., Artigues, M., Avilla, J., Eizaguirre, M., Pons, X., and Sarasua, M. J. "Ensayo de Control de MDMV Mediante Tratamientos Insecticidas." *Actas Do II Congr. Iberico Entomol.* 3 (1985): 281.

166. Singh, C. A. K. "Occurrence of Maize Dwarf Mosaic Virus in India." *Curr. Sci. (Bangalore)* 52 (1983): 818.

167. Daniels, N. E., and Toler, R. W. "Transmission of Maize Dwarf Mosaic by the Greenbug, *Schizaphis graminum.*" *Plant Dis. Rep.* 53 (1969): 59.

168. MacKenzie, D. R., Anderson, P. M., and Wernham, C. C. "A Mobile Air Blast Inoculator for Plot Experiments with Maize Dwarf Mosaic Virus." *Plant Dis. Reptr.* 50 (1966): 363.

169. Toler, R. W., Bockholt, A. J., Walker, H. J., and Leisy, R. "Tractor Equipment for Maize Dwarf Mosaic Virus Inoculation," in *Proc. 8th Biennial Grain Sorghum Res. Util. Conf.* 1973, 41.

170. Toler, R. W., and Miller, F. R. "Effect of Various Combinations of Inoculation Pressure and Concentration on Varietal Disease Response of Sorghum Following Spray Gun Inoculation with Maize Dwarf Mosaic Virus." *Crop Sci.* 23 (1983): 83.

171. Lindner, R. C., and Kirkpatrick, H. C. "The Airbrush as a Tool in Virus Inoculations." *Phytopathology* 49 (1959): 507.

172. Thresh, J. M. "Plant Virus Disease Forecasting," in *Plant Virus Epidemics: Monitoring, Modelling and Predicting Outbreaks,* eds. G. D. McLean, R. G. Garrett, and W. G. Ruesink. (North Ryde, N.S.W.: Academic Press Australia, 1986), 359.

173. Gordon, D. T., Bradfute, O. E., Gingery, R. E., Knoke, J. K., and Nault, L. R. "Maize Virus Disease Complexes in the United States: Real and Potential Disease Problems." *Proc. Annu. Corn Sorghum Res. Conf.* 33 (1978): 102.

174. Slykhuis, J. T. "Virus and Virus-like Diseases of Cereal Crops." *Annu. Rev. Phytopathol.* 14 (1976): 189.

175. Thornberry, H. H. *Plant Pests of Importance to North American Agriculture, Index of Plant Virus Diseases.* U.S. Dept. of Agric., Agric. Handbook, 307 (1966): 446.

176. Milinko, I., Peti, J., Jozsa, S., and Kobza, S. "Kukoricahibridek Kukorica Csikos Mozaik Virus Rezisztencia-Ertekelese." *Novenytermeles* 33 (1984): 147.

177. Knoke, J. K., Louie, R., Madden, L. V., and Gordon, D. T. "Spread of Maize Dwarf Mosaic Virus from Johnsongrass to Corn." *Plant Dis.* 67 (1983): 367.

178. Boothroyd, C. W. "Seed Transmission of Maize Dwarf Mosaic Virus in Sweet Corn and Yield Reduction in Plants from an Infected Seed Lot." *Proc. Am. Phytopathol. Soc.* 4 (1977): 184 (Abstract).

179. Hill, J. H., Martinson, C. A., and Russell, W. A. "Seed Transmission of Maize Dwarf Mosaic and Wheat Streak Mosaic Viruses in Maize and Response of Inbred Lines." *Crop Sci.* 14 (1974): 232.

180. Humaydan, H. S. "Studies on Seed Transmission of Maize Dwarf Mosaic Virus in Sweet Corn, Sorghum and Annual Grass Hosts" (Abstract), in *Proc. 9th Int. Congr. Plant Prot. 1979,* Washington, D. C., 1979 (unpaged).

181. Shepherd, R. J., and Holdeman, Q. L. "Seed Transmission of the Johnson Grass Strain of the Sugarcane Mosaic Virus in Corn." *Plant Dis. Reptr.* 49 (1965): 468.

182. Timian, R. G., Jons, V. L., and Lamey, H. A. "Maize Dwarf Mosaic Virus in North Dakota." *Plant Dis. Rep.* 62 (1978): 674.

183. Paliwal, Y. C., Raychaudhuri, S. P., and Renfro, B. L. "Some Properties and Behavior of Maize Mosaic Virus in India." *Phytopathology* 58 (1968): 1682.

184. Renfro, B. L. "Maize Production and Improvement in the Tropics and Subtropics," in *Proc. Int. Maize Virus Dis. Colloq. and Workshop*, eds. L. E. Williams, D. T. Gordon, and L. R. Nault. (Ohio Agric. Res. Dev. Cent., Wooster, 1977), 9.

185. Knoke, J. K., and Louie, R., unpublished data, 1988.

186. Madden, L. V., Knoke, J. K., and Louie, R. "Classification and Prediction of Maize Dwarf Mosaic Intensity," in *Proc. Int. Maize Virus Dis. Colloq. and Workshop*, eds. D. T. Gordon, J. K. Knoke, L. R. Nault, and R. M. Ritter. (Ohio State Univ., Ohio Agric. Res. Dev. Cent., Wooster, 1983), 238.

187. Blair, B. D. "Aphids Collected from a Scioto County, Ohio, Corn Field and Areas Bordering the Field." *J. Econ. Entomol.* 63 (1970): 1099.

188. Madden, L. V., Knoke, J. K., and Louie, R. "Effect of Source Strength, Distance and Direction on the Spread of Maize Dwarf Mosaic Virus." *J. Phytopathol.* 117 (1986): 92.

189. Knoke, J. K., Louie, R., Madden, L. V., and Gordon, D. T. "The Spread of Maize Dwarf Mosaic Virus from Johnsongrass to Maize." *Phytopathology* 71 (1981): 886 (Abstract).

190. Madden, L. V., Louie, R., Abt, J. J., and Knoke, J. K. "Evaluation of Tests for Randomness of Infected Plants." *Phytopathology* 72 (1982): 195.

191. Madden, L. V., Louie, R., and Knoke, J. K. "Temporal and Spatial Analysis of Maize Dwarf Mosaic Epidemics." *Phytopathology* 77 (1987): 148.

192. Scott, G. E. "Nonrandom Spatial Distribution of Aphid-Vectored Maize Dwarf Mosaic." *Plant Dis.* 69 (1985): 893.

193. Findley, W. R., Josephson, L. M., and Dollinger, E. J. "Breeding for Disease Resistance in Corn," in *Virus and Viruslike Diseases of Maize in the United States*, eds. D. T. Gordon, J. K. Knoke, and G. E. Scott. (Southern Coop. Ser. Bull., 247, Ohio Agric. Res. Dev. Cent., Wooster, 1981), 137.

194. Seifers, D. L., and Harvey, T. L. "Increase Levels of Maize Dwarf Mosaic in Insecticide Treated Sorghum (*Sorghum bicolor* (L.) Moench)." *Phytopathology* 76 (1986): 1076 (Abstract).

195. Rains, B. D., and Christensen, C. M. "Effect of Soil-Applied Carbofuran on Transmission of Maize Chlorotic Dwarf Virus and Maize Dwarf Mosaic Virus to Susceptible Field Corn Hybrid." *J. Econ. Entomol.* 76 (1983): 290.

196. Kuhn, C. W., Jellum, M. D., and All, J. N. "Effect of Carbofuran Treatment on Corn Yield, Maize Chlorotic Dwarf and Maize Dwarf Mosaic Virus Diseases, and Leafhopper Populations." *Phytopathology* 65 (1975): 1017.

197. Pitre, H. N. "Effect of Insecticidal Sprays on Stunt and Mosaic Virus Diseases of Corn in Small Field Plots in Mississippi." *J. Econ. Entomol.* 61 (1968): 585.

198. Szatmari-Goodman, G., and Nault, L. R. "Tests of Oil Sprays for Suppression of Aphid-Borne Maize Dwarf Mosaic Virus in Ohio Sweet Corn." *J. Econ. Entomol.* 76 (1983): 144.

199. Findley, W. R., Louie, R., and Knoke, J. K. "Breeding Corn for Resistance to Corn Viruses in Ohio." *Proc. Annu. Corn Sorghum Res. Conf.* 39 (1985): 52.

200. Findley, W. R., Louie, R., Knoke, J. K., and Dollinger, E. J. "Breeding Corn for Resistance to Virus in Ohio," in *Proc. Int. Maize Virus Dis. Colloq. and Workshop*, eds. L. E. Williams, D. T. Gordon, and L. R. Nault. (Ohio Agric. Res. Dev. Cent., Wooster, 1977), 123.

201. Scott, G. E. "Breeding for and Genetics of Virus Resistance in Field Corn," in *Proc. Int. Maize Virus Dis. Colloq. and Workshop*, eds. D. T. Gordon, J. K. Knoke, L. R. Nault, and R. M. Ritter. (Ohio State Univ., Ohio Agric. Res. Dev. Cent., Wooster, 1983), 248.

202. Josephson, L. M., and Scott, G. E. "Sources of Disease Resistance in Corn," in *Virus and Viruslike Diseases of Maize in the United States*, eds D. T. Gordon, J. K. Knoke, and G. E. Scott. (Southern Coop. Ser. Bull., 247, Ohio Agric. Res. Dev. Cent., Wooster, 1981), 132.

203. Ford, R. E., and Mikel, M. A. "Disease Resistance in Maize," in *Rev. of Trop. Plant Pathol*, Vol. I., eds. S. P. Raychaudhuri and J. P. Verma. (New Delhi: Today and Tomorrow's Printers and Publishers, 1984), 197.

204. All, J. N. "Integrating Techniques of Vector and Weed-Host Suppression into Control Programs for Maize Virus Diseases," in *Proc. Int. Maize Virus Dis. Colloq. and Workshop*, eds. D. T. Gordon, J. K. Knoke, L. R. Nault, and R. M. Ritter. (Ohio State Univ., Ohio Agric. Res. Dev. Cent., Wooster, 1983), 243.

205. All, J. N., Kuhn, C. W., and Jellum, M. D. "Control Strategies for Vectors of Virus and Viruslike Pathogens of Maize and Sorghum," in *Virus and Viruslike Diseases of Maize in the United States*, eds. D. T. Gordon, J. K. Knoke, and G. E. Scott. (Southern Coop. Ser. Bull., 247, Ohio Agric. Res. Dev. Cent., Wooster, 1981), 127.

206. Knoke, J. K. "Control: The Use of Pesticides for Vector Management," in *Proc. Int. Maize Virus Dis. Colloq. and Workshop*, eds. L. E. Williams, D. T. Gordon, and L. R. Nault. (Ohio Agric. Res. Dev. Cent., Wooster, 1977), 129.

207. Roane, C. W., Tolin, S. A., and Genter, C. F. "Inheritance of Resistance to Maize Dwarf Mosaic Virus in Maize Inbred Line Oh7B." *Phytopathology* 73 (1983): 845.

208. Scott, G. E., and Rosenkranz, E. "A New Method to Determine the Number of Genes for Resistance to Maize Dwarf Mosaic in Maize." *Crop Sci.* 22 (1982): 756.

209. Rosenkranz, E., and Scott, G. E. "Determination of the Number of Genes for Resistance to Maize Dwarf Mosaic Virus Strain A in Five Corn Inbred Lines." *Phytopathology* 74 (1984): 71.

210. Misra, K. P. "Inheritance of Resistance to Maize Dwarf Mosaic Virus in Sweet Corn Inbreds Derived from B68 Resistant Dent X Sweet Corn Crosses." (Ph.D. thesis, Cornell University, 1984).

211. Mikel, M. A., D'Arcy, C. J., Rhodes, A. M., and Ford, R. E. "Genetics of Resistance of Two Dent Corn Inbreds to Maize Dwarf Mosaic Virus and Transfer of Resistance into Sweet Corn." *Phytopathology* 74 (1984): 467.

212. Scott, G. E., and Nelson, L. R. "Locating Genes for Resistance to Maize Dwarf Mosaic in Maize Seedlings by using Chromosomal Translocations." *Crop Sci.* 1 (1971): 801.

213. Scott, G. E., and Rosenkranz, E. E. "Use of Chromosomal Translocations to Determine Similarity of Maize Genotypes for Reaction to Maize Dwarf Mosaic." *Crop Sci.* 13 (1973): 724.

214. Eberhart, S. A. "Developing Virus Resistant Commercial Maize Hybrids," in *Proc. Int. Maize Virus Dis. Colloq. and Workshop*, eds. D. T. Gordon, J. K. Knoke, L. R. Nault, and R. M. Ritter. (Ohio State Univ., Ohio Agric. Res. Dev. Cent., Wooster, 1983), 258.

215. Dale, J. L., and York, J. O. "Maize Dwarf Mosaic Virus Ratings of Corn Hybrids and Inbreds Tested in Arkansas in 1966." Arkansas Agric. Exp. Stn. Mimeogr. Ser. 158 (1967), 5.

216. Dale, J. L., and York, J. O. "1985 Yields of Corn Hybrids and Ratings of Inbreds Tested Under Virus Disease Conditions." Arkansas Agric. Exp. Stn. Res. Ser. 329 (1986): 4.

217. Dale, J. L., McFerran, J., Wann, E. V., and Bona, R. L. "Field Evaluation of Sweet Corn Hybrids and Inbreds, with Emphasis on Virus Disease Resistance." Arkansas Agric. Exp. Stn. Rep. Ser. 258 (1981): 12.

218. Dale, J. L., McFerran, J., and Wann, E. V. "Yield and Ear Quality of Sweet Corn Hybrids Grown under Virus Disease Conditions—1987." Arkansas Agric. Exp. Stn. Res. Ser. 362 (1987): 3.

219. Kuhn, C. W. "Corn Viruses in Georgia." Ga. Agric. Exp. Stn. Res. Rep. 23 (1968): 25.

220. Kuhn, C. W., and Jellum, M. D. "Disease Evaluation of Commercial Hybrids," in *1976 Corn and Grain Sorghum Performance Tests.* Ga. Agric. Exp. Stn. Res. Rep. 238 (1977): 52.

221. Poneleit, C. G., and Evans, K. O. "Kentucky Hybrid Corn Performance Test—1986." Ky. Agric. Exp. Stn. Prog. Rep. 177 (1969): 35.

222. Poneleit, C. G., and Evans, K. O. "Kentucky Hybrid Corn Performance Test—1986." Ky. Agric. Exp. Stn. Prog. Rep. 300 (1986): 34.

223. Wallin, J. R., Zuber, M. S., Keaster, A. J., Sheeley, R. D., and Loonan, D. V. "Virus Tolerance Ratings of Corn Strains Grown in Missouri," in *1976 Virus Tolerance Ratings for Corn Strains Grown in the Lower Corn Belt.* (Washington, D.C.: U.S. Dept. Agric., ARS-NC-53, 1977), 1.

224. Wallin, J. R., Darrah, L. L., and Loonan, D. V. "Virus Tolerance Ratings of Corn Strains Grown in Missouri in 1982," in *1982 Virus Tolerance Ratings for Corn Strains Grown in the Lower Corn Belt,* ARR-NC-11, 1983, 1.

225. Zuber, M. S., Keaster, A. J., and Loesch, P. J., Jr. "Virus Ratings of Corn Strains in Missouri—1967." Mo. Agric. Exp. Stn. Spec. Rep. 100 (1968): 26.

226. Zuber, M. S., Calvert, O. H., Keaster, A. J., and Palm, E. W. "Virus Tolerance Ratings for Corn Strains Grown in Missouri," in *1975 Virus Tolerance Ratings for Corn Strains Grown in the Lower Corn Belt.* (Washington, D.C.: U.S. Dept. Agric., ARS-NC-39, 1976), 1.

227. Findley, W. R., Dollinger, E. J., and Williams, L. E. "Reaction of Corn Strains to Virus Disease at Portsmouth, Ohio 1964." (Washington, D.C.: U.S. Dept. Agric., ARS-CR-69-64, 1964), 19.

228. Findley, W. R., Knoke, J. K., Louie, R., and Underwood, J. F. "Performance of Corn Hybrids Exposed to Corn Viruses in Ohio in 1984." Ohio Agric. Res. Dev. Cent. Agron. Dep. Ser. 213 (1985), 9.

229. Jordan, D. M., Louie, R., and Knoke, J. K. "Performance of Corn Hybrids Exposed to Corn Viruses in Ohio, 1985." Ohio Agric. Res. Dev. Cent. Agron. Dep. Ser. 213 (1986): 5.

230. Jordan, D. M., Louie, R., and Knoke, J. K. "Performance of Corn Hybrids Exposed to Corn Viruses in Ohio, 1986." Ohio Agric. Res. Dev. Cent. Agron. Dep. Ser. 213 (1987): 4.

231. Graves, C. R. "1972 Performance of Field Crop Varieties—Corn." Tenn. Agric. Exp. Stn. Bull. 503 (1972): 14.

232. Graves, C. R. "Performance of Corn Hybrids," in *1985 Performance of Field Crop Varieties,* Tenn. Agric. Exp. Stn. Bull., 643 (1986): 16.

233. Josephson, L. M., and Hilty, J. W. "Reaction of Corn Strains to Virus Disease in Tennessee in 1965." Tenn. Agric. Exp. Stn. Mimeogr. Rep. 1965, 20.

234. Josephson, L. M., Graves, C. R., and Kincer, H. C. "Performance of Corn Hybrids Grown under Virus Disease Conditions." Tenn. Farm Home Sci. Prog. Rep 105 (1978): 4.

235. West, D. R., McLaughlin, M. R., and Kincer, H. C. "Reaction of Corn Genotypes to the Corn Virus Disease." Tenn. Agric. Exp. Stn. Res. Rep., RR80-05, 1980, 17.

236. West, D. R., Kincer, H. C., and Kincer, D. R. "Reaction of Corn Genotypes to the Corn Virus Disease Complex in Tennessee in 1982." Tenn. Agric. Exp. Stn. Dep. Plant Soil Sci. RR82-14, (1982): 23.

237. Genter, C. F., Brown, L. W., Camper, H. M., Jr., Carter, M. T., Jones, G. D., Harrison, R. L., Link, L. A., Roane, C. W., Tolin, S. A., and Wilmouth, R. R., "Virginia Corn Performance Trials of Dent, High-Lysine, Waxy and Popcorn Hybrids in 1973." Va. Polytech. Inst. State Univ. Res. Div. Rep. 154 (1973): 26.

238. Genter, C. F., Brann, D. E., Bryant, J., Camper, H. M., Jr., Carter, M. T., Crews, S. L., Harrison, R. L., Jones, G. D., Link, L. A., Roane, C. W., Wilkinson, W. B., and Wilmouth, R. R. "Virginia Corn Performance Trials in 1977." Va. Polytech. Inst. State Univ. Res. Div. Rep. 173 (1977): 15.

239. Persley, D. M., Martin, I. F., and Greber, R. S. "The Resistance of Maize Inbred Lines to Sugarcane Mosaic Virus in Australia." *Aust. J. Agric. Res.* 32 (1981): 741.

240. Deng, T. C. "Identification of Maize Dwarf Mosaic Virus B Strain and Screening for Resistance of Corn." *J. Agric. Res. China* 34 (1985): 195.

241. Heo, N. Y., Kim, D. U., Ryu, G. H., Kang, C. S., and Lee, K. H. "Resistance of Corn to Maize Dwarf Mosaic Virus: Genetic Analysis by Diallel Cross." *Korean J. Plant Pathol.* 1 (1985): 136.

242. Darrah, L. L. "Report of the Inter-regional Maize Inbred Evaluation." Mo. Agric. Exp. Stn. Spec. Rep. 325 (1985): 122.

243. Logrono, M. L., Brewbaker, J. L., and Kim, S. K. "Viral Resistance and Agronomic Traits of Tropical-Adapted Maize inbreds." *Agron. Abstr.* 79 (1987): 70 (Abstract).

244. Keaster, A. J., Zuber, M. S., Fairchild, M. L., and Loesch, P. J., Jr. "Effect of Planting Dates on Incidence and Severity of Corn Virus Diseases." *Agron. J.* 61 (1969): 363.

245. Zuber, M. S. "Date of Planting Studies with Corn in the Missouri Delta Area." Mo. Agric. Exp. Stn. Bull. B862 (1967): 30.

246. Everly, R. T. "Corn Leaf Aphid Populations and Damage in Relation to Planting Date for 12 Selected Dent Corn Inbred Lines." *Proc. North Cent. Branch Entomol. Soc. Am.* 24 (1969): 92.

247. Hackerott, H. L., Harvey, T. L., and Ross, W. M. "Greenbug Resistance in Sorghums." *Crop Sci.* 9 (1969): 656.

248. Snelling, R. O., Blanchard, R. A., and Bigger, J. H. "Resistance of Corn Strains to the Corn Leaf Aphid, *Aphis maidis* Fitch." *J. Am. Soc. Agron.* 32 (1940): 371.

249. Weibel, D. E., Starks, K. J., Wood, A. E., Jr., and Morrison, R. D. "Sorghum Cultivars and Progenies Rated for Resistance to Greenbugs." *Crop Sci.* 12 (1972): 334.

250. Rosenkranz, E. "A New Leafhopper-Transmissible Corn Stunt Disease Agent in Ohio." *Phytopathology* 59 (1969): 1344.

251. Bradfute, O. E., Gingery, R. E., Gordon, D. T., and Nault, L. R. "Tissue Ultrastructure, Sedimentation and Leafhopper Transmission of a Virus Associated with a Maize Dwarfing Disease." *J. Cell. Biol.* 55 (1972): 25a (Abstract).

252. Bradfute, O. E., Louie, R., and Knoke, J. K. "Isometric Virus-Like Particles in Maize with Stunt Symptoms." *Phytopathology* 62 (1972): 748 (Abstract).

253. Pirone, T. P., Bradfute, O. E., Freytag, P. H., Lung, M. C. Y., and Poneleit, C. G. "Virus-like Particles Associated with a Leafhopper-Transmitted Disease of Corn in Kentucky." *Plant Dis. Reptr.* 56 (1972): 652.

254. Nault, L. R., Styer, W. E., Knoke, J. K., and Pitre, H. N. "Semipersistent Transmission of Leafhopper-Borne Maize Chlorotic Dwarf Virus." *J. Econ. Entomol.* 66 (1973): 1271.

255. Ling, K. C. "Nonpersistence of the Tungro Virus of Rice in its Leafhopper Vector, *Nephotettix impicticeps.*" *Phytopathology* 56 (1966): 1252.

256. Ayers, J. E., Boyle, J. S., and Gordon, D. T. "The Occurrence of Maize Chlorotic Dwarf and Maize Dwarf Mosaic Viruses in Pennsylvania in 1977." *Plant Dis. Reptr.* 62 (1978): 820.

257. Damsteegt, V. D. "A Naturally Occurring Corn Virus Epiphytotic." *Plant Dis. Reptr.* 60 (1976): 858.

258. Main, C. E., Nusser, S. M., and Bragg, A. W. *Crop Losses in North Carolina Due to Plant Diseases and Nematodes.* N.C. State University Dept. Plant Pathol. Spec. Publ. 3 (1984), 146.

259. Scott, G. E., Rosenkranz, E. E., and Nelson, L. R. "Yield Loss of Corn Due to Corn Stunt Disease Complex." *Agron J.* 69 (1977): 92.

260. Louie, R., Knoke, J. K., and Gordon, G. T. "Epiphytotics of Maize Dwarf Mosaic and Maize Chlorotic Dwarf of Diseases in Ohio." *Phytopathology* 64 (1974): 1455.

261. Nault, L. R., and Bradfute, O. E. "Corn Stunt: Involvement of a Complex of Leafhopper-Borne Pathogens," in *Leafhopper Vectors and Plant Disease Agents*, eds. K. Maramorosch and K. F. Harris. (New York: Academic Press, 1979), 561.

262. Choudhury, M. M., and Rosenkranz, E. "Differential Transmission of Mississippi and Ohio Corn Stunt Agents by *Graminella nigrifrons*." *Phytopathology* 63 (1973): 127.

263. Gingery, R. E., Gordon, D. T., Nault, L. R., and Bradfute, O. E. "Maize Chlorotic Dwarf Virus," in *Handbook of Plant Virus Infections and Comparative Diagnosis*, ed. E. Kurstak. (Amsterdam, The Netherlands: Elsevier/North Holland Biomedical Press, 1981), 19.

264. Gordon, D. T. "Routine Serological Assays for Diagnosis of Maize Virus Diseases," in *Proc. Int. Maize Virus Dis. Colloq. and Workshop*, eds. L. E. Williams, D. T. Gordon, and L. R. Nault. (Ohio Agric. Res. Dev. Cent., Wooster, 1977), 99.

265. Gingery, R. E. "An Immunofluorescence Test for Maize Chlorotic Dwarf Virus." *Phytopathology* 68 (1978): 1526.

266. Reeves, J. T., Jackson, A. O., Paschke, J. D., and Lister, R. M. "Use of Enzyme-Linked Immunosorbent Assay (ELISA) for Serodiagnosis of Two Maize Viruses." *Plant Dis. Reptr.* 62 (198): 667.

267. Derrick, K. S., and Brlansky, R. H. "Assay for Viruses and Mycoplasmas Using Serologically Specific Electron Microscopy." *Phytopathology* 66 (1976): 815.

268. Nault, L. R., Gordon, D. T., Robertson, D. C., and Bradfute, O. E. "Host Range of Maize Chlorotic Dwarf Virus." *Plant Dis. Reptr.* 60 (1976): 374.

269. Matthews, R. E. F. "Classification and Nomenclature of Viruses. Fourth Report of the International Committee on Taxonomy of Viruses." *Intervirology* 17 (1982): 1.

270. Gordon, D. T. Personal Communication, 1981.

271. Gingery, R. E. "Properties of Maize Chlorotic Dwarf Virus and Its Ribonucleic Acid." *Virology* 73 (1976): 311.

272. Gingery, R. E. Unpublished data, 1986.

273. Choudhury, M. M., and Rosenkranz, E. "Vector Relationship of *Graminella nigrifrons* to Maize Chlorotic Dwarf Virus." *Phytopathology* 73 (1983): 685.

274. Nault, L. R., and Madden, L. V. "Phylogenetic Relatedness of Maize Chlorotic Dwarf Virus Leafhopper Vectors." *Phytopathology* 78 (1988): 1683.

275. Nault, L. R. Personal communication, 1984.

276. Nault, L. R. "Vectors of Maize Viruses," in *Proc. Int. Maize Virus Dis. Colloq. and Workshop*, eds. L. E. Williams, D. T. Gordon, and L. R. Nault. (Ohio Agric. Res. Dev. Cent., Wooster, 1977), 111.

277. Harris, K. F. "Leafhoppers and Aphids as Biological Vectors: Vector–Virus Relationships," in *Leafhopper Vectors and Plant Disease Agents*, eds. K. Maramorosch, K. F. Harris. (New York: Academic Press, 1979), 217.

278. Harris, K. F. "Arthropod and Nematode Vectors of Plant Viruses." *Annu. Rev. Phytopathol.* 19 (1981): 391.

279. Harris, K. F., and Childress, S. A. "Mechanism of Maize Chlorotic Dwarf Virus (MCDV) Transmission by its Leafhopper Vector, *Graminella nigrifrons* (Abstract), in *81st Annu. Meet. Am. Soc. Microbiol. 1981*, Dallas, Texas, 1981, 251.

280. Ammar, E. D., Gordon, D. T., and Nault, L. R. "Ultrastructure of Maize Chlorotic Dwarf Virus Infected Maize and Viruliferous Leafhopper Vectors." *Phytopathology* 77 (1987): 1743 (Abstract).

281. Hunt, R. E., Nault, L. R., and Gingery, R. E. "Evidence for Infectivity of Maize Chlorotic Dwarf Virus and a Helper Component for Its Leafhopper Transmission." *Phytopathology* 77 (1987): 1743 (Abstract).

282. Murant, A. F., Raccah, B., and Pirone, T. P. "Transmission by Vectors of Viruses with Filamentous Particles," in *The Filamentous Plant Viruses*, Vol 4, ed. R. G. Milne, in *The Plant Viruses*, eds. H. Fraenkel-Conrat and R. R. Wagner. (New York: Plenum Press, 1988), 237.

283. All, J. N., Kuhn, C. W., Gallaher, R. N., Jellum, M. D., and Hussey, R. S. "Influence of No-Tillage-Cropping, Carbofuran, and Hybrid Resistance on Dynamics of Maize Chlorotic Dwarf and Maize Dwarf Mosaic Diseases of Corn." *J. Econ. Entomol.* 70 (1977): 221.

284. Gustin, R. D., and Stoner, W. N. "Biology of *Deltocephalus sonorus* (Homoptera: Cicadellidae)." *Ann. Entomol. Soc. Am.* 61 (1968): 77.

285. Stoner, W. N., and Gustin, R. D., "Biology of *Graminella nigrifrons* (Homoptera: Cicadellidae), a Vector of Corn (Maize) Stunt Virus." *Ann. Entomol. Soc. Am.* 60 (1967): 496.

286. Stevens, C., Gudauskas, R. T., and Karr, G. W., Jr. "Seasonal Incidence of Two Viral Diseases of Corn." *J. Ala. Acad. Sci.* 47 (1976): 130.

287. Sedlacek, J. D., and Freytag, P. G. "Field Biology of *Graminella nigrifrons* (Forbes): Vector of Maize Chlorotic Dwarf Virus in Kentucky (Abstract), in *Abstr. 38th Annu. Meet. N. Cent. Br. Entomol. Soc. Am.* No. 98, 1983 (unpaged).

288. Alverson, D. R., All, J.N., and Kuhn, C. W. "Simulated Intrafield Dispersal of Maize Chlorotic Dwarf Virus by *Graminella nigrifrons* with a Rubidium Marker." *Phytopathology* 70 (1980): 734.

289. Knoke, J. K. Unpublished data, 1988.

290. Sedlacek, J. D., and Freytag, P. H. "Seasonal Occurrence of *Graminella nigrifrons* (Forbes) in Corn and Pasture Ecosystems in Kentucky (Abstract), in *Abstr. 37th Annu. Meet. N. Cent. Br Entomol. Soc. Am.* No. 30 (1982) (unpaged).

291. Guthrie, W. D., Tseng, C. T., Knoke, J., and Jarvis, J. L. "European Corn Borer and Maize Chlorotic Dwarf Virus Resistance-Susceptibility in Inbred Lines of Dent Maize." *Maydica* 27 (1982): 221.

292. Bhirud, K. M., and Pitre, H. N. "Bioactivity of Carbofuran and Disulfoton in Corn in Greenhouse Tests, Particularly in Relation to Leaf Position on the Plant." *J. Econ Entomol.* 65 (1972): 1183.

293. Keaster, A. J., and Fairchild, M. L. "Reduction of Corn Virus Disease Incidence and Control of Southwestern Corn Borer with Systemic Insecticides." *J. Econ. Entomol.* 61 (1968): 367.

294. Pitre, H. N. "A Preliminary Study of Corn Stunt Vector Populations in Relation to Corn Planting Dates in Mississippi. Notes on Disease Incidence and Severity." *J. Econ. Entomol.* 61 (1968): 847.

295. Stevens, C., Gudauskas, R. T., Karr, G. W., and Estes, P. M. "Effects of Carbofuran on Incidence of Maize Chlorotic Dwarf and Maize Dwarf Mosaic in Corn." *J. Ala. Acad. Sci.* 48 (1977): 57 (Abstract).

296. Simons, J. N., and Zitter, T. A. "Use of Oils to Control Aphid-Borne Viruses." *Plant Dis.* 64 (1980): 542.

297. All, J. N., Kuhn, C. W., and Jellum, M. D. "The Changing Status of Corn Virus Diseases: Potential Value of a Systemic Insecticide." *Ga. Agric. Res.* 17 (1976): 4.

298. Crawford, J. L., Kuhn, C. W., and Jellum, M. D. "Corn Virus Disease Control." Univ. Ga. College Agric. Coop. Ext. Serv. Circ. 635 (1978): 8.

299. Gordon, D. T., Findley, W. R., Knoke, J. K., Louie, R., Nault, L. R., Bradfute, O. E., Dollinger, E. J., and Gingery, R. E. "Distinguishing Symptoms and Latest Research Findings on Corn Virus Diseases in the United States," in *Proc. 29th Annu. Corn Sorghum Res. Conf.* (Washington, D.C.: Am. Seed Trade Assoc., 1974), 153.

300. Gordon, D. T, Findley, W. R., Knoke, J. K., Louie, R., Nault, L. R., Bradfute, O. E., Dollinger, E. J., and Gingery, R. E. "Maize Dwarf Mosaic and Maize Chlorotic Dwarf Diseases in the United States," in *Proc. Corn Dis. Conf. 1974*, (Lafayette, Ind.: Purdue Univ., 1974), 52.

301. Findley, W. R., Nault, L. R., Styer, W. E., and Gordon, D. T. "*Zea diploperennis* as a Source of Maize Chlorotic Dwarf Virus Resistance: A Progress Report," in *Proc. Int. Maize Virus Dis. Colloq. and Workshop*, eds. D. T. Gordon, J. K. Knoke, L. R. Nault, and R. M. Ritter. (Ohio State Univ., Ohio Agric. Res. Dev. Cent., Wooster, 1983), 255.

302. Naidu, B., and Josephson, L. M. "Genetic Analysis of Resistance to the Corn Virus Disease Complex." *Crop Sci.* 16 (1976): 167.

303. Scott, G. E., and Rosenkranz, E. E. "Location of Genes Conditioning Resistance to the Corn Stunting Disease Complex in Maize." *Crop Sci.* 17 (1977): 923.

304. Grogan, C. O., and Rosenkranz, E. E. "Genetics of Host Reaction to Corn Stunt Virus." *Crop Sci.* 8 (1968): 251.

305. Francki, R. I. B., Milne, R. G., and Hatta, T. "Maize Chlorotic Dwarf Virus Group," in *Atlas of Plant Viruses*, Vol. I. (Boca Raton, Fla.: CRC Press, 1985), 111.

306. Harris, K. F., and Childress, S. A. "Cytology of Maize Chlorotic Dwarf Virus Infection in Corn." *Int. J. Trop. Plant Dis.* 1 (1983): 135.

307. Fuller, C. "Mealie Blight." *Natal Agric. J.* 3 (1900): 389.

308. Storey, H. H. "Streak Disease, An Infectious Chlorosis of Sugar-Cane, not Identical with Mosaic Disease," in *Rep. Imp. Bot. Conf. 1924*, London, 1924, 132.

309. Seth, M. L., Raychaudhuri, S. P., and Singh, D. V. "Bajra (Pearl Millet) Streak: A Leafhopper-Borne Cereal Virus in India." *Plant Dis. Reptr.* 56 (1972): 424.

310. McClean, A. P. D. "Some Forms of Streak Virus Occurring in Maize, Sugar-Cane and Wild Grasses." *Union S. Afr. Sci. Bull.*, 265 (1947): 39.

311. Storey, H. H. "Streak Disease of Sugar-Cane." *S. Afr. Dep. Agric. Sci. Bull.* 39 (1925): 30.

312. Bock, K. R. "Maize Streak Virus." *CMI/AAB Descriptions of Plant Viruses* No. 133, 1974.

313. Bock, K. R. "Geminivirus Diseases in Tropical Crops." *Plant Dis.* 66 (1982): 266.

314. Damsteegt, V. D. "Exotic Virus and Viruslike Diseases of Maize," in *Virus and Viruslike Diseases of Maize in the United States,* eds. D. T. Gordon, J. K. Knoke, and G. E. Scott. (Southern Coop. Ser. Bull., 247, Ohio Agric. Res. Dev. Cent., Wooster, 1981), 110.

315. Francki, R. I. B., Milne, R. G., and Hatta, T. "Geminivirus Group," in *Atlas of Plant Viruses*, Vol. I. (Boca Raton, Fla.: CRC Press, 1985), 33.

316. Goodman, R. M. "Geminiviruses." *J. Gen. Virol.* 54 (1981): 9.

317. Goodman, R. M. "Geminiviruses." in *Handbook of Plant Virus Infections and Comparative Diagnosis*, ed. E. Kurstak. (Amsterdam, The Netherlands: Elsevier/North Holland Biomedical Press, 1981), 879.

318. Harrison, B. D. "Advances in Geminivirus Research." *Annu. Rev. Phytopathol.* 23 (1985): 55.

319. Howarth, A. J., and Goodman, R. M. "Plant Viruses with Genomes of Single-Stranded DNA." *Trends Biochem. Sci.* 7 (1982): 180.

320. Ammar, E. D. "Virus Diseases of Sugarcane and Maize in Egypt," in *Proc. Int. Maize Virus Dis. Colloq. and Workshop.* eds. D. T. Gordon, J. K. Knoke, L. R. Nault, and R. M. Ritter. (Ohio State Univ., Ohio Agric. Res. Dev. Cent., Wooster, 1983), 122.

321. Reyes, G. M. "The Mosaic Disease of Sugar Cane." *Philipp. Agric. Rev.* 20 (1927): 187.

322. Fauquet, C., and Thouvenel, J. C. "Maladies Virales des Plantes Cultivees en Cote d'Ivoire." *Initiations-Documentations Techniques No. 46,* ORSTOM, Paris, 1980, 128.

323. Rossel, H. W., and Thottappilly, G. *Virus Diseases of Important Food Crops in Tropical Africa,* Int. Inst. Trop. Agric., Ibadan, Nigeria, 1985, 61.

324. Johnston, L. M. "The Status of Maize Virus Diseases in Zimbabwe," in *Proc. Int. Maize Virus Dis. Colloq. and Workshop*, eds. D. T. Gordon, J. K. Knoke, L. R. Nault, and R. M. Ritter. (Ohio State Univ., Ohio Agric. Res. Dev. Cent., Wooster, 1983), 155.

325. Von Wechmar, M. B., and Milne, R. G. "Purification and Serology of a South African Isolate of Maize Streak Virus," in *Proc. Int. Maize Virus Dis. Colloq. and Workshop*, eds. D. T. Gordon, J. K. Knoke, L. R. Nault, and R. M. Ritter. (Ohio State Univ., Ohio Agric. Res. Dev. Cent., Wooster, 1983), 164.

326. Edgerton, C. W. "Streak. Cause: A Virus," in *Sugarcane and Its Diseases*, ed. C. W. Edgerton. (Baton Rouge: Louisiana State Press, 1955), 301.

327. Toler, R. W., Cuellar, R., and Ferrar, J. B. "Preliminary Survey of Plant Diseases in the Republic of Panama, 1955–1958." *Plant Dis. Reptr.* 43 (1959): 1201.

328. Draganic, M., Vidakovic, M., and Milosevik, L. "Contribution to the Study of the Resistance of Some Maize Inbred Lines and Hybrids to Corn Streak Virus." *Zast. Bilja* 34 (1983): 33.

329. Le Conte, J. "La Virose du Mais Dahomey." *Agron. Trop.* 29 (1974): 831.

330. Guthrie, E. J. "Virus Diseases of Maize in East Africa," in *Proc. Int. Maize Virus Dis. Colloq. and Workshop*, eds. L. E. Williams, D. T. Gordon, and L. R. Nault. (Ohio Agric. Res. Dev. Cent., Wooster, 1977): 62.

331. Ammar, E. D. "Biology of the Leafhopper *Cicadulina chinai* Ghauri (Homoptera, Cicadellidae) in Giza, Egypt." *Z. Angew. Entomol.* 79 (1975): 337.

332. Fajemisin, J. M., Cook, G. E., Okusanya, F., and Shoyinka, S. A. "Maize Streak Epiphytotic in Nigeria." *Plant Dis. Reptr.* 60 (1976): 443.

333. Fajemisin, J. M., and Shoyinka, S. A. "Maize Streak and Other Maize Virus Diseases in West Africa," in *Proc. Int. Maize Virus Dis. Colloq. and Workshop*, eds. L. E. Williams, D. T. Gordon, and L. R. Nault. (Ohio Agric. Res. Dev. Cent., Wooster, 1977), 52.

334. McKinney, H. H. "Mosaic Diseases in the Canary Islands, West Africa, and Gibraltar." *J. Agric. Res.* 39 (1929): 557.

335. Rose, D. J. W. "Epidemiology of Maize Streak Disease." *Annu. Rev. Entomol.* 23 (1978): 259.

336. Beck, B. D. A. "Sorghum Diseases in Malawi," in *Sorghum Diseases: A World Review,* ed. G. D. Bengston. *Proc. Int. Workshop on Sorghum Diseases 1978,* Hyderabad, India, 1980, 40.

337. De Carvalho, I. "Relacao Preliminar de Dofncas Encontradas em Plantes e Insectos com Amatacoes Fitopatologicas." Colonia de Mocambique, Reparticao de Agricultura, Seccao de Micologia, 1948.

338. Hopkins, J. C. "Suspected 'Streak' Disease of Maize." *Rhod. Agric. J.* 32 (1935): 234.

339. Herd, G. W. "Maize Diseases During the 1954/55 Season." *Rhod. Agric. J.* 53 (1956): 525.

340. Fourie, A. P., and Pienaar, J. H. "Breeding for Resistance to Maize Streak Virus: A Report on the Vaalharts Breeding Programme," in *Proc. 5th S. Afr. Maize Breed. Symp. 1982,* Potchefstroom, S. Afr. Dep. Agric. Tech. Commun. 182 (1983): 44.

341. Gorter, G.J.M.A. "Wheat Stunt—a New Cereal Disease." *Farming S. Afr.* 22 (1947): 29.

342. Seth, M. L., Raychaudhuri, S. P., and Singh, D. V. "A Streak Disease of Bajra [*Pennisetum typhoides* (Burm.f.) Stapf and Hubb.] in India." *Curr. Sci.* 40 (1971): 272.

343. Seth, M. L., Raychaudhuri, S. P., and Singh, D. V. "Occurrence of Maize Streak Virus on Wheat in India." *Curr. Sci. (Bangalore)* 41 (1972): 684.

344. Dastur, J. F. "A Mosaic-like Disease of Sugarcane in the Central Provinces in 1926." *Agric. J. India* 21 (1926): 429.

345. Rhind, D. "Ann. Rept. of the Mycologist," Burma, 1927.

346. Anonymous. "Virus Diseases," in Mauritius Sugar Ind. Res. Inst. Annu. Rep. 1975, Reduit, Mauritius, 1976, 56.

347. Ricaud, C., and Felix, S. "Identification and Relative Importance of Virus Diseases of Maize in Mauritius," in *Proc. 2nd S. Afr. Maize Breeding Symp. 1976.*, Tech. Commu. Dep. Agric. Tech. Serv., S. Afr., No. 142. 1979, 105.

348. Shepherd, E. F. S. "Maize Chlorosis. Notes on Chlorosis of Maize and Other Graminae in Mauritius." *Trop. Agric.* 6, 320, 1929.

349. Dollet, M., Accotto, G. P., Lisa, V., Menissier, J., and Boccardo, G. "A Geminivirus, Serologically Related to Maize Streak Virus, from *Digitaria sanguinalis* from Vanuatu." *J. Gen. Virol.* 67 (1986): 933.

350. Soto, P. E., Buddenhagen, I. W., and Asnani, V. L. "Development of Streak Virus-Resistant Maize Populations through Improved Challenge and Selection Methods." *Ann. Appl. Biol.* 100 (1982): 539.

351. Rose, D. J. W. "Field Studies in Rhodesia on *Cicadulina* spp. (Hem. Cicadellidae), Vectors of Maize Streak Disease." *Bull. Entomol. Res.* 62 (1973): 477.

352. Storey, H. H. "The Transmission of Streak Disease of Maize by the Leafhopper *Balclutha mbila* Naude." *Ann. Appl. Biol.* 12 (1925): 422.

353. Ammar, E. D., Abul-Ata, A. E., El-Sheikh, M. A., and Sewify, G. H. "Incidence of Virus and Viruslike Disease Syndromes on Maize and Sugarcane in Middle and Lower Egypt." *Egypt J. Phytopathol.* 19 (1987): 97.

354. Soto, P. E. "A New Vector of Maze Streak Virus." *E. Afr. Agric. For. J.* 44 (1978): 70.

355. Rossel, H. W., and Thottappilly, G. "Maize Chlorotic Stunt in Africa: A Manifestation of Maize Mottle Virus?" in *Proc. Int. Maize Virus Dis. Colloq. and Workshop*, eds. D. T. Gordon, J. K. Knoke, L. R. Nault, and R. M. Ritter. (Ohio State Univ., Ohio Agric. Res. Dev. Cent., Wooster, 1983), 158.

356. Nagaich, B. B., and Sinha, R. C. "Eastern Wheat Striate: A New Viral Disease." *Plant Dis. Reptr.* 58 (1974): 968.

357. Schindler, A. J. "Insect Transmission of Wallaby Ear Disease of Maize." *J. Aust. Inst. Agric. Sci.* 8 (1942): 35.

358. Ruppel, R. F. "A Review of the Genus *Cicadulina* (Hemiptera, Cicadellidae)." Publ. Mus. Mich. State Univ. Biol. Ser. 2, 1965, 385.

359. Maramorosch, K., Calica, C. A., Agati, J. A., and Pableo, J. "Further Studies on the Maize and Rice Leaf Galls Induced by *Cicadulina bipunctella.*" *Entomol. Exp. Appl.* 4 (1961): 86.

360. Etienne, J., and Rat, B. "Le Stripe: Une Maladie Importante du Mais a la Reunion." *L'Agron. Trop.* 28 (1973): 11.

361. Gorter, G.J.M.A. "Studies on the Spread and Control of the Streak Disease of Maize." Union S. Afr. Dep. Agric. For. Sci. Bull., 341, 1953, 20.

362. Thottappilly, G. "Detection of MSV," in Int. Inst. Trop. Agric. Annu. Rep. 1984, Ibadan, Nigeria, 1985, 52.

363. Ammar, E. D., Kira, M. T., and Abul-Ata, A. E. "Natural Occurrence of Streak and Mosaic Diseases on Sugarcane Cultivars at Upper Egypt and Transmission of Sugarcane Streak by *Cicadulina bipunctella zeae* China." *Egypt J. Phytopathol.* 12 (1980): 21.

364. Bock, K. R., Guthrie, E. J., and Woods, R. D. "Purification of Maize Streak Virus and Its Relationships to Viruses Associated with Streak Diseases of Sugar Cane and Panicum maximum." *Ann. Appl. Biol.* 77 (1974): 289.

365. Damsteegt, V. D. "Maize Streak Virus: Additional Hosts of Virus and Vector." *Phytopathol. News* 12 (1978): 225 (Abstract).

366. Damsteegt, V. D. "Maize Streak Virus: I. Host Range and Vulnerability of Maize Germ Plasm." *Plant Dis.* 67 (1983): 734.

367. Ricaud, C., and Felix, S. "Sources and Strains of Streak Virus Infecting Graminaceous Hosts" (Abstract), in *Proc. 3rd Int. Cong. of Plant Pathol.*, ed. W. Laux. (Berlin: Deutsche Phytomed. Ges., Paul Parey, 1978), 23.

368. Nagaraju, S. V., Channamma, K. A. L., and Reddy, H. R. "Effect of Streak Virus on Growth and Yield of Ragi." *Indian Phytopathol.* 34 (1981): 356.

369. Gelaw, B. "CIMMYT's Maize Improvement Role in East, Central and Southern Africa," in *To Feed Ourselves. A Proceedings of the First Eastern, Central and Southern Africa Regional Maize Workshop*, ed. B. Gelaw. (CIMMYT, Mexico, D. F.: Lusaka, Zambia, 1986), 208.

370. Obi, I. U. "Identified Major Disease Hazards of Maize Production in Southern Nigeria." *Tropenlandwirt* 82 (1981): 137.

371. Choudhary, G. G., Singh, G., and Bhatnagar, G. C. "Losses in Yield Components of Wheat Variety Lal Bahadur Caused by Streak Virus Disease." *Indian Phytopathol.* 33 (1980): 604.

372. Choudhary, G. G., Singh, G., and Bhatnagar, G. C. "Reactions of Pearl Millet Lines to Pennisetum Strain of Maize Steak Virus." *Indian J. Mycol. Plant Pathol.* 10 (1980): 71.

373. Van Rensburg, G. D. J., and Kuhn, H. C. "Maize Streak Disease." Farm. S. Afr. Maize Series, No. E, 1977, 4.

374. Fajemisin, J. M., Kim, S. K., Efron, Y., and Alam, M. S. "Breeding for Durable Disease Resistance in Tropical Maize with Special Reference to Maize Streak Virus," in *FAO Plant Prod. Prot.*, No. 55, Ibadan, Nigeria, FAO-United Nations Publ., Rome, Italy, 1984, 49.

375. Mzira, C. N. "Assessment of Effects of Maize Streak Virus on Yield of Maize." *Zimbabwe J. Agric. Res.* 22 (1984): 141.

376. Mzira, C. N. "Cultural Control of Maize Streak Virus in Wheat by Spacing and Time of Planting." *Zimbabwe Agric. J.* 81 (1984): 189.

377. Guthrie, E. J. "Measurement of Yield Losses Caused by Maize Streak Disease." *Plant Dis. Reptr.* 62 (1978): 839.

378. Whitcomb, R. F., and Davis, R. E. "Mycoplasma and Phytarboviruses as Plant Pathogens Persistently Transmitted by Insects." *Annu. Rev. Entomol.* 15 (1970): 405.

379. Plavsic-Banjac, B., and Maramorosch, K. "Electron Microscopy of African Maize Streak." *Phytopathology* 62 (1972): 671 (Abstract).

380. Sylvester, E. S., Richardson, J., and Nickel, J. L. "An Additional Note on Viruslike Particles Associated with Maize Streak Disease." *Plant Dis. Reptr.* 57 (1973): 414.

381. Bock, K. R., Guthrie, E. J., and Meredith, G. "RNA and Protein Components of Maize Streak and Cassava Latent Viruses." *Ann. Appl. Biol.* 85 (1977): 305.

382. Harrison, B. D., Barker, H., Bock, K. R., Guthrie, E. J., Meredith, G., and Atkinson, M. "Plant Viruses with Circular Single-Stranded DNA." *Nature (Lond.)* 270 (1977): 760.

383. Howell, S. H. "Physical Structure and Genetic Organisation of the Genome of Maize Streak Virus (Kenyan isolate)." *Nucleic Acids Res.* 12 (1984): 7359.

384. Harrison, B. D., and Barker, H. "Nucleic Acid of Maize Streak and Cassava Latent Viruses," in *Annu. Rep. Scott. Hortic. Res. Inst. 1977* (Dundee, Scotland: Invergowrie, 1978), 103.

385. Storey, H. H., and McClean, A. P. D. "The Transmission of Streak Disease Between Maize, Sugar Cane and Wild Grasses." *Ann. Appl. Biol.* 17 (1930): 691.

386. Autrey, L. J. C., and Ricaud, C. "The Comparative Epidemiology of Two Diseases of Maize Caused by Leafhopper-Borne Viruses in Mauritius," in *Plant Virus Epidemiology*, eds. R. T. Plumb, and J. M. Thresh. (Oxford, England: Blackwell Scientific Publ., 1983): 277.

387. Markham, P. G., Pinner, M. S., and Boulton, M. I. "The Transmission of Maize Streak Virus by Leafhoppers, a New Look at Host Adaptation." *Mitt. Schweiz. Entomol. Ges.* 57 (1984): 431.

388. Howell, S. H. "Physical Structure and Genetic Organization of the Genome of Maize Streak Virus (Kenyan isolate). Corrigendum." *Nucleic Acids Res.* 13 (1985): 3018.

389. Mullineaux, P. M., Donson, J., Morris-Krsinich, B. A. M., Boulton, M. I., and Davies, J. W. "The Nucleotide Sequence of Maize Streak Virus DNA." *Eur. Mol. Biol. Organ. J.* 3 (1984): 3063.

390. Clarke, B. A., Kirby, R., and Rybicki, E. P. "Genomic Variation Amongst South African Isolates of Maize Streak Virus." *Phytophylactica* 19 (1987): 127 (Abstract).

391. Donson, J., Accotto, G. P., Boulton, M. I., Mullineaux, P. M., and Davies, J. W. "The Nucleotide Sequence of a Geminivirus from *Digitaria sanguinalis.*" *Virology* 161 (1987): 160.

392. MacDowell, S. W., Macdonald, H., Hamilton, W. D. O., Couts, R. H. A., and Buck, K. W. "The Nucleotide Sequence of Cloned Wheat Dwarf Virus DNA." *Eur. Mol. Biol. Organ.* 4 (1985): 2173.

393. Mullineaux, P. M., Donson, J., Stanley, J., Boulton, M. I., Morris-Krsinich, B. A. M., Markham, P. G., and Davies, J. W. "Computer Analysis Identifies Sequence Homologies Between Potential Gene Products of Maize Streak Virus and Those of Cassava Latent Virus and Tomato Golden Mosaic Virus." *Plant Mol. Biol.* 5 (1985): 125.

394. Morris-Krsinich, B. A. M., Mullineaux, P. M., Donson, J., Boulton, M. I., Markham, P. G., Short, M. N., and Davies, J. W. "Bidirectional Transcription of Maize Streak Virus DNA and Identification of the Coat Protein Gene." *Nucleic Acids Res.* 13 (1985): 7237.

395. Thomas, J. E., Massalski, P. R., and Harrison, B. D. "Production of Monoclonal Antibodies to African Cassava Mosaic Virus and Differences in their Reactivities with Other Whitefly-transmitted Gemini viruses." *J. Gen. Virol.* 67 (1986): 2739.

396. Damsteegt, V. D. "Maize Streak Virus: Effect of Temperature on Vector and Virus." *Phytopathology* 74 (1984): 1317.

397. Markham, P. G., Pinner, M. S., Boulton, M. I., and Plaskitt, K. "Maize Streak Virus—Biology." John Innes Inst. Annu. Rep., 73, 1985, 166.

398. Storey, H. H. "The Transmission of a New Plant Virus Disease by Insects." *Nature (Lond.)* 114 (1924): 245.

399. Storey, H. H. "Report of the Plant Pathologist. Virus Diseases of Plants." East Afr. Agric. Res. Stn., Amani, Annu. Rep., 8 (1936) 11.

400. Storey, H. H. "Virus Diseases of East African Plants. 5. Streak Disease of Maize." *East Afr. Agric. J.* 1 (1936): 471.

401. Fennah, R. G. "A New Species of *Cicadulina* (Homoptera: Cicadellidae) from East Africa." *Ann. Mag. Nat. Hist. (Ser. 13)* 2 (1960): 757.

402. Dabrowski, Z. T. "Comparative Studies of *Cicadulina* Leafhoppers in West Africa," in *Int. Workshop on Leafhoppers and Planthoppers of Economic Importance*, eds. M. R. Wilson and L. R. Nault. (Provo, Utah: Brigham Young Univ., CAB Int. Inst. Entomol., 1897), 35.

403. Nielson, M. W. *The Leafhopper Vectors of Phytopathogenic Viruses (Homoptera Cicadellidae) Taxonomy, Biology, and Virus Transmission.* (Washington, D.C.: U.S. Dep. Agric. Tech. Bull., 1382, 1968), 386.

404. Nielson, M. W. "A Synonymical List of Leafhopper Vectors of Plant Viruses (Homoptera, Cicadellidae)." (Washington, D.C.: U.S. Dep. Agric., Agric. Res. Serv., ARS-33-74, 1962), 12.

405. Van Rensburg, G. D. J. "Laboratory Observations on the Biology of *Cicadulina mbila* (Naude)(Homoptera: Cicadellidae), a Vector of Maize Streak Disease. 1. The Effect of Temperature." *Phytophylactica* 14 (1982): 99.

406. Van Rensburg, G. D. J. "Laboratory Observations on the Biology of *Cicadulina mbila* (Naude)(Homoptera: Cicadellidae), a Vector of Maize Streak Disease. 2. The Effect of Selected Host Plants." *Phytophylactica* 14 (1982): 109.

407. Rose, D. J. W. "Laboratory Observations on the Biology of *Cicadulina* spp. (Hom., Cicadellidae), with Particular Reference to the Effects of Temperature." *Bull. Entomol. Res.* 62 (1973): 471.

408. Okoth, V. A. O., Dabrowski, Z. T., and van Emden, H. F. "Comparative Biology of Some *Cicadulina* Species and Populations from Various Climatic Zones in Nigeria (Hemiptera: Cicadellidae)." *Bull. Ent. Res.* 77 (1987): 1.

409. Amar, E. D. "Biology of *Cicadulina bipunctella zeae* China in Giza, Egypt." *Dtsch. Entomol. Z.* 24 (1977): 345.

410. Grimsley, N., Hohn, T., Davies, J. W., and Hohn, B. "Agrobacterium-Mediated Delivery of Infectious Maize Streak Virus into Maize Plants." *Nature (Lond.)* 325 (1987): 177.

411. Polson, A., and Von Wechmar, M. B. "A Novel Way to Transmit Plant Viruses." *J. Gen. Virol.* 51 (1980): 179.

412. Storey, H. H. "Transmission Studies of Maize Streak Disease." *Ann. Appl. Biol.* 15 (1928): 1.

413. Storey, H. H. "The Inheritance by a Leafhopper of the Ability to Transmit a Plant Virus." *Nature (Lond.)* 127 (1931): 928.

414. Storey, H. H. "The Inheritance by an Insect Vector of the Ability to Transmit a Plant Virus." *Proc. R. Soc. Lond. Biol. Sci.* 112 (1932): 46.

415. Storey, H. H. "Investigations of the Mechanism of the Transmission of Plant Viruses by Insect Vectors. I." *Proc. R. Soc. Lond. Biol. Sci.* 113 (1933): 463.

416. Storey, H. H. "Investigations of the Mechanism of the Transmission of Plant Viruses by Insect Vectors. II. The Part Played by Puncture in Transmission." *Proc. R. Soc. Lond. Biol. Sci.* 125 (1938): 455.

417. Storey, H. H. "Investigations of the Mechanism of the Transmission of Plant Viruses by Insect Vectors. III. The Insect's Saliva." *Proc. R. Soc. Lond. Biol. Sci.* 127 (1939): 526.

418. Boulton, M. I., Markham, P. G., and Davies, J. W. "Nucleic Acid Hybridisation Techniques for the Detection of Plant Pathogens in Insect Vectors," in *Proc. British Crop Protection Conference. Pests and Diseases,* Brighton Metropole, England, 1984, 181.

419. Rose, D. J. W. "The Distribution of Various Species of *Cicadulina* in Different African Countries, Frequency of Their Attack and Impact on Crop Production," in *Proc. Int. Workshop on Biotaxonomy, Classification and Biology of Leafhoppers and Planthoppers (Auchenorrhyncha) of Economic Importance,* eds. W. J. Knight, N. C. Pant, T. S. Robertson, and M. R. Wilson. Commonw. Inst. Entomol., London, 1983, 297.

420. Rose, D. J. W. "Dispersal and Quality in Populations of *Cicadulina* Species (Cicadellidae)." *J. Anim. Ecol.* 41 (1972): 589.

421. Rose, D. J. W. "Times and Sizes of Dispersal Flights by *Cicadulina* Species (Homoptera: Cicadellidae) Vectors of Maize Streak Disease." *J. Anim. Ecol.* 41 (1972): 495.

422. Rose, D. J. W. "Distances Flown by *Cicadulina* spp. (Hem. Cicadellidae) in Relation to Distribution of Maize Streak Disease in Rhodesia." *Bull. Entomol. Res.* 62 (1973): 497.

423. Rose, D. J. W. "The Epidemiology of Maize Streak Disease in Relation to Population Densities of *Cicadulina* spp." *Ann. Appl. Biol,* 76 (1974): 199.

424. Storey, H. H. "Streak Disease of Maize." *Farming S. Afr.* 1 (1926): 183.

425. Okoth, V. A. O., and Dabrowski, Z. T. "Population Density, Species Composition and Infectivity with Maize Streak Virus (MSV) of *Cicadulina* spp. Leafhoppers in Some Ecological Zones in Nigeria." *Acta Oecol. Oecol. Appl.* 3 (1987): 191.

426. Rose, D. J. W. "Pests of Maize and Other Cereal Crops in the Rhodesias." Rhod. Minist. Agric. Br. Entomol. Bull. 2163 (1963): 23.

427. Farina, M. P. W., Channon, P., and Mapham, W. R. "Soil Fertility Effects on the Incidence of Leaf Streak and Ear Rot in Maize." *Gewasprod./Crop Prod.* 5 (1976): 139.

428. Drinkwater, T. W., Walters, M. C., and Van Rensburg, J. B. J. "The Application of Systemic Insecticides to the Soil for the Control of the Maize Stalk Borer, *Busseola fusca* (Fuller) (Lep.: Noctuidae), and of *Cicadulina Mbila* (Naude) (Hem.:Cicadellidae), The Vector of Maize Streak Virus." *Phytophylactica* 11 (1979): 5.

429. Van Rensburg, J. B. J., and Walters, M. C. "The Efficacy of Systemic Insecticides Applied to the Soil for the Control of *Cicadulina mbila* (Naude)(Hem. Cicadellidae), the Vector of Maize Streak Disease, and the Maize Stalk Borer *Busseola fusca* (Fuller) (Lep. Noctuidae)." *Phytophylactica* 10 (1978): 49.

430. Fielding, J. "Field Experimental Work on Rotation Crops." Empire Cotton Growing Assoc. Prog. Rept. 1931–1932, 1933.

431. Rose, M. F. "Rotation Crops," in Emp. Cotton Growing Crop. Prog. Rep. Exp. Stn. 1936–1937, 1938, 21.

432. Gorter, G.J.M.A. "Breeding Maize for Resistance to Streak." *Euphytica* 8 (1959): 234.

433. Storey, H. H., and Howland, A. K. "Transfer of Resistance to the Streak Virus into East African Maize." *East Afr. Agric. For. J.* 33 (1967): 131.

434. Bock, K. R., Guthrie, E. J., Meredith, G., and Ambetsa, T. "Maize Viruses," in East Afr. Agric. For. Res. Organ. Rec. Res. Annu. Rep. 1974, 1975, 121.

435. Bock, K. R., Guthrie, E. J., Meredith, G., and Ambetsa, T. "Maize Viruses," in East Afr. Agric. For. Res. Organ. Rec. Res. Annu. Rep. 1976, 1977, 135.

436. Hainzelin, E. M., and Marchand, J. L. "Registration of IRAT 297 Maize Germplasm." *Crop Sci.* 26 (1986): 1090.

437. Dabrowski, Z. T. "The Biology and Behaviour of *Cicadulina triangula* in Relation to Maize Streak Virus Resistance Screening." *Insect Sci. Applic.* 6 (1985): 417.

438. Bjarnason, M. "Progress in Breeding for Resistance to the Maize Streak Virus Disease," in *To Feed Ourselves. A Proceedings of the First Eastern, Central and Southern Africa Regional Maize Workshop*, ed. B. Gelaw. (CIMMYT, Mexico, D.F.: Lusaka, Zambia, 1986), 197.

439. Dabrowski, Z. T. "Rearing and Releasing *Cicadulina* spp. Leafhoppers for Maize Resistance Screening to the Streak Virus," presented at the National Maize Meeting, Bukavu, Zaire, Jan. 8 to 11, 1985, 7.

12

PEARL MILLET DOWNY MILDEW

D. P. THAKUR

Professor, Pearlmillet Pathology
Haryana Agricultural University, Hissar, India

12-1 INTRODUCTION

Pearl millet crop originally belonged to tropical Africa because a variety of wild species are still abundant in those areas.[1] Nearly 42.35 million hectares of cultivated land in the world are pearl millet crop, out of which 18.50 million hectares are available in India alone.

Downy mildew of pearl millet is most widely present in the temperate and tropical areas of the world, including the United States, Europe, Africa, and Asia (Figure 12-1).

The magnitude of grain yield reduction largely depends on disease severity and stage of crop growth during infection. Total loss may occur in plants exhibiting downy mildew infection in seedling stage and green-ear infection during earhead formation. In East China, nearly 6% yield loss was noted.[2] Mitter and Tandon[3] reported 45% reduction in grain from North India. Nearly 30% reduction in grain yield has been reported in some high-yielding varieties in India.[4]

12-2 SYMPTOMS

Downy mildew symptoms in pearl millet appear at two phases of plant growth:

1. Typical downy mildew symptoms on the vegetative parts during seedling stage, and

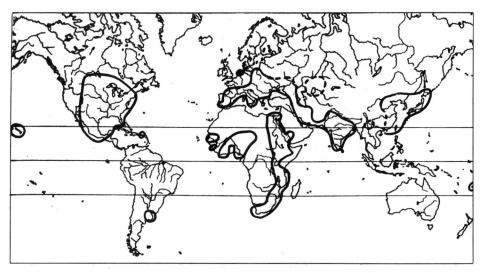

Figure 12-1 CMI geographical distribution map of pearl millet downy mildew.

2. Characteristic green-ear symptom, which brings about different types of distortions and malformations on the affected earheads.

In the downy mildew phase, infected plants initially turn pale yellow [Figure 12-2 (a, b)] from dark green color with variable degrees of stunting. During the advanced stage of downy mildew phase the entire foliage becomes white due to heavy deposition of fungal growth. In a majority of the cases chlorotic streaks with downy growth of the fungus appear on the undersurface of infected leaves. The pale yellow to whitish color of the affected plants continues for nearly 3 weeks after initiation of the disease, and later on, these chlorotic streaks [Figure 12-2 (a, b)] gradually turn into light brown to dark brown lesions on the affected leaves. In severe cases, drying/death of young seedlings takes place resulting in patchy appearance or complete failure of the crop. If the infection is due to soilborne or seedborne oospores in the very early stage, a majority of such seedlings remain stunted without any tillering. Normally, such plants remain dwarf, barely attaining 50 cm in height with typical succulent stems spreading laterally instead of growing upright, thus imparting a bushy appearance to the infected plants.

The green-ear phase normally is the result of systemic infection where different forms of abnormalities occur during the heading stage of the affected plants. The nodal buds of lateral shoots develop into a bushy appearance. The main tiller of the affected plants also develops abnormalities due to transformation of floral parts into wavy or twisted green structures, imparting the name of green-ear phase. In the typical green-ear affected plants the following types of deformations are observed.

1. The earheads become abnormal in all the tillers which are replaced by a bunch of leafy structures.

2. The ear length remains normal, but the entire ear is converted into giant leafy structures.

Figure 12-2 (a), (b)   Symptoms of downy mildew of pearl millet (downy mildew stage on foliage). (Courtesy of CRISAT)

3. Only the basal part of the earhead gets converted into leafy structure and the remaining parts bear the normal grains or only the upper part, or sometimes the middle part of the affected earheads is transformed into a green leafy structure while the remaining part bears normal grains (Figure 12-3).

12-3 THE PATHOGEN

Sclerospora graminicola was originally named as *Protomyces graminicola* by Saccardo[5] who observed this fungus on *Setaria italica*. Schroeter[6] renamed *Protomyces graminicola* as *Sclerospora graminicola*. The reason for this change was formation of multilayered dark brown epispore around the fertilized oogonium. In the beginning of the present century, Ito[7] put *Sclerospora* into subgenera *Eusclerospora* with a type species of *S. graminicola*. Shaw[8] considered *S. graminicola* as a primitive species. Coenocytic and branched mycelium of *S. graminicola* is present in the roots, stems, leaves, and inflorescence of infected plants. From the mycelium, both asexual and sexual structures develop. The asexual phase appears first as whitish downy growth. The asexual stage is followed

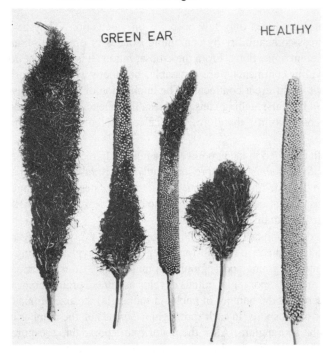

GREEN EAR   HEALTHY

Figure 12-3 Symptoms of downy mildew
of pearl millet (Green-ear stage)

by sexual spore production. The culmination of the vegetative mycelial growth results in
the formation of sex organ initial. According to Kenneth,[9] sporangiophores measures
100.6–221.2 × 15.3–23.7 μm and sporangium, 19.0–31.6 x 15.8–23.7 μm. Oospores
with oogonia measure 47.4–55 μm in diameter, whereas the diameter of the oospore
generally ranges between 31.6 and 39.5 μm on pearl millet. In the presence of water each
sporangium burst, releasing unequal, 3–13 biflagellate zoospores. The zoospores swim
in water for about 30 min before finally encysting. Encysted zoospores immediately germinate. In some cases zoospores germinate within sporangia even before they are liberated.[10] Both oogonia and antheridia develop within infected host cells. The oogonium is
terminal or intercellular, and after fertilization it develops within thick, brown to dark
brown irregular walls. The product of sexual reproduction is the formation of round to
oval oospores which are produced in large numbers and remain scattered in the mesophyll
cells. Oospores have three distinct layers in the wall, i.e., exosporium, mesosporium, and
endosporium. The oogonial wall is persistent. Oospores vary in diameter from 26–32 μm
having an average diameter of 35 μm, and they do not germinate immediately but have a
prolonged resting period.

 Primary infection and the survival of the fungus in nature are affected by oospores.
Some of the hybrids recently developed as resistant to downy mildew in India are susceptible in West Africa. Similarly, there are other cultivars that have displayed variation in
their reaction at different locations within India and West African countries. This indicates that the fungus population has the potential to change pathogenicity. Idris and Ball[11]
have demonstrated the presence of two equally proportioned compatibility types of
downy mildew reaction in pear millet which have been designated as G-1 and G-2. On the
basis of oospore collections from diverse locations in West Africa and India, they indi-

cated that some of the isolates produced only few oospores on inoculation. They have further made the tests for cross-compatibility by combining isolates of opposite sexual compatibility types from different collections. From these observations it is clear that not only the collections from within continents were crossable, but there was also cross-compatibility between isolates of different continents. The implication of such extensive outbreeding in *S. graminicola* has far-reaching consequences. The phenomenon of inter- and intracontinental sexual compatibility, therefore, is to be considered while breeding for downy mildew resistance.

Heterothallism has been recorded in many oomycetous fungi.[12] In recent years there have been indications that *S. graminicola* is predominantly heterothallic in nature. Michelmore et al.[13] have demonstrated heterothallism in different isolates of the fungus collected from India. The existence of heterothallism has also been observed by Kenneth[9] in the isolates of *S. graminicola* from different species.

Safeeulla[14] has described cytological details in the development of *S. graminicola*. Initiation of sporangiophores takes place in the substomal regions. The hyphae, having a number of nuclei, migrate into the young sporangiophores immediately after initiation. When branches in sporangia are laid, sporangial initials develop as minute protruberances on apical sterigmata. Irrespective of the number of nuclei in sporangia, the size of nuclei remain uniform. The number of zoospores in each sporangium depends on the number of nuclei originally present in the sporangium. After the maturity of sporangia, zoospores are released through an opening in papilla one by one and the whole process of zoosporosis is completed within a few minutes.

A detailed study of cytoplasmic behavior has been conducted by Lange, Olson, and Safeeulla.[15] They have reported the cleavage in the zoosporangial cytoplasm of *S. graminicola*. However, the cleavage of *S. graminicola* is different from Saprolegniales where zoosporogenesis is brought about by the expansion of central vacuole or where the plasmalemma alone is used.

12-4 EPIDEMIOLOGICAL STUDIES

12-4-1 Pathogen Under Infection Process

Nutrition of pathogen.   *S. graminicola* happens to be a strict obligate pathogen and as such, the effect of nutrition has only been studied under dual culture conditions. Callus isolated from healthy and proliferated earheads of pearl millet plants infected by the fungus grew best on a medium having basal mineral salt, agar, and several growth-promoting substances between pH 5 and 7. Maximum growth was obtained with 9 mg/1 2, 4-D without which diseased tissues failed to grow even in the White's basal medium.[16] Bhat et al.[17] have also reported the growth of this fungus on host tissue culture. Mycelium was normally coenocytic but occasionally septate too. Both asexual and sexual spores were formed on callus tissue. It is interesting to note that mycelium on the callus tissue retained its infectivity throughout 5 years of subculturing.

Viability of pathogen.   A good amount of literature is available on the viability of fungus both in nature and under control conditions in the axenic culture. Bhat et al.[17]

reported that mycelium of *S. graminicola* formed on callus tissue remained infective through 5 years of subculturing. Similarly, Prabhu et al.[18] reported that mycelium of *S. graminicola* invading seed embryo remained viable even in dry seed, whereas the viability of hyphae present in pericarp decreased as the moisture content of seed declined. When seeds from partially malformed ears were cultured on Murashige and Skoog's[19] medium, thin mycelial growth was observed on callus tissue of 0.1% of the seed. It is interesting to note that mycelium can remain dormant during unfavorable periods but will become active when suitable conditions are restored.

Oospores remain viable in the soil for 3–4 years.[20–21] However, in the infected host tissue the viability of oospores has been reported to be 8 years.[22] Thakur[23] reported that though oospores survive for longer periods their viability was lost after 28 months. Experiments on survival, viability, and infectivity of oospores revealed that they remain pathogenic in soil for 28 months. Deeper placement of oospores in soil decreases the disease incidence.[23]

It is also known that oospores require some weathering period for germination and infection. The overwintering of oospores 25 cm below soil level under natural conditions increased the pathogenic potentiality of oospores in comparison to oospores stored at ambient conditions. One year, weathered oospores either mixed with soil or seed retained highest infectivity.[23] Thus, oospores buried in field soil were more infective than those stored in the laboratory. This indicated that weathering stimulates the germination and infectivity of oospores. It is widely believed that oospores of *S. graminicola* undergo dormancy.[24] However, Bhat[25] claimed that newly formed oospores also germinate satisfactorily.

The survival of oospores, however, largely depends on various soil factors like temperature, pH, soil salinity, water holding capacity of soil, soil organic matter content, and so on. These factors might be operating either alone or in combination to influence the survival of oospores.[26] They have further opined that low levels of ribosomal RNA and ribosomal protein are found in the ungerminated oospores indicating the presence of precursors. The absence of this component of protein synthesis could explain the dormancy of oospores.

Variability of pathogen.   There are several reports which indicate that considerable variability exists in *S. graminicola*. Kenneth[9,27] has reported the occurrence of physiological races. Rasheed et al.[28] and Shetty and Rasheed[29] have also clearly demonstrated the existence of races of *S. graminicola*. Ball[30] has demonstrated host cultivar reaction to infection by different isolates of *S. graminicola*. According to Michelmore et al.[13] *S. graminicola* is predominantly heterothallic with two distinctly compatible types viz., G1 and G2, which occur almost in equal proportion in a given population. In this context it is pertinent to mention that evidences of cross-compatibility between isolates of Indian and African countries have been established. The isolates of West Africa are more pathogenic than Indian isolates.[31] Recent studies confirmed the earlier report of Uppal and Desai[32] when they failed to infect pearl millet with oosporic inoculum from *Setaria* or *Panicum*, the other two major hosts and vice versa, indicating the existence of physiological specialization. Two years later Takasugi[22] reported that *S. graminicola* from *S. italica* could not infect *S. viridis* and vice versa. Another early indication of the existence of physiologic races came from Girard[33] when he reported that some pearl millet varieties resistant at certain places were found susceptible at others. Similarly, Bhat[25] reported hybrid HB-3 as highly resistant at Mysore but it proved susceptible in many other pearl millet growing states of India.

Germination studies.   Techniques for successful germination of oospores have been given by several scientists.[22,34–37] However, reproducible and consistent results have not yet been achieved. The Hiura[35] technique involves incubation of oospores on filter paper in a petri dish at 27–30°C for nearly 48 hr. Pande[37] observed that oospores treated with gibberellic acid (0.2–0.5 ppm) followed by incubation at 15–20°C under alternate light and darkness induced gemination resulting in production of sporangiophores and sporangia. Indirect evidence has been obtained by the author for germination of soilborne oospores embedded in diseased plant debris after weathering the sterile soil under congenial environmental conditions. Sterilized soil was inoculated with oospores at the rate of 100 g oosporic powdered plant debris for every kg of soil. There was successful infection of young seedlings between 2–3 leaf stage development.

Host spectrum.   *S. graminicola* has been reported from a few more plants such as *Agrostis alba, Echinochloa crusgalli, Euchlaena mexicana, Panicum miliaceum, Pennisetum* spp., *Saccharum officinarum, Setaria* spp., and *Zea mays*.[25–26]

Singh and Williams[38] further studied the host range of *S. graminicola*. They studied the reactions of 23 hosts belonging to 11 genera and also included earlier reported host species. *S. graminicola* failed to produce downy mildew symptoms on any of these selected host species. It was later confirmed by Singh and Luther[39] that ICRISAT strain of *S. graminicola* is specific only to *Pennisetum americanum*.

12-4-2 Role of Environmental Factors

On asexual propagules.   According to Safeeulla and Thirumalachar[40] the following factors were required for sporangial production:

1. A period of 15–20 hr since the previous sporangial crop.
2. Free film of moisture on the surface of the leaves.
3. A temperature of about 25°C.

Sporangia germinated at 18–29°C, the optimum being 24–25°C.[41] Melhus et al.[42] and Suryanarayana[43] considered temperature as low as 10°C favorable for sporulation. Sporulation did not occur at temperatures higher than 28°C.[43] For sporangial production, in general, a saturated atmosphere with a film of water on the leaves is best. Under favorable conditions, 35,000 sporangia are produced/m² leaf area of infected pearl millet plants/sporangial crop and as many as 11 crops on successive nights are formed.[14] Wind direction in general affected sporangial spread but spread was considerably more during the rainy season than in the dry postrainy season.[44]

On sexual propagules.   Oospores survive for a minimum of only 8 months to as long as 10 years.[26] Results of experiments conducted by the author for 5 years revealed that the soilborne oospores remained pathogenic for 28 months under extreme semiarid conditions. Higher infection occurs with 1-year-old oospores as compared with fresh oospores or old oospores of 2 or more years of storage.[14] According to Safeeulla,[14] profuse germination of oospore occurs at 20–25 °C and greater germination with overwintered oospores have been observed. The author has also observed that weathered oospores 6 months after overwintering were more pathogenic.[24] They further observed that weathered oospores exhibited more pathogenicity in field soil or through external seed coat in comparison to sterilized soil.

12-4-3 Role of Biotic Factors

Planting date.   Singh and Williams[44] in their study at ICRISAT reported interplot movement of sporangia from earlier planted plots to the later plantings. Thakur[45] conducted experiments in 3 consecutive years to determine the effect of sowing dates on downy mildew incidence in five test cultivars. It was observed that disease incidence was less in early sown crop (till July 5th sown crop) than in late sown ones (July 15th and 25th sown crops).

Inoculum placement.   Pearl millet is a shallow rooted plant with plenty of subsurface feeder root systems. It is expected that deep placement of oospores in soil should not produce much dent on the disease index. In exhaustive experiments, Thakur[45] observed that even deeply placed oospores remained infective enough to initiate a sufficient level of primary infection.

Ratooning.   To take advantage of early rain and also to avoid the risk of pollen washing, many cultivators plant their crop in early rain and when the crop is 4 to 6 weeks old they cut the crop nearly 10 cm above the ground level and the stubble is allowed to regenerate. This practice also gives an additional crop of fodder. However, such regenerated crop or ratooned crop has proved to be more prone to downy mildew infection.[45]

Planting methods.   In a comparative experiment repeated over 3 years,[45] it has been observed that direct seeded crop contracted more than 65% incidence in comparison to only 22.5% infection in transplanted crop of highly susceptible hybrid NHB-3. The logic behind transplanting may be an automatic rejection of the downy mildew infected seedlings before transplanting. In addition, even if 100% downy mildew infected seedlings are transplanted, only 74% of adult plants show typical downy mildew infection. Because of the fact that downy mildew fungus, being a biotroph, gets lost or eradicated during the process of senescence, when the seedlings are uprooted and transplanted, it takes 7–10 days for regeneration of transplanted seedlings. Under such stress conditions due to negligible physiological activities, congenial conditions for continued colonization of pathogen does not exist, which results in the eradication of the pathogen.[45]

Monoculture.   Primary infection comes through hybernating oospores in the soil or oospores adhered on seed coat. The repeated sowings of the susceptible cultivar in the same field is bound to increase the inoculum buildup in soil. The phenomenon has been found true in the case of hybrid BJ-104, which became susceptible to downy mildew within 7 years after its release. This resulted in the denotification of BJ-104 hybrid from the list of cultivable varieties.

12-5 TRANSMISSION

S. graminicola invades the apical meristem of the host and the symptoms appear after the differentiation of infected tissue and colonization of the pathogen within the host system.[46] In general, pearl millet plant remains more susceptible because of the continuous production of tillers. This results in the continuous presence of young susceptible meristem in tillers which are more prone to infection.[44] More often, such early infection eventually leads to systemic infection. It is evident that various types of infective propagules are involved in transmission or perpetuation of pearl millet downy mildew. In case of *S. graminicola*, the infective propagules for the successful transmission of disease are oospores and internally seedborne mycelium acting as primary sources of infection and sporangia/zoospores responsible for secondary spread of the disease.

12-5-1 Role of Soilborne Oospores in Primary Infection

During the recently concluded detailed experiments on the infectivity of oospores over a period of 5 years by the author, several important and basic information(s) have emerged. These are:

Pathogenic viability of soilborne oospores.   Once the weathered oospores are placed in the soil they retain the pathogenic/infective capability for 28 months even under the very dry conditions of Hisar.[45] However, these oospores lost infectivity after 30 months.

Pathogenic potentiality of soilborne oospores in relation to soil depth.   Pearl millet is a shallow rooted crop with a subsurface feeder root system and it is expected that if the oospores are placed deep in soil, beyond 7 cm level, the crop should remain free. However, experimental results revealed that oospores placed even 15 cm deep could cause infection.[45]

Pathogenic potentiality of weathered/unweathered soilborne oospores.   Weathering through overwintering increases the infective capability of soilborne oospores. However, soilborne oospores differ in their pathogenic potentiality depending on the type of soil. It has been proved by the author that field soil rich in organic matter increased the infective potentiality of oospores; contrary to this, oospores placed in sterilized soil without organic matter could cause little infection.

12-5-2 Role of External Seedborne Oospores

Oospores of *S. graminicola* adhering to the seed coat has been considered a potential source for transmission and perpetuation of inoculum from one season to anther or from one place to another.[26,43,47-48] However, the impact of external seedborne oospores depends on the threshold number or quantum of oosporic seed load. Thakur[49-50] observed gradual increase of infection levels from 7% under 0.5 g oosporic material/kg seed to 42.2% under 4 g oosporic material/kg seed. Positive correlation was also observed between the oosporic seed load of individual seed samples and corresponding downy mildew incidence under natural field conditions.

12-5-3 Role of Internal Seedborne Mycelium in Primary Infection

Suryanarayana[51] was first to observe intercellular aseptate mycelium in seed embryo. Tiwari and Arya[53] could trace mycelial fragments inside the seed coat as well in embryo in the seeds collected from partially infected earheads. Sundaram et al.[54] reported the presence of hyphae inside embryo of 4% seeds from infected earhead of hybrid HB-4. They reported that symptoms of downy mildew appeared when such infected seeds were sown at 90% RH and 25–28°C. Thakur and Kanwar[55] recorded an interesting observation that seeds from healthy heads of healthy plants from infected field also produced 1–2% downy mildew infected seedlings. It may be presumed that flower infection may take place like that of loose smut of wheat. They also observed that the extent of infection in seedlings from partially green ear-affected heads was maximum. Shetty et al.[48] showed that transmission of S. graminicola takes place from dormant mycelium carried within the seed after storage for 10 months under dry conditions. The mycelium of fungus invades all parts of seed tissue such as pericarp, aleurone layer, and endosperm, as well as scutellum. However, growing on tests seem to indicate that more likely the mycelium present in the embryo in scutellum takes part in causing systemic infection in seedlings.

12-5-4 Role of Sporangia in Secondary Infection

Suryanarayana[43] observed sporangial germination in the early morning if there is enough dew on leaf surface. Thakur and Kanwar[56] have reported that downy mildew phase of the disease initiated from sporangial infection was maximum in young seedlings. The author, through detailed studies on various aspects of sporangial infection, observed that inoculation of sporangia produced downy mildew seedlings within 6 days under high humidity ($\geq 90\%$) and temperature range of 18–20°C. Use of sporangia by contact method, there was comparatively less infection than when sporangial suspension was used on young seedlings. According to Subramanyam and Safeeulla,[57] two atmospheric processes, namely turbulent diffusion and thermal convex atmosphere, disturb the process of sporangia sedimentation and keep them in dispersion phase from few minutes to several hours resulting in loss of viability up to some extent. Subramanyam et al.[58] observed that as the crop matures the sporangial density in the air is reduced, resulting in a decline of disease incidence. Airborne zoospores released from sporangia can move in water saturated soil and may infect pearl millet seedlings through roots as well.[59] The sporangia also get disseminated from early sown disease plant to late sown crop.[44] According to Thakur and Kanwar[56] and Thakur[45] sporangia play a significant role in secondary spread of downy mildew. Their study showed 13.2% of plants getting downy mildew symptoms only through secondary sporangial infection. Wind direction had an effect on the extent of secondary spread.

Seedling age and secondary infection.   It has been demonstrated that pearl millet seedlings remained highly susceptible till the age of 15 days; thereafter, downy mildew development decreased. Ten-day old seedlings got maximum infection whereas it was least in 20 day old seedlings.[56]

Plant parts vulnerable to sporangial infection.   Symptoms of downy mildew always initiate from whorl of the seedling. This is because of vulnerable young apical cells, function of whorl to accommodate free water for longer duration and thus, acting as an incubation chamber which facilitates movement and germination of zoospores. The author, after concluding an elaborate experiment, found that the whorl inoculation could induce downy mildew on 15.2% seedlings whereas ligule inoculation gave only 4% infected seedlings. There was no infection on other plant parts such as leaf tip and base.

12-5-5 Disease Cycle

As explained earlier, oospores produced in abundance in infected crops perpetuate the disease either through soil or on the seed coat. Such soilborne oospores or sometimes zoospores present in the free film of water near the root zone of host initiate the colonization of organisms at root zone which subsequently produce downy mildew symptoms. After production of millions of asexual sporangia/zoospores in several cycles of sporulation, these infected plant parts convert into the favorable sites for development of sexual spores (oospores) following the mating between antheridia and oogonia. The sporangia/zoospores on the other hand again act as a secondary source of infection, repeating the same process of production of asexual spores followed by sexual spores. Some of these sporangia/zoospores, when they fall on the ground and if free moisture is present in the soil, further infect the root zone of young seedlings, thereby repeating the same cycle again. Ultimately the infected plants are converted into completely or partially malformed earheads, which is essentially systemic in nature. Besides malformations, the systemic infection also makes the various floral parts sterile. In such systemically infected plant parts numerous oospores are present which find their way either to external seed coat or in the soil, thus repeating the cycle again and again.[46]

12-6 CONTROL MEASURES

12-6-1 Cultural Methods

Alteration in planting date.   Delayed planting of pearl millet results in higher downy mildew incidence. The reason for less disease in early planted crop may be the lack of initial inoculum level, which in the later planted situation increases because of the large area under cultivation of susceptible host, and thereby many fold increase in the inoculum buildup.[45]

Modification in planting methods.   Downy mildew incidence could significantly be reduced if the crop is transplanted instead of directly seeded. The direct seeded hybrid HB-3 contracted 65% downy mildew incidence against only 22% infection in transplanted crop. There seem to be two plausible reasons for less disease incidence in transplanted crop. First, the infected seedling is automatically culled out in nursery before transplanting. Second, the abruptly reduced physiological activities or the presence of senescence in the uprooted/transplanted seedlings for a few days during the period of reestablishment may not be favorable for sustained colonization of the pathogen in the infected seedling. This fact has been observed even in the transplanting of 100% downy mildew infected seedlings.[45]

Avoidance of ratooning.   In certain areas, ratooning of the crop is a common phenomenon to take advantage of early rain. It is cautioned that at least susceptible culti-var should not be seeded as ratooned crop because during the regeneration phase high disease infection takes place.[45]

Roguing and gap filling.   Removal of infected plant has two beneficial effects, viz., reduction of asexual spore and thereby restricting the spread of secondary infection, and second, the subsequent restriction in the production of oospores which ultimately reduce the level of primary infection. It has been found that pearl millet downy mildew has significantly been reduced along with a proportionate increase in harvest index if the infected seedlings are eradicated at the pretillering stage. The effect of roguing can be further increased by transplanting healthy seedlings in the gap created by removal of infected plants so that the plant population is maintained simultaneously.[24,45]

Application of balanced nutrients.   Experiments on host nutrition have been extensive but results obtained so far seem to be contradictory and inconclusive. Sivapra-kasam et al.[52] did not observe appreciable effect on downy mildew in three pearl millet hybrids when nitrogen level was increased from 0–200 kg/ha, whereas Deshmukh et al.[61] observed reduced downy mildew when nitrogen was applied increasingly from 0–100 kg/ha. However, Singh[62] recorded increased downy mildew incidence due to increased level of nitrogen from 0–40 kg/ha. Thakur[45] also observed increased level of downy mildew incidence from 33% to 58% under zero level to 40 kg N/ha.

Deep ploughing and sun baking of soil.   Deep ploughing of the field has been suggested by Tuleen et al[63] to reduce oosporic inoculum level in soil in case of sorghum downy mildew. This method is equally justifiable in pearl millet downy mildew also. This practice could be more effective if it is done continuously for 3–4 years.

Removal of infected crop residue.   According to Williams[46] the removal of downy mildew infected crop residue has two beneficial effects. In addition to the reduc-tion of sexual spore production initially during primary infection, it will further reduce the infection foci in the subsequent crop.

Cultivation of trap crop.   In a heavily infected field where oosporic inoculum buildup is high, it has been found beneficial to reduce soilborne inoculum level by raising highly susceptible cultivar in between the main crop season. As soon as the off-season crop comes to knee-height stage, appearance of downy mildew symptoms with abundant production of inoculum takes place. Exactly at this stage the crop should be harvested so that formation of sexual spores does not take place. If this practice is repeated at least 2–3 times between the main crop seasons it will reduce the soilborne inoculum level below the threshold value.

Avoidance of monoculture.   According to Thakur[45] the soilborne inoculum increased gradually with repeated sowing of the same susceptible cultivar year after year.

Thus, avoidance of monoculture, particularly of susceptible cultivar, is the most desirable feature for obtaining good harvest index by reduction of inbuilt soilborne inoculum.

12-6-2 Physical Methods

Hot water seed treatment.   Exposure of infected seeds at 55°C for 10 min reduces disease incidence considerably.[45,55]

12-6-3 Chemical Control

Seed treatment.   Though downy mildew of pearl millet can be reduced by seed treatment and foliar spray either independently or in combination, normally such treatments were generally not economical and often not effective under ever-increasing soilborne inoculum buildup.[64–66] Thakur and Kanwar[67] and Thakur[45] used several fungitoxicants for control of pearl millet downy mildew and observed that seed treatment with Thiram and Captafol gave satisfactory disease control under experimental conditions.

Foliar spray.   Fungitoxicants of different chemical groups have been used as foliar spray alone or in combination with seed dressing fungicides. It has been observed[45] that foliar spray with copper oxychloride or ziram did not prove very effective, but Captan and Mancozeb controlled the disease by more than 67% in the susceptible hybrids.

Use of fungicidal root dip.   It has also been observed[45] that dipping the roots of pearl millet seedlings in water suspension (2,000 ppm) of Aureofungin and Difolatan for 1 hr just before transplanting reduced downy mildew infection in adult plants of highly susceptible hybrid HB-3.

Role of metalaxyl.   With the introduction of acylalanines particularly metalaxyl (methyl N-(2-methoxyacetyl)-N-(2,6-xylyl)-DL-ananinate) formulations such as Ridomil-25 WP, Apron SD-35, etc., control of downy mildew either through seed treatment or through foliar spray has been very effective.[45,68–70] According to Dang et al.[70] and Thakur[45] Ridomil-25 WP and Apron SD-35 as seed treatment at 2 g ai/kg completely protected susceptible pearl millet plants (HB-3) up to 30 days. As foliar spray, Ridomil-25 WP (2,000 ppm) used once at 20 days or twice after 20 and 40 days of plant growth gave least downy mildew incidence at harvest time. Seed treatment followed by one spray of 2,000 ppm Ridomil-25 WP was found to be very effective in controlling the disease even in sick plot. The effect was also curative. These treatments improved the plant growth and yield significantly over nontreated crop. However, because of resistance problem encountered in several plant pathogens Ridomil-25 WP has been replaced by mixed formulations of metalaxyl with protective fungicides like mancozeb, maneb, folpet and captan for the foliar application.

12-6-4 Host Plant Resistance

Today downy mildew resistant cultivars of maize, sorghum, and pearl millet are contributing to the food production efforts in several countries, including India.[46]

Nature of downy mildew inheritance.   Resistance of pearl millet to *S. graminicola* has been considered a dominant attribute over susceptibility.[20] Gill et al.[71] reported that the inheritance of downy mildew resistance or susceptibility is complex in nature. They suggested that inheritance to downy mildew in pearl millet is governed by two duplicated dominant factors. Basavaraju et al.[72] concluded that more than one pair of genes are responsible for resistance against *S. graminicola* and it is quantitative in nature. Both additive and nonadditive gene effects were found to be significant. It has been suggested that upgrading of materials could be done through reciprocal recurrent selection.[73–74] A review of work done[45,75] on the nature of inheritance is given in Table 12-1.

12-6-5 Integrated Control Measures

Incidence of downy mildew in relation to the cultural practices along with the nominal use of chemicals in terms of seed treatment alone or in combination with foliar spray could be taken into account for the formulation of an integrated approach for the economical control of pearl millet downy mildew through minimal use of cash inputs.

The essential ingredients for effective control of *S. graminicola* in pearl millet has been achieved through the use of tolerant cultivars, crop rotation, seed treatment, transplanting, and so on, and the combined effect of all these methods has been found very effective and economical for reduction of pearl millet downy mildew.[45,49]

12-7 FUTURE STRATEGIES

Certain gaps in our knowledge of various aspects of epidemiology and management of downy mildew still exist. There is a need to initiate research along the following lines.

TABLE 12-1   REVIEW OF WORK DONE ON NATURE OF INHERITANCE OF DOWNY MILDEW OF PEARL MILLET

Contributors*	Nature of resistance
Singh	Resistance dominant, qualitative, two dominant genes
Appadurai et al.	Resistance dominant, one or two major genes
Gill et al.	Complex gene interaction, cytoplasmic gene interaction or both
Safeeulla	Resistance dominant
Pethani	Resistance dominant
Kumar et al.	Resistance dominant, nonadditive component predominant
Basavaraju et al.	Resistance dominant, quantitative nature, nonadditive genes most effective
Gill et al.	Resistance dominant, duplicate gene factor
Singh et al.	Susceptibility dominant, quantitative in nature
Dang and Thakur	Resistance partially dominant, quantitative nature, nonadditive gene effect
Dass	Resistance dominant, quantitative nature, nonadditive gene predominant, highly heritable

*For details of references see Thakur[45]

12-7-1 For the Pathogen

Gene for gene studies.   It has been observed that gene for gene hypothesis holds good to several other diseases caused by obligate parasites or biotrophs, i.e., rusts, powdery mildews, and downy mildew. It is assumed that gene for gene theory could also be applicable to pearl millet downy mildew.

Physiologic races.   So far, detailed research work on the existence of physiological races of *S. graminicola* has not been able to clearly indicate the existence of races or mechanism of variability in *S. graminicola*. Only some sketchy information is available on the variability of this fungus with respect to sexual compatibility. Therefore, emphasis is to be placed on defining the races on the basis of their pathogenic behavior against differentials or standard cultivars.

12-7-2 For Host–Pathogen Interaction

Horizontal resistance.   So far very little work has been done on horizontal resistance in pearl millet downy mildew. It is imperative therefore that research work be initiated for development of cultivars having horizontal resistance, because it is considered to be stable, flexible, and efficient.

Estimate for resistance durability.   At present certain multilocational trials are being conducted in India to find out the durability of resistance of various cultivars against *S. graminicola*. However, more uniformity regarding the techniques and methodology is not being followed under field conditions, especially with respect to inoculum load and other cultural and agronomical operations. As a result, the reproducibility of the results of such experimental trials normally becomes difficult and doubtful. Moreover, the parameters of durable resistance are yet to be selected through mutual exchange of views of various pearl millet pathologists and breeders working in India and abroad.

Standardization of rating scale.   Several disease rating scales have been given by various workers from time to time in India and abroad. However, so far the relevance of these scales in the study of epidemic development of disease and consequent yield losses has been overlooked. This aspect of the disease also needs careful investigation.

12-7-3 For Measures to Control Disease

Biological control through microorganisms.   Oospores of *S. graminicola* are the most potential infective propagules for the primary infection of downy mildew of pearl millet. Literature showed that the phenomenon of microparasitism has been used for biological control of several downy mildews. Among these, the presence of mycoparasitic fungi belonging to Chytridiales is very common. The duration of survival of soilborne oospores indicates that these infective propagules are capable of withstanding the loss in their population due to natural mycoparasitism. It has been reported that the downy mildew infected areas of leaves which later on get secondarily infected/colonized by other fungi normally reduce or totally inhibit the colonization and sporangial production by downy mildew fungus. Thus, this phenomenon could be exploited for biological control of pearl millet downy mildew through microorganisms.

Improvement in metalaxyl formulations.   Metalaxyl derivatives are effective against downy mildew disease both in relation to reduction in disease intensity and proportionate increase in harvest index. Though this fungicide possesses both preventive and curative properties, it has still been observed that symptoms of downy mildew/malformations appear even on the seed-treated plants. Though the phenomenon does not attribute to the yield reduction, because of such later symptoms, confusion arises regarding the efficacy of this fungicide. The later symptoms appear because the residual activity of metalaxyl stays only up to 30–35 days after sowing. It is suggested metalaxyl should be modified in such a way that its residual effect could be extended further. Even a new compound may be developed keeping this criterion in mind.

12-8 REFERENCES

1. Krishna Swami, N. "Origin and Distribution of Cultivated Crops of South Asia, Millets." *Indian J. Genet. Pl. Breed.* 11 (1957): 67.

2. Porter, R. H. "A Preliminary Report of Surveys for Plant Diseases in East China." *Plant Dis. Reptr.* Suppl. 46 (1926): 153.

3. Mitter, J. H., and Tandon, R. N. "A Note on *Sclerospora graminicola* (Sacc.) Schroet in Allahabad." *J. Indian Bot. Soc.* 9 (1930): 243.

4. Anonymous. "Annual Report of *All India Co-ordinated Millets Improvement Project (ICAR)* for the Year 1971–72."

5. Saccardo, P. A. "Fungi Veneti novi vel. Critia, Ser. V. No. 91 Nuova Giorn." *Bot. Ital.* 8 (1876): 161.

6. Schroeter, J. "*Protomyces graminicola* Saccardo." *Hedwigia* 18 (1879): 83.

7. Ito, S. "Kleine Notizen uber Parasitische Pilze Japana." *Bot. Mag. Tokyo* 27 (1913): 217.

8. Shaw, C. G. "The Genera of the Peronosporaceae." *Phytopathology* 40 (1950): 25 (Abstract).

9. Kenneth, R. "Studies on Downy Mildew Disease Caused by *Sclerospora graminicola* (Sacc). Schroet and *S. sorghii* Weston and Uppal Scr." *Hierosol* 18 (1966): 143.

10. Weston, W. H., and Weber, G. F. "Downy Mildew (*Sclerospora graminicola*) on Everglade Millet in Florida." *J. Agric. Res.* 36 (1928): 935.

11. Idris, M. O., and Ball, S. L. "Inter- and Intra Continental Sexual Compatibility in *Sclerospora graminicola.*" *Plant Pathol.* 33 (1984): 219.

12. Michelmore, R. W., and Ingram, D. S. "Heterothallism in *Bremia lactucae.*" *Trans. Br. Mycol. Soc.* 75 (1980): 47.

13. Michelmore, R. W., Pawar, M. H., and Williams, R. J. "Heterothallism in *Sclerospora graminicola.*" *Phytopathology* 72 (1982): 1368.

14. Safeeulla, K. M. "Downy Mildew of Bajra. Current Status." Progress Report of *All India Co-ordinated Millet Improvement Programme* (1975–76). Chapter VII. 53–65, 1976.

15. Lange, L., Olson, L. W., and Safeeulla, K. M. "Pearl Millet Downy Mildew (*Sclerospora graminicola*): Zoosporogenesis." *Protoplasma* 119 (1984): 178.

16. Arya, H. C., and Tiwari, M. M. "Growth of *Sclerospora graminicola* on Callus Tissues of *Pennisetum typhoides* and in Culture." *Indian Phytopathol.* 22 (1969): 446.

17. Bhat, S. S., Safeeulla, K. M., and Shaw, C. G. "Growth of *Sclerospora graminicola* in Host Tissue Cultures." *Trans. Br. Mycol. Soc.* 75 (1980): 303.

18. Prabhu, M. S. C., Safeeulla, K. M., and Shetty, H. S. "Tissue Culture Technique to Demonstrate the Viability of Downy Mildew Mycelium in Pearl Millet Seeds." *Curr. Sci.* 52 (1983): 1027.

19. Murashige, T., and Skoog, P. "A Revised Medium for Rapid Growth and Bioassays with Tobacco Tissue Cultures." *Physiol. Plant.* 15 (1962): 473.

20. Safeeulla, K. M. "Downy Mildew of Pearl Millet." Proc. Ann. Rep. of *All India Co-ordinated Millet Improvement Project 1975–76.* 53, 1975.

21. Suryanarayana, D. "Infectivity of Oospore Material of *Sclerospora graminicola* (Sacc.) Schroet., the Bajra (*Pennisetum typhoides*) Green Ear Pathogen." *Indian Phytopathol.* 15 (1963): 247.

22. Takasugi, H., and Akaishi, Y. "Studies on the downy mildew (*Sclerospora graminicola sitariae italicae*) on Italian Millet in Manchuria I. About the Germination of Oospores." *S. Manchri Ry. Co. Agric. Exp. Sta. Res. Bull.* 11 (1933): 1.

23. Thakur, D. P. "Epidemiological Studies on *Sclerospora graminicola* the Pearl Millet Downy Mildew Fungus." *3rd International Symposium on Plant Pathology,* IARI, New Delhi, December 14 to 18, 1981, 11–12 (Abstract).

24. Thakur, D. P., and Gangopadhyay, S. "Biology and Epidemiology of Downy Mildew of Pearl Millet," in *Vistas in Plant Pathology*, eds. A. Varma, and J. P. Varma. (New Delhi: Malhotra Publishing House, 1986), 293.

25. Bhatt, S. S. "Investigations on the Biology and Control of *Sclerospora graminicola* on Bajra." (Ph.D. thesis, Univ. Mysore, India, 1973), 165.

26. Nene, Y. L., and Singh, S. D. "A Comprehensive Review of Downy Mildew and Ergot of Pearl Millet." *Proceedings Consultative Group Meeting on Downy Mildew and Ergot of Pearl Millet.* ICRISAT, Patancheru, India 1–3 Oct., 1975, 16–53.

27. Kenneth, R. "Downy Mildews of Graminae in Israel." *Indian Phytopathol.* 23 (1970): 371.

28. Rasheed, A., Shetty, H. S., and Safeeulla, K. M. "Existence of Pathogenic Races in *Sclerospora graminicola* (Sacc.) Schroet Attacking Pearl Millet (*Pennisetum typhoides* (Burm.) Stapf and Hubb.)." Presented at *3rd International Congress of Plant Pathology,* Munich, 1978.

29. Shetty, H. S., and Rasheed, A. "Physiological Specialization in *Sclerospora graminicola*." *Indian Phytopathol.* 34 (1981): 307.

30. Ball, S. L. "Pathogenic Variability of Downy Mildew (*Sclerospora graminicola*) on Pearl Millet. I. Host Cultivar Reactions to Infection by Different Pathogen Isolates." *Ann. Appl. Biol.* 102 (1983): 257.

31. Ball, S. L., and Pike, D. J. "Intercontinental Variation of *Sclerospora graminicola*." *Ann. Appl. Biol.* 104 (1984): 41.

32. Uppal, B. M., and Desai, M. K. "Physiologic Specialization in *Sclerospora graminicola* (Sacc.) Schroet." *Indian J. Agric. Sci.*, 2 (1932): 667.

33. Girard, J. C. "Three Years of Study on Millet and Sorghum Diseases at C.N.R.A. of Bombay, Senegal (1972–74)." 1–5, 1974.

34. Evans, M. H., and Harrar, J. G. "Germination of Oospores of *Sclerospora graminicola*." *Phytopathology* 20 (1930): 993.

35. Hiura, M. "A Simple Method of Germination of Oospores of *Sclerospora graminicola*." *Science* N. S. 72, 95, 1930.

36. Suryanarayana, D. "Oospore Germination of *Sclerospora graminicola* (Sacc.) Schroet. on Bajra (*Pennisetum typhoides*)." *Indian Phytopathol.* 9 (1956): 182.

37. Pande, A. "Germination of Oospores in *Sclerospora graminicola*." *Mycologia* 64 (1972): 426.

38. Singh, S. D., and Williams, R. J. "A Study on the Host Range of *Sclerospora graminicola*." *International Working Group on Graminaceous Downy Mildew. Newsletter* 1 (1979): 3.

39. Singh, S. D., and Luther, K. D. M. "Further Studies on the Host Range of *Sclerospora graminicola*." *International Working Group on Graminaceous Downy Mildew. Newsletter* 3 (1981): 3.

40. Safeeulla, K. M., and Thirumalachar, M. J. "Periodicity Factor in the Production of Asexual Phase in *Sclerospora graminicola* and *Sclerospora sorghii* and the Effect of Sporangiophores." *Phytopathol. Z.* 26 (1956): 41.

41. Safeeulla, K. M., Shaw, C. G., and Thirumalachar, M. J. "Sporangial Germination of *Sclerospora graminicola* and Artificial Inoculation of *Pennisetum glaucum.*" *Plant Dis. Reptr.* 47 (1963): 679.

42. Melhus, I. E., Van Haffern, F., and Bliss, D. E. "A Study of the Downy Mildew, *Sclerospora graminicola* (Sacc.) Schroet." *Phytopathology* 17 (1927): 57 (Abstract).

43. Suryanarayana, D. "Studies on the Downy Mildew Disease of Millets in India." *Indian Phytopath. Soc. Bull.* 3 (1965): 72.

44. Singh, S. D., and Williams, R. J. "The Role of Sporangia in the Epidemiology of Pearl Millet Downy Mildew." *Phytopathology* 70 (1980): 1187.

45. Thakur, D. P. "Management of Downy Mildew Disease of Pearl Millet in India." *Adv. Biol. Res.* 4 (1986): 17.

46. Williams, R. J. "Downy Mildew of Tropical Cereals," in *Advances in Plant Pathology*, Vol. 3, eds. D. S. Ingran, and P. H. Williams. (London: Academic Press, 1984), 1.

47. Ramakrishnan, T. S. *Diseases of Millets,* Publ. Indian Council of Agricultural Research, New Delhi, 152, 1963.

48. Shetty, H. S., Neergard, P., and Mathur, S. B. "Demonstration of Seed Transmission of Downy Mildew or Green Ear Disease. (*Sclerospora graminicola*) in Pearl Millet (*Pennisetum typhoides*)." *Proc. Indian Nat. Sci. Academy* Series B 43 (1977): 201.

49. Thakur, D. P. "Present Status of External Seed Borne Nature of *Sclerospora graminicola* in Downy Mildew Incidence in Pearl Millet and its Management." Paper presented in the International Seed Pathology Symposium, Mysore, Dec., 1984.

50. Thakur, D. P. "Quantum of Oosporic Seed Load of *Sclerospora graminicola* on Pearl Millet Downy Mildew." *Proc. Natl. Symp. Adv. Front. Pl. Sci.* Jodhpur 43 (1983).

51. Suryanarayana, D. "Occurrence of an Unknown Fungal Mycelium Inside the Sound Grains Produced on Partly Formed Green Ears of Bajra Plants." *Sci. Cult.* 28 (1962): 536.

52. Sivaprakasam, K., Pillayrsamy, K., and Rajagopalan, C. K. S. "Influence of Nitrogen on the Incidence of Downy Mildew Disease of Pearl Millet." *Plant Soil* 41 (1975): 677.

53. Tiwari, M. M., and Arya, H. C. "Studies on Green-Ear Disease of Bajra (*Pennisetum typhoides* Stapf and Hubb.) Caused by *Sclerospora graminicola* (Sacc.) Schroet." *Indian Phytopathol.* 19 (1966): 125.

54. Sundaram, N. V., Ramasastry, D. V., and Nayer, S. K. "Note on the Seed-Borne Infection of Downy Mildew (*Sclerospora graminicola* (Sacc.) Schroet.) of Pearl Millet." *Indian J. Agric. Sci.* 43 (1973): 215.

55. Thakur, D. P., and Kanwar, Z. S. "Internal Seed-Borne Infection and Heat Therapy in Relation to Downy Mildew of *Pennisetum typhoides* Stapf and Hubb." *Sci. Cult.* 43 (1977): 432.

56. Thakur, D. P., and Kanwar, Z. S. "Infectivity of Sporangia of *Sclerospora graminicola* Causing Downy Mildew of Pearl Millet." *Indian J. Mycol. Plant Pathol.* 7 (1977): 104.

57. Subramanyam, S., and Safeeulla, K. M. "Aerosol Viability of Sporangia of *Sclerospora graminicola.*" *3rd Int. Symp. on Plant Pathology*, IARI, New Delhi, Dec. 1981, 12–13 (Abstract).

58. Subramanyam, S., Safeeulla, K. M., Shetty, H. S., and Kumar, R. U. "Importance of Sporangia in the Epidemiology of Downy Mildew of Peare Millet." *Indian Nat. Sci. Acad.* 48 (1982): 824.

59. Safeeulla, K. M. "Secondary Spread of Pearl Millet Downy Mildew." *International Working Group on Greminaceous Downy Mildews. Newsletter* 1 (1979): 6.

60. Dang, J. K., and Thakur, D. P. "Effect of Metalaxyl Treatment on Different Cultivars of Pearl Millet for Control of Downy Mildew Infection." *MILWAI* 4 (1985): 9.

61. Deshmukh, H. G., Mose, B. B., and Utikai, P. G. "Screening Germplasms of Pearl Millet Against Downy Mildew (*Sclerospora graminicola*)." *Indian J. Agric. Res.* 12 (1978): 85.

62. Singh, S. D. "Studies on the Downy Mildew Disease (*Sclerospora graminicola* (Sacc.) Schroet.) of Bajra (*Pennisetum typhoides* (Burm. f) Stapf and C.E. Hubb). (Ph.D. thesis, IARI, New Delhi, India, 1974), 126.

63. Tuleen, D. M., Frederiksen, R. A., and Vudhivanich, P. "Cultural Practices and the Incidence of Sorghum Downy Mildew in Grain Sorghum." *Phytopathology* 70 (1980): 905.

64. Exconde, O. R. "Chemical Control of Maize Downy Mildew." *Trop. Agric. Res. Ser.* 8 (1975): 157.

65. Frederiksen, R. A., and Renfro, B. L. "Global Status of Maize Downy Mildew." *Ann. Rev. Phytopathol.* 15 (1977): 249.

66. Williams, R. J. "Seed Transmission of Some Graminiceous Downy Mildews—Position." Paper presented in Conference on Strategies for Control of Graminceous Downy Mildew, Bellagio, Italy, 1979.

67. Thakur, D. P., and Kanwar, Z. S. "Reducing Downy Mildew (*Sclerospora graminicola*) of Pearl Millet (*Pennisetum typhoides*) by Fungicidal Seed, Soil Treatments." *Pesticides* 11 (1977): 53.

68. Venugopala, M. N., and Safeeulla, K. M. "Chemical Control of the Downy Mildew of Pearl Millet, Sorghum and Maize." *Indian J. Agric. Sci.* 48 (1978): 537.

69. Williams, R. J., and Singh, S. D. "Control of Pearl Millet Downy Mildew by Seed Treatment with Metalaxyl." *Ann. Appl. Biol.* 97 (1981): 263.

70. Dang, J. K., Thakur, D. P., and Grover, R. K. "Control of Pearl Millet Downy Mildew Caused by *Sclerospora graminicola* with Systemic Fungicides in an Artificially Contaminated Plot." *Ann. Appl. Biol.* 102 (1983): 99.

71. Gill, K. S., Phul, P. S., Singh, H. N., and Chahal, S. S. "Inheritance of Resistance to Downy Mildew in Pearl Millet. A Preliminary Report." *Crop Improvement* 2 (1975): 128.

72. Basavaraju, R., Safeeulla, K. M., and Murty, M. R. "Centres of Resistance to Downy Mildew in Pearl Millet" (Abstract). *National Seminar on Genetics of Pennisetum* held at P.A.U., Ludhiana, March 27–29, 1978.

73. Dang, J. K. "Studies on Downy Mildew (*Sclerospora graminicola* (Sacc.) Schroet) of Pearl Millet (*Pennisetum typhoides* (Burm f.) Stapf and Hubb.)." (Ph.D. thesis, Haryana Agril. University, Hisar, India,) 1981, 163 pp. 1981.

74. Dang, J. K., and Thakur, D. P. "Genetic Analysis of Downy Mildew Resistance in Pearl Millet." *International Working Group on Graminaceous Downy Mildew. Newsletter* 3 (1981): 4.

75. Thakur, D. P., and Dang, J. K. "Breeding for Disease Resistance in Pearl Millet." Paper presented in the *AICMIP Annual Workshop held at M.P.A.U.*, Rahuri, April 26–29, 1985.

13

ERGOT
OF PEARL MILLET

R. P. THAKUR AND S. B. KING

Senior Plant Pathologist and Principal Plant Pathologist, Cereals Program, International Crops Research Institute for the Semi-Arid Tropics (ICRISAT), Patancheru, A. P. 502 324, India

13-1 INTRODUCTION

Ergot (*Claviceps fusiformis* Lov.) is an important disease of pearl millet [*Pennisetum glaucum*(L.) R. Br.]. The disease is known to occur in most parts of the pearl millet growing areas of the world (Figure 13-1). Occasionally ergot in severe form is reported in countries of Africa, including Burkina Faso, Ghana, Nigeria, Senegal, Tanzania, Zambia, and Zimbabwe.[1-3] In India the disease occurs almost every year to at least some extent in states of Gujarat, Haryana, Karnataka, Maharashtra, Punjab, Rajasthan, and Tamil Nadu, and the union territory of Delhi; in some areas in some years the disease is severe.

Ergot of pearl millet was first observed in the early 1940s in India,[4] but it was considered to be of minor importance. The first reported ergot outbreaks of significance were reported for Satara district of Maharashtra.[5-6] After this the disease continued to be sporadic until 1967 and 1968 when it appeared in an epidemic form, especially in northern India, on HB 1, the first F1 hybrid released for commercial cultivation.[7] These epidemics drew considerable attention. With the continuing widespread cultivation of hybrids, ergot has become endemic in most pearl millet growing states in India.[8] Local land races and open-pollinated varieties are generally less affected than hybrids.[9]

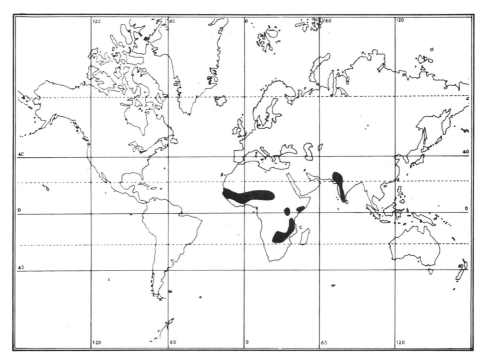

Figure 13-1   Geographical distribution of ergot disease (*Claviceps fusiformis*) of pearl millet

13-2 ECONOMIC IMPORTANCE

Ergot is a double-edged problem. The disease directly reduces crop yield by replacing grain with sclerotia of the causal fungus, and it lowers quality of produce by contaminating grain with sclerotia. Ergot sclerotia contain toxic alkaloids of the clavine group which pose a health hazard to humans and animals.[10]

In a test involving 18 pearl millet cultivars, Natarjan et al.[11] reported a mean grain yield loss of 58% due to ergot. At ICRISAT Center, grain yield losses of 65% in the hybrid, BJ 104, and 55% in the variety, WC-C75, were estimated.[12]

Ergot occurs in a number of cereal crops. Ergotism, a disease of humans and animals, is known to occur following ingestion of grain mixed with ergot sclerotia. Classical ergotism, which is of two characteristic forms, gangrenous and convulsive, is no longer a problem.[13] In India another type of ergotism, manifested mainly as gastrointestinal disturbances, has been reported in parts of Maharashtra, Gujarat, and Rajasthan, as a result of ingestion of ergoty pearl millet grain. The general symptoms of poisoning in man are nausea, repeated vomiting, and giddiness, followed by prolonged sleepiness sometimes extending to 48 hr.[14] Diarrhea and dehydration may also occur, and in extreme cases ingestion may be fatal,[15] although there seems to be some question about the latter.

Ergot of pearl millet contains agroclavine, elymoclavine, chanoclavine, penni-clavine, and setoclavine alkaloids, which are different from those found in rye and wheat ergot.[15] Sundaram et al.[16] isolated ergometrine, another water soluble alkaloid from ergot sclerotia of pearl millet. The total amounts of alkaloid content in honeydew and sclerotia were estimated to be 5 mg/100g and 56 mg/100g, respectively.[17]

In India, the Central Committee of Food Standards of the Directorate General of Health Services has fixed a safe limit of 0.05% of contaminated grain of pearl millet on the basis of the safe limit fixed by western countries for ergot of rye.[10] However, because studies indicate that symptoms of pearl millet ergot poisoning in man are different from those in experimental animals, it may be difficult to fix safe levels in man based on experimental animals.[18]

Loveless[19] reported inhibition of mammary gland development in sows that were fed pearl millet grain contaminated with ergot sclerotia during the last few weeks of pregnancy, which resulted in the death of newborn piglets by starvation. Mantle[29] reported similar observations with mice. Agroclavine was determined as a major alkaloid component suppressing lactation and inhibition of development of mammary glands.

Young monkeys, when administered intraperitoneally with varying doses of alkaloids extracted from sclerotia of pearl millet (5–10 ml/kg body weight), exhibited symptoms of toxicosis within 10 min of administration.[21]

It is evident from the above reports that the alkaloid content in ergot sclerotia in pearl millet varies both qualitatively and quantitatively. Further research is needed to clearly define the toxic contents of pearl millet sclerotia and their toxicity effects.

13-3 SYMPTOMS

Ergot first appears as a viscous, turbid fluid oozing out from the infected florets of pearl millet panicles (Figure 13-2). This carbohydrate-rich fluid contains numerous conidia of the causal fungus, and is called the "honeydew" phase of ergot symptoms. Depending on temperature and host genotype, the honeydew can be initially cream-colored, later becoming pink or brown. The honeydew phase lasts for about 4–6 days, when it may flow down the panicle onto the leaves or the ground. Within 10 days of honeydew initiation, sclerotia become visible in infected florets instead of seeds. Sclerotia, which are initially whitish, elongated, and larger than seed, become hard and brown to dark brown within 10–15 days (Figure 13-3). Sclerotial shape and size vary, depending on weather conditions and host genotype. In some cases sclerotia do not develop and florets have a blasted appearance, with tips of florets becoming black. These florets do not bear seeds.

13-4 CAUSAL ORGANISM

The causal fungus of ergot in pearl millet was known as *Claviceps microcephala* (Waller.) Tul. until 1968 when Loveless[19] proposed a new species name, *Claviceps fusiformis* Loveless. In India, Thomas et al.[4] first described the fungus as *C. microcephala* from *Pennisetum hohenackeri,* which was later confirmed by other workers.[22–23] Based on studies of a large number of ergot samples from pearl millet in Africa, Loveless[19] provided morphological data which differed from those of *C. microcephala.* Siddiqui and Khan[24] studied the ergot fungus on pearl millet in India and their observations were similar to those of Loveless.[19] Other studies by Thakur et al.[25] and Sharma and Chauhan[26] further confirmed

Figure 13-2 Honeydew symptoms of ergot of pearl millet showing a viscous, turbid fluid exuding from infected florets. The turbid fluid contains both micro- and macroconidia.

C. fusiformis Lov. as the causal fungus of ergot on pearl millet in India. Conidia of *C. fusiformis* are hyaline, unicellular, fusiform, broadly fulcate, and measure 13–18 × 3–4 μm.[19,24–25]

Reproduction in *C. fusiformis* is both asexual and sexual. Asexual reproduction is through conidia, which are primarily of two types, macro- and microconidia. These are also called primary, secondary, and tertiary conidia, the latter being produced by secondary conidia.[27] Macroconidia germinate by producing one to three germ tubes from their ends or sides; microconidia usually produce only one germ tube. Both macro- and microconidia are produced on the tips of the germ tubes, and microconidia are also produced by budding in chains.

The sexual phase initiates with sclerotia, which replace grains in the infected florets. Sclerotia germinate by producing one to several fleshy, purplish stipes, 6–26 mm long, that bear light to dark brown, globular capitula, which have numerous perithecial projections (Figure 13-4). Asci are interspersed with paraphyses in perithecia and emerge through ostioles. Asci are long and hyaline with apical pores and narrow ends. Ascospores released from asci are filiform, hyaline, nonseptate, and measure 103–176 × 0.5–0.7 μm.[25]

Chahal et al.[28] studied variation among eight isolates of *C. fusiformis* collected from

Figure 13-3 Ergot-infected panicle of pearl millet showing the dark, pointed sclerotia projecting from florets

different pearl millet growing regions of India. Although differences in virulence among isolates were not clear-cut, some appeared to be more virulent than others. There were, however, distinct morphological and cultural differences among the isolates. The International Pearl Millet Ergot Nursery (IPMEN), conducted every year by the International Crops Research Institute for the Semi-Arid Tropics (ICRISAT), has not clearly demonstrated the existence of different pathotypes in almost 10 years of multilocational testing in India and countries in West Africa. Nevertheless, a more precise study is needed to determine if pathogenic variation occurs in *C. fusiformis*.

13-5 DISEASE CYCLE

The primary source of inoculum is sclerotia left in the field from the previous crop or sown with seed; in some cases, ergot-infected, collateral hosts serve as a source of primary inoculum. Airborne ascospores from germinating sclerotia can infect pearl millet inflorescences at the protogyny stage under favorable weather conditions. Conidia from collateral hosts may also cause infection. Pearl millet flowers are susceptible only after stigma emergence and before pollination and fertilization.[9] Under favorable weather con-

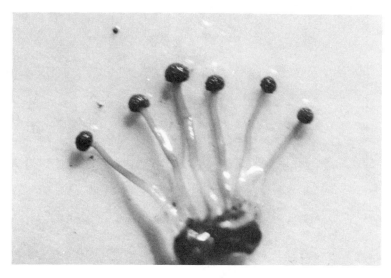

Figure 13-4 Germinated ergot sclerotium of pearl millet showing elongated stipes which bear globular capitula containing perithecia

ditions, particularly high relative humidity (more than 80%) and moderate temperature (20–30°C), honeydew symptoms appear within 4–6 days after infection. Secondary infection can be caused by conidia produced by honeydew. Conidia are disseminated to healthy inflorescences by physical contact with infected inflorescences, splashing rain, air currents, and insects. The role of insects in disease transmission has been demonstrated,[29–30] but their effectiveness in the presence of pollen needs further investigation. Sclerotia develop to a full size about 15–20 days after infection and are harvested with grain or drop to the ground in the field, and the cycle is repeated (Figure 13-5).

13-6 EPIDEMIOLOGY

The important parameters in ergot epidemiology are floral biology and weather conditions.

Flowering in pearl millet is protogynous. Ergot is strictly a floral disease and infection occurs through young, fresh stigmas. Once the stigma withers due to pollination or aging, the infection is prevented.[9] Following pollination, infection, and/or aging, a constriction develops in the stylar tissue enclosed within glumes, which leads to withering of the stigma.[31] The period of protogyny is variable among and within pearl millet genotypes. The susceptibility period in pearl millet to ergot is the period between stigma emergence and pollination, which under natural conditions could last a few hours to several days depending on the genotypes and availability of pollen in the field. Inoculation of panicles with a conidial suspension of *C. fusiformis* at the maximum fresh stigma stage produces maximum infection compared with inoculation before or after this stage.[9] Pollination of stigmas at the same time or as much as 16 hr after inoculation with *C. fusiformis* conidia significantly reduces ergot infection. Under natural conditions in pearl millet fields when pollen availability is abundant, ergot infection is very low, but when

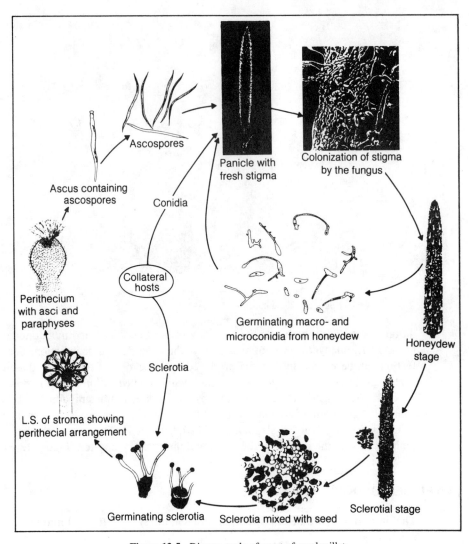

Figure 13-5 Disease cycle of ergot of pearl millet

flowering occurs during wet weather ergot infection increases. This may be due to pollen wash or poor anther dehiscence. Open-pollinated varieties of pearl millet have heterogenous plant populations in which abundant pollen is available for a longer time than in the near homogenous plant populations of F_1 hybrids, where flowering is more synchronous and lasts for a short period. This characteristic of hybrids makes them more vulnerable to ergot than open-pollinated varieties.[9] Ergot susceptibility of F_1 hybrids of pearl millet has been shown to be closely associated with the susceptibility of its female parent, the cytoplasmic male sterile lines.[32]

Among weather factors affecting ergot infection and development, temperature and relative humidity are more important than others. Remaswamy[7] reported the environmental conditions in Delhi for September 1–10, 1967, the period during which pearl millet flowered and high levels of ergot developed. His observations included: relative humidity of 85% to 95% in the morning for the 1st to 10th, and evening relative humidity, which was normally 45% to 50%, of 75% to 90% between the 1st and 5th and 60% to 70% between the 6th and 10th; total cloud amount of 6 to 8 octa (i.e., the sky was 75–100% covered) both morning and evening between the 1st and 8th, which was considerably higher than normal; total number of hours sunshine per day of only 1 to 5 hr, compared to the normal of 7 hr; and daily showers from the 1st to the 6th. Siddiqui and Khan[27] observed high infection of ergot when the mean temperature ranged from 18–20°C (minimum) and 28–30°C (maximum), relative humidity was 90%, and skies were overcast for 5–6 days during flowering. Gupta et al.[33] found a high positive correlation between relative humidity and temperature, and ergot infection. Based on multilocational ergot nursery, Huda and Thakur[34] identified daily rainfall (≥ 5 mm), relative humidity ($\geq 90\%$ in morning, $\geq 70\%$ in afternoon) and sunshine hours (≤ 1 hr) during the 4–day preanthesis stage as most conducive for ergot development in pearl millet.

The occurrence of two collateral hosts of *C. fusiformis*, *Panicum antidotale* in Haryana[35] and *Cenchrus ciliaris* in Rajasthan,[36] may be of important epidemiological significance. These perennial grasses harbor the pathogen and conidial inoculum is available from them when pearl millet flowers.

13-7 CONTROL MEASURES

13-7-1 Chemical Control

Since pearl millet is a food crop of resource-poor farmers of the arid and semi-arid tropics and ergot is strictly an airborne floral disease, the control of this disease by chemical means is neither economical nor feasible. However, there are reports where various degrees of success have been obtained in reducing ergot at experiment stations using chemical sprays. Several spray fungicides, including Aureofungin,[37] ziram, and copper oxichloride + zineb,[38] Cosan-80[39], Duter and Cuman-L,[40] Difolatan,[41] and Bavistin, Benlate, and Brestanol[42] have been reported to provide partial to complete control of ergot. However, none of these are recommended for use by farmers.

13-7-2 Cultural Control

Cultural practices generally aim at reducing the primary inoculum load in the soil and delay the development of disease. The various cultural practices reported to control ergot in pearl millet are deep ploughing, adjustment of sowing dates, balanced soil fertilization, intercropping, and planting sclerotia-free seed.

Deep ploughing helps to bury sclerotia deep in the soil and thus presumably prevent production and liberation of ascospores into the air.[43] This practice, to a certain extent, reduces the primary inoculum load of ergot.

Late-sown crops usually are more infected with ergot that early-sown crops. Crops sown during the second fortnight of July remained free from ergot, but those sown in August and September showed more ergot infection.[44] Pearl millet hybrids sown during June and July showed less ergot infection than when they were sown in later months.[8]

There are two factors that favor ergot development in the late-sown crop, wetter weather at flowering and lower temperature.

Sivaprakasham et al.[45] reported significantly higher ergot incidence with 160–200 kg N/ha. The balanced application of 80 kg N and 20 kg P or 40 kg K/ha was reported effective in maintaining low levels of ergot.[46] Sharma et al.[42] reported application of 100 kg N, 50 kg P, and 40 kg K/ha reduced ergot and increased grain yield. Farmers in the semi-arid tropics usually do not apply fertilizers to pearl millet; therefore, such recommendations are of limited value.

In the northwestern states of India, pearl millet is sometimes intercropped with mungbean, a practice that seems to reduce ergot incidence. A reduction of 18–25% in ergot incidence has been reported by intercropping with mungbean.[47] The thick crop canopy of mungbean apparently intercepts ascospores released from germinating sclerotia in the soil, thus reducing the level of infection.

The use of sclerotia-free seed for planting may be an important means of reducing primary inoculum load in the field. Sclerotia can be removed from seed lots by hand picking, separating sclerotia with the help of a Specific Gravity Table[48] (an instrument used to separate sclerotia from seed), gravity separator,[49] and by immersing contaminated seed in 10% common salt solution[50] to float sclerotia at the surface. However, there is no efficient method available to clean seed lots or grain at a larger scale.

13-7-3 Control by Pollen Management

Based on the finding that rapid pollination reduces ergot infection,[9] experiments were conducted at ICRISAT Center to control ergot in F_1 hybrids of pearl millet by pollen management. In several experiments at ICRISAT Center, when an ergot less-susceptible, early-flowering line was strategically planted as a pollen donor for susceptible hybrids, ergot severity was significantly reduced by 13% to 51% and grain yields were increased in different hybrids.[51] Similar results were obtained in an experiment at Hisar in North India.[52] This approach to the control of ergot in pearl millet hybrids, which is based on pollination-induced resistance, is quite promising but needs wider testing before it can be recommended for use by farmers.

13-7-4 Biological Control

Fusarium sambucinum,[53] *F. semitectum* var *majus,*[54] and *Cerebella andropogonis*[55] have been reported to parasitize both honeydew and sclerotia of pearl millet ergot, and to interfere with sclerotial formation and development under field conditions. However, the potential of these fungi as effective agents in the biocontrol of ergot of pearl millet remains to be demonstrated.

13-7-5 Control by Host Resistance

Use of resistant varieties is often the most economical and effective means of disease control. Ergot resistance research at ICRISAT Center has involved the following activities:

   1. Development of an effective field screening technique;

2. Identification of sources of resistance by screening large and varied germplasm accessions;

3. Determination of stability of resistance through multilocational testing;

4. Understanding the genetics and mechanisms of resistance; and

5. Utilization of resistance to breed disease-resistant cultivars.

An effective field-based screening technique for ergot resistance has been developed.[56] The technique involves bagging pearl millet panicles at the boot stage, inoculating them at the full protogyny stage with an aqueous suspension of conidia obtained from honeydew and rebagging immediately after inoculation, providing overhead sprinkler irrigation 1–2 times per day to maintain the high relative humidity necessary for ergot infection and development, and scoring panicles for ergot severity at about 20 days after inoculation using a standard ergot severity assessment key.[9] This technique is precise, effective, and easily adaptable, and it can be used in one or two crops a year depending on weather conditions and available resources.

Resistance to ergot in pearl millet is difficult to find. In nature, selection for ergot resistance has not been effective, likely because of pollen interference. Indication of this comes from the absence of adequate levels of resistance in the several thousand germplasm accessions screened at ICRISAT Center. Resistance, however, was developed by intercrossing less-susceptible plants and pedigree-selecting plants with higher levels of resistance for several generations.[56] Several inbred lines with high levels of resistance are available at ICRISAT.[57–58] Recently Thakur et al.[59] showed that an ergot severity threshold level of 20–30% in field screening should provide adequate levels of functional field resistance under natural ergot epidemic conditions.

Resistance that is stable, determined by multilocational testing, is desirable for a resistance breeding program. There are several inbred lines that have shown stable resistance across locations over several years of testing[60–61] and probably these are the best sources currently available for use in breeding programs.

Resistance breeding programs to produce ergot-resistant hybrids and varieties are in progress at ICRISAT Center[62] and at the Punjab Agricultural University, Ludhiana.[63–64]

Genetics of ergot resistance is relatively complex. Resistance has been found to be recessive and multigenic.[65] Resistance may be dependent on a pollination-induced, stigmatic constriction that prevents the infection hyphae from reaching the ovary.[31,66]

13-8 CONCLUSIONS

In the absence of ergot-resistant cultivars, economical control of ergot is not possible. Various other control measures described have limited application in experimental plots at research stations. At present, ergot-resistant cultivars are not available and it may take some time before such cultivars become available to farmers. Pathologists and breeders will have to work together to achieve this goal. Breeding ergot-resistant hybrids is no doubt a difficult task because of the complexity of inheritance of resistance. Our current knowledge suggests that both the female and male parents of hybrids must be resistant for the hybrid to be resistant. Successful breeding will require patience and a better understanding of the genetics and mechanisms of resistance.

Although there are no data available to support the existence of pathogenic variation in *C. fusiformis,* more detailed studies on this subject are needed. Mechanisms of resistance, other than pollination-induced resistance, need to be investigated.

13-9 REFERENCES

1. Thakur, R. P., and King, S. B. "Ergot Disease of Pearl Millet." Information Bull No. 24. ICRISAT, Patancheru, 1988, 24.

2. Ramakrishnan, T. S. *Disease of Millets.* New Delhi: ICAR, 1971, p. 61.

3. Rachie, K. O., and Majmudar, J. V, *Pearl Millet.* University Park and London: Pennsylvania State University, Press, 1986, p. 307.

4. Thomas, K. M., Ramakrishnan, T. S., and Srinivasan, K. V. "The Occurrence of Ergot in South India." *Proc. Indian Acad. Sci.* B 21 1945: 93.

5. Bhide, V. P., and Hedge, R. K. "Ergot of Bajra (*Pennisetum typhoides* (Burm.) Stapf. & Hubbard) in Bombay State." *Curr. Sci.* 26 (1975): 116.

6. Shinde, P. A., and Bhide, V. P. "Ergot of Bajri (*Pennisetum typhoides*) in Bombay State." *Curr. Sci.* 27 (1958): 499.

7. Ramaswamy, C. "Meteorological Factors Associated with the Ergot Epidemic of Bajra (*Pennisetum*) in India During the *Kharif* Season–1967—a Preliminary Study." *Curr. Sci.* 37 (1968): 331.

8. Thakur, D. P. "Ergot Disease of Pearl Millet." *Rev. Trop. Plant Pathol.* 1 (1984): 297.

9. Thakur, R. P., and Williams, R. J. "Pollination Effects on Pearl Millet Ergot." *Phytopathology* 70 (1980): 80.

10. Bhat, R. V., Roy, D. N., and Tulpule, P. G. "Ergot Contamination of Bajra." *Proc. Nut. Soc. India* 10 (1975): 7.

11. Natarajan, U. S., Guruswamy Raja, V. B., Selvaraj, S., and Parambaramani, C. "Grain Loss Due to Ergot Disease in Bajra Hybrids." *Indian Phytopathol.* 27 (1974): 254.

12. Thakur, R. P. "Disease of Pearl Millet and Their Management." *Plant protection in field crops:* lead papers of the National Seminar on Plant Protection in Field Crops, 29-31 Jan 1986, Hyderabad, India, eds. M. V. Rao and S. Sithanantham. (Plant Production Association of India, 1987), 147.

13. Bove, F. J. *The Story of Ergot.* Basel and New York: S. Karger, 1970, p. 297.

14. Krishnamachari, K. A. V. R., and Bhat, R. V. "Poisoning by Ergoty Bajra (Pearl Millet) in Man." *Indian J. Med. Res.,* 64 (1976): 1624.

15. Bhat, R. V., Roy, D. N., and Tulpule, P. G. "The Nature of Alkaloids of Ergoty Pearl Millet of Bajra and Its Comparison with Alkaloids of Ergoty Rye and Ergoty Wheat." *Toxicol. Appl. Pharmacol.* 36 (1976): 11.

16. Sundaram, N. V., Bhowmik, T. P., and Khan, I. D. "Water Soluble Alkaloid Content of Ergot [*Claviceps microcephala* (Wallr.) Tul.] Sclerotia of Pearl Millet [*Pennisetum typhoides* (Burm.) Stapf & Hubb]." *Indian J. Agric. Sci.* 40 (1970): 569.

17. Anonymous "Studies on the Outbreak of Ergotism in Humans in Rajasthan." *Ann. Rep. National Institute of Nutrition, Hyderabad, India,* 1976, pp. 25–27.

18. Tulpule, P. G., and Bhat, R. V. "Food Toxins and Their Implications in Human Health." *Indian J. Med. Res.* 68 (1978): 99.

19. Loveless, A. R. "*Claviceps fusiformis* sp. nov. "The Causal Agent of an Agalactia of Sows." *Trans. Br. Mycol. Soc.* 50 (1976): 15.

20. Mantle, P. G. "Inhibition of Lactation of Mice Following Feeding with Ergot Sclerotia [*Claviceps fusiformis* (Loveless)] from Bulrush Millet [*(Pennisetum typhoides)* Staph & Hubbard)] and an Alkaloid Component." *Proc. Roy. Soc. B.* 170 (1968): 423.

21. Anonymous. "Ergot Contamination of Bajra," *in Ann. Rep. National Institute of Nutrition, Hyderabad, India,* 1977, pp. 25–27.

22. Thirumalachar, M. J. "Ergot of *Pennisetum hohenackeri* Hochst." *Nature, London* 156 (1945): 754.

23. Ramakrishnan, T. S. "Observation on Ergot of *Pennisetum* and other Grasses." *Proc. Indian Acad. Sci. B.* 36 (1952): 97.

24. Siddiqui, M. R., and Khan, I. D. "Renaming *Claviceps microcephala*. Ergot Fungus on *Pennisetum typhoides* in India as *Claviceps fusiformis.*" *Trans. Mycol. Soc. Japan.* 14 (1973): 195.

25. Thakur, R. P., Rao, V. P., and Williams R. J. "The Morphology and Disease Cycle of Ergot, Caused by *Claviceps fusiformis,* in Pearl Millet." *Phytopathology* 74 (1984): 201.

26. Sharma, O. P., and Chauhan, R. K. S. "Identity of Pathogen Causing Ergot of Pearl Millet." *Indian Phytopath.* 37 (1984): 539.

27. Siddiqui, M. R., and Khan, I. D. "Dynamics of Inoculum and Environment in Relation to Ergot Incidence on *Pennisetum typhoides* (Burm.) Stapf. and Hubbard." *Trans, Mycol. Soc. Japan* 14 (1973): 280.

28. Chahal, S. S., Rao, V. P., and Thakur, R. P. "Variation in Morphology and Pathogenicity in *Claviceps fusiformis,* the Causal Agent of Pearl Millet Ergot." *Trans. Br. Mycol. Soc.* 84 (1985): 325.

29. Sharma, Y. P., Singh, R. S., and Tripathi, R. K. "Role of Insects in Secondary Spread of the Ergot Disease of Pearl Millet (*Pennisetum americanum*)." *Indian Phytopathol.* 36 (1983): 131.

30. Verma, O. P., and Pathak, V. N. "Role of Insects in Secondary Spread of Pearl Millet Ergot." *Phytophylactica* 16 (1984): 257.

31. Willingale, J., and Mantle, P. G. "Stigma Constriction in Pearl Millet, a Factor Influencing Reproduction and Disease." *Ann. Bot.* 56 (1985): 109.

32. Thakur, R. P., Rao, V. P., and King, S. B. "Ergot Susceptibility in Relation to Cytoplasmic Male Sterility in Pearl Millet." *Plant Dis.* 73 (1989): 676.

33. Gupta, G. K., Subba Rao, G. V., and Saxena, M. B. L. "Relationship Between Meteorological Factors and the Occurrence of Ergot Disease (*Claviceps microcephala)* of Pearl Millet." *Trop. Pest Manag.* 29 (1983): 321.

34. Huda, A. K. S. H., and Thakur, R. P. "Characterization of Environments for Ergot Disease Development in Pearl Millet. Agrometeorological Information for Planning and Operation in Agriculture with Special Reference to Plant Protection." Workshop 22–26 Aug 1988, Calcutta, India. eds. V. Krishnamurthy and G. Mythys. Geneva, Switzerland, WMO, 197, 1989.

35. Thakur, D. P., and Kanwar, Z. S. "Ability of Naturally Incident *Claviceps microcephala* from *Panicum antidotale* to Produce Ergot Symptoms in *Pennisetum typhoides.*" *Indian J. Agric. Sci.* 48 (1978): 540.

36. Singh, G., Vyas, K. L., and Bhatt, B. N. "Occurrence of Pearl Millet Ergot on *Cenchrus ciliaris* Pers. in Rajasthan." *Indian J. Agric. Sci.* 53 (1983): 481.

37. Sulaiman, M. G., Lukade, G. M., and Dawkhar, G. S. "Effect of Some Fungicides and Antibiotics on Sclerotial Development and Germination of Ergot on *Pennisetum typhoides.*" *Hindustan Antibiotic Bull.* 9 (1966): 94.

38. Sundaram, N. V. "Ergot of Bajra," *in Advances in Mycology and Plant Pathology.* eds. S. P. Raychaudhuri, A. Varma, K. S. Bhargava, and B. S. Mehrotra. New Delhi, IARI, 1975, 343.

39. Reddy, K. D., Govindaswamy, C. V., and Vidhyasekaran, P. "Studies on Ergot Disease of Cumbu *(Pennisetum typhoides).*" *Madras Agric. J.* 56 (1969): 367.

40. Sivaprakasham, K. "A Note on the Control of Ergot Disease of Pearl Millet." *Andhra Agric. J.* 18 (1973): 213.

41. Brar, G. S., Chand, J. N., and Thakur, D. P. "Fungicidal Control of Ergot of Bajra." *Haryana Agric. Univ. J. Res.* 6 (1976): 1.

42. Sharma, Y. P., Singh, R. S. and Tripathi, R. K. "Management of Pearl Millet Ergot by Integrating Cultural Practices and Chemical Control." *Indian J. Mycol. Pl. Pathol.* 14 (1984): 69.

43. Sundaram, N. V. "Ergot Disease of Bajra, Its Symptoms and Control." *Indian Farming* 17 (1967): 56.

44. Singh, R., and Singh, S. N. "A Note on Effects of Different Dates of Sowing Hybrid Bajra - 1 on Grain Yield and Incidence of Ergot [*C. microcephala* (Wallr.) Tul.]. *Madras Agric. J.* 56 (1969): 140.

45. Sivaprakasham, K., Pillayarswamy, K. and Ramu, S. "Role of Nitrogen on the Incidence of Ergot Disease of Pearl Millet [*Pennisetum typhoides* (Burm. f.) Stapf and Hubb]." *Madras Agric. J.* 62 (1975): 574.

46. Brar, G. S. "Studies on Ergot Disease of Bajra." M. S. thesis submitted to Haryana Agril. Univ., Hisar, 1975.

47. ICAR (Indian Council of Agricultural Research). *Progress report of the All India Coordinated Millets Improvement Project, 1980–81.* Indian Counc. Agric. Res., New Delhi, 1981.

48. Pathak, V. N., Yadav, K. R., Langkilde, N. E., and Mathur, S. B. "Removal of Ergot Sclerotia from Seeds of Pearl Millet." *Seeds Res.* 12 (1984): 70.

49. Nicholas, I. "Removal of Ergot from Grain or Seed Lots of Bajra by Gravity Separators." *Seeds and Farms.* 1 (1975): 4.

50. Nene, Y. L., and Singh, S. D. "Downy Mildew and Ergot of Pearl Millet." *Pest Artic. News Summ. (PANS)* 22 (1976): 366.

51. Thakur, R. P., Williams, R. J., and Rao V. P. "Control of Ergot in Pearl Millet through Pollen Management." *Ann. Appl. Biol.* 103 (1983): 31.

52. Thakur, D. P. "Epidemiology and Control of Ergot Disease of Pearl Millet." *Seed Sci. Technol.* 11 (1983): 797.

53. Tripathi, R. K., Kolte, S. J., and Nene, Y. L. "Mycoparasite of *Claviceps fusiformis,* the Causal Fungus of Ergot of Pearl Millet." *Indian J. Mycol, Pl. Pathol.* 11 (1981): 114.

54. Rao, V. P. and Thakur, R. P. "Fusarium semitectum var. *majus*—a Potential Biocontrol Agent of Ergot (*Claviceps fusiformis*) of Pearl Millet." *Indian Phytopathol.* 41 (1988): 567.

55. Kulkarni, C. S., and Moniz, L. "Occurrence of Hyperparasite *Cerebella* on Ergot of Bajra." *Curr. Sci.* 43 (1974): 803.

56. Thakur, R. P., Williams, R. J., and Rao, V. P. "Development of Ergot Resistance in Pearl Millet." *Phytopathology* 72 (1982): 406.

57. ICRISAT (International Crops Research Institute for the Semi-Arid Tropics). "Pearl Millet," in *Annual Report 1983*, ed. A. P. Patancheru, 502 324, India: ICRISAT, 1984, pp. 65–117.

58. ICRISAT(International Crops Research Institute for the Semi-Arid Tropics). "Pearl Millet." In *Annual Report 1984,* ed. A. P. Patancheru, 502 324, India: ICRISAT, 1985, pp. 81–130.

59. Thakur, R. P., King, S. B., and Rao, V. P. "Expression of Ergot Resistance in Pearl Millet under Artificially Induced Epidemic Conditions." *Phytopathology* 79 (1989): 1323.

60. Thakur, R. P., Rao, V. P., Williams, R. J., Chahal, S. S., Mathur, S. B., Pawar, N. B., Nafade, S. D., Shetty, H. S., Singh, G., and Bangar, S. G. "Identification of Stable Resistance to Ergot in Pearl Millet." *Plant Dis.* 69 (1985): 982.

61. Thakur, R. P., and King, S. B. "Registration of Four Ergot Resistant Germplasms of Pearl Millet." *Crop Sci.* 28 (1988): 382.

62. Andrews, D. J., King, S.B., Witcombe, J. R., Singh, S. D., Rai, K. N., Thakur, R. P., Talukdar, B. S., Chavan, S. B., and Singh, P. "Breeding for Disease Resistance and Yield in Pearl Millet." *Field Crops Res.* 11 (1985): 11.

63. Chahal, S. S., Gill, K. S., Phul, P. S., and Singh, N. B., "Effectiveness of Recurrent Selection for Generating Ergot Resistance in Pearl Millet." *SABRO Journal.* 13 (1981): 184.

64. Gill, K. S., Chahal, S. S. and Phul, R.S. "Strategy to Develop Ergot Resistance in Pearl Millet," in *Trends in Genet. Res. Pennisetums,* eds. V. P. Gupta and J. L. Minocha. Punjab India: Punjab Agric. Univ. Ludhiana, 1980, 159.

65. Thakur, R. P. , Talukdar, B. S., and Rao, V. P. "Genetics of Ergot Resistance in Pearl Millet," in Abstracts of contributed papers of the XV International Congress of Genetics (Part 2), 12–21 Dec. 1983, New Delhi, India. Oxford and IBH Publishing Co., 1983.

66. Willingale, J., Mantle, P. G., and Thakur, R. P. "Post Pollination Stigmatic Constriction, the Basis of Ergot Resistance in Selected Lines of Pearl Millet. *Phytopathology* 76 (1986): 536.

14

ANTHRACNOSE OF SORGHUM

A. P. SINHA

*Senior Research Officer, Research Station, G.B. Pant University of Agri. & Tech.,
Nagina-246762, Bijnor, U.P., India*

14-1 INTRODUCTION

Sorghum anthracnose, also known as red stalk rot (*Colletotrichum graminicola* (Cesati) Wilson, was first reported separately in North Carolina and Texas in 1911 and 1912, though this disease had been known for sometime in India and elsewhere.[1-3] The disease has since been recorded from most areas where sorghum is cultivated. This disease is considered as one of the primary factors limiting the cultivation of sorghum.

14-2 DISTRIBUTION AND ECONOMIC IMPORTANCE

Anthracnose has been reported from all the sorghum-growing regions of the world including many countries in North, Central, and South America, West Indies, Asia, and Africa.[1] It is far more important in humid regions or during the rainy seasons. It is the most important sorghum disease in Brazil and is a major threat in most of the Latin American countries,[4,5] and in some areas it is one of the main yield-limiting factors, especially in Brazil, Venezuela, and Guatemala.[5-7] The damaging effect of this disease on sorghum in Georgia and other humid areas are well discussed.[8-10] It has also been a destructive disease on grain sorghum and broom-corn in Texas and has caused heavy loss in broom-corn in Illinois.[11-13] The stalk rot aspect of the pathogen on sorghum species has been reported from the United States, Mexico (caused mostly by *Colletotrichum graminicola* and partly by *Gleosporium* species),[14] Australia,[15] Senegal,[16] French Equatorial Africa,[17] and possibly Java (attributed to *C.* falcatum)[18] and Bihar (India).[3] Anthracnose and red rot of sorghum in Netherlands, New Guinea were attributed to *Colletotrichum* sp.[19]

Several workers have done extensive studies on grain loss, disease incidence and grain loss relationship, deterioration in juice/sugar quality, and the like. It has been observed that diseased leaves lost 30% to 40% in weight as compared with the healthy leaves of the same age and position on the stem.[10,20.] Lebeau et al.[21] reported that leaf destruction by the fungus markedly reduced the yield of syrup or sucrose per ton of stalks by reducing total sugar content and at the same time bringing about partial inversion of the sucrose present in the sap. A substantial amount of stalk rot can be present without greatly affecting the total solids and purity of the stalk juice up to the ripe stage, but beyond this stage the normal loss in total solids and purity may be aggravated by severe stalk rot. The crop grown for syrup production plants that are badly damaged by rot yield a syrup of inferior quality. Stokes and co-workers[22] pointed out that losses in syrup yield due to stalk rot can usually be avoided by harvesting in the early dough stage. Such early harvesting involves no remarkable loss in yield or quality. Massive yield loss and lodging due to anthracnose in sorghum at 1 and 2 weeks after the normal harvest date has been observed.[23] Mishra and Siradhana[24] reported that the maximum loss in yield was 16.4%, which was due to anthracnose phase only. They have also noticed that percentage loss in grain yield varied from 1.2% to 16.4% depending on the visible disease intensities. There was a decreasing trend in the grain yield with the increase in disease intensities.

Powell et al.[25] observed that the disease reduced grain yield per head by 70.0%. More than half the yield loss resulted from the incomplete grain fill as shown by 42% decrease in 1,000 seed weight and 17.2% decrease in seed density. Harris and Cunfer[26] observed more than 50% reduction in grain yield on susceptible hybrids. By the use of correlation coefficients and linear regression formulae, Harris et al.[27] made an attempt to evaluate the importance of the disease and importance of the three phases of the diseases Sharma[28] reported that losses due to anthracnose ranged between 41% and 60%. However, Chohan[29] reported that infected grains lose 51% in weight as compared with healthy grains from plants of the same age. Recently, Ferreira and Warren[30] estimated grain losses caused by anthracnose to reach as high as 88.7% in the highly susceptible cultivar. IS-4255, RS-671 had yield losses of 42% where resistant material such as IS-9789 and IS-9569 were essentially free from losses.

14-3 SYMPTOMS

The characteristic symptoms caused by *C. graminicola* in sorghum have been adequately described.[3–4,31–32] Several workers have recognized three phases of anthracnose, which include a foliar phase, stalk rot, and colonization of the panicle including the grain. Many times, anthracnose symptoms are evident in all three or only two plant parts. If symptoms develop only in two plant parts, it is more common on the head and peduncle as compared to leaf.[4]

The characteristic symptoms of anthracnose appear as small, circular-elliptical spots up to about 1/4 in. in diameter that develop on the leaves and leaf sheaths [Figure 14-1(a–b)]. They are well defined, and color ranges from tan orange-red to blackish purple, depending on the variety of the sorghum attacked. The center of older spots

Figure 14-1 Symptoms of anthracnose of sorghum: (a) and (b) Many lesions may develop close together and coalesce. Midrib infection as elongate-elliptical red or purple
lesion. (c) red rot phase (Courtesy of ICRISAT)

become greyish or straw-colored with reddish borders and bear few or several minute
black specks, the fruiting body (acervuli). Mid-rib infection generally occurs as elliptical
discolored areas which may coalesce to affect most of the mid rib [Figure 14-1(a–b)]. The
color of the lesions depends mostly on variety of the host plant. Weakened, chlorotic,
stressed, or senescent leaves of susceptible varieties are quickly colonized by the pathogen. Sometimes leaf spot may develop together, causing extensive necrosis. Infection on
the leaves can occur at all stages of plant growth. Under favorable climatic conditions its
rapid development may even cause death of young plants.

Symptoms of stalk rot can be recognized by their irregularly mottled or marbled
pattern of colonization. The marbling may result from either multiple infection or colonization of stalk tissue from a single inoculation site.[13] In the beginning, the interior of the
stalk is water soaked and discolored but no external symptoms are visible. The discolor-

ation depends on variety of sorghum, ranging from yellowish to blackish purple and appearing more or less uniform or mottled with whitish patches [Figure 14-1(c)]. Discoloration may be continuous through the infected internode or in the form of apparently isolated spots. Nodal tissue is rarely discolored despite its invasion by the pathogen. On the highly susceptible cultivar, young lesions develop as elliptical pockets or bars immediately beneath the epidermal tissue, particularly in the upper internode of the plant. Stalk rot is generally preceded by leaf anthracnose. It is understood that the pathogen can enter the stalk by direct penetration of the rind, and afterwards spreads into the interior, and by growing into the vascular tissues interferes with movement of water and food materials, which result in poor development of heads and grains. The upper internodes and peduncles seem to be especially liable to invasion; the dry rotted internode tissue shrinks and may lead to breaking of the stalks, often near ground level. Surface lesions commonly develop on infected stalks of highly susceptible cultivars as elleptic or oblong spots up to about 1/4 in. in length, with pale centers and reddish purple borders. These spots coalesce and form irregular discolored areas on which several aceruvli develop. In several sorghums the pathogen apparently penetrates the cortex of the stalk near the internode base without producing visible surface lesions. In these cases the rot is usually confined to the more pithy positions of the stalk and rarely involves the more juicy cortical and subcortical layers, thereby causing less damage. Le Beau et al.[21] observed that as *C. graminicola* invades the plant, the latter is stimulated to produce pigments that limit mycelium development of the pathogen. In highly susceptible cultivars, areas of tissue of 1 cm or more in diameter may be invaded by the mycelium before the pigment is produced, thus resulting in lightly colored pockets of invaded tissue filled with mycelium and surrounded by highly colored areas practically devoid of mycelium.[3] On grains, symptoms can be seen as small black dots on its surface. When infected grains are seen under binocular microscope, fruiting structures are visible.

14-4 CAUSAL ORGANISM

Colletotrichum graminicola (Ces.) Wilson has been recognized as the causal agent of anthracnose, which is characterized by dark acervuli [Figure 14-2(a)]. The disc-shaped fruiting structures are present in the necrotic part of the lesions. Acervuli are erumpent stromatic bodies formed on both surfaces of the leaf, on the leaf mid rib, peduncle, rachis branches, and grains, and has dark spines or setae at the edge or among the conidiophores. Conidiophores are numerous, short, crowded, hyaline measuring $1-2 \times 6-12$ μm in size, each of which bears a single terminal conidium.[3] In Georgia (United States), conidiophore dimensions have been given as $1-2 \times 6-12$ μm[20] and as $3.5-4 \times 20-30$ μm in French Equatorial Africa.[17] Mature aervuli are dark brown or black, of different size, and mostly somewhat elongate. The conidia [Figure 14-2(b)] are hyaline, unicellular, terminal, sickle shaped and borne singly. The conidia, which generally have one or two oil drops present, are held together in an orange to salmon to pinkish gelatinous matrix. The conidia have dimensions usually falling within the range $3-7 \times 16-34$ μm^3; e.g., $3.5-5 \times 21-33.5$ $(4.2 \times 25.3)\mu$m in French Equatorial Africa,[17] $4.2-6.4 \times 16.8-28$ $(5.2-24)\mu$m in the United States.[20] Some cells of the acervuli develop in long, rigid, unbranched, dark setae with many transverse septa of about 175 μm in length and 4–5 μm

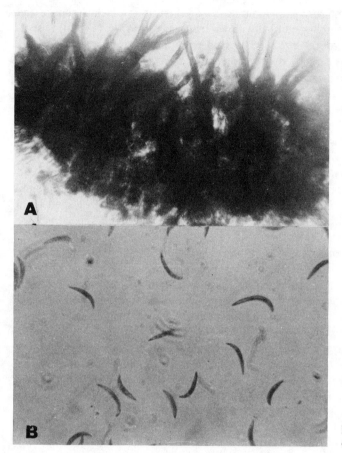

Figure 14-2   (a) Acervuli of pathogen causing sorghum anthracnose (b) Conidia

in width. They may be few or many in number, and taper toward their tips, which are lighter in color, and often rounded. Tarr[3] has pointed out wide variation in conidial shape and dimensions within *C. graminicola* which may possibly be associated with different strains of the pathogens. Mishra[33] observed thin, loose cottony growth of mycelium at initial stage and after 3–4 days the colonies became thick and greenish in color. Later, the color changed to greyish black to black. Mycelium is septate, branched, hyaline to greenish black and measure 3.5 to 4.0 μm in width.[34–35] Chohan[29] reported appressed and fluffy growth of aerial mycelium. The amount of mycelium ranges from fair to very good, to greyish white.

Mishra and Siradhana[36] conducted a detailed study on the effect of light exposures on the growth and sporulation of *C. graminicola*. They observed that maximum colony diameters of isolates I_1, I_2, and I_3 were formed under the diurnal light and minimum in the 8-hr light and 16-hr dark treatments. Significant differences in colony diameter of I_1, I_2, and I_3 were recorded in all the light exposures. I_1 and I_2 produced maximum aerial growth of the mycelium, spores, acervuli, and setae in diurnal light. The same was observed in the case of I_3 with regard to the production of spores and acervuli in diurnal light; however, the aerial growth of mycelium and setae production was better in continuous light.

14-5 PATHOGEN VARIABILITY AND PHYSIOLOGIC SPECIALIZATION

Colletotrichum graminicola is a highly variable species. Reports on races are available from the United States and from other parts of the world.[37] Frederiksen and Rosenow[38] observed specific differences in reactions among selected sorghum lines in Texas, Mississippi, and Georgia. They demonstrated that the majority of sorghum genotypes tested as resistant or susceptible in one area of the United States tended to react similarly in another. An International Sorghum Virulence Nursery (ISAVN) was initiated by the Texas Agricultural Experiment Station in the year 1975 to study the virulence of *C. graminicola*, and to monitor the possible existence or appearance of different or new physiological races of the pathogen. The information gathered strongly suggested that physiologic races exist within the isolates of *C. graminicola* that infect sorghum.[4] The variety "widely" was infected by *C. graminicola* in Venezuela but it is highly resistant elsewhere. Other varieties like Mn-960 and TAM-428 were highly susceptible to the pathogen in Nigeria but these were resistant in the Americas. On the other hand, the variety TAM-428, resistant in the United States in the year 1976, was infected by *C. graminicola* in Puerto Rico.[4] Gradual erosion of resistance is noticed by the changes in reaction of resistance of T x 2536 and other sorghum lines in Georgia and Puerto Rico.[13] Nakamura[5] noticed five races of *C. graminicola* using five differential cultivars of sorghum. These races were classified from 1983 single spore isolates gathered from a number of diseased plants throughout Brazil. Thus, Nakamura's work confirms the hypothesis of many workers that several physiologic forms of the fungus are present not only within an area, but between locations as well.[13] Other races of *C. graminicola* probably exist in India because all the sorghum in the International Sorghum Anthracnose Virulence Nursery (ISAVN) showed susceptible reaction to anthracnose at Pantnagar in northern India.[13] However, other sorghum varieties screened at Pantnagar are resistant. Frederiksen[13] pointed out the significance of these observations; first the species is dynamic and affected by directional selection pressure by host resistance genes, and secondly profoundly various races exist in different regions of the world. He further concluded that these facts pose challenging problems if host resistance is used as the sole measure of controlling the disease in those areas where it develops in severe form. Ferreira and Casela[39] noted differential interactions between isolates and cultivars when 12 differential sorghum cultivars were inoculated with conidial suspensions of 7 monosporic isolates of the pathogen.

Ali and Warren[40] tested 9 sorghum isolates of *C. graminicola* from various geographical areas for pathogencity on 6 sorghum lines. These lines were evaluated for reaction type and disease severity. They observed that sorghum lines IS 4225 and IS 8361 were susceptible, whereas BR 64 and 954206 were resistant to all isolates. Based on the differential responses of sorghum lines 954130 and 954062, the isolates were grouped into 3 physiological races designated 1, 2, and 3. These results suggest that a resistant sorghum cultivar in one region may succumb to leaf anthracnose in another region because of the prevalence of a different virulent race.

14-6 EPIDEMIOLOGY

14-6-1 Host Range and Pathogenicity

When Wilson[41] studied the species of *Colletotrichum* causing anthracnose of grasses, he combined many of them into single species which he erected as *C. graminicolum* (Ces.) Wilson 1914. In this new broad species, he included *Colletotrichum lineola* Corda, *C. bromi* Jennings, *C. cereale* Manns, *Dicladium graminearum* Cesati, *Psilonia apalospora* Berk and Curt., *Vermicularis culmigera* Cooke, *V. holci* Sydon, *V. melicae* Fuckle, *V. sanguinea* Ellis and Halsted, *V. gaminicola* Westend, *V. affinis* Socc. and Briard.[41]

The pathogens have been recorded on several wild species of cereals and grasses including barley, corn, oats, rye, sorghum, Sudan grass, johnsongrass, wheat, and a large number of grasses.[42-46] However, isolates from one species may or may not infect other species. The capacity of an anthracnose isolate to infect several host species is not generally agreed by the workers.

Selby and Manns[47] noticed typical anthracnose symptoms on wheat and emmer in the field, with spore suspensions obtained from these two hosts. They also reported infection of emmer with isolate from wheat. In the United States, Edgerton[48] was unable to infect sugarcane stalks with either *C. cereale* obtained from johnsongrass and broom-corn or *C. lineola* obtained from wheat and other grasses. In India, *C. falcatum* (sugarcane) and *C. lineola* (sorghum) were morphologically closely allied, but the latter did not attack sugarcane and the former was apparently restricted to it.[49] An isolate from oats did not infect wheat, barley, flax.[44] An isolate of *C. graminicola* cultured from maize readily caused infections on sorghum and infected to a lesser extent other hosts including barley, wheat, oat, finger millet, Italian millet, feosinte, *Panicum frumentaceum*, *Pennisetum typhoideum*, and *Paspalum scrobiculatum*.[34] The infection of maize with an isolate from sorghum has also been recorded.

Abbott[50] found that the *C. falcatum* from sugarcane was morphologically similar to the *Colletotrichum* affecting other grass species in Louisiana. He also pointed out that *Colletotrichum* isolates from certain grass species in Louisiana were parasitic to sugarcane and concluded that these isolates represented *C. falcatum* rather than a different species. In his studies, inoculation of sugarcane with the culture of *C. graminicola* failed to produce red rot. Lohman and Stokes[10] described a disease of sorghum caused by *Colletotrichum* sp. pointing out the close similarity between the causal agent and *C. falcatum*. They also noticed red rot of sorghum was produced by cultures of *C. falcatum* from sugarcane, *Pennisetum typhoideum* and *Paspalum scrobiculatum*.[34] The infection of maize with an isolate from sorghum has also been recorded.

LeBeau[51] observed that cultures from sugarcane were highly pathogenic to sugarcane, but rarely pathogenic to sorghum, while those isolated from sorghum, johnsongrass, Sudangrass, *Erianthus*, and broom-corn were nonpathogenic on sugarcane but were usually highly pathogenic on sorghum. Bruehl and Dickson[42] observed host specificity among various isolates of *C. graminicola* from various cereals and grasses with symptoms produced only by the isolates from the same or closely related species. It has been pointed out that *C. graminicola* comprised more than one pathogenic race, and that the isolates from sorghum, sudan grass, johnsongrass, broom-corn, and *Erianthus* seem to form one or more races characterized by their high pathogenicity to sorghum, whereas isolates from 14 other grass species appear to belong to one or more races which do not attack sorghum.[21] Certain isolates of maize anthracnose are capable of infecting sorghum in the glasshouse.[52] In Sudan, Tarr[3] had not been able to observe the pathogen on maize or other cultivated cereals, although it is very common on sorghum species. A local isolate associated with maize anthracnose in Ohio did not infect wounded leaves of wheat, oats, or barley.[53] Similarly, Dale[54] observed that a maize isolate from Arkansas failed to infect sorghum and an isolate from sorghum did not infect maize. The results obtained by Chohan[29] indicate that isolate G of *C. graminicola*, which is highly virulent on sorghum, is pathogenic to barley, Sudan grass, baru grass, sugarcane, and *Sorghum nitidum*, but not to wheat, bajra, or maize. It has been demonstrated that some isolates of *Colletotrichum* resembling *Glomerella tucumanensis* (Isolate F) occur on sorghum in Punjab, are capable of infecting sugarcane, and cause typical red rot.[55] Bergquist[56] reported that *C. graminicola* recovered from sorghum was weakly pathogenic to sugarcane. *C. graminicola* lesions on sugarcane foliage were superficial reddened lesions on mid rib and leaf as contrasted to deep, reddened lesions from *C. falcatum*. *C. falcatum* was not pathogenic to sorghum seedlings or older sorghum plants.

Thirty isolates of *C. graminicola* from six states (United States) readily infected maize and several species of *Sorghum*(*S. bicolor*, *S. sudanense* and *S. halepense*).[57] It was further reported that isolates from *Avena, Medicago, Hordeum, Bromus, Triticum, Calamagrostis, Festuca, Sorghum,* and *Danthonia* were nonpathogenic to maize. However, Nicholson[58] noticed that a *C. graminicola* isolate from Indian maize failed to infect wild cane, johnsongrass, sorghum, fall panicum, and six different foxtails. Isolates from maize were pathogenic to maize but not sorghum.[59]

Shahnaz and Nicholson[60] isolated *C. graminicola* from infected corn, sorghum, and shatter cane and tested for host range and pathogenicity. Plant (15 days old) of corn (Mo 940, 33-16; and Mo 15 Ht x B73Ht), sorghum (P72 1N and Br - 54) and field selection of shattercane were inoculated with each isolate. It was observed that Mo 940 and Mo17 Ht x B73 Ht were susceptible to the corn isolate and 33-16 was hypersensitively resistant. Corn only showed chlorotic flecks in response to sorghum (So) and shattercane (SH) isolate.

Knowledge of host range of the causal organism is essential to develop control measures for anthracnose of sorghum or other cereals or grasses. This is particularly important in the case of sorghum anthracnose which has a wide host range. Paster-Corrales and Frederiksen[4] have rightly pointed out that more information is needed on the ability of different geographical anthracnose isolates to cross-infect one host species to another.

14-6-2 Survival

The fact that there are many reports of the destructive nature of anthracnose indicates the existence of efficient mechanism for the survival of the fungus from one season to another. Several scientists have investigated this aspect and reported that the fungus survives mainly in the diseased crop debris, seeds from infected plants, and on alternate hosts.

Crop debris and soil.   Mishra and Siradhana[61] reported that cultures kept in sterilized and unsterilized soil survived from 9 to 6 weeks, respectively. Misra[33] observed that the survival ability of the fungus declined sharply over the period of time, when infected leaf pieces were incubated at room temperature. He concluded that high temperature (40°C) during the months of April to June did reduce the survival of the pathogen but the fungus was not altogether eliminated, as the recovery of the colonies from infected leaves continued up to June. When infected leaf pieces were buried in sterilized and natural soils and incubated at different temperatures (15°C to 45°), the pathogen did not survive for more than 8 months.

Seed. Luthra[62] and Chilton[63] reported that *C. graminicola* is carried on sorghum seeds. Misra and Siradhana[61] observed that the pathogen remained viable for 2 1/2 years on seeds, whereas Siddiqui et al.[64] noticed that *C. graminicola* survived for 7 years in seeds of sorghum stored at 5°C. They further suggested that the presence of pathogen in the embryo may be a reason for its long survival. Bergquist[56] noticed that *C. graminicola*-infected seeds had a significantly lower germination and may cause a post-emergence damping-off. Infected seed may also serve as a source of primary inoculum for spread of the disease to healthy plants during the growth period of the crop.

Site of Infection. Prasad et al.[65] reported that the pathogen, when plated on PDA, grew from all parts of the grain naturally infected with *C. graminicola*. Infection was greater in pericarp (55%) than in the endosperm (40%) or embryo (5%). These workers further observed a number of acervuli with setae over the seed surface. Mycelium was observed in the endosperm tissue just below the pericarp. Acervuli with setae were also present in the endosperm and embryo. Basuchaudhary and Mathur[66] detected *C. graminicola* mainly in the pericarp, occasionally in the endosperm, and rarely in the embryo of sorghum grains. These authors were of the opinion that the pathogen is thus primarily extraembryolar in sorghum and probably may follow the type of disease cycle involving extraembryolar infection followed by systemic infection. Siddique et al.[64] demonstrated that in component planting of sorghum seeds from 1975 and 1976, *Colletotrichum* was detected in all parts of the seeds, including the embryo. They concluded that this may be a reason for the long survival of the pathogen in seeds of sorghum, especially wherever low seed infection survived long storage periods.

Transmission of the Pathogen. Basuchaudhary and Mathur[66] reported that 14% to 16% of seedling developed reddish brown lesions on the cole-optile, older lesions turned blood red to almost black and were surrounded by a reddish brown periphery.

14-6-3 Seasonal Persistance and Spread

Seasonal persistence occurs through infected crop residues or on colletral hosts, especially perennial grasses. Sporulation has been noticed on sorghum stalks and stubbles after overwintering in the field.[21] Bergquist[56] was of the opinion that infected seed may serve as a source of primary inoculum for spread of the disease to healthy plants during the growth period of the crop.

Low and unbalanced soil fertility favored the disease. More damage was noticed on open coarse soil.[67] Wheeler et al.[57] were of the opinion that low light intensity and prolonged periods of high humidity seem to be important environmental factors that enhance disease severity.

14-7 INOCULATION AND DISEASE RATING

There are many published techniques for the inoculation: (a) use of leaf debris from previous year's or season's crop, (b) artificially infested grain dropped in whorls of young plants, (c) hypodermic placement of conidia in whorls, and (d) toothpicks inserted into stalks, a method that will work in dry environments.

Coleman and Stokes[68] reported the inoculation technique by the use of infected leaves. They obtained spores of *C. graminicola* from leaves of the susceptible variety Collier. Several infected leaves were placed in a moist chamber at room temperature at about 4 p.m. At 7 a.m. the next morning, these leaves were washed in tap water and the resulting spore suspension was strained through cheese cloth to eliminate trash large enough to plug the hypodermic needle. The concentration of spores in the suspension was adjusted by dilution to not less than 30 per low microscopic field. Every plant was inoculated for leaf anthracnose in the leaf whorl when the plants were still in the jointing stage and about 3 ft high. The main stalk of every plant was inoculated for red rot at about heading time. LeBeau[69] developed a method for large-scale inoculation of sorghum with *Colletotrichum*. He designed a special needle which was used with the compressed air sprayer for red-rot inoculation. Harris and Sowell[70] and Bergquist[56] sprayed spore suspension on the foliage and/or head. In another experiment, Lebeau[51] punctured the medium internode of the experimental stalks with a special cork-borer like instrument and inoculated a spore and mycelial suspension of *C. graminicola*. The inoculated stalks, in an upright position in metal cans, were incubated at 65°F to 80°F. Dean[71] inoculated 'Tracy' sorgo with *C. graminicola* by using a tractor-mounted sprayer normally used for applying herbicides.

Several techniques were employed by various workers to produce a large quantity of the fungus spores for artificial inoculation. In one petri plates containing steam sterilized pieces of sorghum leaves placed over moist filter paper were inoculated with spores and mycelium of *C. graminicola*. Plates were kept in a chamber illuminated with fluorescent lights at room temperature. In another procedure spores were obtained by growing the fungus in autoclaved V-8 broth. Flasks containing broth and fungus were placed on a rotatory shaker for about 1 week. The conidial suspension obtained from the plates and from the V-8 broth was filtered through many layers of cheese cloth and adjusted for the suitable number of spores per ml of water. Toothpicks were soaked in PDA broth, autoclaved, and the excess broth poured off. Mycelium or spores of *C. graminicola* were added to the toothpicks which were colonized by the fungus. Inoculum was also obtained by growing the fungus on sorghum seeds. Sorghum seeds soaked in water were autoclaved and inoculated with *C. graminicola*. Flasks containing seeds and the fungus were shaken daily to avoid the formation of heavy mats of fungal mycelium and to facilitate better sporulation.

Basically, the determination of plant disease resistance is based on measurement of the disease intensity and translating this into categories of resistance. Therefore, identification and detection of resistance will depend on the accuracy of the disease scale.

ICRISAT recommends a 1–5 disease scale[28] (Table 14-1). Disease rating 2 is descriptive and does not demarcate between 2 and 3 and thus may suffer from error. The disease rating 5 is very wide and does not specify the maximum disease point. All India Coordinated Sorghum Improvement Programme (AICSIP) suggested a scale which is

TABLE 14-1   DISEASE RATING SCALE RECOMMENDED BY ICRISAT

Rating value	Percentage area infected	Description
1	—	No symptoms
2	—	Few scattered lesions/spots
3	up to 25	Typical lesions developing on the leaves covering up to 25% leaf area
4	26 to 40	Coalescing spots covering about 26–40% leaf areas
5	40	Symptoms more severe covering more than 40% leaf area

similar to ICRISAT (Table 14-2). It is based mainly on percentage diseased leaf area. The disease rating 2 is very wide and ignores a marking at 5% a point recognized in general as "epidemic outbreak point" for leaf diseases.[72]

A detailed and descriptive scale for measuring intensity of foliar diseases of sorghum was developed at AICSIP, JNKVV, Indore. The scale is based on percentage leaf area infected.[73] Visual standards 1–9 represent (in percent) 0, 2.5, 5, 10, 20, 35, 50, 75, and 100% disease affected area. Disease indices 0 to 5 are marked at 0, 5, 20, 35, 50 and 100% on the disease scale to denote important conversion point for disease reaction and epidemic force.[28]

Harris and Sowell[70] used a 0–50 symptoms index for recording percentage leaf area affected (Table 14-3).

TABLE 14-2   DISEASE INDEX SCALE RECOMMENDED BY AICSIP

Rating value	Percentage area infected	Description
1	—	Free from disease
2	Trace–10	Slight symptoms
3	11–25	Moderate symptoms
4	26–50	Moderately severe
5	50	Very severe

TABLE 14-3   0 TO 5.0 SYMPTOMS INDEX FOR RECORDING PERCENTAGE LEAF AREA AFFECTED (HARRIS AND SOWELL[70])

Symptoms index	Percentage leaf area affected
0	0
0.5	Trace–10
1.0	11–20
1.5	21–30
2.0	31–40
2.5	41–50
3.0	51–60
3.5	61–70
4.0	71–80
4.5	81–90
5.0	91–100

Sharma[74] obtained data on degree of infection from 10 varieties which differed markedly in response to *C. graminicola* during intensive and continuous screening from 1974 to 1977. These were used to construct an infection index for each variety based on percentage leaf area infected. The indices of a candidate variety and a susceptible control were then compared to give a susceptible ratio. This ratio is believed to remain stable under varying conditions of natural infection. Susceptibility ratios of 0 to 0.14 and 0.15 to 0.35 are regarded as resistant and moderately resistant, respectively.[75] He has also presented model illustrations to show immune, moderately resistant, and susceptible reactions.[74]

14-8 MANAGEMENT

14-8-1 Cultural

Clean cultivation to eliminate susceptible hosts (weeds) such as johnsongrass, deep plowing during summer, destruction of infected crop and weed residues, and crop rotation may help in controlling the disease by minimizing primary inoculum. Harvesting of crop should not be delayed because the disease is known to cause severe damage following physiological maturity of sorghum.[4] Planting should be so adjusted that maturity does not concide with rainy weather.[75]

In view of the wide host range of *C. graminicola* among grasses, it is advisable to alternate sorghum (or other cereals) with noncereal crops wherever practicable.[3]

14-8-2 Chemical

Seed treatment.   Benlate and Ceresan wet @ 2 g/kg seed gave most effective control followed by Bavistin, Thiram, and Agrosan G.N.[76] Plantvax and Agrimycin also reduced disease intensity significantly.[77]

Foliar spray.   Sanden and Gorbet[77] evaluated thiophanate-methyl (1.12 kg/ha) maneb (2.24 kg/ha) triphenyl hydroxide (0.56 kg/ha) captafol (1.50 kg/ha) and chlorothalonil (1.13 and 2.24 kg/ha). Treatments initiated at boot stage and continued at biweekly intervals for 4 applications. Leaf infection was reduced from 8.0 in the unsprayed plots to 2.8 in plots sprayed with captafol. Yield increases over the check ranged from 1258 kg/ha fro thiophanate-methyl to 1370 kg/ha for captafol. The use of chlorothalonil resulted in a reduction in seed set and a corresponding reduction in yield. Another report indicates that of the several chemicals used as foliar spray, only Plantvax could reduce the disease intensity significantly.[78] Spray of Agrimycin was not effective.

Mishra and Siradhana[76] obtained significant control with Benlate, Difolatan, and Bavistin. Dithane Z-78, Dithane M-45, and Aureofungin were also effective. Sharma[28] reported that zineb and ziram, used as foliar sprays (@ 0.2%) were highly effective in reducing disease severity. In another report,[33] Bavistin, Topsin M, and Difolatan offered better control than Dithane Z-78 and Dithane M-45.

14-8-3 Use of Resistant Genotypes

Five American grain sorghums viz., Texas Milo, Club Kafir, Hegari, Pink Kafir, Western Blackhull Kafir were resistant to stalk rot and leaf anthracnose.[21] Most of the African varieties were highly resistant to leaf anthracnose and a few were to stalk rot. Texas Milo is said to be highly resistant to anthracnose in Georgia[20] and Sart is also resistant.[79] In Netherlands New Guinea Early White and Bird-Proof were relatively resistant to anthracnose and stalk rot.[19] Harris and Johnson[80] observed KS652, KS701, De Kalb E-57, and De Kalb F-64 as resistant hybrids. Of the 12 varieties tested, CSH-1, CSH-2, Khedi BK-1-1, Improved Sauner, and Ramkel were immune and Khedi BK 2-2-10, Mahamadapuri, P.J.8.K showed moderately resistant reaction.[81] Sharma and Jain[82] observed sorghum cultivars CSV-4, CSV-5, and CSH-5 as moderately resistant to the disease.

Sorghum cultivars like CS3541, CSH-4, CSH-5 were reported to be resistant against *C. graminicola*.[83–84] However, Mishra and Siradhana[85] observed CSH-1 as susceptible to anthracnose. The reaction to anthracnose of the sorghum cultivars was tested for three crop seasons at more than three locations in India. Nine sorghum cultivars were designated as resistant (SPV3, SPV5, SPV9, SPV12, SPV13, SPV23, SPV33, SPV34, and SPV35), five as moderately resistant (SPV1, SPV4, SPV29, SPV59, and SPV70), and one as susceptible (SPV69).

Rawal and his co-workers[86] screened 21 sorghum entries against anthracnose. Only SPV192 showed highest degree of resistance to disease while SPV106 and SPV229 were only next to SPV192 in their tolerance to the disease. As against this, SPV233 scored the highest grade (3.1) among all the entries included in the trial. BR500, BR501, BR503, and BR504 were found to be resistant to *C. graminicola*.[87] SL35, PC28, SPV98 and 73/53 were completely free from the disease.[88] Mathur and Naik[89] studied the reaction of 15 sorghum varieties against anthracnose. These authors reported SPV224 as highly resistant, SPV-128 and SPV220 as resistant, eight varieties as moderately resistant, and SPV105, SPV106, SPV192 as susceptible against anthracnose. In another report, sorghum cultivars viz., SC 326-6, SC 414-12E, R3338, R4244, 82CS444, SC748-5 showed highly resistant reaction to the disease.[89] Sinha et al.[90] reported that SPV615, SPV669, SPV707, and CSH5 showed a high degree of resistance and SPV710, CSV10, CSV11, and CSH1 were susceptible to anthracnose. Several other reports are available on the screening of sorghum genotypes against anthracnose.[91–92]

Inheritance of resistance.   LeBeau and Coleman[93] studied inheritance of resistance and reported that resistant to anthracnose was dominant and governed by a single major factor. Development of plant color after injury by disease or wounding was inherited independently. Lebeau et al.[21] observed that resistance to leaf anthracnose appeared to be independent of reaction to the stalk rot aspect of the disease, and inheritance of the latter was more complex. Resistance to stalk rot seemed to be relatively unstable and dependent not only on the inheritance of the plant but largely on such factors as temperature in relation to date of maturity, inoculum potential from the time of heading to maturity, and so on. Coleman and Stokes[68] pointed out that reaction was controlled by the single factor pair in which resistance was dominant. Reaction to stalk rot and anthracnose are closely linked with 9.57% crossing over and inherited independently of seed coat color.

Mechanism of resistance.   Little is known about the mechanism of resistance to spread of *C. graminicola* in stalk tissue of sorghum. Katsanos and Pappelis[94-95] demonstrated through a series of experiments that in the stalk tissues of sorghum, *C. graminicola* spreads through areas of dead cells and is inhibited by living cells or factors associated with them.

Biochemical basis of resistance.   A number of reports indicate that leaf exudates facilitate or inhibit spore germination of the pathogen. Sharma and Sinha[96] observed that the exudate of resistant variety TADA IS-3975 caused effective inhibition of the conidial germination. The exudates of both young and old sorghum leaves collected from midnight to morning showed maximum inhibition of conidial germination indicating inhibitory nature of exudates and thereby rendering the variety to show resistance. The exudates from young leaves were more inhibitory. Sharma[97] observed that out of 12 sugars detected in the leaf leachates, dextrose and raffinose seem to be associated with resistance. Likewise, out of 13 amino acids present in the leachate, DL-tryptophan and DL-alanine were found exclusively in the leachate of the resistant variety.

Mishra and Siradhana[98] have also observed significant reduction in germination in the leaf diffusate of resistant plants collected 8 hr after inoculation and in leaf extracts of resistant plants. Chiranjeevi and Tripathi[99-100] noticed that the anthracnose infection caused a 7% and 13% decrease in total chlorophyll b contents. In contrast, carotenoid content was more than four times higher. The diseased leaves had lower contents of ethanol extractable material, total sugars, reducing sugars, starch, lipid fraction, and deoxyribonucleic acid (DNA). However, nonreducing sugars, total nitrogen, soluble proteins, ribonucleic acid (RNA), and phenols had increased. Shahaz et al.[101] observed that PQ7, the most resistant variety, showed the least loss in chlorophyll. Carotanoid contents were affected only in the moderately resistant variety JS 1.

Mishra et al.[102] observed that total phenols and *o*-dehydroxphenols increased in resistant sorghum (CSH-5) leaves following inoculation. They concluded that phenols may be responsible for the resistance. Sharma[103] indicated that certain chemicals viz., DL-tryptophan, IAA, and HCN, may be responsible for bringing about the biochemical resistance in some sorghum varieties to *C. graminicola*. Leaves of the susceptible host plants were sprayed separately with these chemicals in different combinations of concentrations before inoculation with conidia of the pathogen. Spraying of HCN, IAA, or tryptophan + zinc (but not tryptophan alone) induced host resistance.[103]

14-9 CONCLUSION

Information on the relationship between disease severity and yield loss, pathogenic variability and physiological races, and host range of sorghum isolates of *C. graminicola* are inconclusive. Epidemiological aspects, particularly survival and spread of the pathogen, need to be extensively studied. Whether a perfect stage of the fungus exists is also a question. Information is required to establish the physiology of host response to infection, biochemistry of pathogenesis, and nature and mechanism of resistance. Effective and relevant laboratory and field-screening techniques need to be developed to identify sources of resistance. Methods for cultural and biocontrol of anthracnose are also very important areas for future investigation particularly in view of the fact that chemicals are usually too expensive to be used by the marginal farmers.

14-10 REFERENCES

1. Heald, F. D. "A Plant Disease Survey in the Vicinity of San Antonio, Texas." *U.S. Bureau of Plant Industry* Bulletin 226, (1912).

2. Stevens, F. L., and Hall, J. H. "Notes on the Plant Diseases Occurring in North Carolina." *North Carolina: Agricultural Experiment Station* 33 (1911): 59.

3. Tarr, S. A. J. *Diseases of Sorghum, Sudangrass, and Broom Corn.* Kew, Surrey: The Commonwealth Mycological Institute, 1962, 380.

4. Pastor-Corrales, M. A., and Frederiksen, R. A. "Sorghum Anthracnose," in Proceedings of the International Workshop on Sorghum Diseases, 11–15 December, 1978, Patancheru, Andhra Pradesh, India: ICRISAT, 1980, 289.

5. Nakamura, K. "Especializacao Fisiologica em *Colletotrichum graminicola* (Ces.) Wils. (Sensu Arx. 1957) Agente Causal da Anthracnose do Sorgho (Sorghum spp.)." (Ph.D. thesis, Universidade Estadual Paulista, Joboticabal, Brazil, 1943), 1082.

6. Maunder, B. "Potential for Sorghum Production in Latin America, Commercial Point of View," in Proceedings International Sorghum Workshop, University of Puerto Rico, Maya Gucz, PR, 3, 1975.

7. Sharvelle, E. C. "Sorghum Diseases in Brazil," in Proceedings International Sorghum Workshop, University of Puerto Rico, Mayaguoz, PR, 212, 1975.

8. Harris, H. B., and Fisher, D. "Yield of Grain Sorghum in Relation to Anthracnose Expression at Different Developmental Stages of Host," in 8th Grain Sorghum Research Utilization Conference Biennial Program, Lubbock, Texas: Grain Sorghum Producers' Association, 1973, 44.

9. Leukel, R. W., Martin, J. H., and Lefebvre, C. L. "Sorghum Diseases and Their Control." *USDA Farmers Bulletin*, 1059, 1951.

10. Lohman, M. L., and Stokes, I. E. "Stem Anthracnose and Red Rot of Sorgo in Mississippi." *Plant Dis. Reptr.* 28 (1944): 76.

11. Koehler, B. "Disease Threatening Broomcorn Production in Illinois." *Plant Dis. Reptr.* 27 (1943): 70.

12. Toler, R. W., and Frederiksen, R. A. "Sorghum Diseases, Grain Sorghum Research in Texas." Texas Agricultural Experiment Station, Texas A and M University, College State, Texas, 1970.

13. Frederiksen, R. A. "Anthracnose Stalk Rot, in Sorghum Root and Stalk rots—A Critical Review," Proceedings of the Consultative Group Discussion on Research Needs and Strategies for Control of Sorghum Root and Stalk Rot Diseases, Nov. 27–Dec. 2, 1983, Patancheru, ICRISAT, India, 1984, 37.

14. Hsi, C. H. "Stalk Rots of Sorghum in Eastern New Mexico." *Plant Dis. Reptr.* 40 (1956): 369.

15. Noble, R. J. "Australia: Notes on Plant Diseases Recorded in New-South Wales for the Year Ending June 30, 1937." *Internat. Bull. Plant Prot.* 11 (1937): 246.

16. Jaubert, P. "Listeannotee des Principles Affections Parasitaires (Mycoses, Bacterioses, Viroses) ainsi que des Affections de Causes mal Definies a des Plants Nuisibles aux Cultures du Senegal." *Centre Rech. Agron. Bombay, A.O.F. Bull.* 7 (1953): 2.

17. Saccas, A. M. "Les Champignons Parasites des Sorghos (*Sorghum vulgare*) et des Penicillaires (*Pennisetum typhoideum*) en Afrique Equatoriale Francaise." *Agron Trop. Nogent.* 9 (1954): 135, 263, 647.

18. Van Hoof, H. A. "Attack of Sorghum Vulgare Pers by *Colletotrichum falcatum* Went." *Landbouw* 21 (1949): 267.

19. Moll, H. W. "Outbreaks and New Records. Netherlands New Guinea." *F.A.O. Plant Prot. Bull.* 2 (1954): 121.

20. Luttrell, E. S. "Grain Sorghum Disease in Georgia." *Plant Dis. Reptr.* 34 (1950): 45.

21. LeBeau, F. J., Stokes, I. E., and Coleman, O. H. "Anthracnose and Red Rot of Sorghum." *Tech. Bull. U.S. Dept. Agr.* 1935 (1951): 21.

22. Stokes, I. E., Anderson, W. S., and Ferris, E. S. "Growing Sorgo for Syrup Production." *Miss. Agric. Exp. Sta. Circ.* (1944): 117.

23. Bockholt, A. J., Toler, R. W., and Frederiksen, R. A. "Measuring the Effect of Disease on Grain Sorghum Performance," in Proceedings of the Seventh Biennial Grain Sorghum Research and Utilization Conference, Sponsored by the Grain Sorghum Producers Association (GSPA) and Sorghum Improvement Conference of North America, Texas, 13, 1971.

24. Mishra, A., and Siradhana, B. S. "Estimation of Losses due to Anthracnose of Sorghum." *Indian J. Mycol. Plant Path.* 9 (1975): 257.

25. Powell, P., Ellis, M., Alaheda, M., and Sotomayor, A. "Effect of Natural Anthracnose Epiphytotic on Yield, Grain Quality, Seed Health and Seedborne Fungi in *Sorghum bicolor.*" *Sorghum Newsletter* 20 (1977): 77.

26. Harris, H. B., and Cunfer, B. M. "Observations on Sorghum Anthracnose in Georgia." *Sorghum Newsletter* 19 (1976): 100.

27. Harris, H. B., Johnson, B. J., Dobson, J. W., Jr., and Luttrell, E. S. "Evaluation of Anthracnose on Grain Sorghum." *Crop Sci.* 4 (1964): 460.

28. Sharma, H. C. "Screening of Sorghum for Leaf-Disease Resistance in India," in Proceedings of the International Workshop on Sorghum Diseases, 11–15 December 1978, Hyderabad, India. Patancheru, Andhra Pradesh, India. ICRISAT, 1980, 249.

29. Chohan, J. S. "Anthracnose Disease of Jowar (*Sorghum vulgare* Pers.) Caused by *Colletotrichum graminicolum* (Ces.) Wils, in the Punjab." *J. Res. Ludhiana,* 4 (1967): 394.

30. Ferreira, A. S., and Warren, H. L. "Resistance of Sorghum to *Colletotrichum graminicola.*" *Plant Dis. Reptr.* 44 (1982): 773.

31. Dickson, J. C. *Diseases of Field Crops.* New York: McGraw-Hill, 1956, p. 517.

32. Williams, R. J., Frederiksen, R. A., and Girard, J. C. "Sorghum and Pearl Millet Disease Identification Handbook." ICRISAT Information Bulletin No. 2, ICRISAT, Patancheru, A. P., India, 1978, 98.

33. Misra, A. N. "Studies on Anthracnose of Sorghum Caused by *Colletotrichum graminicalum.*" (M.S. Ag. thesis, G. B. Pant University of Agriculture and Technology, Pantnagar, India, 1987), 99.

34. Chowdhury, S. C. "A Disease of *Zea mays* Caused by *Colletotrichum graminicolum* (Ces.) Wils." *Ind. J. Agric. Sci.* 6 (1936): 833.

35. Ali, M. M. "Comparison of the Physiology of Three Isolates of *Colletotrichum graminicola.*" *Mycopath. Mycol. Appl.* 17 (1962): 261.

36. Mishra, A., and Siradhana, B. S. "Sorghum Anthracnose and the Growth and Sporulation of the Causal Fungus at Different Light Exposures." *Philippine Agricult.* 63 (1980): 67.

37. Foster, J., and Frederiksen, R. A. "Anthracnose and Other Sorghum Diseases in Brazil," in Proceedings of the Eleventh Biennial Grain Sorghum Research and Utilization Conference, sponsored by the Grain Sorghum Producers Association (GSPA) and Sorghum Improvement conference of North America, Texas, 76, 1979.

38. Frederiksen, R. A., and Rosenow, D. T. "Disease Resistance in Sorghum," Proceedings, 26th Annual Corn and Sorghum Conference 26, 71, 1971.

39. Ferreira, A. S., and Casela, C. R. "Pathogenic Races of *Colletotrichum graminicola*, Causal Agent of Sorghum (*Sorghum bicolar*) Anthracnose (Pt.)." *Fitopatologia Brasileira* 11 (1986): 83.

40. Ali, M. E. K., and Warren, H. L. "Physiological Races of *Colletotrichum graminicola* on Sorghum." *Plant Dis.* 71 (1987): 402.

41. Wilson, G. W. "The Identify of the Anthracnose of Grasses in the United States." *Phytopathology* 4 (1914): 106.

42. Bruehl, G. W., and Dickson, J. B. "Anthracnose of Cereals and Grasses." *Tech. Bull. U.S. Dept. Agr.* 1005 (1950): 37.

43. Luke, H. H., and Sechler, D. T. "Rye Anthracnose." *Plant Dis. Reptr.* 47 (1963): 936.

44. Sanford, G. B. "*Colletotrichum graminicola* (Ces.) Wils. or a Parasite of the Stem and Root Tissues of *Avena sativa.*" *Science Agric.* 15 (1935): 370.

45. Shurtleff, M. C. *A Compendium of Corn Diseases.* St. Paul, Minn.: The American Phytopathological Society, 1973, p. 64.

46. Wiese, M. V. *Compendium of Wheat Diseases.* St. Paul, Minn.: The American Phytopathological Society, 1977, 106.

47. Selby, A. D., and Manns, T. F. "Study on Diseases of Cereals and Grasses." *Ohio Agricultural Experiment Station Bull* 19 (1909): 475.

48. Edgerton, C. W. "The Red Rot of Sugarcane. A Report of Progress." *Louisiana Agricultural Experiment Station Bulletin* (1911): 133.

49. Butler, E. J., and Hafiz Khan, A. "Red Rot of Sugarcane." *Mem. Dep. Agric., India (Bot. Ser.)* 6 (1913): 151.

50. Abbott, E. V. "Red Rot of Sugarcane." *U.S. Dep. Agric. Tech. Bull.* (1938): 641.

51. LeBeau, F. J. "Pathogenicity Studies with *Colletotrichum* from Different Hosts on Sorghum and Sugarcane." *Phytopathology* 40 (1950): 430.

52. Muntnola, M. "Parasitos Criptogamicos de los Sorgo en la Provincia de Tricuman." *Revista Argentina, Agronomica* 19 (1952): 220.

53. Williams, L. E., and Willis, G. M. "Diseases of Corn Caused by *Colletotrichum graminicola.*" *Phytopathology* 53 (1963): 364.

54. Dale, J. L. "Corn Anthracnose." *Plant Dis. Reptr.* 47 (1963): 243.

55. Chohan, J. S. "Variability in *Colletotrichum graminicolum* (Ces.) Wils., the Pathogen Causing Anthracnose of Sorghum caused by *Glomerella tucumanensis* speg. Arx. and Muller (*Colletotrichum falcatum* Went.)." *J. Res. Ludhiana* 5 (1968): 220.

56. Bergquist, R. R. "*Colletotrichum graminicolum* in *Sorghum bicolor* in Hawaii." *Plant Dis. Reptr.* 57 (1973): 272.

57. Wheeler, H., Politis, D. J., and Poneleit, C. G. "Pathogenicity, Host Range, and Distribution of *Colletotrichum graminicola* on Corn." *Phytopathology* 64 (1974): 293.

58. Nicholson, R. L. "The Potential of Corn Anthracnose," Corn Disease Conference, Purdue University, 41, 1974.

59. Hooker, H. L. "Corn Anthracnose Leaf Blight and Stalk Rot," Proceedings of the 31st Annual Corn and Sorghum Research Conference, 31, 167, 1974.

60. Shahnaz, F. F., and Nicholson, R. L. "Colletotrichum graminicola: Host Specificity and Plant Age." *Phytopathology* 69 (1979): 542.

61. Mishra, A., and Siradhana, B. S. "Studies on the Survival of Sorghum Anthracnose (*Colletotrichum graminicolum*)." *Philippine Agricult.* 62 (1979): 149.

62. Luthra, J. C. "Some Fungal Diseases of Farm Crops Recently Discovered in the Punjab (India)." *Internat. Bull. Plant Prot.* 8 (1932): 181.

63. Chilton, S. J. P. "The Occurrence of *Helminthosporium turcicum* in the Seed and Glumes of Sudan Grass." *Phytopathology* 30 (1940): 553.

64. Siddiqui, M. R., Mathur, S. B., and Neergaard, P. "Longevity and Pathogenicity of *Colletotrichum* spp. in Seed Stored at 5°C." *Seed Sci. Technol.* 11 (1983): 353.

65. Prasad, K. V. V., Khare, M. N., and Jain, A. C. "Site of Infection and Further Development of *Colletotrichum graminicola* (Ces.) Wilson in Naturally Infected Sorghum Grains." *Seed Sci. Technol.* 13 (1985): 37.

66. Basuchaudhary, K. C., and Mathur, S. B. "Infection of Sorghum Seeds by *Colletotrichum graminicola.* 1. Survey, Location in Seed and Transmission of the Pathogen." *Seed Sci. Technol.* 7 (1979): 87.

67. Ramakrishnan, T. S. *Diseases of Millets*, New Delhi: Indian Council of Agricultural Research, 1963, 152.

68. Coleman, O. H., and Stokes, I. E. "The Inheritance of Resistance to Stalk Red Rot in Sorghum." *Agron. J.* 46 (1954): 61.

69. LeBeau, F. J. "A Method for Large Scale Inoculation of Sorghum with *Colletotrichum.*" *Phytopathology* 41 (1951): 378.

70. Harris, H. B., and Sowell, O., Jr. "Incidence of *Colletotrichum graminicola* on *Sorghum bicolor* Introduction." *Plant Dis. Reptr.* 54 (1970): 60.

71. Dean, J. L. "Anthracnose Inoculation of Sorgo with a Tractor-Mounted Sprayer." *Phytopathology* 54 (1964): 621.

72. Vanderplank, J. E. *Plant Disease Epidemics and their Control.* New York and London: Academic Press, 1963.

73. Coleman, O. H., and Labeau, F. J. "A Comparison of Three Methods of Rating Red Rot (*Colletotrichum graminicolum*) Infection in Sorghum." *Agron. J.* 45 (1953): 377.

74. Sharma, H. C. "A Technique for Identifying and Rating Resistance to Foliar Diseases of Sorghum under Field Conditions." *Proc. Indian Acad. Sci. (Plant Sciences)* 92 (1983): 271.

75. Dhaliwal, J. S., Sidhu, M. S., and Singh, G. "Effect of Different Dates of Sowing of Sorghum on the Incidence of Pests in Punjab." *J. Res., Punjab Agricultural University* 18 (1981): 400.

76. Mishra, A., and Siradhana, B. S. "Chemical Control of Anthracnose of Sorghum." *Indian Phytopathol.* 31 (1978): 225.

77. Sanden, G. E., and Gorbet, D. W. "Effect of Various Foliar Fungicides for the Control of Anthracnose of Grain Sorghum." *Proc. American Phytopathol. Soc.* 4 (1977): 154.

78. Agrawal, S. C., and Kotasthane, S. R. "Control of Leg Spot of Jowar by Systemic Fungicides and Antibiotics." *Food Farming Agric.* 8 (1976): 8.

79. Karper, R. E. "Registration of Sorghum Varieties." *Agronomy J.* 45 (1953): 322.

80. Harris, H. B., and Johnson B. J. "Sources of Anthracnose Resistance in Sorghum." *Sorghum Newsletter* 8 (1965): 17.

81. Kulkarni, N. B., and Kalekar, A. R. "Varietal Resistance in Sorghum to *Colletotrichum graminicolum* (Ces.) Wilson in Maharashtra State." *Research J. Mahatma Phule Agricultural University* 2 (1971): 159.

82. Sharma, H. C., and Jain, N. K. "A Note on Screening and Certifying Sorghum Varieties Against Foliar Diseases of a Region." *JNKVV Research J.* 12 (1978): 105.

83. Mathur, R. S., and Prakash, J. "Field Screening of Grain Sorghum Against Certain Leaf Spotting fungi." *Indian Photopathol.* 28 (1975): 423.

84. Prakash, J., Sainger, D. K., Mathur, R. S., and Sundaram, N. V. "Resistance of Five Leaf Spotting Fungi in Forage and Grain Sorghums in India." *Plant Dis. Reptr.* 59 (1975): 179.

85. Mishra, A., and Siradhana, B. S. "Reaction of Sorghum Germplasm to *Colletotrichum graminicolum.*" *Indian J. Mycol. Pl. Path.* 12 (1983): 314.

86. Rawal, P. P., Vadher, P. V., and Desai, K. B. "Screening of Breeder's Material in Advanced Yield Trials for Resistance to Sorghum Grain Mold, Leaf Blight, Anthracnose and Ergot." *Sorghum Newsletter* 23 (1980): 127.

87. Raupp, A. A. A., Cordeiro, D. S., Petcini, J. A., Porto, M. P., Brancao, N., and Santos Filho, B. G. Dos. "Sweet Sorghum Cultivation in the South-eastern Region of Rio Grande do Sul." *EMBRAPA* 12 (1980): 15.

88. Naik, S. M. P., and Mathur, K. "Varietal Reaction of Forage Sorghum Against Major Leaf-spot Diseases." *Indian Phytopathol* 34 (1981): 122.

89. Mathur, K., and Naik, S. M. P. "Field Evaluation of Advanced Sorghum Varieties Against Leaf Spot Diseases." *Sorghum Newsletter* 26 (1983): 116.

90. Sinha, A. P., Singh, R., and Shrotia, P. K. "Screening of Breeder's Material in Advanced Varietal Trials for Resistance to Sorghum Anthracnose, Zonate Leaf Spot and Grain Mold." *Sorghum Newsletter* 29 (1986): 12.

91. Ducan, R. R. "Reaction of Selected Sorghum Genotypes to Four Pathotypes of (*Colletotrichum graminicola*)." *Sorghum Newsletter* 28 (1985): 97.

92. Pastor-Corrales, M. A., and Frederiksen, R. A. "Sorghum Reaction in Anthracnose in the United States, Guatemala and Brazil." *Sorghum Newsletter* 22 (1979): 127.

93. Lebeau, F. J., and Coleman, O. H. "The Inheritance of Resistance in Sorghum to Leaf Anthracnose (*Colletotrichum graminicolum*)." *Agron. J.* 42 (1950): 33.

94. Katson, R. A., and Pappelis, A. J. "Relationship of Cell Death Patterns and Spread of *Colletotrichum graminicola* in Sorghum Stalk Tissue." *Phytopathology* 56 (1966): 468.

95. Katsonos, R. A., and Pappelis, A. J. "Relationship of Living and Dead Cells to Spread of *Colletotrichum graminicola* in Sorghum Stalk Tissue." *Phytopathology* 59 (1969): 132.

96. Sharma, J. K., and Sinha, S. "Effect of Leaf Exudates of Sorghum Varieties Varying in Susceptibility and Maturity on the Germination of Conidia of *Colletotrichum graminicola* (Ces.) Wilson Causing anthracnose." *Ecology of Leaf Surface Micro-organism*, eds. T. F. Preece and C. H. Dickinson. London: Academic Press, 1971, 597.

97. Sharma, J. K. "Conidial Germination and Growth Studies of *Colletotrichum graminicola* in the Leaf Extracts of Sorghum Varying in Susceptibility to Anthracnose." *Sorghum Newsletter* 16 (1973): 94.

98. Mishra, A., and Siradhana, B. S. "Inhibition of *Colletotrichum graminicolum* Spore Germination by Leaf Diffusates and Extracts from Sorghum." *Indian J. Myco. Pl. Path.* 10 (1980): 194.

99. Chiranjeevi, V., and Tripathi, R. K. "Changes in Chlorophyll and Carotenoid Contents in Sorghum Leaves Due to Zonate Leaf Spot and Anthracnose." *Indian J. Myco. Pl. Path.* 5 (1975): 98.

100. Chiranjeevi, V., and Tripathi, R. K. "Changes in Chemical Constituents of Anthracnose Infected Sorghum Leaves." *Indian J. Myco. Pl. Path.* 6 (1976): 171.

101. Shahaz, F. F., Azmi, A. R., and Mirza, J. H. "Effect of Sorghum Anthracnose on the Chlorophyll a, Chlorophyll b, and Total Carotenoids of Different Varieties of Sorghum," Proceedings of the XXVI/XXVIII Pakistan Science Conference, Lahore 1979. Part III. Abstracts of papers, Lahore, Pakistan; Pakistan Association for the Advancement of Science, 103A, 34, 1979.

102. Mishra, A., Siradhana, B. S., and Shivpuri, A. "Phenols in Relation to Resistance of Sorghum to Anthracnose." *Philippine Agricult.* 63 (1980): 71.

103. Sharma, J. K. "Effect of Certain Chemicals in Inducing Resistance to Anthracnose of Sorghum Caused by *Colletotrichum graminicola.*" *Sorghum Newsletter* 17 (1974): 67.

15

EYESPOT
OF CEREALS

B. D. L. FITT

AFRC Institute of Arable Crops Research,
Rothamsted Experimental Station, Harpenden, Herts., AL5 2JQ, UK

15-1 INTRODUCTION

Eyespot, caused by *Pseudocercosporella herpotrichoides* (Fron) Deighton is a serious disease on winter cereals and also infects a wide range of grasses.[1-3] The disease is known by a variety of names, associated with the symptoms it causes: eyespot (UK), footrot, strawbreaker (U.S.), piétin-verse (France), and Halmbruchkrankheit (Germany). Eyespot was probably responsible for the "straggling" of winter cereal crops which has been known for centuries, but this disease was first recognized in North America and France in the early twentieth century. Fron[4] described the causal fungus as *Cercosporella herpotrichoides*, but it was not proved to be the cause of the disease until the 1930s[5] because it was difficult to induce sporulation in culture. The disease was first described in the UK by Glynne[6] [Figure 15-1(a)], who studied it extensively over the next 30 years. Deighton[7] reclassified the causal fungus as *Pseudocercosporella herpotrichoides*, on the basis of conidiophore morphology. The teleomorph *Tapesia yallundae* Wallwork & Spooner was first discovered in Australia.[8]

15-2 DISTRIBUTION AND ECONOMIC IMPORTANCE

Eyespot is a widespread disease on wheat, barley, and rye in temperate climates, occurring in Europe, the USSR, South Africa, and parts of North America and Australasia.[9-10] In many of these countries, eyespot is present on most autumn-sown cereal crops, but it

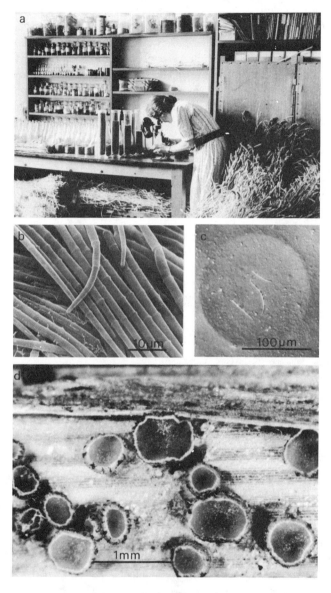

Figure 15-1 (a) Dr Mary D. Glynne; (b) Conidia of *Pseudocercosporella herpotrichoides*; (c) Spore-carrying droplet on photographic film; (d) Apothecia of *Tapesia yallundae* (F. R. Sanderson & A. C. King).

decreases yields only when epidemics become severe.[11-12] Yield losses of up to 50% can occur in individual crops, particularly when severe eyespot lesions weaken the stem bases of plants so that widespread lodging occurs.[11] The importance of eyespot has increased greatly in Europe in recent years, since populations of *P. herpotrichoides* are now largely resistant to the carbendazim-generating (MBC) fungicides which previously gave effective control of the disease.[13-15]

In the UK the Agricultural Development and Advisory Service (ADAS) has estimated losses caused nationally by eyespot in most years since 1975 for winter wheat and since 1981 for winter barley, from surveys of crops sampled at growth stage[16] (GS)

75.[12,15,17] Financial losses have been estimated from disease survey data and compared with fungicide costs estimated from fungicide use survey data. Thus it was estimated that eyespot caused yield losses in winter wheat and winter barley crops of £32M, £33M, and £19M, respectively, in 1987, 1988, and 1989, despite an expenditure of £23M, £16M, and £17M on fungicides for control of eyespot (Table 15-1). However, these loss figures are probably underestimates since they ignore both losses from eyespot-induced lodging which may occur after GS 75 and losses in yield quality, and hence value, which may be caused by eyespot.

TABLE 15-1   ESTIMATED COSTS OF EYESPOT ON WINTER WHEAT AND WINTER BARLEY CROPS IN ENGLAND AND WALES IN THE PERIOD 1985–1989; LOSS ASSESSMENTS AND FUNGICIDE COSTS CALCULATED FROM AGRICULTURAL DEVELOPMENT AND ADVISORY SERVICE SURVEY DATA[1]

	Wheat			Barley		
	Yield[2] loss (%)	Financial[3] loss (£M)	Fungicide[4] costs (£M)	Yield loss (%)	Financial loss (£M)	Fungicide costs (£M)
1985	2.0	25.7	14.6	—	—	—
1986	0.7	10.3	17.3	0.9	5.0	8.3
1987	2.4	29.0	15.6	0.7	3.3	7.2
1988	1.6	19.2	11.6	3.3	14.6	4.5
1989	1.0	11.6	12.8	1.3	5.5	4.6

[1]ADAS survey data from Polley and Thomas (personal communication)
[2]Yield loss (%) estimated from ADAS survey data by the method of Clarkson (1981):

$$L = [(0.1n_2 + 0.36n_3/n_t].100$$

where n_2 and n_3 are the number of stems with moderate or severe eyespot lesions and n_t is the total number of stems in the survey.
[3]Financial loss estimates: $[y/(1-L)].L.a.p.$ where y is mean national yield (t ha^{-1}), L is estimated loss (proportion), a is the area grown (M ha), and p is the price (£ t^{-1}).
[4]Total costs of fungicides applied at GS 30/31, excluding application and wheeling damage costs.

15-3 SYMPTOMS

Eyespot takes its name from the characteristic oval brown-bordered lesions, 15–30mm in length, which form at the base of cereal stems [Figure 15-2(b)]. These lesions typically have black dots (pupils) of fungal stroma in the center. Frequently *P. herpotrichoides* invades the stem base at ground level via the coleoptile,[18] but leaf sheaths may be infected directly.[19] Initially eyespot lesions are rather diffuse with a brown indistinct appearance but, as the fungus penetrates deeper into the outer leaf sheath and subsequently into the tissues below, lesions become more defined. Eyespot lesions are generally more difficult to recognize on barley and rye than on wheat. Eyespot lesions can be distinguished from lesions of sharp eyespot (*Rhizoctonia cerealis* van der Hoeven) by their indistinct, as opposed to sharp, edge and because they are generally confined to the lowest internode of the stem.[20] Furthermore, when infected leaf sheaths are peeled back, black dots of fungal stroma which cannot easily be rubbed off are observed on eyespot but not sharp eyespot lesions.

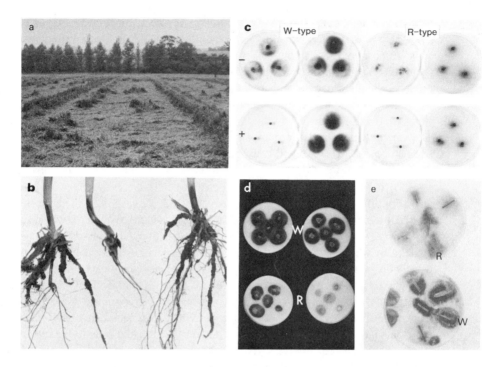

Figure 15-2   (a) Eyespot-induced lodging in winter wheat (G. L. Bateman); (b) Slight, moderate, and severe eyespot lesions on wheat; (c) MBC-sensitive and MBC-resistant W-type and R-type isolates of *Pseudocercosporella herpotrichoides* on potato dextrose agar, with (+) or without (-) carbendazim (1μg ml⁻¹); (d) W-type and R-type isolates on maize-meal agar; (e) Identification of W-type and R-type isolates on maize-meal agar; (c, d, & e: N. F. Creighton).

After stem extension, eyespot lesions can spread from the innermost leaf sheath on to the stem. Slight stem lesions cause little damage but severe lesions weaken the stem at this point so that it bends or breaks. If severe eyespot is widespread in a crop, it is likely to fall over or "lodge" before harvest [Figure 15-2(a)], particularly if stormy weather occurs. Less severe lesions may cause plants to die prematurely and produce shriveled ears known as "whiteheads," often scattered throughout the crop. Such ears may be colonized by secondary molds and turn black, particularly in wet weather. However, in crops with no symptoms of the disease visible on shoots, damaging lesions are often present at the stem bases when plants are pulled up and inspected.

15-4 CAUSAL ORGANISM

15-4-1 Pseudocercosporella herpotrichoides

The anamorph *P. herpotrichoides* is a member of the Moniliales, reproducing asexually from conidia produced abundantly on straw debris after harvest[21-22] and from hyphal fragments. In culture on potato dextrose agar (PDA) the vegetative mycelium is yellow to dark brown and some colonies have a fluffy grey appearance. The fungus survives on straw debris as stroma, composed of dark, thick-walled, polygonal pseudoparenchyma cells. Almost colorless sympodial conidiophores are produced on the ends of the olivaceous branched hyphae and are up to 20 × 3–3.5 μm with 2–3 distinct septa. Conidia are produced singly or in pairs and are colorless, acicular, widest at about a third of the length from the base with a truncate, unthickened hilum [Figure 15-1(b)]. They are straight or slightly curved, smooth, 3–7 septate, blunt or pointed at the apex, 30–80 × 1–3 μm.[23-24] They are produced in large numbers in mucilage composed of sugars and proteins on stromatic tissue. The two end cells may have opposite polarity since conidia migrate to the oil-water interface when mineral oil is shaken with aqueous conidial suspensions.[25]

Differences in cultural characteristics between isolates of *P. herpotrichoides* associated with differences in pathogenicity to wheat and rye have been recognized for a long time.[9] However, Lange-de la Camp[26] first distinguished the W-type (or wheat type) and R-type (or rye-type) of *P. herpotrichoides* on the basis of cultural characteristics and pathogenicity to wheat and rye in glasshouse experiments. In current practice, isolates are usually distinguished in culture. On PDA, W-type isolates produce colonies with smooth, even margins whereas R-type isolates produce colonies with feathery margins and generally grow at about half the rate of W-type isolates.[1] In glasshouse pathogenicity experiments (mean temperature *c.*7°C, range 4–19°C), W-type isolates have generally been more pathogenic to wheat than to barley and much less pathogenic to rye, whereas R-type isolates have been equally pathogenic to wheat, barley, and rye.[1,26-27] However, in recent controlled environment experiments at 10–15°C both W-type and R-type isolates have generally been more pathogenic to wheat than to rye, although R-type isolates have been more pathogenic than W-types to rye.[28-31] Furthermore, there has been a wide variation in pathogenicity between isolates, suggesting that pathogenicity is not a suitable criterion for use in classifying individual isolates as W-type or R-type. A third type, C-type, which is pathogenic to wheat, barley, and couch grass (*Elymus repens*) was distinguished by Cunningham.[3] There is no evidence that isolates of *P. herpotrichoides* exhibit physiologic specialization to particular cereal varieties. Pathogenic adaptation seems to be expressed at the host species level, as with other necrotrophic pathogens.[2]

Nirenberg[32] distinguished two varieties of *P. herpotrichoides, viz., P. herpotrichoides* var. *herpotrichoides* and *P. herpotrichoides* var. *acuformis,* on the basis of conidial characteristics. Conidia of *P.h.* var. *herpotrichoides* were curved or straight, with 4 septa, and were 35–80 (mean 52) μm long, whereas conidia of *P.h.* var. *acuformis* were straight, had 4–6 septa, and were 43–120 (mean 65) μm long. It has been suggested that W-type isolates can be identified as *P.h.* var. *herpotrichoides* and R-type isolates as *P.h.* var. *acuformis.*[28] However, it is not easy to classify individual isolates of *P. herpotrichoides* on the basis of spore characteristics[29] and there is now evidence that the varieties identified by Nirenberg may not be directly equivalent to W-types and R-types.[30] Isolates of both W-type and R-type *P. herpotrichoides* may be either sensitive or resistant to MBC fungicides [Figure 15-2(c)]. There are apparently no differences between MBC-sensitive and MBC-resistant isolates in cultural or conidial characteristics or in their pathogenicity to cereal seedlings.[27-29]

15-4-2 *Tapesia yallundae*

The teleomorph *T. yallundae* was first discovered in Australia on infected wheat and grass straw which was incubated for 6 months on sterile moist sand at 10°C in a dark room with occasional exposure to light.[8,33] Apothecia [Figure 15-1(d)] were 0.5–1.5 μm in diameter, sessile, produced on a subiculum, and had a grey hymenium. Asci were unitunicate, 40–55 × 4–6 μm, and contained eight spores. Ascospores were nonseptate, hyaline, fusiform, and 7–11 × 1.5–2.0 μm. The first report of *T. yallundae* on crops in the field was from Southland, New Zealand, in the spring when apothecia were abundant on wheat and barley stubble from previous crops.[34] Apothecia have now been found on overwintering stubble in Belgium,[35] the UK,[36] and West Germany.[37] The occurrence of the teleomorph may allow rapid changes in eyespot populations through sexual recombination and airborne dispersal of ascospores. However, the role of *T. yallundae* and of the ascospores in the epidemiology of eyespot is unclear since little stubble is left to overwinter in many cereal-growing countries and the distribution of *T. yallundae* is unknown.

15-5 EPIDEMIOLOGY

Epidemics of eyespot on cereals are composed of four overlapping phases, namely sporulation, spore dispersal, infection, and lesion development [Figure 15-3]. Eyespot may be considered as a monocyclic disease, *sensu* Van Der Plank[38] since there is little evidence

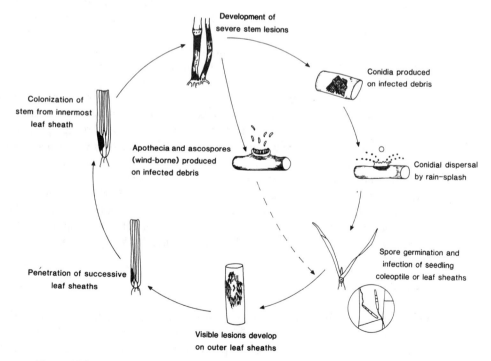

Figure 15-3 Life cycle of the eyespot fungus (anamorph *Pseudocercosporella herpotrichoides*; teleomorph *Tapesia yallundae*) on cereals (modified from Fitt et al.[17]).

that secondary disease spread is important and the increase in incidence of lesions fits the equation for a monocyclic disease, under ideal conditions.[39]

15-5-1 Sporulation

In wet, mild winters *P. herpotrichoides* produces large numbers of conidia on infected stem bases. On naturally infected straws sporulation occurs over the temperature range 1–20°C with an optimum at 5°C.[21,40] In pure culture the optimum temperature is 10–15°C[41] but temperatures above 10°C stimulate growth of competing fungi and bacteria on naturally infected stem bases.[40] Sporulation on stem bases occurs only after water has been absorbed[21,41] and in pure culture conidia were produced under near-ultra-violet (NUV) light (wavelength *c.* 350 nm) but not in darkness,[42] suggesting that both water and NUV are essential for sporulation.

Viable conidia (primary inoculum) were produced on infected stem bases throughout the growing season with a maximum in March-April followed by a decline as temperatures increased in field experiments in the UK,[19,21,22,40] France,[43] and North America.[44] By comparison with this primary inoculum, the secondary inoculum produced by sporulation on infected plants seems unimportant.[44–45]

15-5-2 Spore Dispersal

Conidia of *P. herpotrichoides* are normally dispersed from infected stem bases to plants of the new crop by rain splash. No spores are removed from infected straws by dry air because they are produced in mucilage. However, when drops are allowed to fall on to the straws they dissolve the mucilage and the resulting splash droplets contain large numbers of conidia [Figure 15-1(c)].[21,46,47] Furthermore, slides and pieces of fixed photographic film[48] exposed near infected stem bases in the field collected no spores during dry weather or light rain (<0.2mm h[-1]) but collected many spores in heavier rain (>3mm h[-1]), even in periods as short as 15 min.[22] When several spore samplers were exposed together for weekly periods, most spores were collected by samplers appropriate for splash-dispersed spores and few were collected by samplers appropriate for wind-dispersed spores.

Eyespot exhibits steep disease gradients, with a rapid decrease in disease with distance away from the source.[44] Such steep disease gradients reflect the steep spore deposition gradients[22,47] which are typical for a splash-dispersed pathogen.[49] However, both in rain tower experiments on splash dispersal of *P. herpotrichoides*[46] and above infected crop debris,[22] small numbers of airborne *P. herpotrichoides* conidia have been collected by suction samplers. These airborne conidia may be important in the long-distance spread of eyespot if they remain viable, but their significance in the epidemiology of eyespot is unknown.

15-5-3 Infection

On seedlings the coleoptile is the tissue that is most susceptible to infection[18] and many leaf sheath infections are probably initiated by coleoptile infections in the autumn. However, plants remain susceptible to eyespot throughout the growing season. When Hollins and Scott[19] exposed wheat plants of comparable age to the surrounding crop on an infected site, they developed lesions throughout the season provided they were incubated under conditions favorable for infection. Plants transplanted from an uninfected to an infected site from November to April developed lesions but those transplanted in May–June did not, suggesting that by then environmental conditions in the field were unfavorable for infection.

15-5-4 Lesion Development

In many countries where eyespot occurs, the weather is often favorable for sporulation, spore dispersal, and infection during the autumn and winter, and the severity of epidemics is more dependent on conditions influencing the rate of lesion development in the plants. Lesion development may be divided into four phases, namely leaf sheath lesion establishment, leaf sheath lesion development, stem lesion establishment, and stem lesion development.[39,50] The leaf sheath lesion establishment phase is the period between infection and the appearance of a visible lesion. The coleoptile infection is generalized and not visible to the eye; the appearance of stromata between the coleoptile and the outer leaf sheath is the first evidence that successful penetration has occurred. However, by the time a localized lesion becomes visible on the outer leaf sheath, the fungus may already have penetrated the leaf sheath below.

Leaf sheath lesion development progresses as the fungus penetrates successive leaf sheaths. It has been suggested that the rate of penetration is a function of temperature and a formula has been developed that attempts to relate the number of leaf sheaths penetrated to accumulated temperature in day-degrees.[43,51] In glasshouse experiments, the number of leaf sheaths penetrated increased with increasing temperature over the range 6–18 °C[52–53] and in some seasons the rate of leaf sheath lesion development in wheat crops fitted this formula well.[39] However, the formula did not fit well in other seasons, because it does not take into account factors that may restrict the rate of leaf sheath penetration. Lesions may fail to develop further if the rate of death of outer leaf sheaths is greater than the rate of penetration so that outer leaf sheaths die and decay before the fungus has colonized the leaf sheaths below them. Periods of cold weather, with frosts in winter[39] and warm, dry weather in spring[50] may both enhance the rate of death of outer leaf sheaths and thus impede lesion development.

A crucial stage in the development of eyespot lesions is the period of stem lesion establishment, after stem extension has begun, when the fungus is spreading from the innermost leaf sheath on to the stem.[50] In crops with a high incidence of leaf sheath lesions at GS 30-31, epidemics may fail to become severe if the weather is hot and dry during this period so that infected leaf sheaths die before the fungus has become established on the stems. Conversely, late epidemics may occur in crops with a low incidence of visible lesions at GS 30-31, if weather in this period favors the development of late, cryptic infections so that they become established on the stems subsequently. After they have become established, stem lesions may develop steadily, gradually increasing in severity, until harvest. A formula relating stem lesion development to accumulated temperature[43,51] fits well in some seasons but not others because it does not allow for failure of lesions to become established in stems.[39]

Field experiments in winter wheat and winter barley crops have suggested that there are differences between W-type and R-type isolates in the lesion development phase of eyespot epidemics.[54–55] The results from these experiments suggest that cold winters may favor the early development of R-type isolates in leaf sheaths. This is in agreement with results of seedling pathogenicity tests where R-type isolates grew comparatively better than W-types at lower temperatures.[27,29,30–31] However, in the spring W-type isolates generally became established on stems more rapidly than R-type isolates. In most seasons R-type isolates grew more rapidly than W-type isolates at the end of the season so that there was no difference between W-type and R-type plots in the incidence or severity of eyespot by harvest.

15-6 TECHNIQUES

15-6-1 Inoculation

Infected stem bases collected from the field can be used as a source of inoculum for both field and glasshouse or controlled environment experiments; on rewetting, conidia are produced abundantly, even after several months of storage. However, pure cultures of known isolates are frequently required as inoculum. Such isolates may be stored successfully as mycelial cultures on sterile soil at 4°C.[56] When such inoculated soil is spread on distilled water agar (DWA), abundant sporulation occurs within 1 week under NUV at 17°C. Even greater numbers of spores can be produced by inoculating potato dextrose agar (PDA) with conidial suspensions and incubating them under NUV.

An epidemic of eyespot can be expected if crops are grown after two or more previous consecutive susceptible cereal crops, or if they are inoculated with stem bases taken from infected crops. For inoculation of field experiments with known isolates the method of Bruehl and Nelson[57] has been widely used. Oat grains, which have been autoclaved, colonized by *P. herpotrichoides,* and air-dried, are spread over the site and provide an excellent source of inoculum throughout the season. Scott and Hollins[11] used colonized straws placed vertically in the ground, but this method is unsuitable for inoculating large areas.

Most methods for inoculating seedlings in glasshouse and controlled environment experiments use mycelial inoculum. Inoculated lengths of internodal straw placed over emerging coleoptiles[58] or annuli cut out from mycelial agar cultures and placed in contact with coleoptiles[58] were covered with sand to maintain humidity at the site of infection, whereas colonized filter paper pieces with attached agar plugs were placed within glass tubes over the emerging seedlings.[60] Since plants receive inoculum as conidia in the field, Higgins and Fitt[40] used spores on filter paper discs sealed between the coleoptile and the first leaf sheath with parafilm, and Bateman and Taylor[18] used spores applied with a soft brush to plants then covered with plastic tubes. No method for inoculating plants with ascospores has been described.

15-6-2 Diagnosis

Experienced observers can often distinguish symptoms of eyespot from those of sharp eyespot and brown footrot (*Fusarium* spp) in both seedlings and adult plants.[20] However, these symptoms are more difficult to distinguish on some cultivars and species than others. Confirming the identity of eyespot by isolation can take 6–8 weeks. Because *P.*

herpotrichoides grows slowly by comparison with other stem base fungi,[59] mycelial isolation is not as reliable as the induction of sporulation by placing infected material under NUV. However, growers need to be able to identify eyespot rapidly and reliably so that they can decide whether or not they need to apply a control fungicide spray. There have been several recent attempts to develop rapid serologically based diagnostic methods. Polyclonal antisera are generally unsuitable for detecting fungi since they frequently cross-react with other fungi. Polyclonal antisera produced in West Germany[61–62] were apparently not tested against *R. cerealis*, the fungus that cross-reacted most strongly to many of the monoclonal antisera produced by Dewey.[63] However, some monoclonal antisera did not cross-react and have been used to develop a "dipstick" assay (Farmers Weekly, 22 Sept. 1989).

In pure culture W-type and R-type isolates of *P. herpotrichoides* may be distinguished by their growth habit on agar[1,26] or by isozyme polymorphisms using polyacrylamide gel electrophoresis.[64] Isolates may also be classified on the basis of conidial production within sugar nutrient agar (SNA)[65] but not by observing spores produced on lesions under NUV.[39] However, Creighton[66] developed a maize-based agar on which W-type isolates produce a greenish-black color, whereas R-type isolates produce a pink or pale brown color [Figure 15-2(d)] in the medium under NUV at c. 13°C. This method was successful in mixed cultures produced by plating out naturally infected stem bases[67] [Figure 15-2(e)].

15-6-3 Disease Assessment

Eyespot infection normally occurs in patches and samples should consist of at least 25 plants per plot or 100 plants per field. Plants can be sampled along transects across fields[12] or plots[55] or from the edge in a W-shaped pattern.[68] The incidence of eyespot in the samples is frequently assessed as the percentage of plants with lesions until GS 30/31 and as the percentage of stems with lesions subsequently. A scale for assessing the severity of leaf sheath lesions was developed by Scott:[52] 1, coleoptile infected; 2, coleoptile penetrated; 3, first leaf sheath infected, etc. To incorporate incidence and severity data, a leaf sheath penetration index has been developed, based on the severity of lesions (0–3 scale) and the number of leaf sheaths infected.[40] However, Higgins et al.[50] showed that the mean number of leaf sheaths penetrated is equally reliable. Scott and Hollins[11] devised a 0–3 scale for assessing the severity of stem lesions (1, slight, lesions occupying less than half the stem; 2, moderate, lesions occupying more than half the stem; 3, severe, stem girdled by lesion with tissue softened) and calculated a stem severity index. More complicated assessments of the proportion of the cross-sectional area of the stem occupied by lesions[51] are, in practice, more difficult to use and probably do not greatly improve the accuracy of assessments. Eyespot-induced lodging can be assessed on a scale that takes account of both the angle and the area of lodging.[69]

15-7 MANAGEMENT

Strategies for control of eyespot include cultural practices, varietal resistance, and fungicides, but their relative importance differs greatly between countries. In many countries where cereals are not grown intensively it is uneconomic to use fungicides, and control relies on cultural practices and varietal resistance. However, in countries where cereals are grown intensively, the use of fungicides has increased greatly since the mid-1970s and the contribution to control of cultural practices, such as rotations, and varietal resistance has declined.

15-7-1 Cultural Practices

Cultural methods for control of eyespot, especially crop rotation, have been reviewed by Glynne and Salt.[9] Two-year breaks with nonsusceptible crops decreased eyespot damage in the following wheat crop to negligible levels,[70] provided that volunteer cereals were controlled to prevent carryover of the disease. However, winter cereals are often grown more than one year in three and one-year breaks are insufficient to control eyespot.[71] Furthermore, eyespot can be a serious problem in second cereal crops, particularly in wetter areas. Nevertheless, the opportunities to decrease eyespot by rotation are increasing because the increasing cultivation of alternative crops means that more winter cereal crops are being grown after 2-year breaks of nonsusceptible crops.

Glynne and Salt[9] concluded that eyespot is less likely to be severe in late-sown crops than in early-sown crops, and the trend toward early sowing of winter cereal crops has been an important factor contributing to the increase in the importance of eyespot in Europe.[14] Furthermore, this effect of sowing date on eyespot severity is probably even more important now than previously in countries such as the UK[13,15] and West Germany[65] where W-type populations have been largely replaced by R-type populations, because R-type isolates generally develop more slowly than W-type isolates in winter wheat or winter barley crops.[54-55] However, in many countries, early sowing of winter cereal crops will continue because of other agronomic and management reasons, although delayed sowing may be appropriate in others such as Canada.[72] Nevertheless, growers should be aware that if crops are late-sown (for example, because of a late previous harvest) the risk that serious eyespot epidemics will develop is much reduced. Excessive use of nitrogen fertilizers and very high seed rates both produce lush crops that favor eyespot.[70] In the Pacific northwest of the United States, where cultural methods for control of eyespot are very important, because use of fungicides is often uneconomic, less eyespot developed in crops with minimal cultivation before sowing than in crops sown in ploughed land.[73] This suggests that, although there has been an increasing use of minimal cultivation in some areas, it is unlikely that this will increase the severity of eyespot.

In Europe straw residues are now frequently incorporated rather than baled (uneconomic) or burnt (environmentally undesirable). In experiments, eyespot has typically been less severe where straw has been chopped and incorporated than where it has been burnt despite the fact that burning destroys at least some of the infected stem bases.[74] The reasons for this effect are not known but it may be because burning partially eliminates the competition by other microorganisms growing on the straw (Jenkyn, personal communication). One consequence of the ban on straw burning, which operates in much of western Europe, and the increase in alternative noncereal crops, many of which are spring-sown, may be that more stubble is left to overwinter. This may increase the occurrence of the teleomorph *Tapesia yallundae* which does not currently appear to be widespread in Europe because little stubble has been left to overwinter.

15-7-2 Varietal Resistance

Varietal resistance to eyespot is both agriculturally and environmentally desirable. It is seen by the grower as an inexpensive method of control that imposes no great constraints on methods for growing the crop and may decrease the need for the application of chemicals. Resistance to eyespot may operate directly on the growth of the pathogen or indirectly through increased resistance to lodging. Winter wheat varieties differ markedly in their resistance to eyespot but winter barley varieties are, in general, less severely attacked. Lodging resistance, which is widespread in modern varieties with short, stiff straws, reduces damage from eyespot. Tall-strawed, lodging-susceptible varieties yielded more when lodging was prevented by netting, but nets had no effect on yield of short-strawed, lodging-resistant varieties.[11] Lodging resistance does not necessarily confer resistance to growth of the eyespot fungus. However, stem base characteristics, like thickening of the hypodermis and secondary cell walls, which are important in lodging resistance, may also be important in eyespot resistance.[75]

Most of the current commercially available UK varieties of winter wheat incorporate resistance to eyespot, discovered in the 1950s,[76] derived from the French variety, Cappelle-Desprez.[77] The Cappelle-Desprez resistance to eyespot is readily introduced into new varieties and appears to be durable since it has not been overcome, despite widespread exploitation, for over 30 years. A new variety, Rendezvous, incorporates more effective resistance to eyespot than that of Cappelle-Desprez. This new resistance is from a wild goat grass, *Aegilops ventricosa*, and was first transferred into the French breeding line VPM 1.[78] Although Rendezvous can be infected by the eyespot fungus, symptoms at the start of grain-filling (GS 71) are less severe, and in the presence of the disease lodging is less and yields are greater than in moderately resistant varieties such as Norman.[79] The durability of this resistance will become apparent only when it is tested by extensive exposure to the pathogen in commercial agriculture.

15-7-3 Fungicides

Until the mid-1970s no effective fungicides had been developed against eyespot on cereal crops. However, since the introduction of the effective MBC fungicides in the mid-1970s the usage of fungicides on cereals against eyespot has increased greatly in countries where cereals are grown intensively. When they were introduced, MBC fungicides gave excellent control of eyespot and good yield increases, even when amounts of eyespot in crops were small. Consequently, they were often used routinely and resistance to MBC fungicides developed in *P. herpotrichoides*. This resistance was first identified in West Germany[80] but was thought to be unimportant because the incidence of resistance apparently increased little with the continued use of MBC fungicides.[81] MBC-resistance was detected in the UK in 1981 and was widespread by 1983.[13] By 1987 over 70% of isolates were MBC-resistant.[15] Field experiments have shown that repeated applications of MBC fungicides can increase the proportion of MBC-resistant isolates from <3% to >90% in two seasons, although they have little or no effect on the balance between W-type and R-type isolates.[68,82] Evidence from Europe suggests that if MBC fungicides were to become widely used against eyespot in countries where MBC-resistance is not yet a problem,[72-73] these populations would also become MBC-resistant.[13,84-84] Furthermore, when MBC fungicides are used against MBC-resistant populations of *P. herpotrichoides* they may increase the severity of eyespot and decrease yields,[13] possibly because they control stem base fungi that are antagonistic to *P. herpotrichoides.*

The demethylation-inhibiting (DMI) fungicide prochloraz, first recommended for control of eyespot in 1980, has now replaced MBC fungicides as the principal fungicide for control of eyespot on winter cereals. Although it may be less effective than MBC fungicides against MBC-sensitive isolates of *P. herpotrichoides,*[82] prochloraz is effective against both MBC-sensitive and MBC-resistant populations.[68,82] There is no evidence that resistance to prochloraz has developed in populations of *P. herpotrichoides* despite extensive use over the last few years.[85] Repeated use of prochloraz over several seasons did select for R-type isolates and reduced the proportion of W-type isolates in a population; it can also select against MBC resistance.[68] However, the use of a mixture of carbendazim plus prochloraz selects for MBC-resistant R-type isolates.[68,82] Another broad-spectrum triazole fungicide that has some activity against eyespot is flusilazole, although it has been found to be less effective than prochloraz.[86]

For both economic and environmental reasons it is important that fungicide sprays are applied at optimum times and only when crops need treatment against eyespot. Therefore, several forecasting schemes, reviewed by Fitt et al.,[17] have been developed to predict the occurrence of severe eyespot at a time when spray decisions need to be made. Weather-based forecasting schemes[51,87] have proved unsuitable for the UK, where ADAS schemes use a combination of risk assessment and disease assessments in winter wheat[88] and winter barley crops.[89] It has never been possible to predict satisfactorily the magnitude of eyespot-related yield losses at the time when spray decisions are made because these losses are very dependent on the occurrence of factors (e.g., lodging) late in the season.[90–91] However, when UK eyespot populations were predominantly W-type and MBC-sensitive, disease assessments at GS 30/31 gave a reasonably accurate prediction of eyespot severity during grain-filling.[91] Because the R-type isolates that now predominate develop more slowly than W-type isolates in winter wheat and winter barley crops,[54–55] it is now more difficult to predict the occurrence of severe epidemics. However, the wide period over which prochloraz can give effective control of eyespot[92] has enabled ADAS guidance to be changed.[88–89] If the spray threshold is not exceeded at GS 30, growers are advised to re-examine their crops up to GS 32, thus improving the accuracy with which severe epidemics during grain-filling can be predicted.

15-8 CONCLUSIONS

Eyespot remains a damaging disease on winter wheat and winter barley crops in temperate climates. The disease has increased in importance during the 1980s in many countries, because changes in cropping practices, such as earlier sowing, have favored the development of severe eyespot epidemics, and pathogen populations have become resistant to MBC fungicides that previously gave effective control. There are no varieties of wheat or barley with complete resistance to eyespot, and fungicides currently in use are less effective against eyespot than MBC fungicides were previously. There is a need to improve the estimates of losses caused by eyespot by extending disease surveys to new areas and by recalculating the yield loss formula with modern varieties of wheat and barley. Estimates of losses caused by eyespot-induced lodging and of losses in quality caused by eyespot should also be calculated.

Since it is not always easy to distinguish between eyespot and other stem base pathogens, and rapid diagnostic methods would be of great benefit to the agricultural industry, there is an urgent need to develop rapid methods for identifying eyespot and for distinguishing between the different types of the fungus. There is considerable scope for applying modern serological and molecular biological methods to this problem and for extending the techniques currently being developed to improve their accuracy and reliability.

It is likely that there will be a decrease in the use of fungicides against eyespot, even in countries where cereals are grown intensively. Nevertheless, new fungicides with different modes of action to prochloraz need to be developed. Furthermore, methods for predicting the severity of eyespot epidemics should be improved so that fungicides are applied only when they are needed. The increase in the number and area grown of non-susceptible crops should improve opportunities for reducing the risk of eyespot by crop rotation. Crop rotation and other cultural methods for eyespot control need to be reinvestigated in the light of recent changes in cropping practices. Furthermore, there is an urgent need for new varieties with improved resistance to eyespot combined with good agronomic qualities appropriate to the different countries where they are to be grown.

15-9 REFERENCES

1. Scott, P. R., Hollins, T. W., and Muir, P. "Pathogenicity of *Cercosporella herpotrichoides* to Wheat, Barley, Oats and Rye." *Trans. Brit. Mycol. Soc.* 65 (1975): 529.

2. Scott, P. R., Defosse, L., Vandam, J., and Doussinault, G. "Infection of Lines of *Triticum, Secale, Aegilops* and *Hordeum* by Isolates of *Cercosporella herpotrichoides.*" *Trans. Brit. Mycol. Soc.* 66 (1976): 205.

3. Cunningham, P. C. "Occurrence, Role and Pathogenic Traits of a Distinct Pathotype of *Pseudocercosporella herpotrichoides.*" *Trans. Brit. Mycol. Soc.* 76 (1981): 3.

4. Fron, G. "Contribution a l'Etude de la Maladie de Piéd Noir des Céreales ou Maladie du Piétin." *Ann. Sci. Agron. Fr. Etr.* 4 (1912): 1.

5. Sprague, R. A. "*Cercosporella herpotrichoides* Fron, the Cause of the Columbian Basin Footrot of Winter Wheat." *Science* 24 (1931): 51.

6. Glynne, M. D. "Some New British Records of Fungi on Wheat." *Trans. Brit. Mycol. Soc.* 20 (1936): 120.

7. Deighton, F. "Studies on *Cercosporella* and Allied Genera. IV. *Cercosporella* Sacc., *Pseudocercosporella* gen. nov. and *Pseudocercosporidium* gen. nov." Mycol. Papers 133. Commonwealth Mycological Institute, Kew, 1973.

8. Wallwork, H. "A *Tapesia* Teleomorph for *Pseudocercosporella herpotrichoides,* the Cause of Eyespot of Wheat." *Austral. Pl. Path.* 16 (1987): 92.

9. Glynne, M. D., and Salt, G. A. "Eyespot of Wheat and Barley." *Rothamsted Exp. Sta., 1957* (1958): 231–241.

10. Anonymous. "Distribution Maps of Plant Diseases 74." Commonwealth Mycological Institute, Kew., 1981.

11. Scott, P. R., and Hollins, T. W. "Effects of Eyespot on the Yield of Winter Wheat." *Ann. App. Biol.* 78 (1974): 269.

12. Clarkson, J. D. S. "Relationship Between Eyespot Severity and Yield Loss in Winter Wheat." *Pl. Path.* 30 (1981): 125.

13. King, J. E., and Griffin, M. J. "Survey of Benomyl Resistance in *Pseudocercosporella herpotrichoides* on Winter Wheat and Barley in England and Wales in 1983." *Pl. Path.* 34 (1985): 272.

14. Yarham, D. J. "Change and Decay—the Sociology of Cereal Footrots." *Proc. 1986 Brit. Crop Prot. Conf.—Pests and Diseases,* Thornton Heath, U.K.: British Crop Protection Council, 1986, pp. 401–410.

15. Fitt, B. D. L. Eyespot Disease of Cereals." *HGCA Research Review* 1. Home-grown Cereals Authority, London, 1988, 88 pp.

16. Tottman, D. R., and Broad H. "The Decimal Code for the Growth Stages of Cereals, with Illustrations." *Ann. App. Biol.* 110 (1987): 441.

17. Fitt, B. D. L., Goulds, A., and Polley, R. W. "Eyespot *(Pseudocercosporella herpotrichoides)* Epidemiology in Relation to Prediction of Disease Severity and Yield Loss in Winter Wheat; A Review." *Pl. Path.* 37 (1988): 311.

18. Bateman, G. L., and Taylor, G. S. "Seedling Infection of Two Wheat Cultivars by *Pseudocercosporella herpotrichoides.''* *Trans. Brit. Mycol. Soc.* 67 (1976): 95.

19. Hollins, T. W., and Scott, P. R. "Epidemiology of Eyespot *(Pseudocercosporella herpotrichoides)* on Winter Wheat, with Particular Reference to the Period of Infection." *Ann. App. Biol.* 95 (1980): 19.

20. Goulds, A., and Polley, R. W. "Assessment of Eyespot and Other Stem Base Diseases of Winter Wheat and Winter Barley." *Mycol. Res.* 94, 1990: 819.

21. Glynne, M. D. "Production of Spores by *Cercosporella herpotrichoides.''* *Trans. Brit. Mycol. Soc.* 36 (1953): 46.

22. Fitt, B. D. L., and Bainbridge, A. "Dispersal of *Pseudocercosporella herpotrichoides* Spores from Infected Wheat Straw. *Phytopathol. Z.* 106 (1983): 214.

23. Booth, C., and Waller, J. M. *Pseudocercosporella herpotrichoides. C.M.I. Descriptions of Pathogenic Fungi and Bacteria.* Kew, Surrey, U.K.: Commonwealth Mycological Institute 386, 1973.

24. Sprague, R. A., and Fellows, H. "*Cercosporella* Footrot of Winter Cereals." *Tech. Bull. U.S. Dept. Agr.* 428, 1934.

25. Fitt, B.D.L., and Bainbridge, A. "Recovery of *Pseudocercosporella herpotrichoides* Spores from Rain-Splash Samples." *Phytopathol. Z.* 106 (1983): 177.

26. Lange-de la Camp, M. "Die Wirkungsweise von *Cercosporella herpotrichoides* Fron., dem Erreger der Halmbruchkrankheit des Getreides. I. Feststellung der Krankheit. Beschaffenheit und Infektionsweise ihres Erregers." *Phytopathol. Z.* 55 (1986): 34.

27. Hollins, T. S., Scott, P. R., and Paine, J. R. "Morphology, Benomyl Resistance and Pathogenicity to Wheat and Rye of Isolates of *Pseudocercosporella herpotrichoides. Pl. Path.* 34 (1985): 369.

28. Sanders, P. L., De Waard, M. A., and Loerakker, W. M. "Resistance to Carbendazim in *Pseudocercosporella herpotrichoides* from Dutch Wheat Fields." *Neth. J. Pl. Path.* 92 (1986): 15.

29. Fitt B. D. L., Creighton, N. F., and Bateman, G. L. "Pathogenicity to Wheat Seedlings of Wheat-Type and Rye-Type Isolates of *Pseudocercosporella herpotrichoides." Trans. Brit. Mycol. Soc.* 88 (1987): 149.

30. Mauler, A., and Fehrmann, H. "Erfassung der Anfälligkeit von Weizen gegenüber *Pseudocerosporella herpotrichoides.* I. Untersuchungen zur Pathogenität verschiedener Formen des Erregers." *Z. Pflkrankheiten Pflschutz* 94 (1987): 637.

31. Creighton, N. F., Cavelier, N., and Fitt, B. D. L. "Pathogenicity to Wheat and Rye of *Pseudocercosporella herpotrichoides* Isolates from France and the U.K." *Mycol. Res.* 92 (1989): 13.

32. Nirenberg, H. I. "Differenzierung der Erreger der Halmbruchkrankheit. I. Morphologie." *Z. Pflkrankheiten Pflschutz* 88 (1981): 241.

33. Wallwork, H., and Spooner, B. *"Tapesia yallundae*—the Teleomorph of *Pseudocercosporella herpotrichoides.'' Trans. Brit. Mycol. Soc.* 91 (1988): 703.

34. Sanderson, F. R., and King, A. C. "The Occurrence in the Field of *Tapesia yallundae,* the Teleomorph of *Pseudocercosporella herpotrichoides." Austral. Pl. Path.* 17 (1988): 20.

35. Moreau, J. M., Van Schingen, J. C., and Maraite, H. "Detection of *Tapesia yallundae,* the Teleomorph of *Pseudocercosporella herpotrichoides,* on Wheat Stubbles in Belgium." *Med. Fac. Land. Rijk., Gent.* 54 (1989): 555.

36. Hunter, T. "Occurrence of *Tapesia yallundae,* Teleomorph of *Pseudocercosporella herpotrichoides,* on Unharvested Wheat Culms in England." *Pl. Path.* 38 (1989): 598.

37. King, A. C. "First Record of *Tapesia yallundae* as the Teleomorph of *Pseudocercosporella herpotrichoides* var. *acuformis,* and Its Occurrence in the Field in the Federal Republic of Germany." *Pl. Path.* 39 (1990): 44.

38. Van der Plank, J. E. *Plant Diseases: Epidemics and Control.* New York: Academic Press, 1963.

39. Fitt B. D. L., and White, R. P. "Stages in the Progress of Eyespot Epidemics in Winter Wheat Crops," *Z. Pflkrankheiten Pflschutz.* 95 (1988): 35.

40. Higgins, S., and Fitt, B. D. L. "Production and Pathogenicity to Wheat of *Pseudocercosporella herpotrichoides* conidia." *Phytopathol. Z.* 111 (1984): 222.

41. Rowe, R. C., and Powelson, R. L. "Epidemiology of *Cercosporella* Footrot of Wheat: Spore Production." *Phytopathology* 63 (1973): 984.

42. Chang, E. W. P. and Tyler, L. J. "Sporulation by *Cercosporella herpotrichoides* on Artificial Media." *Phytopathology* 54 (1964): 729.

43. Ponchet, J. K. "La Maladie du Piétin-Verse des Céréales: *Cercosporella herpotrichoides* Fron. Importance agronomique, Biologie, Épiphytologie." *Ann. Épiph.* 10 (1959): 45.

44. Rowe, R. C., and Powelson, R. L. "Epidemiology of *Cercosporella* Footrot of Wheat Disease Spread." *Phytopathology* 63 (1973): 984.

45. Nelson, K. E., and Sutton, J. C. "Epidemiology of Eyespot on Winter Wheat in Ontario." *Phytoprotection* 69 (1988): 9.

46. Fitt, B. D. L., and Nijman, D. J. "Quantitative Studies on Dispersal of *Pseudocercosporella herpotrichoides* spores from Infected Wheat Straw by Simulated Rain." *Neth. J. Pl. Path.* 89 (1983): 198.

47. Fitt B. D. L., and Lysandrou, M. "Studies on Mechanisms of Splash Dispersal of Spores, Using *Pseudocercosporella herpotrichoides* Spores." Phytopathol. Z. 111 (1984): 323.

48. Fitt, B. D. L., Lysandrou, M., and Turner, R. H. "Measurement of Spore-Carrying Splash Droplets Using Photographic Film and an Image-Analyzing Computer." *Pl. Path.* 31 (1982): 19.

49. Fitt, B. D. L., McCartney, H. A., and Walklate, P. J. "The Role of Rain in Dispersal of Pathogen Inoculum." *Ann. Rev. Phytopath.* 27 (1989): 241.

50. Higgins, S., Fitt, B. D. L., and White, R. P. "The Development of Eyespot (*Pseudocercosporella herpotrichoides*) Lesions in Winter Wheat Crops." *Z. Pflkrankheiten Pflschutz* 93 (1986): 210.

51. Rapilly, F., Laborie, Y., Eschenbrenner, P., Choisnel, E., and Lacroze, F. "La Prévision du Piétin-Verse sur Blé d'Hiver." *Persp. Agric.* 23 (1979): 30.

52. Scott, P. R. "The Effect of Temperature on Eyespot (*Cercosporella herpotrichoides*) in Wheat Seedlings." *Ann. App. Biol.* 68 (1971): 169.

53. Higgins, S., and Fitt, B. D. L. "Effects of Water Potential and Temperature on the Development of Eyespot Lesions in Wheat." *Ann. App. Biol.* 107 (1985): 1.

54. Goulds, A., and Fitt, B. D. L. "The Comparative Epidemiology of Eyespot (*Pseudocercosporella herpotrichoides*) Types in Winter Cereal Crops." *Proc. 1988 Brighton Crop Prot. Conf.—Pests and Diseases,* Thornton Heath, U.K.: British Crop Protection Council, pp. 1035–1040, 1988.

55. Goulds, A., and Fitt, B. D. L. "The Development of Eyespot on Seedling Leaf Sheaths in Winter Wheat and Winter Barley Crops Inoculated with W-type or R-type Isolates of *Pseudocercosporella herpotrichoides.*" *J. Phytopath.* 130 (1990): 161.

56. Reinecke, P., and Fokkema, N. J. "*Pseudocercosporella herpotrichoides:* Storage and Mass Production of Conidia." *Trans. Brit. Mycol. Soc.* 72 (1979): 329.

57. Bruehl, G. W., and Nelson, W. L. "Techniques for Mass Inoculation of Winter Wheat in the Field with *Cercosporella herpotrichoides.*" *Pl. Dis. Rep.* 48 (1964): 863.

58. Macer, R. C. F. "Resistance to Eyespot Disease (*Cercosporella herpotrichoides* Fron.) Determined by a Seedling Test in Some Forms of *Triticum, Aegilops, Secale* and *Hordeum.*" *J. Agric. Sc. Cambridge* 67 (1966): 389.

59. Bateman, G. L., Smith, C., Creighton, N. F., Li, K. Y., and Hollomon, D. W. "Characterization of Wheat Eyespot Populations Before Development of Fungicide Resistance." *Trans. Brit. Mycol. Soc.* 85 (1985): 335.

60. Evans, M. E., and Rawlinson, C. J. "A Method for Inoculating Wheat with *Cercosporella herpotrichoides.*" *Ann. App. Biol.* 80 (1975): 339.

61. Bolik, M., Casper, R. and Lind, V. "Einsatz serologischer und gelektrophoretischer Verfahren zum Nachweis von *Pseudocercosporella herpotrichoides.*" *Z. Pflkrankheiten Pflschutz* 94 (1987): 449.

62. Unger, J. G., and Wolf, G. "Detection of *Pseudocercosporella herpotrichoides* (Fron.) Deighton in Wheat by Indirect ELISA." *J. Phytopathol.* 122 (1988): 281.

63. Dewey, F. M. "Development of Immunological Diagnostic Assays for Fungal Plant Pathogens." *Proc. 1988 Brighton Crop Prot. Conf., Pests and Diseases,* Thornton Heath, U.K.: British Crop Protection Council 1988, pp. 777–786.

64. Julian, A. M., and Lucas, J. A. "Isozyme Polymorphism in Pathotypes of *Pseudocercosporella herpotrichoides* and Related Species from Cereals." *Pl. Path.* 39 (1990): 178.

65. Schreiber, M. T., and Prillwitz, H. G. "Verkommen von *Pseudocercosporella*—Taxa an Wintergetreide in Rheinland-Pfalz." *Nach. Pfl. (Braunschweig)* 37 (1985): 145.

66. Creighton, N. F. "Identification of W-type and R-type Isolates of *Pseudocercosporella herpotrichoides.*" *Pl. Path.* 38 (1989): 484.

67. Creighton, N. F., and Bateman, G. L. "Improved diagnosis of Eyespot Pathotypes: Application of a New Method." *Proc. 1988 Brighton Crop Prot. Conf., Pests and Diseases,* Thornton Heath, U.K.: British Crop Protection Council 1988, pp. 1023–1027.

68. Bateman, G. L., Fitt, B. D. L., Creighton, N. F., and Hollomon, D. W. "Changes in Populations of *Pseudocercosporella herpotrichoides* in Successive Crops of Winter Wheat in Relation to Initial Populations and Fungicide Treatments." *Crop Prot.* 9 (1990): 135.

69. Caldicott, J. J. B., and Nuttall, A. M. "A Method for the Assessment of Lodging in Cereal Crops." *J. Nat. Inst. Agric. Bot.* 15 (1968): 88.

70. Glynne, M. D., and Slope, D. B. S. "Effects of Previous Wheat Crops, Seed-Rate and Nitrogen on Eyespot, Take-All, Weeds and Yields of Two Varieties of Winter Wheat: Field Experiment 1954–56." *Ann. App. Biol.* 47 (1959): 187.

71. Maenhout, C.A.A.A. "Eyespot in Winter Wheat: Effects of Crop Rotation and Tillage, and the Prediction of Incidence." *E.P.P.O. Bull.* 5 (1975): 407.

72. Nelson, K. E., and Sutton, J. C. "Fungicides, Fungicide Timing and Sowing Date in Relation to Eyespot Incidence in Ontario Winter Wheat." *Phytoprotection* 68 (1987): 111.

73. Herrman, T., and Wiese, M. V. "Influence of Cultural Practices on Incidence of Footrot in Winter Wheat." *Pl. Dis.* 69 (1986): 948.

74. Jenkyn, J. F., Gutteridge, R. J., and Thomas, M. R. "Effects of Straw Incorporation and Cultivations on Cereal Diseases." *Asp. App. Biol.* 17 (1988): 181.

75. Murray, T. D., and Bruehl, G. W. "Role of Hypodermis and Secondary Cell Wall Thickening in Basal Stem Internodes in Resistance to Straw-Breaker Footrot in Winter Wheat." *Phytopathology* 73 (1983): 261.

76. Vincent, A., Ponchet, J., and Koller, J. "Recherche de Variétés de Blés Tendres peu Sensibles au Piétin-Verse: Résultats Préliminaires." *Ann. Amél. Pl.* 3 (1952): 459.

77. Batts, C. C. V., and Fiddian, W. E. H. "Effect of Previous Cropping on Eyespot in Four Varieties of Winter Wheat." *Pl. Path.* 4 (1955): 25.

78. Doussinault, G., Koller, J., Touvin, H., and Dosba, F. "Utilisation de Géniteurs VPM 1 dans l'Amélioration de l'État Sanitaire du Blé Tendre." *Ann. Amél. Pl.* 24 (1974): 215.

79. Hollins, T. W., Lockley, K. D., Blackman, J. A., Scott, P. R., and Bingham, J. "Field Performance of Rendezvous, a Wheat Cultivar with Resistance to Eyespot (*Pseudocercosporella herpotrichoides*) Derived from *Aegilops ventricosa.*" *Pl. Path.* 37 (1988): 251.

80. Rashid, T., and Schlösser, E. "Resistenz von *Cercosporella herpotrichoides* gegenüber Benomyl." *Z. Pflkrankheiten Pflschutz* 82 (1975): 765.

81. Fehrmann, H., and Schrödter, H. "Control of *Cercosporella herpotrichoides* in Winter Wheat in Germany." *Proc. 7th Brit. Insect. Fung. Conf. 1973,* (1973), 119–126.

82. Hoare, F. A., Hunter, T., and Jordan, V. W. L. "Influence of Spray Programmes on Development of Fungicide Resistance in the Eyespot Pathogen of Wheat *Pseudocercosporella herpotrichoides.*" *Pl. Path.* 35 (1986): 506.

83. Cavalier, N., Leroux, P., Hanrion, M., and Curé, B. "Résistance de *Pseudocercosporella herpotrichoides* aux Benzimidazoles et thiophanates Chez le Blé d'Hiver en France." *EPPO Bull.* 15 (1985): 495.

84. Schreiber, B., and Schlesinger, W. "Contribution to the MBC-Resistance of *Pseudocercosporella herpotrichoides.*" *Med. Fak. Land. Rijk., Gent.* 50 (1985): 1181.

85. Gallimore, K., Knights, I. K., and Barnes, G. "Sensitivity of *Pseudocercosporella herpotrichoides* to the Fungicide Prochloraz." *Pl. Path.* 36 (1987): 290.

86. Bateman, G. L. "Comparison of the Effects of Prochloraz and Flusilazole on Footrot Diseases and on Populations of the Eyespot Fungus." *Pseudocercosporella herpotrichoides,* in winter wheat." *Z. Pflkrankheiten Pflschutz,* 97, (1990): 508.

87. Fehrmann, H., and Schrödter, H. "Control of *Cercosporella herpotrichoides* in Winter Wheat in Germany." *Proc. 7th Brit. Insect. Fung. Conf. 1973,* (1973), pp. 119–126.

88. Anonymous. "Winter Wheat—Managed Disease Control." *Agric. Dev. Adv. Serv. Leaflet* 831 (revised). Ministry of Agriculture, Fisheries and Food, Alnwick, 1986.

89. Anonymous. "Winter Barley—Managed Disease Control." *Agric. Dev. Adv. Serv. Leaflet* 843 (revised). Ministry of Agriculture, Fisheries and Food, Alnwick, 1987.

90. Polley, R. W., and Clarkson, J. D. S. "Forecasting Cereal Disease Epidemics," in *Plant Disease Epidemiology,* eds. P. R. Scott and A. Bainbridge. (Oxford: Blackwell Scientific Publications, 1978), pp. 141–150.

91. Scott, P. R., and Hollins, T. W. "Prediction of Yield Loss Due to Eyespot in Winter Wheat." *Pl. Path.* 27 (1978): 125.

92. Marshall, J., and Ayres, R. J. "Timing of Prochloraz Sprays for Control of Cereal Stem Base Diseases." *Proc. 1986 Brit. Crop Prot. Conf., Pests and Diseases,* Thornton Heath, U.K.: British Crop Protection Council 1986, pp. 1185–1192.

16

THE CEREAL CYST NEMATODES

FRANÇOISE PERSON—DEDRYVER

Laboratoire de Zoologie INRA
BP 29 - 35650 - Le Rheu - France

16-1 INTRODUCTION

Nematode damage in cereal crops was first recorded in eastern Germany by Kühn.[1] At the time this was incorrectly attributed to a beet cyst nematode (*Heterodera schachtii,* Schmidt),[2] albeit considered a different strain. It was also referred to as *H. schachtii* "oat" or "wheat strain."

Wollenweber[3] identified it as a new species (*H. avenae*) whereas others still treated it as a race of *H. schachtii* sub. sp. major.[4] The first attribution was subsequently accepted, although the nematode has also been referred to as the cereal cyst eelworm, cereal cyst nematode (CCN), cereal root eelworm, oat cyst nematode, oat root eelworm, and wheat root eelworm.

Since 1955, a cyst nematode species complex has been recognized on cereals and grasses, ten morphologically distinct forms have been described, and seven of these may cause limited cereal damage. Since cyst nematodes are sedentary parasites, with restricted host range, Stone and Hill[5] have suggested that parasite host coevolution could account for the complex.

Four different species are associated with rice cultures, but have been less studied (*H. oryzae,* Luc & Berdon;[6] *H. sacchari,* Luc & Merny;[7] *H. elachista,* Oshima;[8] *H. oryzicola,* Seshagri Rao & Jayaprakash[9]).

Only *H. avenae* is considered a major pest because of its wide distribution and pronounced effects on host plants.

16-2 SYMPTOMS

Nematode attack is detected at tillering, sometimes provoking reddening of leaf tips. Later there may be local areas of poor growth, open to weed invasion (Figure 16-1). Infected plants are dwarfed and frail, with a reduced tillering flower head. The root system is shallow with numerous short lateral roots conferring a characteristic coralliform aspect (Figure 16-2).

In north temperate regions, summary washing of roots in May or June is sufficient to reveal females as small (approx 0.5 mm) white balls, which turn brown as cereals mature. They transform to cysts containing several hundred larvae, which are viable for 5 years or more.

16-3 SPECIES AND PATHOTYPES

16-3-1 Cyst Nematode Complex Species Associated with Cereals and Grasses

Ten species have been identified according to morphology of dead females or cysts, together with host range-barleys, oats, wheats, cultivated or wild grasses.[10] Sturhan[11] developed a classification key, based on differences in second stage larvae infesting J2. Mulvey and Golden[12] suggested that *Bidera* should be considered equivalent to *Heterodera*.

Six species have been found only in Europe, three in grasses only (*H. mani,* Mathews;[13] *H. iri,* Mathews;[13] and *H. arenaria,* Cooper[14]). *Heterodera hordecalis,* Andersson,[15] has only been found in Sweden, Germany, Poland, and Bulgaria.[10] *Heterodera*

Figure 16-1 An infected wheat crop showing patches of poor growth caused by *Heterodera avenae* (Photograph: INRA Framu; Rivoal, R.)

Figure 16-2 Characteristic coralliform
aspect of wheat roots attacked by
Heterodera avenae. (Photograph: INRA,
France, Rivoal, R.)

bifenestra, Cooper,[14] is probably synonymous with *H. longicaudata*, Seidel.[15,16] It is almost certain that the *H. avenae* pathotype 3[17] and *B. filipjevi*, Madzidov,[18] are a same
species.[19] *Heterodera avenae* pathotype 3 may also be related to *H. avenae*, Gotland type,
recorded in southern Sweden.

   Heterodera graminophila, Golden and Birchfield,[20] has only been found in Louisi-
ana (US) on grasses. *Heterodera avenae*, *H. latipons*,[21] and *Punctodera punctata*,
Thorne,[22] are mainly associated with cereal crops, and their distribution extends beyond
Europe.

   Species identification from soil samples is often difficult and more particularly so
for pathotype 3.[5-11] This led Rumpenhorst[23] to examine total protein with electrophoretic
techniques, for six species in the cyst nematode complex. There was some biochemical
variation between *H. avenae*, *H. mani*, *H. hordecalis*, *H. latipons*, *H. bifenestra*, and *P.
punctata*. The analyses also revealed four populations of *H. avenae*, distinct from six
populations of *H. mani*, but two groups of *H. mani* were also biochemically distinguishable. Two-dimensional protein electrophoresis for several populations of *H. avenae* and
H. avenae Gotland type showed differences between the two strains.[24]

16-3-2 Pathotype Identification

For each cyst nematode species, pathotypes or populations can be found that may or may not be able to overcome host gene resistances.

To detect such pathotypes, *H. avenae* has been presented with a range of hosts consisting of a number of wheat, oat, and barley cultivars. Andersen and Andersen[25] proposed a unified classification for *H. avenae* pathotypes throughout the world, and distinguished three groups (Table 16-1).

Group 1. This currently includes eight pathotypes, avirulent on Drost and Ortolan cultivars with the *Ha1* resistance gene.

Ha11 has been found in seven European countries: Denmark, Sweden, Britain, Holland, Germany, France, and Eire. It has also been described as pathotype 1,[26] A,[27] or FR3.[28]

Ha21 was recorded in Holland in 1964,[27] and since then in Germany[29] and in India (Jaitpura and Govindpura in Rajasthan province).[30]

Ha31 is an Indian pathotype from Bhagru (Rajasthan province).[30]

Ha41 has been recorded in France and named FR1.[28] The population in Shrimadhopur province, India, can also be considered as the Ha41 pathotype.[30]

Ha51 or pathotype B, first found in Holland,[27] has since been reported in Germany[29] and Norway.[31]

Ha61 is known as pathotype E in Germany[29] and Holland.

Ha71 and Ha81 have recently been described from Spain[32] at Toledo and Seville respectively.

Group 2. This includes pathotypes virulent on Drost and Ortolan, but not on cultivars with the resistance gene *Ha2* (KVL 191 and Siri).

Ha12 has been detected by Andersen[26] in Denmark, who called it pathotype 2, and has also been described as pathotype C.[27] Using Andersen and Andersen's[25] criteria, the French pathotypes FR2 and FR4 can be considered equivalent and as *Ha12*. *Ha12* has also been identified in Eire.[33]

Ha22 is a new pathotype, found in Spain.[32]

Group 3. This includes pathotypes virulent to genes *Ha1* and *Ha2* but not to the *Ha3* gene of Morocco cultivar.

Ha13 is the Australian pathotype.[34] Ha23 has only been found in Britain.[17] Ha33 has been found in Denmark.[35]

Only *H. avenae*, Gotland type, considered to be different species to *H. avenae*, also posesses pathotypes. It is distinguished from *H. avenae* by its ability to multiply on the 640328-40-2-1 oat line, and avirulence to the Selma variety. Ireholm[36] has distinguished two *H. avenae* Gotland pathotypes, one of which does not multiply on Ortolan, Emir, and Siri barleys, and occur in western Sweden.

TABLE 16-1 REACTION OF CULTIVARS IN THE TEST COLLECTION TO PATHOTYPES OF *HETERODERA AVENAE, H. HORDECALIS,* AND *H. BIFENESTRA. S* AND *R* INDICATE SUSCEPTIBILITY OR RESISTANCE; (R) MODERATE RESISTANCE; - NO OBSERVATION (FROM ANDERSEN & ANDERSEN[25]).

Group of pathotypes	1						2	3				
Pathotype no. (Dutch classification)	Ha11	Ha21	Ha31	Ha41	Ha51	Ha61	Ha12	Ha13	Ha23	Ha33	Hh1	Hb1
Dutch classification	A	D			B	E	C					
French classification	Fr3			Fr1			(Fr2-Fr4)					
Authors[x]	1,2,5	3	4	5	3,6	7	1,2,5	8,9	10,11	12	13	13
Barley												
Varde/Emir	S	S	—	S	—	S	S	S	S	S	S	S
Drost/Ortolan (*Ha1*)[xx]	R	R	R	R	R	R	S	S	S	S	S	S
KVL 191/Siri (*Ha2*)	R	R	R	S	S	S	R	S	(R)	S	R	S
Morocco C.I. 3902 (*Ha3*)	R	R	R	R	R	R	R	R	(R)	R	R	S
Marocaine C.I. 8341	R	R	—	—	—	—	R	R	(R)	—	—	(R)
Bajo Aragon 1-1	R	—	R	R	R	R	S	S	S	R	S	S
Herta	S	S	—	R	—	—	S	R	S	S	S	S
Martin 403-2	R	—	—	R	R	R	S	S	(R)	S	(R)	S
Dalmatische	(R)	—	—	S	—	R	—	—	R	S	(R)	S
La Estanzuela 750-1-15	—	—	—	—	—	—	R	—	(R)	S	—	—
Harlan 43	R	—	—	—	—	—	R	—	(R)	S	—	—
Oats												
Nidar II	—	—	—	S	—	S	S	S	S	S	R	S
Sun II	S	R	R	R	R	S	S	S	S	S	R	S
640318-40-2-1 (C.I.3444)	R	R	—	R	R	R	R	S	S	S	R	S
Silva	(R)	—	—	R	—	(R)	(R)	(R)	S	S	R	S
A. sterilis I. 376	R	R	—	R	R	R	R	R	R	R	R	S
IGV. H72-646 (*A. sterilis*)	R	—	—	R	—	R	R	S	S	S	—	S
Ty. 72-42-1 (*A. sterilis*)	R	—	—	R	—	—	R	—	S	S	R	—
Wheat												
Capa	S	S	—	S	—	S	S	S	S	S	R	S
63/1-7-15-12 (Loros)	R	R	—	R	—	(R)	R	(R)	S	S	R	(R)
Iskamisch K-2-light	S	—	—	R	—	(R)	(R)	S	S	S	R	R
AUS 10894	R	—	—	R	—	R	R	(R)	S	S	R	(R)
Psathias	—	—	—	S	—	—	S	S	R	S	R	S

[x](1) Andersen, 1959,[26] 1961;[81] (2) Nielsen, unpubl.; (3) Kort et al., 1964;[27] (4) Mathur et al., 1974;[30] (5) Rivoal, 1977;[28] (6) Stoen, 1971;[31] (7) Lücke, 1976;[29] (8) Brown & Meagher, 1970;[34] (9) O'Brien, P. C. & Fisher, J. M. *Nematologica* 25, 1979: 261, (10) Cook, 1975;[17] (11) Cook, 1982;[10] (12) Jakobsen, 1981;[35] (13) Andersson, 1976.[29]

[xx]Symbols for genes for resistance or original source in ().

359

16-4 DISTRIBUTION AND ECONOMIC IMPORTANCE OF CYST NEMATODES

16-4-1 Economic importance

A total of 180 references between 1982 and 1986 in Helminthological Abstracts deal with cereal cyst nematodes. Of these, 160 concern *H. avenae*, which alone is considered to be an economically serious crop pest. The eight other species, *H. hordecalis, H. latipons, H. mani, H. bifenestra, B. filipjevi, H. iri, H. arenaria,* and *P. punctata* are only cited from one to five times over the same interval.

It remains difficult to quantify the economic impact of *H. avenae*. It has direct effects on crop production. Losses can attain or exceed 50% when nematode proliferation coincides with development of young crops. This is so for Australian[37] and Indian[38] wheat crops, for hardwheat in the Lauragais, France[39] and spring cereals in northern Europe. Plant infestation can be favored by certain climatic conditions, such as wet springs in Siberia. If summer is dry, damage to spring cereals may account for up to 70% of lost production.[40]

Heterodera avenae may also affect production costs because, when present, growers tend to use nematicides or increase fertilizer applications to offset damage. In Australia 250,000 ha were treated against *H. avenae* in 1984.[41]

In Australia, where *H. avenae* damage to wheat is the highest recorded, costs of treatment have been estimated at 72 million $A per annum for 2 million ha infected in Victoria and southern Australia.[41]

16-4-2 Distribution

Heterodera avenae has been reported from 32 temperate countries (Figure 16-3). First reports throughout the world are:

Europe. Belgium,[42] Bulgaria,[43] Czechoslovakia,[44] Denmark,[45] East Germany,[1] Eire,[46] France,[47] Great Britain,[48] Greece,[49] Holland,[50] Ireland,[51] Italy,[52] Norway,[53] Poland,[54] Portugal,[55] Spain,[56] Sweden,[57] Switzerland,[58] Yugoslavia.[59]

Asia. In India in the Rajasthan,[60] then in the Haryana the district of Mohindergarh, Gurgaon, Rohtak, and Bhiwani,[62] the Punjab and the Himachal Pradesh,[61] the Jammu and Kashmir,[63] and finally in the Uttar Pradesh.[64]
In Japan.[65]

Africa. Algeria,[66] Morocco,[67] Tunisia.[68]

Pacific. Australia,[69] New Zealand.[70]

America. Canada,[71] Oregon (United States),[72] Peru.[73]

USSR. In the Bakhir, where damage to maize was first reported,[74] in the Ukraine,[75] in the plain of the Volga and the pre-Ural region,[76] and lastly in Estonia.[77]

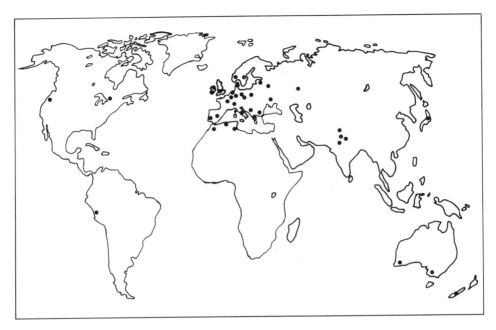

Figure 16-3   Distribution map of Cereal Cyst Nematode: *Heterodera avenae*

Heterodera avenae is not found in Asia Minor, where wheat and barley originated. Meagher[78] therefore concluded that the original plant hosts would have been the oats and rye cultivated in Europe at the time of its discovery in eastern Germany.[1] The nematode has then probably been introduced to Australia, New Zealand, and America.[78]

Wind and water favor nematode dispersion, and human activity even more so (transport of tools and plants). Soil type may be the most important factor limiting dispersion. Heavy soils with poor physical structure restrict larval mobility and nematode multiplication.[78]

16-5 BIOLOGICAL CYCLE AND ASSESSMENT OF DAMAGE

16-5-1 Biological Cycle

Heterodera avenae only reproduces once a year, and larvae are usually retained inside chitinous cysts inside dead females until the second stage of development (J2). The active phase of the cycle coincides with that of cereal plant hosts. The sinous juvenile larvae emerge from cysts as these break down, and invade plants by penetrating just behind root tips and then migrating toward the central cylinder (Figure 16-4). Plants react by producing large multinucleate cells (syncitia) which nourish nematodes and allow them to complete development.[79] Larvae (J2) undergo three molts in transformation either to long (approx. 2 mm) wiry males, or hypertrophied (approx 0.30 to 0.25 mm diam) females. The white females can then be detected on roots after washing. Males leave roots in search of females, and cross-fertilization is essential for reproduction.[80] Females produce from 100 to 600 eggs internally,[81] and larvae pass from first to second stage while in eggs.

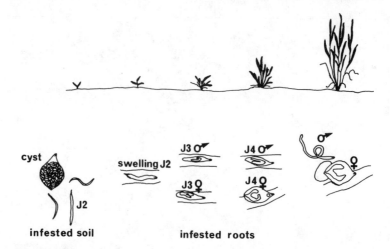

Figure 16-4   Biological cycle of Cereal Cyst Nematode: *Heterodera avenae.*
Juvenile larvae: J2, J3, and J4.
Adult nematode: ♀ female;
               ♂ male.

Where diapause is facultative, eggs may hatch at once, as in certain French populations with the pathotype Ha12 (or FR4),[82] in Czechoslovakia,[83] and India.[84] More often, the nematode goes through diapause, which may be obligatory as for the French pathotype Ha41 or FR1.[82] Cessation of diapause is essentially dependent on temperature and not on host conditions, although high humidity is also favorable.[85]

Two periods of activity have been found according to geographic location of *H. avenae* populations studied. Mediterranean populations are active during winter (in cold), in France for the Ha41 (FR1) pathotype,[86] in Spain for all *H. avenae* populations tested,[87] and in Australia.[88-89] Activity occurs in spring with rise in temperature in Canada,[90] and in northern European countries, Britain,[91-92] Denmark,[81] Czechoslovakia,[93] and in France for the Ha12 (FR4) pathotype.[86] Spanish populations from Toledo and Teruel are an exception, showing two activity periods, one in winter and the other in spring, winter cold probably provoking a second diapause.[87] In general 60% of cysts hatch in the first year.[81,82,94] Hatching may extend over three to four cycles.[95]

16-5-2 Assessment of Damage

Heterodera avenae attack on wheat and barley results in significant long-term reduction in length of primary roots.[96-98] Nematode penetration also provokes formation of abundant lateral roots.[97] In general terms, nematode presence restricts uptake of water and mineral salts, and cell division and root apical development are also affected.[99] This leads to a drop in cereal production, with highly significant ($r = -0.86$; $p < 0.001$) negative correlations between root damage and yield. Furthermore, root damage is positively correlated with number of white females on roots at flowering.[100]

Two methods for assessing *H. avenae* damage are commonly used.

1. Determination of *H. avenae* soil populations before planting, and comparison between parasite densities and fall in crop production.

2. Comparision of production from infected plots with that from plots cleared of nematodes.

Negative correlations have been found between soil nematode populations before hard wheat Bidi 17 planting, and production, at Castelnaudary in the south France.[101] Extent of damage attributable to nematodes here is, however, variable, and depends notably on quality of soil preparation prior to sowing, autumn rainfall, intensity of winter freezing, and spring climatic conditions.[102]

Treatment with dibromochloropropane at 45L/ha has resulted in 83% increase in wheat production in India in an area with initial *H. avenae* infestation levels of 12.8 eggs and J2 larvae per g soil.[103] In a second experiment a greater improvement in production (95%) was obtained for a lower initial infestation level. Assessment of nematicide use may then be affected by a variety of factors, soil type and condition, humidity, methods of nematicide application, duration of application, time of sowing.[103]

16-6 EPIDEMIOLOGY

16-6-1 Role of Plants

Host plants.   Female *H. avenae* can develop in graminaceous plant hosts. Cereal crops are, however, more favorable in this respect than forage grasses, and reproductive performance depends on both nematode pathotype and plant species or cultivar. Several authors working on northern pathotypes concur in ranking host quality in decreasing order as: winter–oats, spring–oats, spring–barley, spring–wheat, winter–barley, winter–wheat, rye.[104–107] The order appears to be reversed for mediterranean climate pathotypes, such as those found in Australia or southern France. Here most cysts occur in wheat, followed by barley, then oats, considered to be a poor host.[33,106]

Forage grasses also carry high densities of *H. avenae*, and *Festuca pratensis, Huds, Lolium perenne,* L, and *L. multiflorum,* Lam, are among the best hosts.[81]

Nonhost plants.   Nonhost plants other than graminaceae are not penetrated by *H. avenae,* and their use reduces soil nematode population levels. Extent of decrease varies, and depends on annual J2 larval emergence from cysts which itself depends on pathotype.[104,106]

Resistant plants.   Plants resistant to *H. avenae* are penetrated by second stage larvae, and only resist development of females,[108] thus inhibiting proliferation. This too can result in reduction of nematode numbers, favorable to subsequent growth and production of cereal crops in a rotation.[109] Annual nematode decline of between 61% and 85% has been obtained in northern regions in this way.[104,109,111–113] It is lower (of the order of

48% to 67%) for pathotypes in the south of France which have low annual hatching levels.[113]

There are numerous known efficient sources of resistance to nematodes in wheat and related species.[101-114] Only a few of these have been exploited in selection for resistant genitors. Apart from genes *Ha1* and *Ha2* in barley, the oat variety cv. Nelson, *Avena sterilis* 1–376 and the wheat lines Loros and AUS 10894 provide the basis for current selection for *H. avenae* resistant cereals.[114] These sources of resistance are not always efficient against all nematode pathotypes, and it is important to identify nematode pathotypes before choosing the resistant cereal cultivar.

16-6-2 Role of Fungal Parasitism

In northern Europe *H. avenae* populations decline in intensive cereal cultures that are supposed to favor the nematode.[115-118]

Examination of females in these conditions shows that they are often parasitized by fungi.[107,118-120]

The fungi *Nematophtora gynophila* and *Verticillium chlamydosporium* parasitize females on roots, and can sometimes either destroy them completely, or stimulate the formation of small cysts which are frequently empty. Parasitized females often have lowered fertility, and may be fragile, only extracted from soil with difficulty. Because of this, level of parasitism can be underestimated.[107-121] In field conditions it has been shown that 60% of females with lowered fertility may be attacked by these fungi.[121] *Nematophtora gynophila* is the most widespread of the two in Britain, and its activity is reduced in dry summers.[120] *Verticillium chlamydosporium* is widespread in southern Australia and Victoria, but only 1% of females examined were infected. *Verticillium chlamydosporium* spore density approaches 50 clamydospores/g of soil under Australian conditions, and seems insufficient to limit *H. avenae* populations.[122]

Verticillium chlamydosporium cultures on ground oat seed added to soil have caused between 26% and 80% reductions in *H. avenae* populations.[123]

16-6-3 Effects of Abiotic Factors

Soil type.   This clearly affects *H. avenae* distribution and intensity of damage caused to crops. In Victoria and South Australia, *H. avenae* has only been found on sandy and clay soils. It is not found on neighboring areas with heavy and poorly structured soils.[41] Highest *H. avenae* populations are found where soil spores range from 30 to 100 µm.[124]

Rainfall.   Under British weather conditions a positive correlation has been found between rainfall (6 to 64 mm/month) and numbers of invading *H. avenae* larvae in root systems of oat and barley cultivars, number of immature females formed in July, and levels of *V. chlamydosporium* in nematode eggs.[125]

Fertilizer applications.   Superphosphate addition may favor increase in numbers of *H. avenae* females formed on wheat plant root systems.[126] High phosphate concentrations appear to stimulate first and second order lateral roots production and extension in wheat, which would then increase favorable nematode invasion sites.[126]

Nitrogen fertilizer application may also be responsible for increased cyst number in cereals. However, results vary with studies. Maximal effects have been obtained with 100 kg N/ha for wheat in Australia,[127] and 46.5 kg N/ha for oats in Denmark.[128]

16-7 CONTROLS

16-7-1 Crop Rotations

Many nematologists believe that the most effective counter-nematode measure is a system of crop rotation including nonhost plants, or fallow, and the effectiveness of the procedure is confirmed by numerous examples.

When nonhost plants are grown two years out of three, *H. avenae* densities do not attain critical levels.[41,104,129] Schönrock-Fischer and Sachse[105] report no damage with 60% to 80% cereals in rotations.

Triennial rotations, such as are practised with beet (a nonhost), wheat, and oats, are most favorable to nematode development in northern regions and should be avoided. Damage under this cropping pattern can be severe.[113,130]

Only economic considerations set limits to control based on crop rotations, since wheat is often more profitable and therefore introduced more often.

16-7-2 Resistant Plants

Use of resistant plants restricts nematode development, while not disturbing agricultural practices, and provides a simple nonpolluting anti-parasitic action.

Use of resistant cultivars in French cropping systems has been more effective than that of nonhost plants.[106] Increase in production resulting from this practice may attain 20 q/ha for a winter wheat resistant to *H. avenae*, derived from Loros.[106] Use of resistant cultivars has been practiced notably in northern European countries. In 1981, in Denmark, about 25% of registered barley seed was *H. avenae* resistant. In 1980 nearly 65% of barleys were resistant in the Swedish province of Haland. This no doubt accounts for the decline in *H. avenae* in these regions over recent years.[104] Resistant oats have been grown in Australia since 1930.[41] Australians were the first to specifically breed a wheat for *H. avenae* resistance (cv. Katyil) from AUS 10894 genitor. Large-scale use in France of such cultivars requires breeding programs adapted to French pedo-climatic conditions.[41]

Although resistant barley cultivars tolerate presence of *H. avenae*,[101,106,112,131] resistant oat production may fall appreciably in presence of the nematode.[104,106] Breeders have therefore attempted to introduce tolerance factors to the cultivars. Experimental programs are currently under way in Britain, France, and Australia.[132]

Repeated use of a same resistant cultivar strain, with a same resistance gene on the same ground, has never been overcome by nematode adaptations.[104,106] Genetically dependent resistance to *H. avenae* (whether mono- or oligo-genic) can thus be considered durable, and chances of appearance, establishment, and maintenance of new, more virulent pathotypes are low.[133]

16-7-3 Timing of Sowing

In the Australian districts of Mallee and Wimmera (Victoria), a peak emergence of J2 *H. avenae* larvae has been found in July. Here early sowing at the beginning of May is advisable. This allows better plant establishment before larval emergence, and also confers some kind of tolerance. Production losses should be minimized, as compared with those for later sowing in June.[41]

16-7-4 Biological Control

Despite the important effects of *N. gynophila* or *V. chlamydosporium* parasitism on nematode populations, biological control has never been attempted on a large scale. Stirling and Kerry[122] did, however, suggest increasing fungal parasitism by altering rotation systems, or increasing natural fungus levels with artificial innocula.

16-7-5 Chemical Control

Chemical control measures have been enthusiastically applied in Australia.[41] The period of nematode emergence coincides with plant establishment and protection is thus essential. The use of nematicides under these conditions frequently results in a 200% yield increase.[41] Ethylene dibromide (EDB), at 3.7 L/ha, was the first nematicide registered for application to cereals in Australia. Oxamyl and carbofuran treatment of seed, and aldicarb, terbufos, and carbofuran applied at sowing as granules followed. In 1984, at least 250,000 ha were treated in Australia.[41] In northern Europe, nematicide application is often unprofitable in the usual cropping systems.[104] Use of resistant cultivars is in fact sufficient to resolve the *H. avenae* problem. Nevertheless, in the south of France, where the nematode shows precocious activity, aldicarb applications at 1 kg active ingredient/ha currently permitted for cereal crops by legislation, offers real protection in early stages of crop growth.[39]

   It must be stressed, however, that wherever a systemic nematicide is economically viable, it still does not provide satisfactory control of parasite populations.[104] At the application doses, the products do not cause direct destruction, but only a temporary and reversible inhibition of hatching and root penetration by nematodes.[134]

16-8 CONCLUSIONS

Heterodera avenae is alone among cyst nematodes in causing serious damage to cereal crops. Efficient methods of controlling the parasite exist. Some require more detailed study if they are to be fully exploited. This is so for biological control, which has not yet been applied on a large scale. Use of nonhost plants in rotations, though effective, is often incompatible with economic requirements. Chemical control, when economically profitable, does not otherwise ensure more general or long control over parasite populations. The ideal solution remains use of *H. avenae*-resistant cultivars, which possess durable resistance in field.[106] This persistence of resistance seems to be attributable both to low frequency of any establishment and maintenance of more virulent nematode pathotypes, which might otherwise overcome the resistance.

Thorough exploitation of resistance genes in plants used in rotations also requires genetic study of the host resistance mechanisms, and nematode virulence. It has been found that for a same genitor, genes responsible for resistance to *H. avenae* can differ from one pathotype to another.[135] Alternation of resistant cultivars with different resistance genes in rotations should counter any eventual possibility of the parasite overcoming host resistance through selection. An effort should, however, be made to discover new sources of resistance, notably in wheat and related wild species. Selection for resistant cultivars should also include selection for tolerance, to avoid damage to resistant varieties due to nematodes.

Studies of *H. avenae* biology itself may also be worthwhile, since they should clarify the way in which nematodes adapt to regional climatic conditions. Distinct characteristics have in fact been reported for regional nematode populations.

16-9 REFERENCES

1. Kühn, J. "Über das Vorkommen von Rübennematoden an den Wurzeln der Halmfrüchte." *Z. viss. Landw. Arch. Kgl. Preuss. Landes—ÖKon. Kolleg* 3 (1874): 47.

2. Schmidt, A. "Über die Rübennematoden (*Heterodera schachtii*)." *Z. Ver. Rübenzucherindustrie* 21 (1871): 1.

3. Wollenweber, H. W. "Zur Kenntnis der Kartoffel-Heteroderen." *Illustr. Landwirtsch. Ztg.* 12 (1924): 100.

4. Schmidt, O. "Sind Ruben und Hafernematoden identisch?" *Archiv. Pflbau* 3 (1930): 420.

5. Stone, A. R., and Hill, A. J. "Some Problems Posed by *Heterodera avenae Complex.*" *EPPO Bull.* 12 (1982): 317.

6. Luc, M., and Berdon Brizuela, R. "*Heterodera oryzae* n.sp. (Nematoda : Heteroderidae) Parasite du Riz en Cote d'Ivoire." *Nematologica* 6 (1961): 272.

7. Luc, M., and Merny, G. "*Heterodera sacchari* n.sp. (Nematoda: Tylenchoidea) Parasite de la Canne à Sucre au Congo-Brazaville." *Nematologica* 9 (1963): 31.

8. Oshima, Y. "*Heterodera elachista* n. sp. an Upland Rice Cyst Nematode from Japan." *Jap. J. Nematol.* 4 (1974): 51.

9. Seshagiri Rao, Y., and Jayaprakash, A. "*Heterodera oryzi cola* n.sp. (Nematoda: Heteroderidae) a Cyst Nematode on Rice (*Oryza sativa* L.) from Kerala Stage India." *Nematologica* 24 (1978): 341.

10. Cook, R. "Cereal and Grass Hosts of Some Gramineous Cyst Nematodes." *EPPO Bull.* 12 (1982): 399.

11. Sturhan, D. "Species Identification of European Cereal and Grass Cyst Nematodes by Larval Characters." *EPPO Bull.* 12 (1982): 335.

12. Mulvey, R. H., and Golden, A. M. "An Illustrated Key to the Cyst-Forming Genera and Species of Heteroderidae in the Western Hemisphere with Species Morphometrics and Distribution." *J. Nematol.* 15 (1983): 1.

13. Mathews, H. J. P. "Two New Species of Cyst Nematode *Heterodera mani* n.sp. and *H. iri* n.sp. from Northern Ireland." *Nematologica* 17 (1971): 553.

14. Cooper, B. A. "A Preliminary Key to British Species of *Heterodera* for Use in Soil Examination," in *Proc. University of Nottingham Second Easter School in Agricultural Sciences, Soil Zoology,* ed. D. K. McE.Kevan. London: Butterworths, 1955, 269.

15. Andersson, S. "*Heterodera hordecalis* n.sp. (Nematoda: Heteroderidae) a Cyst Nematode of Cereals and Grasses in Southern Sweden." *Nematologica* 20 (1974): 445.

16. Seidel, M. "*Heterodera longicaudata* n.sp. Ein and Gramineen Vorkommendes Zystenälchen von Grünlandflächen in Norden der DDR." *Nematologica* 18 (1972): 31.

17. Cook, R. "Observations on the Relationship Between Morphological Variation and Host Range in Populations of Cereal Cyst Nematode." *Ann. Appl. Biol.* 81 (1975): 199.

18. Madzidov, A. R. "(*Bidera Filipjevi* n.sp. (*Heteroderina*: Tylenchidae) in Tadzhikistan)." *Izv. Akad. Nauk. Tadzh. SSR. Biol. Nauki* 2 (1981): 40.

19. Andersson, S. Personal Communication, 1984.

20. Golden, A. M., and Birchfield, W. "*Heterodera graminophi la* n.sp. (Nematoda: Heteroderidae) from Grass with a Key to Closely Related Species." *J. Nematol.* 4 (1972): 147.

21. Franklin, M. T. "*Heterodera latipons* n.sp., a Cereal Cyst Nematode from the Mediterranean Region." *Nematologica* 15 (1969): 535.

22. Thorne, G. "*Heterodera punctata* n.sp. a Nematode Parasitic on Wheat Roots from Saskatchewan." *Sci. Agric.* 8 (1928): 707.

23. Rumpenhorst, H. J. "Vergleichende Elektrophoretische Untersuchungen von Proteinen einiger Zystennematoden von Getreide und Gräsern." *Land und Fortwirtschaft,* Berlin Dahlem 226 (1985): 67.

24. Ferris, V. R., Ferris, J. M., and Faghihi, J. "Two Dimensional Protein Patterns of Strains and Pathotypes in *Heterodera avenae*," in *Proc. 26th meeting Soc. Nematologists, J. Nematol.*, 19 (1987): 522.

25. Andersen, S., and Andersen, K. "Suggestions for Determination and Terminology of Pathotypes and Genes for Resistance in Cyst-Forming Nematodes, especially *Heterodera avenae.*" *EPPPO Bull.* 12 (1982): 379.

26. Andersen, S. "Resistance of Barley to Various Populations of the Cereal Root Eelworm (*Heterodera avenae*)." *Nematologica* 4 (1959): 91.

27. Kort, J., Dantuma, G., and Vanessen A. "On Biotypes of Cereal-Root Eelworm (*Heterodera avenae*) and Resistance in Oats and Barley." *Neth. J. Pl. Path.* 70 (1964): 9.

28. Rivoal, R. "Identification des Races Biologiques du Nématode à Kyste des Céréales, *Heterodera avenae* Woll. en France." *Ann. Zool. Ecol. Anim.* 9 (1977): 261.

29. Lücke, E. "Pathotype Studies on *Heterodera avenae* Populations (1966–1975)." *Z. Pflkrankh. Pflschutz.* 83 (1976): 647.

30. Mathur, R. H., Arya, H. C., Mathur, R. L., and Handa, D. K. "The Occurrence of Biotypes of the Cereal Cyst Nematode (*Heterodera avenae*) in the Light Soils of Rajasthan and Haryana, India." *Nematologica* 20 (1974): 19.

31. Stoen, M. "*Heterodera avenae*, Race and Resistance Studies." *Nord. Jordbrugst.* 53 (1971): 308.

32. Sanchez, A., and Zancada, M. C. "Characterization of *Heterodera avenae* Pathotypes from Spain." *Nematologica* 33 (1987): 55.

33. Brown, R. H. "Report on Overseas Study Tour Victoria Dept of Agric." Victorian Plant Res. Inst. Burneley, Victoria, 30, 1974.

34. Brown, R. H., and Meagher, J. W. "Resistance in Cereals to the Cyst Nematode (*Heterodera avenae*) in Victoria." *Aust. J. Expl. Agric. Anim. Husb.* 10 (1970): 360.

35. Jakobsen, J. "(*Heterodera avenae*—Occurrence, Pathotypes and Importance in Denmark)." *Nordisk Väsctskyd d skonferens*, 76, 1981.

36. Ireholm, A. Personal communication, 1984.

37. Meagher, J. W. "World Dissemination of the Cereal-Cyst-Nematode (*Heterodera avenae*) and Its Potential as a Pathogen of Wheat." *J. Nematol.* 9 (1977): 9.

38. Swarup, G., and Singh, K. "Molya Disease of Wheat and Barley in Rajasthan." *Indian Phytopathol.* 14 (1961): 127.

39. Esmenjaud, D., Marzin, H., and Rivoal, R. "Fortes Attaques du Nématode *Heterodera avenae* sur Blé dur dans le Lauragais." *Phytoma* 390 (1987): 25.

40. Shiabovat, T. N. "Pathogenicity Factors of *Heterodera avenae.*" *Nauchno-tekhnicheshii Byulleten 'VASKHNIL,* Sibirskoe Otdelenie, 22 (1982): 32.

41. Brown, R. H. "Ecology and Control of Cereal Cyst Nematode (*Heterodera avenae*) in Southern Australia." *J. Nematol.* 16 (1984): 216.

42. Brande, J. Van Den, and Gillard, A. "Importance et Répartition en Belgique des Nématodes de la Sous-Famille des Heteroderinae." *Z. Pflansenkr. Pflanzensch.* 64 (1957): 493.

43. Stoyanov, D. "Resultati ot Izsledvanija za Ustanovjavane na Zistoobrazuvǎusǔte Nematodi v Bulgarija. 1. Zveklova nematoda (*Heterodera schachtii* Scmidt, 1871), *Rasteni-evudni nauki.*" *Sofija* 4 10 (1967): 111.

44. Klumpar, J. "Škodlivost a Hubeni hád'Átek v Polnich Plodinách." *Za Vysokori Urodu* 15 (1967): 26.

45. Hansen, K. "Nematoder i Havre." *Vyesker. Landm.* (1987): 17.

46. Duggasn, J. J. "The Occurrence and Control of Plant Parasitic Nematodes in Ireland." *Irish Crop Protection Conf.* Dublin, 1962, 180.

47. Ritter, M. "Biologie, Variation de Populations et Importance Pratique dans le Midi de la France d'*Heterodera avenae* Filipjev (1934) Nématode Parasite des Céréales." *C.R. 86e Congr. Soc. Savantes,* Montpellier, 657, 1961.

48. Theobald, F. I. "An Attack of *Heterodera* on Oats." *J.S.-E. Agric. Coll. Mye* 17 (1908): 150.

49. Hirschmann, H., Paschkali-Kour-Tzi, N., and Triantaphyllou, A. C. "A Survey of Plant-Parasitic Nematodes in Greece." *Annu. Inst. Phytopath.,* Benaki, N.S. 7 (1966): 144.

50. Ritzema Bos, J. *Meded. Nederl. Phytopath. Ver. Het Nederlandsche Tuinbouwblat* 7 (1891): 216.

51. Anonymous. Res. Exp. Rec. Minist. Agric. Northern Ireland 12 (1964): 152.

52. Mezetti, A. "Osservazioni sull'Anguillulosi Radicule dei Cereali in Italia." *Ann. della Sperimentazione Agraria* N.S. 7 (1953): 743.

53. Schoyen J. H. "Beretning on Skadeinsektenes Optreden i Lang-og Habruket i Arene 1924 og 1925." Oslo, Grondahl and Sons, (1926), 31.

54. Wilski, A. "Przycynek do Wystepowania Matwika Zbozowego (*Heterodera avenae* Wollenweber, 1924: H. major), *na terenie województwa Bydgoskiego. Biul. Ior.''* 10 (1960): 185.

55. Macara, A. M. "Algumas Consideracões Sobre Nemátodes, Nomeadamente os de Interesse Agricola: Sua Disseminacão e Importância." *Agros* 46 (1963): 99.

56. Tobar, J. "Especies del Genero *Heterodera* A. Schmidt 1871 (Heteroderidae: Nematoda) de la Provinica de Granada, con Description de un Suevo Procedimiento para el Recuento de los Guistes y sus Larvas Infectivas, en Casos Desinfecciones Multiples, *En analyses rutinarios del suelo.*" *Revista Iber-Parasit.* 23 (1963): 325.

57. Kellgren, A. "Nematoder pa Havre." *Landmannen* 9 (1897): 210.

58. Savary, A. "Le Problème des Nématodes dans les Cultures de Céréales." *Revue Romande Agric. Vitic. Arboric.* 13 (1957): 93.

59. Grujicic, G. "Phytoparasitic Nematodes on the Territory of the S.R. of Serbia with Special Reference to Bioecologic Investigations of *Heterodera schachtii* and *Anguina tritici* (STEIN-BUCH), Filipjev." *Bull. Sci. Sect.* A.T. 11, 11/12 (1966): 272.

60. Prasad, N., Mathur, R. L., and Sehgal, S. P. "Molya Disease of Wheat and Barley in Rajasthan." *Curr. Sci.* 28 (1959): 453.

61. Koshy, P. K., and Swarup, G. "Distribution of *Heterodera avenae, H. zeae, H. cajani,* and *Anguina tritici.*" *Indian J. Nematol.* 1 (1971): 106.

62. Bhatti, D. S., Dahiya, R. S., Gupta, D. C., and Malhan, I. "Plant-Parasitic Nematodes Associated with Various Crops in Haryana." *Haryana Agricultural University J. Research* 8, 1980.

63. Swarup G., Sethi C. L., Kaushal, K. K., and Nand, S. "Distribution of *Heterodera avenae* the Causal Organism of "Molya" Disease of Wheat and Barley in India." *Curr. Sci.* 18 (1982): 896.

64. Siddiqui, Z. A., Siddiqui, M. R., and Hussain, S. I. "First Report on the Occurrence of *Heterodera avenae* Wollenweber 1924, in Utar Pradesh, India." *International Nematology Network Newsletter* 3 (1986): 11.

65. Ichinoe, M. "Heterodera avenae is Detected in Japan." *A. Rep. Soc. Pl. Prot. North Jap.* 5 (1954): 83.

66. Scotto La Massese, C. "Aperçu sur les Problèmes Posés par les Nématodes Phytoparasites en Algérie," in *"Les Nématodes,"* *C.R. Journ. Etudes Inf. FNGPC-ACTA,* Versailles, 16-17 Nov. 1961, 83.

67. Franklin, M. T. "The Cyst-Forming Species of *Heterodera.*" Commonwealth Agricultural Bureaux, 145, 1951.

68. Delanoue, P. "L'Anguillulose des Céréales. Moyens Susceptibles d'en Limiter les Dégâts en Tunisie." *Tunisie Agric.* 23 février, 1953.

69. Davidson, J. "Eelworms (*Heterodera schachtii* Schm.) Affecting Cereals in South Australia." *J. Dep. Agric. South Aust.* 34 (1930): 378.

70. Grandison, G. S., and Halliwell, H. G. "New Pest of Cereals." *N.Z.J. Agric.* 130 (1975): 64.

71. Putnam, D. F., and Chapman, L. J. "Oat Seedling Diseases in Ontario 1. The oat Nematode *Heterodera schachtii* Schm." *Sci. Agric.* 15 (1935): 633.

72. Jensen, H. J., Estiaghi, H., Koepsell, P. A., and Goetze, N. "The Oat Cyst Nematode, *Heterodera avenae* Occurs in Oats in Oregon." *Plant Dis. Reptr.* 59 (1975): 1.

73. Krusberg, L. R., and Hirschmann, H. "A Summary of Plant Parasitic Nematodes in Peru." *Plant Dis. Reptr.* 42 (1958): 599.

74. Mamonova, Z. M. "*Heterodera avenae* Filipjev, 1934 in Bashkir." *A.S.S.R. Tr. Gelmintol. Lab. Akad. Nauk* SSR9 (1959): 188.

75. Ladygina, N. M. "Etude Comparative des Nématodes de la Betterave et des Céréales," in *Nematody Urednye v Sel'skom Khozyriste i Bor'ba s Nimi,* Samarkand, 152, 1962.

76. Popova, M. B. "Distribution et Caractéristiques Écologiques des Nématodes à Kystes dans les Régions de la Volga et de l'Oural," in *Nematodnye bolezni i mery bor'by s nimi,* Moscou, 1972, 27.

77. Krall, E. L., and Krall, Kh. A. "*Heterodera avenae* en Estonie." *Zashch. Rast.* 50, 1974.

78. Meagher, J. W. "World Dissemination of the Cereal-Cyst Nematode (*Heterodera avenae*) and its Potential as a Pathogen of Wheat." *J. Nematol.* 9 (1977): 9.

79. Johnson, P. W., and Fushtey, S. G. "The Biology of the Oat Cyst Nematode, *Heterodera avenae*, in Canada. II. Nematode Development and Related Anatomical Changes in Roots of Oats and Corn." *Nematologica* 12 (1966): 630.

80. Andersen, S. "Heredity of Race 1 or Race 2 in *Heterodera avenae.*" *Nematologica* 11 (1965): 121.

81. Andersen, S. "Resistens Mod Havreal," Dissertation, Kongelige Veterinaer-og Landbrugets Plantekultur, Copenhagen, 1961, 179 pp.

82. Rivoal, R. "Biologie d'*Heterodera avenae* Wollenweber en France. II. Etude des Différences dans les Conditions Thermiques de'Éclosion des races FR1 et FR4." *Revue Nematol.* 2 (1979): 233.

83. Liskova M., Sabova, M., and Valocka, B. "Development of the Oat Cyst-Nematode *Heterodera avenae* Wollenweber, 1924 under Experimental Conditions." *Helminthologia* 20 (1983): 53.

84. Swarup, R., and Swarup, G. "Dormancy in Cereal Cyst Nematode, *Heterodera avenae.*" *Indian J. Nematol.* 15 (1985): 118.

85. Dixon, C. M. "The Effect of Spring Rainfall on the Host–Parasite Relationship Between the Cereal Root Eelworm (*Heterodera avenae* Woll.) and the Oat Plant (*Avena sativa* L.)." *Nematologica* 9 (1963): 521.

86. Rivoal, R. "Biologie d'*Heterodera avenae* Wollenweber en France. I. Différences dans les Cycles d'Éclosion et de Développement des Deux Races FR1 et FR4." *Revue Nematol.* 1 (1978): 171.

87. Valdeolivas, A., and Romero, M. D. "The Biology of *Heterodera avenae* in Spain," in *Cyst Nematodes,* eds. F. Lamberti and C. E. Taylor. (New York: Plenum Press, 1986), p. 287.

88. Meagher, J. W. "Seasonal Fluctuations in Numbers of Larvae of the Cereal Cyst Nematode (*Heterodera avenae*) and of *Pratylenchus minyus* and *Tylenchorhynchus brevidens* in Soil." *Nematologica* 16 (1970): 333.

89. Banyer, R. J., and Fisher, J. M. "Seasonal Variation in Hatching of Eggs of *Heterodera avenae.*" *Nematologica* 17 (1971): 225.

90. Fushtey, S. G., and Johnson, P. W. "The Biology of the Oat Cyst Nematode *Heterodera avenae* in Canada. I. The Effect of Temperature on the Hatchability of Cyst and Emergence of Larvae." *Nematologica* 12 (1966): 313.

91. Duggan, J. J. "Seasonal Variations in the Activity of Cereal Root Eelworm (*Heterodera major,* O. Schmidt, 1930)." *Scient. Proc. R. Dubl. Soc.* Series B 1 (1961): 21.

92. Kerry, B. R., and Jenkinson, S. C. "Observations on Emergence, Survival and Root Invasion of Second Stage Larvae of the Cereal Cyst-Nematode, *Heterodera avenae.*" *Nematologica* 22 (1976): 467.

93. Sabova, M., Lisková, M., and Valocká, V., "Ontogenesis of the Cereal Cyst Nematode, *Heterodera avenae* Wollenweber, 1924 on Oat Under the Climatic Conditions of Slovakia." *Helminthologia* 22 (1985): 285.

94. Meagher, J. W., and Rooney, D. R. "The Effect of Crop Rotations in the Victorian Wimmera on the Cereal Cyst Nematode (*Heterodera avenae*), Nitrogen Fertility and Wheat Yield." *Aust. J. Exp. Agric. Anim. Husb.* 6 (1966): 425.

95. Rivoal, R. "Biologie d'*Heterodera avenae* Wollenweber en France. III. Evolution des Diapauses des Races FR1 et FR4 au Cours de Plusieurs Années Consécutives; Influence de la Température." *Revue Nematol.* 6 (1983): 157.

96. Meagher, J. W. "Yield Loss Caused by *Heterodera avenae* in Cereal Crops Grown in a Mediterranean Climate." *EPPO Bull.* 12 (1982): 325.

97. Rawthorne, D., and Hague, N. G. M. "Relationship Between Cereal Cyst Nematode Inoculation Techniques and Root Growth." *Sup. Ann. Appl. Biol.* 102 (1983): 20.

98. Price, N. S., Clarkson, D. T., and Hague, N. M. "Effect of Invasion by Cereal Cyst Nematode (*Heterodera avenae*) on the Growth and Development of the Seminal Roots of Oats and Barley." *Plant Pathol.* 32 (1983): 377.

99. O'Brien, P. C., and Fisher, J. M. "Ontogeny of Spring Wheat and Barley Infected with Cereal Cyst Nematode (*Heterodera avenae*)." *Aust. J. Agric. Res.* 32 (1981): 553.

100. Simon, A., and Rovira, A. D. "The Relation Between Wheat Yield and Early Damage of Roots by Cereal Cyst Nematode." *Aust. J. Exp. Agric. Anim. Husb.* 22 (1982): 201.

101. Rivoal, R., Besse, T., Morlet, G., and Penard, P., "Nuisibilité du Nématode à Kyste *Heterodera avenae* et Perspectives de Lutte," in *Les Rotations céréalières intensives,* dix années d'études concertées INRA-ONIC-ITCF, 1973-1983, INRA, Paris, 1986, 151.

102. Ritter, M. "Incidence Économique des Nématodes sur la Production Agricole." *EPPO Bull.* 3 (1973): 37.

103. Bhatti, D. S., Dalal, M. R., and Malhan, I. "Estimation of Loss in Wheat Yield due to the Cereal-Cyst Nematode *Heterodera avenae*." *Trop. Pest Manag.* 27 (1981): 375.

104. Andersson, S. "Population Dynamics and Control of *Heterodera avenae*. A Review with Some Original Results." *EPPO Bull.* 463, 1982.

105. Schönrock-Fischer, R., and Sachse, B. "Auftreten von *Heterodera avenae* bei Monokultur von Getreidearten Sowie in Fruchtfolgen mit Unterschiedlichem Spezialisierungsgrad." *Tag.-Ber., Akad. Landwirtsch.-Wiss. DDR,* Berlin 181 (1980): 97.

106. Rivoal, R. "Recherches sur le Nématode *Heterodera avenae* Woll. en Relation avec la Résistance des Variétés de Céréales et leur Utilisation au Champ." (Thèse de Doctorat es Sciences présentée devant L'Université de Rennes I le 17 décembre 1987), 121 pp.

107. Kerry, B. R., Crump, D. H., and Mullen, L. A. "Studies of the Cereal Cyst-Nematode, *Heterodera avenae* under Continuous Cereals, 1974–1978. I. Plant Growth and Nematode Multiplication." *Ann. Appl. Biol.* 100 (1982): 477.

108. Cook, R. "Nature and Inheritance of Nematode Resistance in Cereals." *J. Nematol.* 6 (1974): 165.

109. Williams, T. D., and Beane, J. "The Effects of Nematode Resistant and Susceptible Spring Oat Cultivars and Aldicarb on the Cereal Cyst Nematode *Heterodera avenae* and Yields in Contrasting Soil Types." *Ann. Appl. Biol.* 95 (1980): 115.

110. Andersen, K, and Andersen, S. "Decrease of Cereal Cyst Nematode Infestation after Growing Resistant Barley Cultivars or Grasses." *Tidsskr. Planteavl.* 74 (1970): 559.

111. Cotten, J. "Field Experiments with Spring Barley Resistant to Cereal Cyst Nematode, 1965–1968." *Ann. Appl. Biol.* 65 (1970): 165.

112. Graham, C. W., and Stone, L. E. W. "Field Experiments on the Cereal Cyst-Nematode (*Heterodera avenae*) in South-east England." *Ann. Appl. Biol.* 80 (1975): 61.

113. Rivoal R. "Biology of *Heterodera avenae* Wollenweber in France. IV. Comparative Study of Hatching Cycles of Two Ecotypes after their transfer to Different Climatic Conditions." *Revue Nematol.* 9 (1986): 405.

114. Cook, R., and York, P. A. "Resistance of Cereals to *Heterodera avenae*: Methods of Investigation, Sources and Inheritance of Resistance." *EPPO Bull.* 12 (1982): 423.

115. Gair, R., Mathias, P. L., and Harvey, P. N. "Studies of Cereal Cyst Nematode Populations and Cereal Yield under Continuous or Intensive Culture." *Ann. Appl. Biol.* 63 (1969): 503.

116. Jakobsen, J. "The Importance of Monocultures of Various Host Plants for the Population Density of *Heterodera avenae*." *Tidsskrift for Planteavl.* 78 (1974): 697.

117. Ohnesorge, B., Freidel, J., and Oeterlin, O. "Investigations on the Distribution Pattern of *Heterodera avenae* and Its Changes in a Field under Continuous Cereal Cultivation." *Z. Pflkrankh. Pflschutz.* 81 (1974): 356.

118. Schönhammer, A., and Fischbeck, G. "Untersuchungen zur Populations Dynamik von *Heterodera avenae* (Woll.) in Getreidereichen Fruchtfolgen und Getreidemonokulturen. II. Die Antogonistische Wirkung pilzlicher Nematodenfeinde." *Bayerisches Land wirtschaftliches Jahrbuch* 62 (1985): 85.

119. Juhl, M. "Cereal Cyst Nematode, Soil Heating and Fungal Parasitism." *EPPO Bull.* 12 (1982): 505.

120. Kerry, B. R. "The Decline of *Heterodera avenae* Populations." *EPPO Bull.* 12 (1982): 491.

121. Kerry, B. R., Crump, D. H., and Mullen, L. A. "Studies of the Cereal Cyst-Nematode, *Heterodera avenae* under Continuous Cereals, 1975–1978. II. Fungal Parasitism of Nematode Females and Eggs." *Ann. Appl. Biol.* 100 (1982): 489.

122. Stirling, G. R., and Kerry, B. R. "Antagonists of the Cereal Cyst Nematode *Heterodera avenae* Woll. in Australian Soils." *Aust. J. Exp. Agric. Anim. Husb.* 23 (1983): 318.

123. Kerry, B. R., Simon, A., and Rovira, A. D. "Observations on the Introduction of *Verticillium chlamydosporium* and Other Parasitic Fungi into Soil for Control of the Cereal Cyst-Nematode *Heterodera avenae*." Ann. Appl. Biol. 105 (1984): 509.

124. Håkansson, J., and Videgård, G. "Resistance Studies on Cereal Root Eelworm. I. Test Method Considering the Physical Properties of the Soil." *Nematologica* 11 (1965): 601.

125. Graham, C. M. "The Effects of Rainfall and Soil Type on the Population Dynamics of Cereal Cyst Nematode (*Heterodera avenae*) on Spring Barley (*Hordeum vulgare*) and Spring Oats (*Avena sativa*)." *Ann. Appl. Biol.* 94 (1980): 243.

126. Simon, A., and Rovira, A. D. "The Influence of Phosphate Fertilizer on the Growth and Yield of Wheat in Soil Infested with Cereal Cyst Nematode (*Heterodera avenae* Woll.)." *Aust. J. Exp. Agric.* 25 (1985): 191.

127. Barry, E. R., Brown, R. H., and Elliott, B. R. "Cereal Cyst Nematode (*Heterodera avenae*) in Victoria: Influence of Cultural Practices on Grain Yields and Nematode Populations." *Aust. J. Exp. Agric. Anim. Husb.* 14 (1974): 566.

128. Juhl, M. "Cereal Cyst Nematode and N Fertilization." *EPPO Bull.* 5 (1975): 437.

129. Andersson, S. "Plant Protection Problems in Crop Rotations: Pests). Rapporter och avhadlingar 41, *Inst. för växtodling, Lantbrukshögskolan,* Uppsala 1976, K 1-9, 1976.

130. Vallotton, R., and Vez, A. "Essais de Lutte Contre le Nematode à Kyste des Céréales (*Heterodera avenae*) par la Rotation de Cultures, la Désinfection Chimique du Sol et l'Utilisation de Variétés Résistantes." *Revue Suisse Agric.* 13 (1981): 171.

131. Andersson, S. "The Cereal Cyst Nematode in a Series of Field Experiments in Haland." *Hallands läns hushållnings sällskaps Tidskr.* 18 (1975): 9.

132. Fisher, J. M., Rathjen, A. J, and Dube, A. J. "Tolerance of Commercial Cultivars and Breeders' Lines of Wheat to *Heterodera avenae* Woll." *Aust. J. Agric. Res.* 32 (1981): 545.

133. Person-Dedryver, F. "Etude de la Variabilité dans les Relations Hôtes-Parasites Liant les Espèces ou Variétés de Céréales à Paille ou (et) de Graminées Fourragères au Némätode à Kyste *Heterodera avenae* Woll. et au Némätode à Galle *Meloidogyne naasi* Franklin." (Thèse

de Docteur d'Etat es Science présentée devant l'Université de Paris-Sud, centre d'Orsay le 16 Juillet 1987, 176, 1987).

134. Cavelier, A. "Le Mode d'Action des Nématicides non Fumigants." *Agronomie* 7 (1987): 747.

135. Person-Dedryver, F., and Doussinault, G. "Interactions Génétiques entre Pathotypes Français d'*Heterodera avenae* Woll. et Variétés d'Orge: I. Aspect Variétal." *Agronomie* 4 (1984): 763.

17

PHYTOPHTHORA BLIGHT OF PIGEON PEA

U. P. SINGH AND V. B. CHAUHAN

Department of Mycology and Plant Pathology, Institute of Agricultural Sciences
Banaras Hindu University, Varanasi 221 005, India

17-1 INTRODUCTION

Phytophthora blight of pigeon pea (*Cajanus cajan* (L.) Millsp.) was reported for the first time from India in 1966 by Williams et al.[1] Pal et al.[2] described this disease as a stem rot of pigeon pea. They observed this disease in serious form during 1968–69 on pigeon pea cultivar Type-21. The incitant of this disease was identified as *Phytophthora drechsleri* var. *cajani* Pal, Grewal and Sarbhoy. Five years later, Williams et al.[3] described a "Phytophthora stem blight" on pigeon pea from the same areas. However, they could not identify the species at that time but the same group of workers identified the pathogen as *P. cajani* Amin, Baldev and Williams (Amin et al.).[4] Kaiser and Melendez[5] reported a "stem canker" of pigeon pea caused by *P. parasitica* Dast. from Puerto Rico. The morphology and genetics of the pathogen causing Phytophthora blight of pigeon pea was studied by Kannaiyan et al.[6] at ICRISAT, India and also at the University of California, United States. Based on their results it was concluded that the pathogen should be designated as *P. drechsleri* f. sp. *cajani* (Pal et al.) Kannaiyan et al. because of its host specificity.

17-2 DISTRIBUTION AND ECONOMIC IMPORTANCE

The disease was first reported from IARI, New Delhi. Later, it was found to occur at the Agricultural Farm of Kanpur. Nene et al.[7] reported its occurrence from ICRISAT, Hyderabad, India. The authors recently observed it in mild to severe form at the experimental plot of Institute of Agricultural Sciences, Banaras Hindu University, India. Now the disease is known to occur at most of the pigeon pea growing areas of India.

Phytophthora blight of pigeon pea is a sporadic disease but occasionally assumes an epidemic proportion at places of heavy and frequent rains. The disease is becoming more severe every year, ranking high among the destructive diseases of pigeon pea. Since the affected plants dry up rapidly, the yield loss due to Phytophthora blight has been estimated to be up to 98%.[2-3] High incidence is usually associated with poor surface drainage. However, in 1969, the disease was epidemic even in an apparently well drained field at New Delhi.[3] Reports from most of the pigeon pea growing areas reveal a progressively increasing trend toward excessive severity of Phytophthora blight which causes heavy damage to the crop. About 10–80% of the plants were killed in the field during 1981–82 and 1982–83 trials at the experimental farm of Banaras Hindu University. A similar situation occurred in a farmer's fields.[8]

17-3 SYMPTOMS

P. drechsleri f. sp. *cajani* produces different types of symptoms on infected pigeon pea plants depending on the age and the plant parts infected.

17-3-1 Seeding Infection

The appearance of symptoms commences with the onset of heavy rains in the months of July and August. Initially purple to dark brown necrotic lesions girdle the basal portion of the stem (Fig. 17-1) and later may occur on aerial parts of the seedlings. At first the lesions are small and smooth, later enlarging and becoming slightly depressed. The infected tissues become soft and the whole plant wilts.

17-3-2 Mature Plant Infection

The Infection is mostly confined to the basal portions of the stem. During favorable environmental conditions the pathogen can also infect upper parts of the stem and the branches. The infected bark first becomes brown, the tissues softening causing the plant to collapse. The infected branches may also break off in wind or simply due to the weight of the foliage. The infected plants do not break if the upper portion is not very luxurient. The upper portions of the infected twigs eventually wilt and dry (Figure 17-2).

If the plants are moderately resistant and infection occurs at the basal portion, the infected area becomes hypertrophied, forming a gall (Fig. 17-3). Although the galls are usually formed on the basal portion, they may also occur on any part of the stem. The infected bark cracks open in most of the cases. More galls than one are seen in some cases. The stem breaks at the infected site (Fig. 17-3). In some cases even after the gall formation the upper part of the plant remains healthy, bearing flowers and fruits similar to healthy plants. The gall formation is attributed mostly to host resistance. Roots are not infected.

Figure 17-1 Dark brown necrotic lesions produced by *Phytophthora drechsleri* f. sp. *cajani* girdling the basal part of the stem of pigeon pea

17-3-3 Leaflet Infection

The infection (leaf spots) on leaflets is observed mostly on the plants that are below 125 cm high growing under water-logged conditions. Localized yellowing starts at the tip and also from the margin of the lamina and gradually extends toward the midrib. The center of the spots later turns brown and tissues become hard. Usually the infected portion curls upward toward the pedice of the leaflet (Fig. 17-4). One or two circular to angular spots are also formed on the lamina of the leaflet. The margin of the infected spots is conspicuously dark brown. The spots increase in size during severe infection and cover a major portion of the lamina. Eventually, the infected tissues become necrotic and hard. There is no leaf fall even after severe infection.

17-4 CAUSAL ORGANISM

Studies on the genus *Phytophthora* (Gr. *Phyton* = a plant, *thora* = destruction) commenced with the work of de Bary[9–10] in Strasbourg, Germany on *P. infestans* followed by the isolated studies of Thaxter[11] in America on *P. phaseoli* and of Van Breda de Haan[12] in

Figure 17-2 Infection by *Phytophthora drechsleri* f. sp. *cajani* on the mature plants of pigeon pea showing breaking and collapse

Figure 17-3 Gall formation by *Phytophthora drechsleri* f. sp. *cajani* on basal part of stem of pigeon pea

Figure 17-4 Leaflets of pigeon pea infected by *Phytophthora drechsleri* f. sp. *cajani* showing curling from the tip

Java on *P. nicotianae* (= *P. parasitica*). Several other workers have studied the fungus extensively.[9-22]

The general characters of the fungus causing blight in pigeon pea resemble that of *P. drechsleri*. Kannaiyan et al.[6] made a detailed study of several isolates of *P. drechsleri* var. *cajani* from different locations in India. Based on the sporangium shape and size, oogonium and oospore formation, temperature requirements, pathogenicity, and specificity to pigeon pea, they proposed that these isolates be designated as *P. drechsleri* f. sp. *cajani*. The use of *forma specialis* was considered appropriate because of the specificity of these isolates to pigeon pea and *Atylosia* species, which are wild relatives of pigeon pea.

P. drechsleri f. sp. *cajani* produces aerial white mycelium on potato dextrose agar (PDA) medium. Hyphae hyaline, cottony, devoid of granular contents, coenocytic, filamentous, smooth, slender, 3–6 μm in diameter, provided with abundant terminal and intercalary, irregular swellings (12–15 μm in diameter) with tubular projections (Fig. 17-5(A)). Low temperature (15–20°C) favors swellings but high temperature (38°C) suppresses their formation.

Sporangiophores are usually hyphalike except for the swellings on the tip which develop into sporangia (Fig. 17-5(B)). Sporangia are not formed on solid media; however, they are produced when the culture is transferred to sterilized distilled water. Sporangia mature within 2–3 hr. They are hyaline, terminal, ovate to pyriform, nonpapillate, and measure 41–78 × 28–45 μm (Fig. 17-5(C)). Sporangial germination is by zoospores which mature within the sporangium and emerge individually usually after dissolution of the apical portion of the sporangium (Fig. 17-5(D)). A sporangium may produce another sporangium after germination (Fig. 17-5 (E-F)). Each sporangium pro-

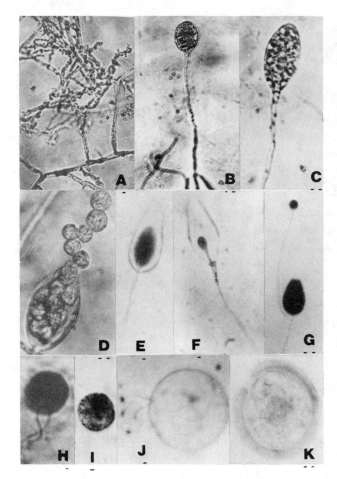

Figure 17-5 Asexual and sexual structures of *P. drechsleri* f. sp. *cajani.*

(a) Mycelium with abundant swellings X 650

(b) Sporangiophore with initiation of sporangium development X 350

(c) Mature sporangium with sporangiophore X 300

(d) Germinating zoosporongium showing the liberation of zoospores X 450

(e) Internal proliferation of sporangium at the base and empty sporangium X 450

(f) Internal proliferation with short sporangiophore X 450

(g) Sporangium formation on externally proliferated sporangiophore X 350

(h) Motile zoospore with two unequal flagella X 1350

(i) Encysted zoospore X 1100

(j) Oogonium with antheridium X 900

(k) Mature oospore X 1100

duces 8–20 zoospores. Only rarely, under adverse conditions, does a sporangium germinate by a germ tube. Extended or nested internal proliferation of sporangia is not uncommon. Rarely, external proliferations with a secondary sporangiophore may also be formed (Fig. 17-5(G)).

Zoospores are hyaline, ovoid to reniform or variable, tapering slightly to the anterior end (Fig. 17-5(H)). Zoospores are biflagellate, swimming for 2–5 hr, becoming nonmotile and forming a spherical cyst (Fig. 17-5(I)) that usually germinates, with one or more germ tubes sometimes ending in a microsporangium.

Oogonia hyaline when immature, becoming thick walled and purple yellow to brown after maturity, smooth, spherical, and measure 24–37 μm in diameter (Fig. 17-5(J)). Antheridia simple, hyaline, amphigynous, persistent, 12.5–19.0 × 10.0–17.0 μm. Oospores spherical to globose, smooth, thick walled (1.2 μm) and 20–32 μm in diameter (Fig 17-5(K)), concolorous with the oogonia.

17-5 DISEASE CYCLE

The fungus probably survives in the soil and plant debris in the form of oospores. With the arrival of favorable environmental conditions it is assumed that oospores germinate by a zoosporangium and/or mycelium, the former producing zoospores. The zoospores encyst and germinate either by one or several germ tubes which infect fresh pigeon pea plants. If the weather is favorable, a large number of sporangia are formed from the mycelial growth on the host which produce the zoospores after germination. They encyst after swimming for 2 hr and germinate by germ tubes. The germ tube develops into mycelium which gives rise to new crops of sporangia. A large number of asexual generations are produced in one growing season if conditions favor the development of the fungus (Fig 17-6).

When the asexual generation stops due to unfavorable conditions, the sexual organs, antheridium, and oogonium develop and produce oospores.

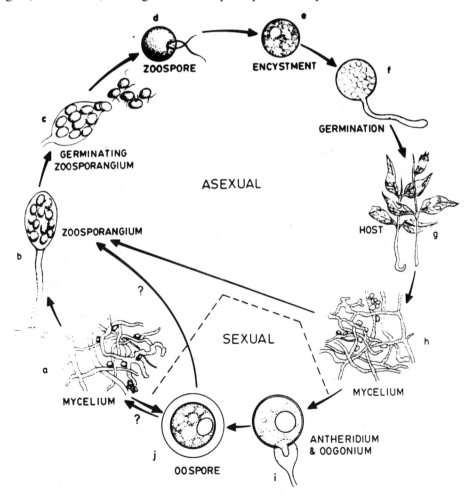

Figure 17-6 Disease cycle of *Phytophthora drechsleri* f. sp. *cajani*.

17-6 DEVELOPMENT OF *P. DRECHSLERI* f. sp. *CAJANI* ON LEAFLETS OF PIGEON PEA

Although the pathogen is suspected to be soilborne, it does not infect roots. Zoospores probably constitute the main inoculum responsible for infection of pigeon pea under suitable environmental conditions. Plant to plant infection also occurs by zoospores transported by rain splashes or air currents.

Leaflets of susceptible pigeon pea cultivar are detached from plants and floated on 5% aqueous sucrose solution in petri dishes.[23] They are inoculated with 0.1 ml zoospore suspension containing approximately 1,000 zoospores and incubated at 28–30°C. The observations on infection processes are taken at 2 hr intervals for the first 12 hr, followed by 3 hourly observations for the next 48 hr.

The inoculated leaflets are immersed in 10 ml of acid-alcohol (equal parts of glacial acetic acid and 95% ethyl alcohol) at 60°C and then gently boiled for a few minutes until the leaflets lose the chlorophyll and become translucent. This is followed by clearing in saturated chloral hydrate solution for 70 min. Leaflets are then mounted in cotton blue prepared in lactophenol for observation under a microscope.[24] The good differentiation is obtained as the fungus mycelium is stained blue and the host tissues remain clear.

Microtomed sections of the infected leaflets are also cut to observe the host–parasite relationship. Infected leaf pieces (4 × 2 mm) are fixed in formalin-acetic acid-alcohol mixture (50% ethyl alcohol 90 ml + glacial acetic acid 5 ml + formalin 5 ml) and after dehydration embedded in paraffin wax (54–56°C mp). Sections (10–12 μm thick) are cut and stained with safranin and fast green.

The zoospores placed in water drops on the surface of the leaflets swim for 2 hr, encyst and after 6 hr germinate by formation of a germ tube. The germ tubes penetrate through the stomata after 5 hr. Direct penetration through the cuticle is apparently rare. After 18 to 21 hr the germ tubes penetrate and establish intracellularly further inside the host tissues, branching profusely within 24 hr to form the mycelium. Hyphalike sporangiophores emerge through the stomata.

Profusely growing mycelium is observed after 30 hr incubation on the upper surface of the leaflets. Oospore formation by the mycelium on the leaflets commences in about 30 hr, and abundant oospores are formed in 36 hr. A typical browning symptom on the leaflets is associated with it. A large area of the leaflets becomes involved within 48 hr.

17-7 OOSPORE FORMATION IN *P. DRECHSLERI* f. sp. *CAJANI* IN VITRO

Oospores are rarely formed in vitro except in very old cultures.[6] A technique has been evolved that enables abundant oospore formation in this fungus under laboratory conditions.

The zoospores of the pathogen are obtained using tomato juice broth described earlier for preparation of spray inoculum. The inoculum contains 8,000 zoospores ml[-1]. Inoculation is done using detached host leaflets floated on 5% solution in 10 cm diameter petri dishes. The petri dishes are incubated in total darkness at 15, 20, 25, 30, and 35°C.

The oospore formation commences after 30 hr. Abundant mature oospores are formed in about 36 hr at 25°C but no oospores are seen above or below this temperature.[25]

The oospores can also be produced by the mycelium when transferred in distilled water on glass slides. They are incubated at 15, 20, 25, 30, and 35°C and abundant mature oospores are formed at 25°C within 36 hr. The technique reveals that the living host is not essential for oospore formation. It is the temperature that is critical. However, it is not yet clear as to why the oospores are not formed in nature in the host tissues in which the fungus grows. This technique will the helpful in studying the life cycle of the fungus and suggesting suitable control measures for Phytophthora blight of pigeon pea.

17-8 A SUITABLE CRITERION FOR IDENTIFYING RESISTANCE IN PIGEON PEA AGAINST *P. DRECHSLERI* f. sp. *CAJANI*

During the course of inoculation studies it was observed that the resistance of cultivars is expressed differently. Galls seen at the base of plants in Phytophthora-sick plots are thought to be a form of host response against the pathogen in escaped plants. Highly resistant plants do not show any symptom. After observing the reaction of inoculation of a number of cultivars it is concluded that the types of host reaction can form a criterion for identifying resistance in pigeon pea.

Three kinds of symptoms are produced on the stems of different cultivars, viz., (a) no symptom of disease development (highly resistant), (b) formation of galls following inoculation (moderately resistant), (c) brown lesions appearance, collapse and death of plants (susceptible). Even under natural conditions the three types of symptoms are observed. This constitutes a suitable criterion to observe different degrees of resistance in pigeon pea cultivars following infection with the pathogen.[26]

17-9 EPIDEMIOLOGY

17-9-1 Influence of Field Topography and Light and Darkness on Disease Development

Phytophthora blight of pigeon pea is most severe during the rainy season (July–September) on seedlings and on plants up to 2 months old. The scoring of plant mortality and lesion size can be done from July to September. It is generally observed that the disease incidence is greater in low-lying areas of the field. To verify it, a contour map of the field was prepared and it was observed that although disease is present all over the field, a maximum number of infected or dead plants were present in low-lying areas where the water stagnates (Figure 17-7).

Light and darkness affect lesion development in this disease. The increase in lesion size is higher in darkness as compared with continuous light in the glasshouse experiments. In the field also the effect of diurnal light and darkness shows significant difference in leaf spot development. Lesion size increases linearly with increase in the inoculation period.[27]

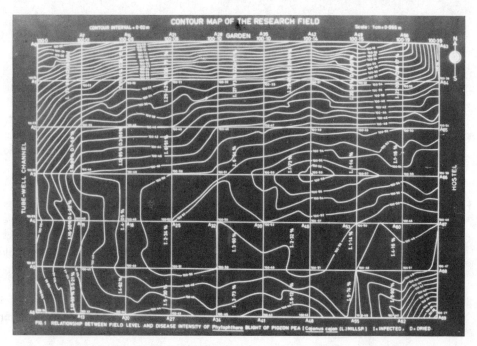

Figure 17-7 Relationship between field level and disease intensity of Phytophthora blight of pigeon pea. I = Infected; D = Dried.

17-9-2 Spread of Inoculum

Spread of fungal inoculum by wind and splashing rain in a diseased nursery of a Phytophthora-sick plot at ICRISAT has been studied by Nene and his co-workers (cited in Anonymous[28–30]). According to them, rain and splashing rain contribute to the dispersal of the fungus inoculum within a crop. They also suspect the possibility of rain water running through a Phytophthora-infested field carrying the inoculum.

17-9-3 Survival of the Inoculum

The survival of the pathogen was studied in the infected stems of pigeon pea.[29] Fungus could survive only up to 3 months in the infected pieces. Since there is more than a 3-month gap between the harvest of pigeon pea and the appearance of the crop in the next season, it is doubtful whether mycelium present in the infected pigeon pea plants could initiate primary infection in the succeeding crop.

Singh and Chauhan[31] collected oospores on the senescent as well as dead leaves of *Cynodon dactylon* Pers. from the Phytophthora-infested field. These morphologically resembled the oospores formed artificially on the leaves of pigeon pea or other substrata. However, since their germination could not be observed it is not possible to establish their identity in relation to *P. drechsleri* f. sp. *cajani*. It is quite likely that the formation of oospores may also occur in nature which will carry over the pathogen from one season to another.

17-10 MANAGEMENT

Metalaxyl (Ridomil) has been reported to control diseases caused by oomycetous fungi. According to results obtained at ICRISAT, foliar spray is ineffective in controlling this disease, but seed treatment (up to 7 g kg⁻¹) gives satisfactory control. Inefficiency of foliar spray of this fungicide has been attributed to its nonpersistence in pigeon pea plants. A combination of the three methods, viz., seed dressing, soil application, and foliar spray was more effective than applied singly. Spray in combination with seed dressing was most effective.[28-30]

Chauhan[8] tried to control the disease in greenhouse as well as under field conditions with comparatively higher doses of metalaxyl and obtained satisfactory results. However, the disease recurred in the field under frequent rains. Chauhan and Singh[31] suspected natural resistance on the part of parasite for this behavior against the fungicide and conducted an experiment to elucidate this phenomenon. They compared several species, in many cases, several *formae specialis* of one species along with four isolates of *P. drechsleri* f. sp. *cajani* for their sensitivity to metalaxyl. The isolates of the above pathogen were found uniformly less sensitive to the commonly used dose of metalaxyl effective against other species of *Phytophthora*.[32]

Since Phytophthora blight of pigeon pea is difficult to control in the rainy season with metalaxyl, Chauhan[8] suggested shifting the sowing from kharif (June–July) to rabi (September–October) season leading the crop to escape the disease.

Several thousand germplasms of pigeon pea have been screened against *P. drechsleri* f. sp. *cajani* at ICRISAT by Nene and his co-workers,[28-29] by Chauhan[8] at Banaras Hindu University, Varanasi, India, at Kanpur and IARI, New Delhi by several other workers.[33] No line has yet been found which shows high degree of resistance against all the four isolates of this fungus. Nene[34] concluded that because of high variability in the pathogen, *P. drechsleri* f. sp. *cajani*, it has been difficult to identify stable and durable resistance against this disease.

17-11 CONCLUSION

Phytophthora blight is an important disease of pigeon pea. More than one type of symptoms, for instance, basal stem blight, stem canker, and leaf blight are discerned depending on the environmental conditions. Perhaps *P. drechsleri* f. sp. *cajani* is unique among all oomycetes that produce such discrete types of symptoms. The pathogen requires relatively high temperature (28–30°C) for its growth in vitro as well as in vivo, which it shares with a few other species of *Phytophthora*. Growth at this temperature relates to its probably physiological evolution from temperate to tropical conditions. The undifferentiated sporangiophores from the normal hyphae and excessive sporangial proliferation in *P. drechsleri* f. sp. *cajani* place it closer to the genus *Pythium*. Hence, evolutionarily, *P. drechsleri* f. sp. *cajani* appears to be the most primitive species of *Phytophthora*.

Nonproduction of oospores either in culture or in the host tissues often poses a question of its mode of survival in nature. Studies of Chauhan[8] and Singh and Chauhan[25] suggest that oospore formation possibly occurs in nature also. However, further study is needed to fill this gap of knowledge of this disease.

Presence of metalaxyl resistance in *P. drechsleri* f. sp. *cajani* makes its control difficult. Exhaustive study is needed to develop suitable fungicides that may be effective against this pathogen under water-logged conditions. Natural resistance in this fungus against metalaxyl might be a result of natural mutation in *P. drechsleri* leading to the acquisition of specificity in pathogenicity for pigeon pea.

More knowledge regarding the in vivo infection, spread of disease from plant to plant, and the factors governing these processes (epidemiology) is needed. It is only presumed, but not confirmed, that the dormant oospores might be found in soil and germinate to initiate primary infection in the next crop by zoospores carried through water splash/air currents. However, detailed study is required to confirm this idea.

17-12 REFERENCES

1. Williams, F. J., Grewal, J. S., and Amin, K. S. "Serious and New Diseases of Pulse Crops in India in 1966." *Plant Dis. Reptr.* 52 (1968): 300.

2. Pal, M., Grewal, J. S., and Sarbhoy, A. K. "A New Stem Rot of Arhar Caused by *Phytophthora.*" *Indian Phytopathol.* 23 (1970): 583.

3. Williams, F. J., Amin, K. S., and Baldev, B. "Phytophthora Stem Blight of *Cajanus cajan.*" *Phytopathology* 65 (1975): 1029.

4. Amin, K. S., Baldev, B., and Williams, F. J. "*Phytophthora cajani*, a new Species Causing Stem Blight of *Cajanus cajan.*" *Mycologia* 70 (1978): 171.

5. Kaiser, W. J., and Melendez, P. L. "A Phytophthora Stem Canker Disease of Pigeon Pea in Puerto Rico." *Plant Dis. Reptr.* 62 (1978): 240.

6. Kannaiyan, J., Ribeiro, O. K., Erwin, D. C., and Nene, Y. L. "Phytophthora Blight of Pigeon Pea in India." *Mycologia* 72 (1980): 169.

7. Nene, Y. L., Sheila, V. K., and Sharma, S. B. "A World List of Chickpea (*Cicer arietinum* L.) and Pigeon Pea (*Cajanus cajan* (L.) Millsp.) Pathogens." ICRISAT Pulse Pathology Progress Report, Patancheru, Andhra Pradesh, India, 1984, 32.

8. Chauhan, V. B. "Studies on Phytophthora Blight of Pigeon Pea (*Cajanus cajan* (L.) Millsp.)." (Ph.D. thesis, Banaras Hindu University, India, 1985).

9. de Bary, A. "Researches into the Nature of Potato Fungus, *Phytophthora infestans.*" *J. R. Agric. Soc. Engl.* 12 (1876): 239.

10. de Bary, A. "Zur Kenntniss der Peronosporeen." *Bot. Zeit*, 39, 33, 1881.

11. Thaxter, R. "Mildew of Lima Beans (*Phytophthora phaseoli* Thaxt.)." *Rep. Coon. Agric. Exp. Stn.*, 161, 1889.

12. Van Breda De Haan, J. "De bibitziekte in de Deli-Tabak Door *Phytophthora nicotianae.*" Meded, Pltuin, *Batavia*, 15 (1896): 1.

13. Blackwell, E. "Terminology in *Phytophthora.*" *Mycol. Pap.* 30 (1949): 1.

14. Butler, E. J. "Potato Diseases of India." *Agric. Ledger* 4 (1903): 112.

15. Dastur, J. F. "The Potato Blight in India." *Mem. Dep. Agr. India Bot.* 7 (1915): 163.

16. Dastur, J. F. "*Phytophthora* Species of Potatoes (*Solanum tuberosum* L.) in Simla hills." *Indian Phytopathol.* 1 (1948): 19.

17. Ho, H. H. "Synoptic Keys to the Species of *Phytophthora.*" *Mycologia* 73 (1981): 705.

18. Newhook, F. J., Waterhouse, G. M., and Stamps, D. J. "Tabular Key to the Species of *Phytophthora* de Bary." *Mycol. Pap.* 143 (1978): 1.

19. Tucker, G. M. "Taxonomy of the Genus *Phytophthora* de Bary." *Univ. Mo. Agric. Exp. Stn. Bull.* 1931, 153.

20. Waterhouse, G. M. "The Genus *Phytophthora*—Diagnoses (or Descriptions) and Figures from the Original Papers." *Kew Misc. Publ., C.M.I.* 1956, 12.

21. Waterhouse, G. M. "Key to the Species of *Phytophthora* de Bary." *Mycol. Pap.* 92 (1963): 1.

22. Waterhouse, G. M. "The Genus *Phytophthora* de Bary." *Mycol. Pap.* 122 (1970): 1.

23. Yarwood, C. F. "Detached Leaf Culture." *Bot. Rev.* 12 (1946): 1.

24. Latch, G. C. M., and Hanson, E. W. "Comparison of Three Stem Diseases of *Melilotus* and Their Causal Agents." *Phytopathology* 52 (1962): 300.

25. Singh, U. P., and Chauhan, V. B. "Oospore Formation in *Phytophthora drechsleri* f. sp. *cajani.*" *Phytopathol. Z.* 123, 1988: 89.

26. Singh, U. P., and Chauhan, V. B. "A Suitable Criterion for Identifying Resistance in Pigeon Pea Against *Phytophthora drechsleri* f. sp. *cajani.*" *Agric. Res. Rural Dev.* 8 (1985): 46.

27. Singh, U. P., and Chauhan, V. B. "Relationship Between Field Levels and Light and Darkness on the Development of Phytophthora Blight of Pigeon Pea (*Cajanus cajan* (L.) Millsp.). *Phytopathol. Z.* 114 (1986): 160.

28. Anonymous. ICRISAT Annual Report, Patancheru, Andhra Pradesh, India, 1982.

29. Anonymous. ICRISAT Annual Report, Patancheru, Andhra Pradesh, India, 1983.

30. Anonymous. ICRISAT Annual Report, Patancheru, Andhra Pradesh, India, 1985.

31. Singh, U. P., and Chauhan, V. B. Unpublished data.

32. Chauhan, V. B., and Singh, U. P. "A Naturally Occurring Resistant Forma Specialis of *Phytophthora drechsleri* f. sp. *cajani* to Metalaxyl." *Phytopathol. Z.* 120 (1987): 93.

33. Pal, M., and Shukla, P. Personal communication.

34. Nene, Y. L. "Multiple-disease Resistance in Grain Legumes." *Ann. Rev. Phytopathol.* 26 (1988): 203.

18

WILT
OF PIGEON PEA

R. S. UPADHYAY AND BHARAT RAI

Department of Botany, Banaras Hindu University
Varanasi - 221 005, India

18-1 INTRODUCTION

Pigeon pea (*Cajanus cajan* (L.) Millsp.), provides staple diet for a large proportion of people in tropical and subtropical countries. It is extensively cultivated in developing countries, particularly in Asia and Africa. In these continents it is usually consumed as staple diet (dal) but in the western hemisphere it is eaten as green pea and canned commercially.[1-2] The wilt disease was first discovered in 1906 in India by E. J. Butler. Later he isolated and identified the causal organism as *Fusarium udum*. Since then various aspects of the disease and its causal organism have been studied but a great loss to pigeon pea still occurs due to this disease. The work done on this disease has been earlier reviewed by Nene et al.,[3] Upadhyay,[4] and Upadhyay and Rai.[5] Dahiya[6] and Nene et al.[7] have compiled an annotated bibliography on pigeon pea diseases. An account of the causal organism has been given by Wollenweber,[8] Wollenweber and Reinking,[9-10] Booth et al.[11], Gerlach and Nirenberg[12] and Upadhyay and Rai.[5] The perfect state of *Fusarium udum* has been recently discovered and described as *Gibberella indica* by Rai and Upadhyay.[13]

18-2 HISTORY

The disease was discovered and described by E. J. Butler in 1906.[14-15] He isolated and identified the causal organism as *Fusarium udum*.[16] A detailed account of the disease and the pathogen was given later.[17] In spite of the fact that *F. udum* was responsible for about

15–25% pigeon pea mortality[4] the disease did not draw much attention till 1920s when W. McRae described it in series of reports of Imperial Mycologists published between 1923–1933.[18–25] During 1939–40 Padwick and his associates pursued research on the disease and published a few papers dealing with *F. udum*.[26–28] Dey[29–31] brought out a series of reports dealing with mixed cropping of sorghum with pigeon pea to reduce the wilt and on breeding resistant varieties. From 1906 until 1948 work on pigeon pea diseases was done mostly at the Imperial Agricultural Research Institute at Pusa, Bihar.

Work on the wilt disease was also initiated at Madras Botany school during 1940s. Sarojini worked on soil conditions and the wilt disease[32–35] and on sporulation of *F. udum*.[36] Other workers at Madras who contributed to work on wilt are Kalyansundaram[37] and Satyanarayana and Kalyansundaram[38] on pathogenic forms of *F. udum* and host specificity; Yogeshwari[39] on the effect of trace elements on the wilt; C.V. Subramanian[40] on the strains of *F. udum*; S. Subramanian[42–44] on symptomatology and infection, changes in host metabolism due to infection, and role of manganese on disease resistance.

Vasudeva and his associates did extensive work on *F. udum* at IARI, New Delhi during 1949–69. The pioneering work done by them was on the effect of associated microflora on *F. udum*.[45–53,189] They isolated *Bacillus subtilis* as a potent antagonist of the pathogen which produces an antibiotic "bulbiformin" and was capable of reducing the wilt disease.[52–54] Singh and Husain initiated work on *F. udum* at Agra during early 1960s. The studies of these workers provided useful information on the role of enzymes and toxins in pathogenesis.[55–59] They proved for the first time that the pathogen produces cellulase, polygalacturonase, and pectin methyl esterase, which are involved in pathogenesis causing degradation of cell wall and disorganization of vessels. They also isolated fusaric acid from infected pigeon pea and implicated its role in the pathogenesis. Prasad and Chaudhary[60–64,187] and Chaudhary and Prasad[65] worked on the wilt pathogen and provided useful information on physiological aspects of the pathogen. Sen Gupta, Kaiser, and associates[66–68] and Shit and Sen Gupta[69–71] worked on infection and pathological histology, physiological races, cross-protection, and enzymatic variability in *F. udum*.

Singh and his associates took up this problem for an in-depth study and worked on the effect of soil amendment with different substances and associated microflora on *F. udum*.[72–81] Rai and Upadhyay made significant contributions on various aspects of ecopathology of *F. udum* and biological control of the disease.[13,82–98] The most important work was the discovery of the perfect state of the pathogen as *Gibberella indica*[98] based on which they have proposed a new disease cycle of the wilt.[91]

As a result of work done on this disease for more than 80 years (1906–1987) a fairly large number of papers have been published on various topics. However, it was realized during the 1970s that despite much information on the disease there is no successful method of its control/management and it remains a serious disease problem in pigeon pea growing countries. In view of the increasing severity and growing need of this pulse crop in third world countries, the wilt of pigeon pea drew the attention of plant pathologists again. As a result, further work was initiated on the disease at the International Crops Research Institute for Semi-Arid Tropics (ICRISAT), Patancheru, Pantnagar, Varanasi, Kalyani, Ranchi and at several other places in India[99–104,185,186] to find a suitable remedy to this disease menace. The information coming from these groups as a result of their renewed interest is important and may provide a basis for developing a suitable method for managing the disease in the years to come.

18-3 DISTRIBUTION AND ECONOMIC IMPORTANCE

The disease widely occurs in Asia and Africa[105] (Figure 18-1). The distribution and occurrence of the disease were earlier doubtful beyond India,[12-17] but Booth[106] stated that the disease is also prevalent in other tropical countries, which was confirmed later through worldwide extensive surveys made by the ICRISAT.[107-109] The disease occurs in Kenya, Tanjania, Uganda, Malavi, Thailand, Indonesia, and Trinidad.[3,105,106]

In India the disease occurs in almost every state in which pigeon pea is grown.[105,184] The average incidence of the wilt in India varies from 0.1% in Rajasthan to 22.6% in Maharashtra.[107-109] The disease is severe in Maharastra (average above 20%), Bihar (10–20%), Uttar Pradesh (5.1–10%), and is comparatively less severe in other states (≤5.0%). The disease is not important in southern states of India.[17,108,111]

Kanniayan et al.[109-110] have reported that the wilt disease is economically very important. They have estimated annual loss in the crop production due to a combined effect of wilt and sterility mosaic in India to be equivalent to US$ 113 million. In Africa the estimated loss due to the wilt is about US$ 5 million every year.

18-4 SYMPTOMS

The symptoms of the disease have been described in detail by Butler,[16-17] Satyanarayana and Kalyansundaram,[38] Subramanian,[42] Chaubey,[112] Amin et al.,[113] Nene et al,[3] and Upadhyay and Rai.[5] The infection is systemic, occurring through fine lateral roots by the germ hyphae, produced commonly either from conidia or chlamydospores[114] or asco-

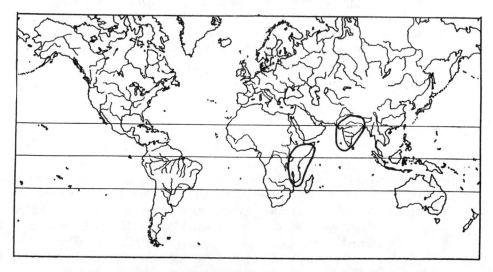

Figure 18-1 World distribution map of the wilt disease of pigeon pea (CMI Disease Distribution Map No. 563, 1985[105])

spores.[13] The pathogen enters the vascular system and traverses all along, producing conidia and chlamydospores within the xylem vessels. The xylem vessels are frequently blocked by clumps of the mycelia of the pathogen (Figures 18-2—18-4). Blackening due to infection frequently appears in vascular tissues of the host.

The wilt appears in early stages of plant growth, i.e., when the plants are about 4–6 weeks old.[3,82,112] The severity has been recorded at a maximum during flowering and podding stages.[3,188] However, the disease may appear at any stage of plant growth from young seedlings up to podding and maturity. The typical symptoms of the disease appear in plants as gradual or sudden withering and drying of green parts exactly as if they were suffering from drought, even though there may be plenty of water in the soil (Figures 18-5,18-6).[17] In the beginning, yellowing of leaves and blackening of portions of the stem starting from collar to the fine branches may appear which gradually result in drying of leaves, stems, fine branches, and finally the death of the whole plant. Partial wilting of the plants is very common in the field during the advanced sages of plant growth. Patches of the wilted plants in infested fields are frequently seen which increase in size with successive cropping of pigeon pea in the same field.

Amin et al.[113] have pointed out that *Phytophthora* blight of pigeon pea can be easily mistaken for the *Fusarium* wilt. The distinction between these can be made as follows: (a) the leaves of the plants affected by *F. udum* frequently turn yellow before drying, whereas

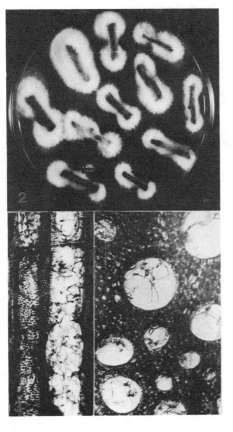

Figure 18-2 Colonies of *Fusarium udum* on root bits of pigeon pea placed on nutrient agar

Figure 18-3 L.S. of roots of pigeon pea showing hyphae of *Fusarium udum* inside the vessels

Figure 18-4 T.S. of pigeon pea roots showing hyphae of *Fusarium udum* inside the vessels

Figure 18-5 A wilted pigeon pea plant in field

the leaves of the plants affected by *Phytophthora* blight remain initially light green with upward rolling and usually dry rapidly, and (b) frequent blackening of the host tissues occurs in case of the wilt but not in case of the blight.

18-5 THE CAUSAL ORGANISM

The causal organism of the wilt disease of pigeon pea is *Gibberella indica*, Rai and Upadhyay[105] (*Fusarium udum* Butler). Its synonyms are *Fusarium butleri*,[115] *Fusarium uncinatum*,[116] *Fusarium lateritium* var. *uncinatum*[8], *Fusarium oxysporum* f.sp. *udum*,[117] *Fusarium lateritium* f. sp. *cajani*,[118] *Fusarium udum* var. *cajani*.[27] The name *Fusarium oxysporum* f. sp. *udum* was frequently used by researchers in the past. However, now *Fusarium udum* is widely accepted as the name of imperfect state of the pathogen.[5,11,12,41,89,93,106] The suggestion was made in the past that *F. udum* is a form of *F. oxysporum*. However, the discovery of its perithecial state by Rai and Upadhyay[13] resolved

this controversy. Because the perfect state *G. indica* belongs to *Gibberella baccata* group, *F. udum* should be retained in the Lateritium group of Fusarium[105] but not in the Elegans group.[41]

A detailed description of the causal organism has been given by Wollenweber,[8] Wollenweber and Reinkings,[9-10] Booth,[106] Booth et al.,[11] Subramanian,[40-41] Gerlach and Nirenberg,[12] and Upadhyay and Rai[5] (Figure 18-7—18-15).

The pathogen is specific to parasitism and is pathogenic to only pigeon pea.[5,12,41,106,119] The causal organism is a soilborne facultative parasite that enters through roots and then becomes systemic. It can be isolated from all parts of the host from lateral fine roots to pedicel and pod hull.[3] The fungus usually occurs more frequently and in high population in the vicinity of the infected and wilted plant roots (Figure 18-16). Its population increases in the infested plots when pigeon pea is grown successively in the same plot.[3-5] The pathogen extends more rapidly from one place to another along the root than across the soil.[16] It is dispersed through irrigation, rain water, and displacement of host debris from one place to another. The propagules of the pathogen are also dispersed by termites that feed frequently on the dead wilted plants.[89] It has also been found to be of a seedborne nature.[111,120,121] Susceptible cultivars of pigeon pea carry seedborne infections that can be eradicated by seed treatment with Benlate + thiram mixture.[121]

Fusarium udum shows a great deal of variation in cultural characteristics.[16,34,40,69,70,122-124,190] Sarojini[34] and Shit and Sen Gupta[70-71] found variation in pathogenicity of different isolates of *F. udum* collected from different regions of India which was correlated with variation in production of polymethyl galacturonase and cellulase enzymes in

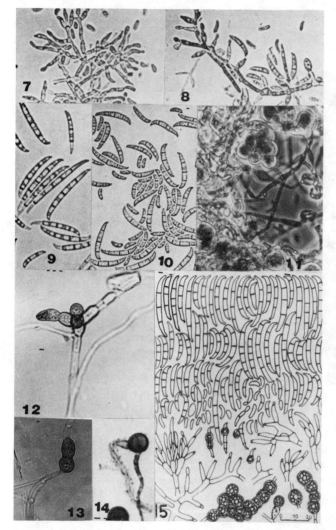

Figures 18-7, 18-8   Conidiophores and conidia (x 520)

Figures 18-9, 18-10   Conidia of *Fusarium udum* (x 520)

Figure 18-11   Clumps and chain of chlamydospores (x 800)

Figure 18-12   Chlamydospore formation

Figure 18-13   Intercalary chlamydospores (x 750)

Figure 18-14   One-celled terminal chlamydospore and germinating chlamydospore (x 750)

Figure 18-15   Digrammatic presentation of various structures of *Fusarium udum*. (Figs. 18-7–18-10, 18-15 from Gerlach and Nirenberg[12]).

different isolates. Based on the cultural characteristics, Nene et al.[3] classified *F. udum* in 12 distinct groups.

Physiological races have also been suggested to exist.[70,111,123,125] Single spore isolates from single strains also vary among themselves with regard to growth pattern, segmentation, and capacity of secreting metabolic products. A number of transitional forms of *Fusarium* pathogenic to pigeon pea have also been recorded.[126]

The pathogen has been reported to produce three enzymes viz., pectin methyl esterase, polygalacturonase, and cellulase in vivo and in vitro[56,58] and a toxin fusaric acid[57,59] which have been shown to play a role in pathogenecity. A comprehensive account of other aspects of *F. udum*, viz., ecology, physiology, competitive saprophytic ability has been given by Upadhyay and Rai.[5]

In the parasitic phase the population of the pathogen remains very high on infected and wilted plants in comparison to the healthy ones (Figure 18-16). During the saprophy-

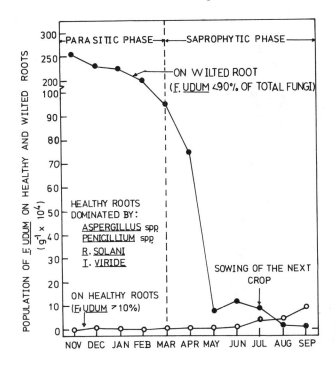

Figure 18-16 Population dynamics of *Fusarium udum* on roots of pigeon pea in parasitic and saprophytic phases during 1977–78. (Based on Upadhyay and Rai[89])

tic phase the healthy and wilted roots are vigorously colonized by other saprophytic microorganisms resulting in a reduced population of the pathogen on decomposing roots. However, at the time of the next sowing of the crop, sufficient inoculum of *F. udum* is still present in field soils, mostly on host debris,[89] to cause the disease.

Fusarium udum can survive saprophytically in soil in the absence of its host for a period of 3–4 years. The fungus passes from decaying roots into the soil and continues to grow and form spores. There are different views on its period of survival in soil. Butler[17] and Mundkur[127] were of the common view that *F. udum* survives saprophytically in soil for an indefinite period. McRae and Shaw[25] and Agnihothrudu[128] suggested that the fungus survives in already invaded host tissues for about 8 years. Singh[114] pointed out that the saprophytic survival of the pathogen in soil is limited to continued presence of dead host roots and other debris. In recent reports Nene et al.[3] and Nene and Reddy[129] have mentioned that the pathogen survives in buried pigeon pea substrates for about 2 1/2–3 years. It has been found that crop sequence,[31,130] use of fertilizers and other micronutrients,[19,33,131] organic amendments with different substance,[23,78–80, 132] types of soils,[3,82] and associated microflora[48,49,89,133] influence the behavior and survival of the pathogen in soil.

The perfect state of *Fusarium udum* has been discovered as *Gibberella indica* and has been described recently (Figures 18-17–18-31).[13] Some unidentified species of *Gibberella* were recorded from pigeon pea in the past,[32,134,135] which were thought to cause a similar disease in pigeon pea. However, their relationship with *F. udum* and their role in the disease development was not established. Recently, ICRISAT has also started a search for the perfect state of *F. udum*.[111] They found that on mating, dark red perthecia-like bodies were formed on potato sucrose agar at lower temperatures (10–15°C) but without asci and ascospores.

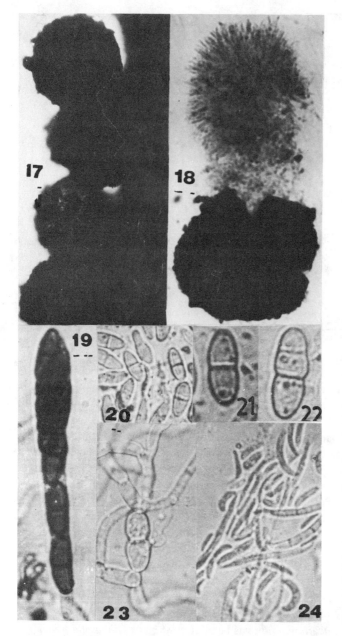

Figures 18-17—18-24 *Gibberella indica* (the perfect state of *Fusarium udum*)
18-17 A group of perithecia on the host surface (x 75)
18-18 A broken perithecium with exposed asci (x 125)
18-19 An ascus with 8 ascospores (x 1000)
18-20 Ascospores (x 822)
18-21 Two-celled ascospore (x 1380)
18-22 Three-celled ascospore (x 1380)
18-23 A germ hypha producing microconidia after germination of an ascospore (x 1248)
18-24 Micro- and macroconidia obtained through single ascospore culture (x 1100). (from Rai and Upadhyay[13])

In fields, perthecia of *G. indica* are formed but infrequently on wilted pigeon pea. They are formed on exposed roots and collar region, which are superficial, commonly aggregated, globose to subglobose, sessile, and smooth walled. *G. indica* is a heterothallic fungus and therefore the formation of perithecia in culture can be induced after mating different strains on nutrient medium or on sterilized host substrate on to the medium at 25 ± 2°C. *G. indica* also initiates the disease development by producing germ hyphae or

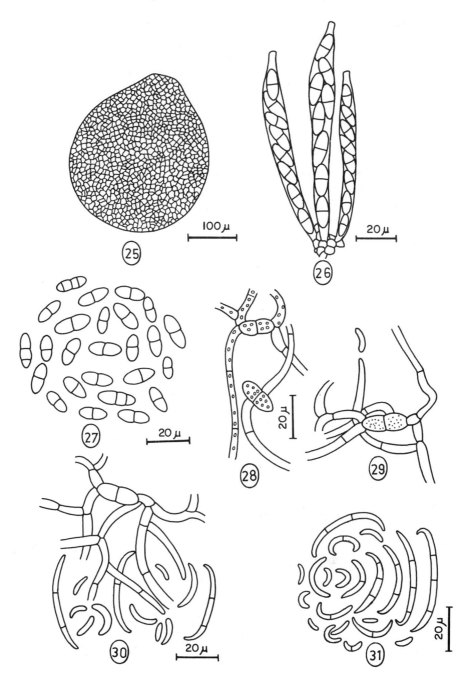

Figures 18-25—18-31 *Gibberella indica* 18-25 A single perithecium 18-26 Asci with ascospores 18-27 Ascospores 18-28 Germinating ascospores 18-29 Microconidia produced on a germ hypha from ascospore 18-30 Germ hyphae of ascospore producing micro- and macroconidia 18-31 Micro- and macroconidia. (From Rai and Upadhyay[13])

conidia from ascospores (Figures 18-23 and 18-32). It seems that the disease could not be controlled in the past due to lack of knowledge on the perfect state of the pathogen. It is therefore necessary to understand the biology of the perfect state and its role in pathogenesis. Further work on the perfect state is in progress at ICRISAT and Varanasi.

18-6 DISEASE CYCLE

The disease cycle was known earlier to occur only through the imperfect state. After discovery of the perfect state it was observed and established by Upadhyay and Rai[91] that the disease cycle of the wilt occurs both through imperfect (*F. udum*) and perfect (*G. indica*) states (Figure 18-32). However, the asexual state is more important and common

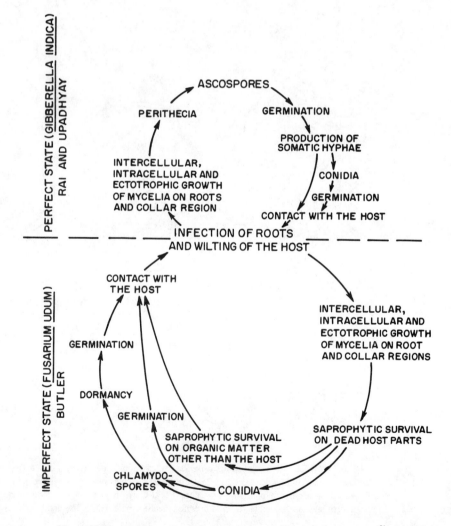

Figure 18-32   Disease cycle of wilt of pigeon pea. (from Upadhyay and Rai[91])

since it is more frequently observed in nature. In both cases the pathogen grows externally as well as internally. It produces a mass of mycelia and conidia on the surface of the host, particularly on the collar region and on the roots. The infection takes place through fine lateral branches of the roots and the fungus continues to grow in the xylem vessels. Consequently, upon wilting the pathogen lives and survives saprophytically for several years, primarily on dead parts of the host either in the imperfect state[3,17,136] or in the perfect state.[91] The longevity of survival of the pathogen through its perfect state is yet to be determined. The perfect state of the pathogen occurs along with the imperfect state simultaneously on the host.[82] The saprophytic survival is confined mostly in the infected dead roots and other host debris. However, the fungus may also survive for a limited period on organic matter other than the host. Mycoparasitic survival of *F. udum* on other fungi in soil has been reported recently.[97] *F. udum* propagules have been recorded from the bodies of termites also feeding on the wilted host roots.[90] Depending on the environmental conditions, chlamydospores are formed both in parasitic and saprophytic phases from the hyphae and conidial cells.[114] These chlamydospores serve as structures of survival during prolonged absence of the host and on return of the favorable conditions, these resting spores germinate to cause infection. The spread of the disease occurs through contact of the infected roots with the healthy ones, irrigation, rain water, and termites. The fungus exists in soil mostly in patches in localized spots with an increase in its size when the same crop is grown successively.[17,114] After death of the infected plants the fungus lives saprophytically in the host substrates for a limited period and then forms chlamydospores to enter the dormant resting stage.

Sometimes a large number of dark violet perithecia are produced on the collar region and exposed roots of the host. These perithecia also serve as resting structures under unfavorable conditions. The perithecia produce a large number of ascospores that remain in soil for a limited period under a physiologically inactive state, and under favorable conditions they produce somatic hyphae on germination either causing infection in the host roots or producing conidia (*F. udum*) which, in turn, may cause infection.

18-7 EPIDEMIOLOGY

The incidence and severity of the disease depend primarily on (a) soil conditions and (b) the type of cultivar used. The disease is very common in northern and central parts of India. It is very severe along the bank of the River Ganges. Slightly acidic to slightly alkaline soils containing 50% or more sand particles favor the disease incidence in susceptible cultivars.[5,82] Shukla[137] noted that a higher proportion of sand in soil favors occurrence of the wilt. Soil temperature and moisture also play a significant role in occurrence of the wilt disease. Upadhyay[82] reported a range of temperature between 20–29°C most suitable for the disease development at moistures 6–16%. However, Mundkur[136] had earlier reported a temperature range of 12–29°C favorable for the disease. Upadhyay and Rai[87] made an attempt to correlate the soil physico-chemical properties with fungistasis and the wilt of pigeon pea and have concluded that less incidence of the wilt disease in soils of southern states of India observed by Kannaiyan et al.[109] might be due to a higher level of soil fungistasis against the pathogen. Based on the incidence of the disease, Upadhyay and Rai[4,5,87] have classified different Indian soils into three categories: (a) conducive soils that support the growth of *F. udum* and promote the wilting, (b) suppressive soils that do not favor the growth of *F. udum* and suppress the incidence of the wilting, and (c) intermediate soils that moderately suppress *F. udum* and the disease.

The disease incidence in a particular soil depends mainly on the saprophytic activity and survival of *F. udum* in soil which are favored by continued presence of the host substrate. The disease becomes increasingly more severe when susceptible varieties of pigeon pea are grown in infested soils successively. The severity of the disease and the size of patches of infections are markedly increased under such conditions.

18-8 MANAGEMENT OF THE DISEASE

18-8-1 Cultural Control

Crop rotation and mixed croppings are traditional practices. The former varies according to the infestation of fields by the pathogen. Mundkur[127] suggested that if long rotations are practiced the disease can be effectively reduced in intensity. The crop should not as a rule be sown in the same soil at least for 3 years in succession. However, the problem with crop rotation in India is that generally short periods of rotations (1–2 years) are adopted because of low holdings of farmers,[17] which is insufficient to eliminate the pathogen from field soils. Upadhyay and Rai[89] have also suggested crop rotation for management of the disease. The type of the crop used in crop rotation greatly influences the effectivity of the rotation. Bose[130] found that rotation of pigeon pea with tobacco after every 3 to 4 years significantly suppresses the disease.

Intercropping of pigeon pea with *Sorghum* reduces the wilt significantly.[29,138,139] It was found that pigeon pea intercropped with sorghum had only 24% wilt against 85% in the sole crop treatment which was consistent across 14 susceptible pigeon pea genotypes.[139] Breaks of sorghum and fallowing fields for one or more years also reduce the wilt considerably. It has also been recorded that a one-year break between pigeon pea crop by fallowing or by growing sorghum or tobacco reduces the wilt to 22%, 20%, and 44% respectively.[111] A break with cotton or maize was not very effective. Mixed cropping of pigeon pea with *Crotalaria madicaginea* also suppresses the disease to a good extent.[88] Management of the disease by amendments of soil has also been reported. A considerable reduction in the incidence of wilt disease was found by green manuring[19–22] and by addition of compost or the leaves of *C. medicaginea*.[88] Amendments of autoclaved or natural soils with *Azadirachta indica* or *Ricinus communis* oil cakes, rice husk, or saw dust favor the lytic effect of *Bacillus subtilis* on *F. udum* which affect the population of the pathogen in soil.[72,73] However, soil amendment with these substances for control in the wilt is little practiced.

Another precautionary measure suggested as early as 1918 is the systematic removal of wilted plants from fields and their subsequent burning. However, this practice is neither of much use nor practically feasible because *F. udum* also colonizes and lives on healthy plants, though in low population.[89] Therefore, even if the wilted plants are removed the root systems of the healthy plants which remain in the soil after harvest are vigorously colonized and inhabited by *F. udum* (Figure 18-16).

18-8-2 Chemical Control

There are several examples of use of fungicides and other chemicals for the disease management.[88,102,140–143] Sinha[143] observed a significant control of the disease by Bavistin applied as a soil drench at 2,000 ppm 10 days before inoculation of pigeon pea with *F. udum*. In laboratory experiments Bavistin and BASF38601 F were highly effective in suppressing mycelial growth. The spore germination was suppressed by Benlate and Campogran-M at 50 ppm. Haider et al.[102] reported the disease control over three years by captan, Brassicol (quintozene), and phenyl mercury acetate. Upadhyay and Rai[88] found a considerable reduction in the wilt incidence by Phygon XL, Dithane-78, and Zincop when applied as soil treatment. Kotasthane et al.[192] recently reported that seed treatment with Benlate + thiram (1:3) considerably reduced the wilt incidence. However, the effect of the fungicide applied at the time of sowing did not persist for the whole cropping season and thus a single treatment did not provide a remedy for the disease control.

Antibiotics griseofulvin[144] and bulbiformin[52] have been reported effective in controlling the disease. However, these antibiotics do not provide an economical approach to disease management. Therefore, it may be concluded that fungicides and antibiotics do not hold well for the disease control, and other methods of the disease management should be explored.

The effect of superphosphate on the wilt incidence was studied by McRae.[19–23] Superphosphate added in the green manured soil reduced the wilt incidence by 25%. The disease was however favored when superphosphate alone was added in soil. Rai and Upadhyay[98] in a similar observation found urea to promote the activity of *F. udum* in soil. Therefore, use of urea in infested fields should be avoided.

18-8-3 Biological Control

Biological control remains an attractive possibility for management of soilborne plant pathogens. Some encouraging results have been reported for wilt causing Fusaria.[5,145–149] Attempts to identify antagonists of *F. udum* were made in the past.[49,52,53,72–74,83,94,150–152] *Bacillus subtilis* has been reported to possess strong antagonistic property against *F. udum*[49,52,53,153–155] which produces an antibiotic bulbiformin and is able to reduce the wilt incidence. The production of antibiotics is stimulated by magnesium sulphate, manganese, and iron and soil amendment with roots of certain leguminous crops, molasses and oil-cakes.[132,152,153] A suitable medium for the growth of *B. subtilis* favoring production of antibiotics against *F. udum* has also been described[50]. Inoculation of *B. subtilis* into autoclaved soil amended with molasses and sweet clover roots and groundnut-cake reduced 88% of the pigeon pea wilt. Unfortunately, work has not been pursued to use *B. subtilis* for biological control of the disease. Therefore, further work is needed on this bacterium as a biocontrol agent. *B. subtilis* has already been proved to have biological potential for many soilborne plant pathogens.[156–161]

Another antagonist reported to have potential for biocontrol of *F. udum* is *Micromonospora globosa*[84] which kills and destroys *F. udum* even in its resting stage. Destruction of chlamydospore wall, which resists even the most adverse conditions, is caused by the antagonist.

Aspergillus niger, A. flavus, A. terreus, Penicillium citrinum, Trichoderma har-

zianum, T. viride, and *Streptomyces griseus*[94] and *Aspergillus nidulans*[83] also posses the ability of biological control of wilt disease. Upadhyay and Rai[94] and Upadhyay[83] have tested these antagonists for biological control of the wilt disease. The antagonists significantly suppress *F. Udum* in the soil and root region of the host.

Recently, workers at ICRISAT isolated *P. pinophilum* from soil solarized by polythene mulching[111,121] which shows antagonism against *F. udum* and controls the wilt incidence on inoculation in soil, seeds or on roots of the host. Solarization-induced control of the disease has been observed at ICRISAT[111,121] and Varanasi (Upadhyay and Sinha, unpublished data).

Cross protection of the wilt by other *Fusarium* spp. has also been reported.[66,68,120,152] Prior inoculation of the host with *Fusarium oxysporum* f. spp. *ciceri* and *vasinfectum* provided a considerable reduction in the disease. Cross-protection through avirulent strains of pathogens has been reported in the case of many soilborne diseases.[162–163]

18-8-4 Resistant Varieties

Work on developing resistant varieties against the wilt disease was started as early as 1908.[15] Butler[17] had remarked that "the only thoroughly satisfactory way of dealing with wilt disease is the selection or breeding resistant varieties." Since then work on this aspect of the disease has been done by several investigators.[16,24,25,30,31,46,111,112,121,125,140,164–181]

A list of the cultivars/varieties showing resistance against the wilt is presented in Table 18-1. Many of these cultivars are under cultivation in India. Several cultivars that earlier claimed to possess resistance against the wilt failed to give uniform performance at ICRISAT.[3] However, cultivars NP (WR)-15, (NP-24 x NP-51), 15-3-3, DDN-1 and 20-1 show a consistently low level of the disease incidence.

Systematic and intensive work on screening and developing resistant varieties of pigeon pea has been taken up by the ICRISAT[111,121] and Indian Council of Agricultural Research (ICAR) with emphasis on (a) trials on resistant varieties already claimed resistant, (b) multilocation uniform trials, and (c) screening for broadbased combined resistance for wilt + sterility mosaic (SM), wilt + SM + *Phytophthora* blight (PB) and wilt + *Heliothis* and other diseases of pigeon pea. Since 1976 more than 11,000 entries have been screened for wilt resistance and many promising lines have been developed (Table 18-1). Attempts are also being made to develop short duration pigeon pea which have been found to show generally less susceptibility to the wilt.[182]

It has been frequently found that due to segregation some resistant varieties lose their resistance during later generations.[3] Mundkur[178] reported that cultivars D-16-12-2, PT-12, and D-33-4-22 lost resistance when grown in field. The cultivars C-11, C-28, C-36, F-18, NP (WR)-15, NP-41, and T-17 earlier reported to be resistant or tolerant later became susceptible to the wilt.[179] A similar observation has been made for many lines at ICRISAT[111,121] during coordinated varietal trials. Another problem associated with this method of control is that many of the resistant lines do not possess resistance against many other diseases of pigeon pea, some of which, like PB and SM are prevalent and serious. For example, out of 52 entries of pigeon pea evaluated for resistance against wilt, SM and PB for multilocation trials, only ICP 8860, ICP 8861 were found highly resistant to these diseases (incidence less 20%) at all the sites.[121] However, many other lines were promising for more than one pathogen. ICP 8860, ICP 8867, BWR 159 were highly

TABLE 18-1   CULTIVARS/VARIETIES OF PIGEON PEA REPORTED TO POSSESS
RESISTANCE AGAINST *F. udum*[a]

	Cultivars/varieties	References
1.[b]	Sabour 2E, Sabour 75, Pusa P	164
2.	17 W/2	168
3.	15-3-3, ET-236-6-3-102, (C.11×N.252), (C.11×N.252) 10, Vita-1, Osmanabad-1-5, Udgir-500	169
4.	NP(WR)-15, S-103	111,123
5.	IP-41, Hybrid-5 (D-419-2-4) IP-80	170
6.	NP-80	30,31
7.	D-16-17-2, NP-41, C-38-1-2, D-419-2-4	172
8.	Type 16,41,50,51,80,82	24,25
9.	A-126-4-1, IP-80, IP⁻ 41, C⁻ 38, C-15, D-16-12-2[c], PT-12[c], D-33-4-22[c]	178
10.[c]	C-11, C-28, C-36, F-18, NP(WR)-15, NP-41, T-17	179
11.[d]	Bori 192-15-2-2-11-41, Bori 192-12-5-1-2	179
12.	NP-15	42
13.	ST-1, ST-2, ST-3	173
14.	S-55	46,47
15.	ICP-8859, ICP-8860, ICP-8861, ICP-8862, ICP-8864, ICP-8865, ICP-8867, ICP-8868, ICP-8869	3
16.	ICPL-83-227	183
17.	C 11	111,123,181
18.	Purple-1 (Malviya Arhar-1)	104
19.	Bhavani Sagar, ICPL 8357, AKT 1, ICPL 227, MTH 9, DA 15, DA 12, PPA 84-8-3, NPRR 1	181
20.[e]	ICP 9174, ICP 11297, ICP 12748, ICPL 8357, BWR 105	111
21.[f]	BWR 159, BWR 332, ICP 12825, ICP 12924, ICP 13235, ICP 13237, ICP 13622, ICP 13625, ICP 13626, ICP 13627, ICP 13641, ICP 13646, ICP 13656, ICP 13668, ICP 13684, ICP 13713, ICP 13753, ICP 13759	111
22.	ICP 1097, ICP 7867, ICP 8863, ICPL 227, ICPL 8357, NPRR 1, DA 12, MA 97, T 7	111
23.[g]	ICP 3461, ICP 8859, ICP 8859, ICP 8863, ICP 9174, ICP 11297, ICP 12731, ICP 12745, ICP 12748, ICP 12758, ICP 12759, ICPL 84007, ICPL 84008, ICPL 84013, ICPL 84017, ICPL 8362, ICPL 8357, ICPL 8358, ICPL 8363, BWR 105, BWR 117	181
24.[h]	ICP 8863 (Maruthi)	3,185,186
25.[i]	NPP 675/2, NPP 675/6, NPP668/4, NPP 690/1, NPP 691/2, NPP 693/3, NPP 696/2, NPP 699/1, NPP 699/2	111
26.[j]	BSMR 65, BSMR 242, BSMR 243, BSMR 270, BSMR 294, BSMR 519, BSMR 520, BSMR 528, BSMR 539, BSMR 540, BSMR 544, BSMR 615, BSMR 697, BSMR 731, BSMR 732, BSMR 736, BSMR 493, BSMR 511, BWR 97, BWR 153, BWR 159, BWR 175, BWR 177, BWR 250, BWR 259, BWR 322, BWR 332	191
27.[k]	BDN-1, BDN-2	176

[a] This is not an exhaustive list. [b] Lost resistance in field. [c] Lines earlier reported resistant proved susceptible. [d] Moderately resistant. [e] Highly resistant and show broadbased resistance on multilocation test. [f] Showed resistance to wilt and sterility mosaic but not to Phytophthora blight. [g] Showed resistance at 4 or more sites on multilocation test (wilt below 20%) of ICAR-ICRISAT Uniform Trial. [h] Highly successful in Karnataka, India. [i] Kenyan lines showing promise for wilt resistance. [j] Resistant to wilt and sterility mosaic both (wilt below 20%). [k] Recognized all over India.

resistant to both SM and wilt.[121] In another trial none of the 140 entries of ICP germplasm of pigeon pea lines at ICRISAT showed resistance against wilt + SM + PB during 1986–87 in multiple disease tests. Many lines that showed no wilt were highly susceptible to either PB or SM or both. The lines that possess less than 20% of both SM and wilt are ICP 13622, ICP 13626, ICP 13632, ICP 13641, ICP 13646, ICP 13656, ICP 13684, ICP 13713, ICP 13715, ICP 13759, BWR 159, BWR 332, ICP 12825, ICP 12924, ICP 13235, ICP 13237. Therefore, it seems very difficult to get lines/varieties resistant to all the pathogens of pigeon pea. Most of these ICRISAT-ICAR lines[111,121,181] are still under multilocation or multiple disease resistance trials and are not released for cultivation however, such a testing has helped in identification of several stable sources of resistance to Fusarium wilt and sterility mosaic[183] (see Table 1 of Chapter 19). Nevertheless it may be concluded that so far we have not been able to fully manage the disease problem by this method too. However, the disease can be minimized to a great extent by the use of some promising varieties like ICP 9174, ICP 11297, ICP 12748, ICPL 8357, BWR 105, C 11, NP (WR)-15, S-103, (NP-24 x NP-25), 15-3-3, BDN-1, 20-1, ICP 8863 (Maruthi), Purple-1 (Malviya Arhar-1), Bhawani Sagar, ICPL 8357, AKT 1, ICPL 227, MTH 9, DA 15, DA 12, DPPA 84-8-3, NPRR 1, etc. The enormous emphasis given by ICRISAT-ICAR provides us with great hope of finding a solution to this serious disease problem by developing resistant varieties.

18-8-5 Integrated Management

Integrated management has been found successful for many soilborne plant pathogens.[5] In recent years this approach has been emphasized to be better than the use of only a single component for disease control. There has been little effort to adopt an integrated approach for managing pigeon pea wilt. Lower dosages of some effective fungicides would be proved effective if used in combination with the identified antagonists. Solarization-induced control of the disease is also a step in this direction where the control of the disease is achieved by solar heating to reduce the fungus propagule in soil or the control by augmentation of population of antagonists active against the pathogen at higher temperature.[111] Recently, Upadhyay[83] suggested an integrated approach to control the disease by selecting higher temperature and alkaline pH which favors growth of an antagonist *A. nidulans*. Intercropping or crop rotation may also be taken into consideration as components in integrated management of the disease by combining these practices with antagonists, solarization, or use of the resistant varieties to make the disease control more effective.

18-9 CONCLUSIONS

Pigeon pea wilt occurs widely in India and many other countries. It causes great destruction of pigeon pea in certain areas, creating major economic loss. No other disease does more damage to pigeon pea than the wilt, as sometimes in natural fields the disease incidence reaches up to 70–80%. In spite of sufficient attention given to this disease, no suitable method of its management has evolved so far. The only way to overcome the disease menace, to some extent, is the use of resistant varieties and crop rotation or intercropping.

Extensive work has been done on ecopathology, antagonism, and biological control, developing resistant varieties, and management of the disease by cultural methods. Fortunately, work on these aspects is still being pursued in India which may lead to a better understanding on management of this disease in future. The discovery of the perfect state of the pathogen has provided substantial knowledge about the pathogen and has opened a new avenue for research on the disease and its causal organism. However, more efforts should be made to study (a) crop rotation, crop sequences, and intercropping for natural suppression of the disease, (b) antagonism and biological control, (c) survival of the pathogen and factors adversely affecting the survival, and (d) developing resistant varieties of pigeon pea. Recent work on solarization at ICRISAT and Varanasi indicates that there is a promising possibility of using this technique for management of the disease. Integrated approach using two or more components of disease management could be more fruitful for control of the disease.

18-10 REFERENCES

1. Sharma, D., and Green, J. M. "Perspective of Pigeon pea and ICRISATS's Breeding Programme," in *International Workshop on Grain Legumes*, International Crops Research Institute for the Semi-Arid Tropics, Patancheru, India, 1975.

2. Abrams, R. "Status of Research on Pigeonpeas in Puerto Rico," in *International Workshop on Grain Legumes*, International Crops Research Institute for Semi-Arid Tropics, Patancheru, Andhra Pradesh, India, 1975.

3. Nene, Y. L., Kannaiyan, J., Haware, M. P., and Reddy, M. V. "Review of Work done at ICRISAT on Soil-borne Diseases of Pigeon-Pea and Chick-Pea," in *Proceedings of the Consultants Group Discussion on the Resistance to Soil-borne Diseases of Legumes*, ICRISAT, Patancheru, A.P., India, 1979, 3.

4. Upadhyay, R. S., "Ecology and Biological Control of *Fusarium udum*," in *Proceedings of the 11th International Congress of Plant Protection*, Manila, 1987, 121 (Abstract).

5. Upadhyay, R. S., and Rai, B. "Wilt Disease of Pigeon-Pea and Its Causal Organism *Fusarium udum*," in *Perspective of Phytopathology*, eds. V. P. Agnihotri, U. S. Singh, H. S. Chaube, N. Singh and T. S. Dwivedi. New Delhi: Today and Tomorrow's Printers and Publishers, 1989.

6. Dahiya, B. S. *An Annotated Bibliography of Pigeon-pea*, International Crops Research Institute for Semi-Arid Tropics, Patancheru, India, 1977.

7. Nene, Y. L., Kaiser, W. J., Grewal, J. S., Kannaiyan, J., and Beniwal, S. P. S. *An Annotated Bibliography of Pigeon-Pea Diseases, 1906–81*. Bibliography 013, International Crops Research Institute for Semi-Arid Tropics, Patancheru, India, 1985, 128 pp.

8. Wollenweber, H. W. "*Fusarium*—Monographie. Fungi Parasitici et Saprophytici." *Z. Parasitenkol*. 3 (1931): 269.

9. Wollenweber, H. W., and Reinking, O. A. *Die Fusarien, ihre Beschreibung, Schadwirkung und Bekampfung*. Berlin: Paul Parey, 1935, p. 335.

10. Wollenweber, H. W., and Reinking, O. A. *Die Verbreitung der Fusarien in der Natur*, Friedlander und Sohn, Berlin, 1935, 80.

11. Booth, C., Mordue, J. E. M., and Gibson, I. A. S. *CMI Descriptions of Pathogenic Fungi and Bacteria Set 58 Nos. 571–580*, 1978, 20.

12. Gerlach, W., and Nirenberg, H. *The Gunus Fusarium—a Pictorial Atlas*. Berlin and Hamburg: Kommissionsverlag Paul Parey, 1982, p. 406.

13. Rai, B., and Upadhyay, R. S. "*Gibberella indica*: The Perfect State of *Fusarium udum*." *Mycologia* 74 (1982): 343.

14. Butler, E. J. "The Wilt Disease of Pigeon-Pea and Pepper." *Agriculture India* 1 (1906): 25.

15. Butler, E. J. "Selection of Pigeon-Pea for Wilt Disease." *Agriculture India* 3 (1908): 182.

16. Bulter, E. J. "The Wilt Disease of Pigeon-Pea and the Parasitism of *Neocosmospora vasinfecta* Smith." *Mem. Dep.Agric. India (Bot. Sec.)* 2 (1910): 1.

17. Butler, E. J. *Fungi and Plant Diseases*. Dehradun and Delhi: Bishen Singh and Mahendra Pal Singh, 1918, p. 547.

18. McRae, W. "Report of the Imperial Mycologist." *Scient. Rep.Agric. Res. Inst. Pusa 1922–23*, 53.

19. McRae, W. "Report of the Imperial Mycologist." *Scient. Rep. Agric. Res. Inst. Pusa 1923–24*, 1924, 41.

20. McRae, W. "Report of the Imperial Mycologist." *Scient. Rep. Agric. Res. Inst. Pusa 1925–26*, 1926, 54.

21. McRae, W. "Report of the Imperial Mycologist." *Scient. Rep. Agric. Res. Inst. Pusa 1927–28*, 1928, 56.

22. McRae, W. "Report of the Imperial Mycologist." *Scient. Rep. Agric. Res. Inst. Pusa 1928–29*, 1930, 51.

23. McRae, W. "Note on Wilt of Rahar in Permanent Plots at Pusa." *Proc. Bd. Agric. India, 1929*, Appendix 3, 236, 1931.

24. McRae, W., and Shaw, F. J. F. "Report on Experiments with *Cajanus indicus* (Rahar) for Resistance to *Fusarium vasinfectum* (Wilt Disease)." *Scient. Rep. Agric. Res. Inst. Pusa. 1925–26*, 1926, 208.

25. McRae, W., and Shaw, F. J. F. "Influence of Manures on the Wilt Disease of *Cajanus indicus* Spreng. and Isolation of Types Resistant to the Disease. Part II. The Isolation of Resistant Types." *Scient. Monogr. Coun. Agric. Res. Pusa* 7 (1933): 37.

26. Padwick, G. W. "Report of the Imperial Mycologist." *Scient. Rep. Agric. Res. Inst. New Delhi. 1938–39*, 105.

27. Padwick, G. W. "Genus *Fusarium* 5: *Fusarium udum* Butler, *F. vasinfectum* Atk. and *F. lateritium* var. *uncinatum* W. R." *Indian J. Agric. Sci.* 10 (1940): 863.

28. Padwick, G. W., Mitra, M., and Mehta, P. R. "The genus *Fusarium*. IV. Infection and Cross Infection Tests with Isolates from Cotton (*Gossypium* sp.), Pigeon-Pea (*Cajanus cajan*) and Sunn-Hemp (*Crotalaria juncea*). *Indian J. Agric. Sci.* 10 (1940): 697.

29. Dey, P. K. *Administration Report Agriculture Department. U.P. 1944–45*, 38, 1947.

30. Dey, P. K. *Administration Report Agriculture Department. U.P. 1945–46*, 43.

31. Dey, P. K. *Administration Report Agriculture Department. U.P. 1946–47*, 39.

32. Sarojini, T. S. *Soil Conditions and Root Diseases with Special Reference to Fusarium udum on Red Gram*. (Ph.D. thesis, University of Madras, Madras, 1946).

33. Sarojini, T. S. "Soil Conditions and Root Diseases, Micro-Nutrient Elements and Disease; Development of *Fusarium udum* on Red Gram (*Cajanus indicus* Linn.)." *J. Madras Univ.* (Sec. B) 19 (1950): 1.

34. Sarojini, T. S. "Soil Conditions and Root Diseases. Part II. *Fusarium* disease of Red Gram (*Cajanus indicus*)." *Proc. Indian Acad. Sci.* (Sec. B) 33 (1951): 49.

35. Sarojini, T. S. "Soil Conditions and Root Diseases. III. *Neocosmospora vasinfecta* Smith. Disease of *Cajanus cajan.*" *J. Madras Univ.* (Sec. B) 24 (1954): 137.

36. Sarojini, T. S., and Yogeswari, L. "Aeration Affecting Growth and Sporulation of Some Fusaria in Liquid Cultures." *Proc. Indian Acad. Sci.* (Sec. B) 26 (1947): 69.

37. Kalyansundaram, R. "Ascorbic Acid and *Fusarium* Wilted Plants." *Proc. Indian Acad. Sci.* (Sec. B) 36 (1952): 102.

38. Satyanarayana, G., and Kalyansundaram, R. "Soil Conditions and Root Diseases. V. Symptomology of Wilted Cotton and Red Gram." *Porc. Indian Acad. Sci.* (Sec. B) 36 (1952): 54.

39. Yogeshwari, L. "The Element Nutrition of Fungi. I. The Effect of Boron, Zinc, and Manganese on *Fusarium* Species." *Proc. Indian Acad. Sci.* (Sec. B) 28 (1948): 177.

40. Subramanian, C. V. "Studies on South Indian Fusaria. IV. 'The Wild Type' in *Fusarium udum* Butler." *J. Indian Bot. Soc.* 34 (1955): 29.

41. Subramanian, C. V. *Hyphomycetes.* Indian Council of Agricultural Research, New Delhi, 1971, 930.

42. Subramanian, S. "*Fusarium* Wilt of Pigeon-Pea. I. Symptomology and Infection Studies." *Proc. Indian Acad. Sci.* (Sec. B) 57 (1963): 134.

43. Subramanian, S. "*Fusarium* Wilt of Pigeon-Pea. II. Changes in the Host Metabolism." *Proc. Indian Acad. Sci.* (Sec. B) 57 (1963): 178.

44. Subramanian, S. "*Fusarium* Wilt of Pigeon-Pea. III. Manganese Nutrition and Disease Resistance." *Proc. Indian Acad. Sci.* 57 (1963): 259.

45. Vasudeva, R. S. "The Effect of Associated Soil Microflora of *Fusarium udum* Butler on the Causing of Wilt of Pigeon-Pea (*Cajanus indicus*)." *Proc. Sixth Inter. Cong. Microbiol.* 5 (1955): 239.

46. Vasudeva, R. S. "Report of the Division of Mycology and Plant Pathology." *Rep. Agric. Res. Inst., New Delhi, 1955–56,* 85.

47. Vasudeva, R. S. "Report of the Division of Mycology and Plant Pathology." *Rep. Agric. Res. Inst., New Delhi, 1956–57,* 86.

48. Vasudeva, R. S., and Govindaswamy, C. V. "Studies on the Effect of Associated Soil Microflora on *Fusarium udum* Butl. The Fungus causing the Wilt of Pigeon-Pea (*Cajanus cajan* (L.) Millsp.) with Special Reference to Its Pathogenecity." *Ann. Appl. Biol.* 40 (1953): 573.

49. Vasudeva, R. S., and Roy, T. C. "The Effect of Associated Soil Microflora on *Fusarium udum* Butler the Fungus Causing Wilt of Pigeon-Pea (*Cajanus cajan* (L.) Millsp.)." *Ann. Appl. Biol.* 38 (1950): 169.

50. Vasudeva, R. S., Jain, A. C., and Nema, K. G. "Investigation of the Inhibitory Action of *Bacillus subtilis* on *Fusarium udum* Butl., the Fungus Causing Wilt of Pigeon-Pea (*Cajanus cajan* (L.) Millsp.)." *Ann. Appl. Biol.* 39 (1952): 229.

51. Vasudeva, R. S., Subbaiah, T. V., Sastry, M. L. N., Rangaswamy, G., and Ayengar, R. S. "'Bulbiformin' an Antibiotic Produced by *Bacillus subtilis.*" *Ann. Appl. Biol.* 46 (1958): 336.

52. Vasudeva, R. S., Singh, G. P., and Iyengar, M. R. S. "Biological Activity of Bulbiformin in Soil." *Ann. Appl. Biol.* 50 (1962): 113.

53. Vasudeva, R. S., Singh, P., Sen Gupta, P. K., and Mahmood, M. "Further Studies on the Biological Activity of Bulbiformin." *Ann. Appl. Biol.* 51 (1963): 415.

54. Mahmood, M. "Factors Governing the Production of Antibiotic Bulbiformin and Its Use in the Control of Pigeon-Pea Wilt. (Ph.D. thesis, IARI, New Delhi, 1962), 101.

55. Singh, G. P. "Studies on Wilt of Arahar." (Ph.D. thesis, Agra University, Agra, India, 1965).

56. Singh, G. P., and Husain, A. "Production of Pectic and Cellulolytic Enzymes by Arhar Wilt Fungus." *Curr. Sci.* 31 (1962): 110.

57. Singh, G. P., and Hussain, A. "Presence of Fusaric Acid in Wilt Affected Pigeon-Pea Plants." *Curr. Sci.* 33 (1964): 287.

58. Singh, G. P., and Husain, A. "Role of Enzymes in Pathogenesis by *Fusarium lateritium* f. sp. *cajani.*" *Indian Phytopathol.* 21 (1968): 361.

59. Singh, G. P., and Husain, A. "Role of Toxic Metabolites of *Fusarium lateritium* f. sp. *cajani* (Padw.) Gord. in the Development of Pigeon-Pea Wilt." *Proc. Natl. Acad. Sci. India* (Sec. B) 40 (1970): 9.

60. Prasad, M., and Chaudhary, S. K. "Effect of Sulphur on Sporulation of *Fusarium udum* Butler." *J. Indian Bot. Sci.* 46 (1967): 45.

61. Prasad, M., and Chaudhary, S. K. "Influence of IAA and Gibberellic Acid on Mycelial Output and Sporulation of Varied Spore Forms of *Fusarium oxysporum* f. sp. *udum.*" *Indian Phytopathol.* 29 (1976): 193.

62. Prasad, M., and Chaudhary, S. K. "Relation of pH Levels and Varied Nutrient Media to Growth and Sporulation of *Fusarium oxysporum* f. sp. *udum* (Butler) Sn. et H." *Res. J. Ranchi Univ.* 13 (1977): 214.

63. Prasad, M., and Chaudhary, S. K. "Effect of Different Concentrations of DL-isoleucine, DL-valine and DL-alanine on Growth and Sporulation of *Fusarium oxysporum* f. *udum* (Butler) Sny. et Hans." *Zent. Bakteriol., Parasit. Infect. Hygiene* 132 (1977): 735.

64. Prasad, M., and Chaudhary, S. K. "Effect of C.C.C. 2-(Chloroethyl), Trimethyl Amonium Chloride on Growth and Sporulation in *F. oxysporum* f. sp. *udum* (Butl.) Sn. et H." *Zent. Bakteriol. Parasiten, Infect. Hygiene* 133 (1978): 86.

65. Chaudhary, S. K., and Prasad, M. "Variations in Sugar Contents of Healthy and *Fusarium oxysporum* f. sp. *udum* Infected Plants of *Cajanus cajan.*" *Phytopathol. Z.* 80 (1974): 303.

66. Kaiser, S. A. K., and Sen Gupta, P. K. "Cross Protection Against Wilt Disease Caused by *Fusarium oxysporum* f. sp. *udum* in Pigeon-Pea." *Indian J. Mycol. Res.* 7 (1960): 38.

67. Kaiser, S. A. K., and Sen Gupta, P. K. "Infection and Pathological Histology of Pigeon-Pea (*Cajanus cajan* (L.) Millsp.) Inoculated with Pathogenic and Non-pathogenic Formae Speciales of *Fusarium oxysporum.*" *Z. Pflanzk. Pflanzs.* 82 (8–9) (1975): 482.

68. Kaiser, S. A. K. M., and Sen Gupta, P. K. "Inhibition of Wilt Symptoms of *Fusarium oxysporum* f. sp. *udum* in Pigeon-Pea (*Cajanus cajan*) Induced by Other Forme Speciales of *Fusarium oxysporum.*" *Phytopathol. Mediterr.* 16 (1977): 1.

69. Shit, S. K. "Studies on Variability in *Fusarium oxysporum* f. sp. *udum* the Incitant of Wilt of Pigeon-Pea." (M. S. thesis, Bidhan Chandra Krishi Viswavidyalaya, Kalyani, West Bengal, 1976), 57.

70. Shit, S. K., and Sen Gupta, P. K. "Possible Existence of Physiological Races of *Fusarium oxysporum* f. sp. *udum*, Incitant of the Wilt of Pigeon-pea." *Indian J. Agric. Sci.* 48 (1978): 629.

71. Shit, S. K., and Sen Gupta, P. K. "Pathogenic and Enzymatic Variation in *Fusarium oxysporum* f. sp. *udum.*" *Indian J. Microbiol.* 20 (1980): 46.

72. Singh, N., and Singh, R. S. "Inhibition of *Fusarium oxysporum* f. sp. *udum* by Soil Bacteria." *Indian Phytopathol.* 33 (1980): 356.

73. Singh, N., and Singh, R. S. "Lysis of *Fusarium oxysporum* f. sp. *udum* Caused by Soil Amended with Organic matter." *Indian J. Mycol. Plant Pathol.* 10 (1980): 146.

74. Singh, N., and Singh, R. S. "Lysis of Mycelium of *Fusarium oxysporum* f. sp. *udum* in Soil Amended with Organic Matter." *Plant Soil* 59 (1981): 9.

75. Singh, N., and Singh, R. S. "Effect of Oil-Cake Amended Soil Atmosphere on Pigeon-Pea Wilt Pathogen." *Indian Phytopathol.* 35 (1982): 300.

76. Singh, N., and Singh, R. S. "Development of *Fusarium udum* (Pigeon-Pea Wilt Pathogen) in Oil Cake Amended Soil." *Indian Botanical Reptr.* 2 (1983): 116.

77. Singh, N., and Singh, R. S. "Chlamydospore Formation in *Fusarium udum* by Soil Bacteria." *Indian Phytopathol.* 36 (1983): 165.

78. Singh, R. S. "Effect of Host and Non-host Crop on Vegetative Growth and Sporulation of *Fusarium udum* in Soil." *Indian Phytopathol.* 27 (1974): 553.

79. Singh, R. S. *Studies on Fusarium. Research Bulletin No. 7*, G.B. Pant University of Agriculture and Technology, Pantnagar, Nainital, U.P., 1975.

80. Singh, R. S., and Chaube, H. S. "Development of Fusaria in Soil Amended with High Carbon Sources," in *Studies on Fusarium*, R. S. Singh. Research Bull. No. 7, College of Agriculture and Experiment Station, G.B. Pant Univ. Agric. & Technology, Pantnagar, 1975, 97.

81. Singh, R. S., and Singh, N. "Effect of Cake Amendment of Soil on Sporulations of Some Wilt Causing Species of *Fusarium.*" *Phytopathol. Z.* 69 (1970): 160.

82. Upadhyay, R. S. "Ecological Studies on *Fusarium udum* Butler Causing Wilt Disease of Pigeon-Pea." (Ph.D. thesis, Banaras Hindu University, 1979).

83. Upadhyay, R. S. "Tolerance to Higher Temperature by *Aspergillus nidulans* and its Possible Implication on biological Control of Wilt Disease of Pigeon-Pea." *Plant Soil* 97 (1987): 273.

84. Upadhyay, R. S., and Rai, B. "*Micromonospora globosa* Krass: A Destructive Parasite of *Fusarium udum* Butler." *Microbios Lett.* 8 (1978): 123.

85. Upadhyay, R. S., and Rai, B. "Tolerance of Higher Temperatures by *Aspergillus nidulans* in Competition with Other Soil Fungi." *Proc. Natl. Acad. Sci. India* 77, 1978.

86. Upadhyay, R. S., and Rai, B. "*Coprinus lagopus* as Potent and Frequent Saprophytic Colonizer of Pigeon-Pea in Soil." *Sci. Cult.* 45 (1979): 171.

87. Upadhyay, R. S., and Rai, B. "Fungistatic Activity of Different Indian Soils Against *Fusarium udum* Butler." *Plant Soil* 63 (1981): 407.

88. Upadhyay, R. S., and Rai, B. "Effect of Cultural Practices and Soil Treatments on Incidence of Wilt Disease of Pigeon-Pea." *Plant Soil* 62 (1981): 309.

89. Upadhyay, R. S., and Rai, B. "Ecology of *Fusarium udum* Causing Wilt Disease of Pigeon-Pea: Population Dynamics in the Root Region." *Trans. Br. Mycol. Soc.* 78 (1982): 209.

90. Upadhyay, R. S., and Rai, B. "A Possible Mode of Dispersion of *Fusarium udum* Butler in Soil by Termites." *Sci. Cult.* 48 (1982): 207.

91. Upadhyay, R. S., and Rai, B. "A New Disease Cycle of Wilt of Pigeon-Pea." *Curr. Sci.* 52 (1983): 978.

92. Upadhyay, R. S., and Rai, B. "Hyphal Interaction and Parasitism Amongst Some Rhizosphere Fungi of Pigeon-Pea." *Indian Phytopathol.* 36 (1983): 302.

93. Upadhyay, R. S., and Rai, B. "Competitive Saprophytic Ability of *Fusarium udum* Butler in Relation to Some Microfungi of Root Region of Pigeon-Pea." *Indian Phytopathol.* 36 (1983): 539.

94. Upadhyay, R. S., and Rai, B. "Antagonistic Behaviour of Root Region Microfungi of Pigeon-Pea against *Fusarium udum*," in *Ecology and Management of Soil-borne Plant Pathogens*, eds. C. A. Parker, A. D. Rovira, K. J. Moore, P. T. W. Wong, and J. F. Kollmorgen. The American Phytopathological Society, (St. Paul, Minnesota: 1985), 131.

95. Upadhyay, R. S., Rai, B. "Studies on Antagonism Between *Fusarium udum* Butler and Root Region Microflora of Pigeon-Pea." *Plant Soil* 101 (1987): 79.

96. Upadhyay, R. S., Rai, B., and Gupta, R. C. "Hyphal Parasitism and Reproductive Body Formation by *Fusarium udum* within *Cunninghamella echinulata* During Parasitism." *Microbios Lett.* 11 (1979): 69.

97. Upadhyay, R. S., Rai, B., and Gupta, R. C. "A Possible Mode of Survival of *Fusarium udum* Butler as a Mycoparasite." *Acta Mycologica* 19 (1983): 115.

98. Rai, B., and Upadhyay, R. S. "Competitive Saprophytic Colonization of Pigeon-Pea Substrate by *Fusarium udum* in Relation to Environmental Factors, Chemical Treatments and Microbial Antagonism." *Soil boil. Biochem.* 15 (1983): 187.

99. Murthy, G. S., and Bagyaraj, D. J. "Free Amino Nitrogen and Amino Acids in *Cajanus cajan* in Relation to *Fusarium* Wilt Resistance." *Indian Phytopathol.* 31 (1978): 482.

100. Murthy, G. S., and Bagyaraj, D. J. "Rhizosphere Mycoflora of *Cajanus cajan* in Relation to *Fusarium* Wilt Resistance." *Plant Soil* 50 (1978): 485.

101. Murthy, G. S., and Bagyaraj, D. J. "Flavanol and Alkaloid Content of Pigeon-Pea Cultivars Resistant and Susceptible to *Fusarium udum*." *Indian Phytopathol.* 33 (1980): 633.

102. Haider, M. G., Singh, R. K., Prasad, H., Nath, R. P., and Sharma, R. N. "Effect of Some Common Fungicides on the Incidence of Pigeon-Pea Wilt." *Indian Phytopathol.* 31 (1978): 511.

103. Venkateshwarulu, S., Reddy, A. R., Singh, R. M., and Singh, R. B. "Reaction of Pigeon-Pea Varieties to Wilt and Mosaic Virus (SMV)." *Trop. Grain Leg. Bull.* 17/18, (1980): 25.

104. Venkateshwarulu, S., Reddy, A. R., Chauhan, V. B., Singh, O. N., Singh, R. M., and Singh, U. P. "A Promising Resistant Line of Pigeon-Pea (Malviya Arar-1) for Wilt and Sterility Mosaic (virus?)." *Trop. Grain Leg. Bull.* 23 (1981): 27.

105. *CMI Distribution Maps of Plant Diseases, Map No.563*, Commonwealth Mycological Institute, Kew, Surrey, England, 1985.

106. Booth, C. *The Genus Fusarium*. Commonwealth Mycological Institute, England, 1971, 137.

107. Kannaiyan, J., Reddy, M. V., Nene, Y. L., and Raju, T. N. "International Survey of Pigeon-Pea Diseases." *Pulse Pathology Progress Report No. 12* (1981): 82.

108. Kannaiyan, J., Reddy, M. V., Nene, Y. L., Ryan, J. G., and Raju, T. N. "Prevalance of Pigeon-Pea Wilt and Sterility Mosaic in India: 1975–1980." *International Pigeon-pea Newsletter* 1 (1981): 24.

109. Kannaiyan, J., Nene, Y. L., Reddy, M. V., Ryan, J. G., and Raju, T. N. "Prevalence of Pigeon-pea Diseases and Associated Crop Losses in Asia, Africa and the Americas." *Trop. Pest Manag.* 30 (1984): 62.

110. International Crops Research Institute for the Semi-Arid Tropics. *Annual Report 1982*, Patancheru, A. P., India, ICRISAT, 1983.

111. International Crops Research Institute for the Semi-Arid Tropics, *Annual Report–Pulse Pathology*, Patancheru, A. P., India, ICRISAT, 1986–87.

112. Chaube, H. S. "Combating Diseases of Arhar and Gram." *Indian Farmers' Digest* 1(7) (1968): 26.

113. Amin, K. S., Baldev, B., and Williams, F. J. "Differentiation of *Phytophthora* Stem Blight from *Fusarium* Wilt of Pigeon-Pea by Field Symptoms." *FAO Pl. Prot. Bull.* 24 (1976): 123.

114. Singh, R. S. *Introduction to Principles of Plant Pathology.* New Delhi: Oxford and I. B. H. Publishing Co., 1975, 282.

115. Wollenweber, H. W. "Studies on *Fusarium* Problem." *Phytopathology* 3 (1913): 24.

116. Wollenweber, H. W. "Fusaria Autographice Delineata." *Ann. Mycol.* 15 (1917): 1.

117. Snyder, W. C., and Hansen, H. N. "The Species Concept in *Fusarium.*" *Am. J. Bot.* 27 (1940): 64.

118. Gordon, W. L. "The Occurrence of *Fusarium* Species in Canada. II. Prevalence and Taxonomy of *Fusarium* Species in the Cereal Seed." *Can. J. Bot.* 30 (1952): 209.

119. Kannaiyan, J., Nene, Y. L., and Raju, T. N. "Host Specificity of Pigeon-Pea with Pathogen *Fusarium udum.*" *Indian Phytopathol.* 38 (1985): 553.

120. Dwivedi, R. S., and Tandon, R. N. "Studies on Some Aspect of Seed Mycoflora of Pigeon-Pea." *Proc. Indian Sci. Cong.* 63 (1975): 63.

121. International Crops Research Institute for Semi-Arid Tropics. *Annual Report–Pulse Pathology*, Patancheru, A. P., India, 1985–86.

122. Roy, T. C. "I. Studies on the Soil Microorganisms with Special Reference to Their Antibiotic effect on *Fusarium udum* Butler, the Wilt Organism of Arhar (*Cajanus cajan* (Linn.) Millsp.). II. Some Aspects of Cultural Variation and taxonomic Considerations of *F. udum* Butler, the Causal Organism of Wilt of Pigeon-Pea." (Thesis, Indian Agricultural Research Institute, New Delhi, 1949), 52.

123. Baldev, B., and Amin, K. S. "Studies on the Existence of Races in *Fusarium udum* Causing Wilt of *Cajanus cajan.*" *SABRAO J.* 6(2) (1974): 201.

124. Jeswani, M. D., Prasad, N., and Gemawat, P. D. "Morphological Variability in *Fusarium lateritium* f. sp. *cajani.*" *Indian J. Mycol. Pl. Path.* 5 (1977): 4.

125. Mukherjee, D., De, T. K., and Parui, N. R. "A Note on Screening of Arhar Against Wilt Disease." *Indian Phytopathol.* 24 (1971): 598.

126. Venkata Ram, C. S. "Soil Fusaria and Their Pathogenicity." *Proc. Indian Acad. Sci.* (Sec. B) 42 (1955): 124.

127. Mundkur, B. B. *Fungi and Plant Diseases.* London: Macmillan & Co. Limited, 1967.

128. Agnihothrudu, V. "Soil Conditions and Wilt Disease in Plants. Rhizosphere Microflora in Relation to Fungal Wilts." (Ph.D. thesis, Madras University, Madras, 1954).

129. Nene, Y. L., and Reddy, M. V. "Survival of Pigeon-Pea Wilt *Fusarium* in Vertisols and Alfisols," in *Proc. Intl. Workshop on Pigeon-Peas,* Vol. 2, Patancheru, ICRISAT, Andhra Pradesh, 1981, 291.

130. Bose, R. D. "The Rotation of Tobacco for the Prevention of Wilt Disease of Pigeon-Pea." *Agric. Live-Stock India* 6 (1938): 653.

131. Mitra, M. "Report of the Imperial Mycologist." *Scient. Rep. Agric. Res. Inst. Pusa. 1924–25*, 45.

132. Mahmood, M. "Factors Governing the Production of Antibiotics Bulbiformin and Its Use in the Control of Pigeon-Pea Wilt." *Sci. Cult.* 30 (1964): 352.

133. Agnihothrudu, V. "Incidence of Fungistatic Organisms in the Rhizosphere of Pigeon-Pea (*Cajanus cajan*) in Relation to the Resistance and Susceptibility to Wilt Caused by *Fusarium udum* Butl." *Naturwiss.* 42 (1955): 1.

134. Hansford, C. G. "Annual Report of the Plant Pathologist 1936." *Rep. Dep. Agric. Uganda*, 1936–37 (Part 2), 43, 1938.

135. Wiehe, P. O. "Division of Plant Pathology." *Rep. Dep. Agric. Mauritius* 1938, 34–39.

136. Mundkur, B. B. "Influence of Temperature and Maturity on the Incidence of Sunn-Hemp and Pigeon-Pea Wilt at Pusa." *Indian J. Agric. Sci.* 5 (1935): 609.

137. Shukla, D. S. "Incidence of *Fusarium* Wilt of Pigeon-Pea in Relation to Soil Composition." *Indian Phytopathol.* 28 (1975): 295.

138. Gupta, S. L. "The Effect of Mixed Cropping of Arhar (*Cajanus indicus* Spreng.) with Jowar (*Sorghum vulgare* Pers.) on Incidence of Arhar Wilt." *Agric. Anim. Husb. U.P.* 3 (1961): 31.

139. Natrajan, M., Kannaiyan, J., Willey, R. W., and Nene, Y. L. "Studies on Cropping System on *Fusarium* Wilt of Pigeon-Pea." *Field Crops Research* 10 (1985): 333.

140. Anonymous. "Testing of Arhar (Pigeon-Pea) Strains Against Wilt Disease." *Pesticides* 10 (1976): 17.

141. Ghosh, M. K. *Control of Fusarium Wilt of Pigeon-Pea by Various Treatments.*" (M. S. thesis, Bidhan Chandra Krishi Viswavidyalaya, Kalyani, West Bengal, 1975).

142. Ghosh, M. K., and Sinha, A. K. "Laboratory Evaluation of Some Systemic Fungicides Against *Fusarium* Wilt of Pigeon-Pea." *Pesticides* 15 (1981): 24.

143. Sinha, A. K. "Control of *Fusarium* Wilt of Pigeon-Pea with Bavistin, a Systemic Fungicide." *Curr. Sci.* 44 (1975): 700.

144. Chakrabarti, S., and Nandi, P. "Effect of Griseofulvin on *Fusarium udum* Butler and Its Host Pigeon-Pea (*Cajanus cajan* (L.) Millsp.)." *Proc. Indian Sci. Acad.* 56 (1969): 228.

145. Alabouvette, C., Rouxel, F., and Louvet, J. "Recherche sur la Resistance des Sols aux Maladies III. Effects du Rayonnement sur la Microflore d'un Sol et sa Resistance a la Fusariose Vascul Aire du Melon." *Ann. Phytopath.* 9 (1977): 467.

146. Alabouvette, C., Rouxel, F., and Louvet, J. "Characteristic of *Fusarium* Wilt Suppressive Soils and Prospects for their Utilization in Biological Control," in *Soil Borne Plant Pathogens*, eds. B. Schippers and W. Gams. (London; Academic Press, 1979), 165.

147. Alabouvette, C., Rouxel, F., and Louvet, J. "Recherche sur la Resistance des Sols aux Maladies. VI. Mise en Evidence de la Specificite de la Resistance d'un Sol vis-a-vis des Fusarioses Vasculaires." *Ann. Phytopath.* 12 (1980): 11.

148. Alabouvette, C., Tramier, I. R., and Grouet, D. "Recherches sur la Resistance des Sols aux Maladies. VIII. Perspective d'Utilization de la Resistance des Sols pour Lutter Contre les Fusarioses." *Ann. Phytopath.* 12 (1980): 83.

149. Alabouvette, C., Couteaudier, Y., and Louvet, J. "Soil Suppressive to *Fusarium* Wilt: Mechanisms and Management of Suppressiveness," in *Ecology and Management of Soil-Borne Plant Pathogens*, eds. C. A. Parker, A. D. Rovira, K. J. Moore, P. T. W. Wong, and J. F. Kollimorgen. (St. Paul, Minnesota: The American Phytopathological Society, 1985), 101.

150. Nema, K. G. "Inhibitory Effect of Certain Soil Microorganisms on *Fusarium udum* Butler, the Pigeon-Pea (*Cajanus cajan* (L.) Millsp.) Wilt Organism." (Thesis, Indian Agricultural Research Institute, New Delhi, 1950), 54.

151. Jain, A. C. "Effect of Certain Microorganisms on the Activity of *Fusarium udum* Butler, the Causal Agent of Pigeon-Pea (*Cajanus cajan* (Linn.) Millsp.) Wilt." (Thesis, Indian Agricultural Research Institute, New Delhi, 1949), 50.

152. Maitra, A., and Sinha, A. K. "Partial Inhibition of *Fusarium* Wilt Symptoms in Pigeon-Pea by Non-Pathogenic Formae of *Fusarium oxysporum*." *Curr. Sci.* 42 (1973): 654.

153. Indian Agricultural Research Institute, Scientific Report of the Indian Agricultural Research Institute, New Delhi for the year 1947–48, 182.

154. Podile, A. R., and Dube, H. C. "Effect of *Bacillus subtilis* on Vascular Wilt Fungi." *Curr. Sci.* 54 (1985): 1282.

155. Podile, A. R., Prasad, G. S., and Dube, H. C. "*Bacillus subtilis* as an Antagonist to Vascular Wilt Pathogens." *Curr. Sci.* 54 (1985): 864.

156. Utkhede, R. S., and Rahe, J. E. "Biological Control of White Onion Rot." *Soil Biol. Biochem.* 12 (1980): 101.

157. Olsen, C. M., and Baker, K. F. "Selective Heat Treatment of Soil, and Its Effect on the Inhibition of *Rhizoctonia solani* by *Bacillus subtilis*." *Phytopathology* 58 (1968): 79.

158. Chang, I. P., and Kommedahl, T. "Biological Control of Seedling Blight of Corn by Coating Kernels with Antagonistic Microorganisms." *Phytopathology* 58 (1968): 1395.

159. Swinburne, T. R., Barr, J. G., and Brown, A. E. "Production of Antibiotics by *Bacillus subtilis* and Their Effect on Fungal Colonists of Apple Leaf Scars." *Trans. Brit. Mycol. Soc.* 65 (1975): 211.

160. Broadbent, P., Baker, K. F., and Waterworth, Y. "Bacteria and Actinomycetes Antagonistic to Fungal Root Pathogens in Australian Soils." *Austral. J. Biol. Sci.* 24 (1971): 925.

161. Cook, R. J., and Baker, K. F. *The Nature and Practice of Biological Control of Plant Pathogens.* St. Paul, Minnesota: The American Phytopathological Society, 1983, p. 539.

162. Deacon, J. W. "Biological Control of Take-All Fungus, *Gauemannomyces graminis* var. *tritici*, by *Phialophora radicicola* and Similar Fungi." *Soil Biol. Biochem.* 8 (1976): 275.

163. Hornby, D. Personal communication.

164. Alam, M. *Administration Report of the Botanical Section for the year ending 31st March 1931. Appendix 1(8): Rep. Dep. Agric. Bihar. Orissa. for the period from 1st April 1930 to 31st March 1931*, 42, 1931.

165. Anonymous. *Review of Agricultural Operations in India. 1928-29. Imp. Coun. Agric. Res. Pusa.* 1931, 251.

166. Anonymous. Pigeon-pea. *Indian Fung.* 1 (1940): 178.

167. Anonymous. *Agriculture and Animal Husbandry in India 1938–39, Imp. Coun. Agric. Res. Delhi*, 1941, 422.

168. Anonymous. *Annual Administration Report of the Department of Agriculture. Uttar Pradesh. for the year ending June 30, 1949*, 125.

169. Ravishanker, "Isolation of Wilt Resistant Tur." *Nagpur Agric. Coll. Mag.* 10 (1936): 162.

170. Dastur, J. F. "Report of the Imperial Mycologist." *Scient. Rep. Agric. Res. Institute. New Delhi. 1944–45*, 66.

171. Indian Council of Agricultural Research. *Annual Report for 1948–49. Delhi*, 1950, 177.

172. Indian Agricultural Research Institute. *Scientific Report of the Indian Agricultural Research Institute. for the year ended 30th June 1951*, 120.

173. Vaheeduddin, S. "Selection of Tur (*Cajanus cajan* (L.) Resistant Varieties Against Wilt (*Fusarium udum* Butler)." *Agric. Coll. J. Osmania Univ.* 3 (1956): 12.

174. Vaheeduddin, S., and Nanjundiah, S. N. "Evolving Wilt Resistant Strains in Tur (*Cajanus cajan* L.)." *Proc. Indian Sci. Congr. Assoc.* 43 (1956): 20.

175. Deshpande, R. B., Jeswani, L. M., and Joshi, A. B. "Breeding Wilt Resistant Varieties of Pigeon-Pea." *Indian J. Genet.* 23 (1963): 58.

176. Raut, N. K., and Bhombe, B. B. "A Review of the Work of Selections of Tur Varieties Resistant to *Fusarium* Wilt at College of Agriculture, Parbani (Maharashtra)." *Magaz. Coll. Agric., Parbani* 12 (1971): 37.

177. Patil, B. G., and Samble, J. E. "A Note on the Screening of Tur Against Wilt Disease." *PKV Res. J.* 2 (1973): 73.

178. Mundkur, B. B. "Report of the Imperial Mycologist." *Scient. Rep. Agric. Res. Inst., New Delhi, forthe Triennium ended 30th June, 1944,* 57, 1946.

179. Singh, D. V., and Mishra, A. N. "Search for Wilt Resistant Varieties of Red Gram in Uttar Pradesh." *Indian J. Mycol. Pl. Pathol.* 6 (1976): 89.

180. Nene, Y. L., and Kannaiyan, J. "Screening Pigeon-Pea for Resistance to *Fusarium* Wilt." *Plant Disease* 66 (1981): 306.

181. Amin, K. S. *Consolidated Report on Kharif Pulses, Plant Pathology* (All India Co-ordinated Pulses Improvement Project), Directorate of Pulses Research, ICAR, Kanpur, India, 1986–87.

182. Reddy, M. V., Raju, T. N., and Nene, Y. L. "Low Wilt Incidence in Short Duration Pigeon-peas." *Intl. Pigeonpea Newsletter* 7, 1986.

183. Nene, Y. L. "Multiple-disease Resistance in Grain Legumes." *Ann. Rev. Phytopathol.* 26 (1988): 203.

184. International Crops Research Institute for Semi-Arid Tropics, Annual Report—Pulse Pathology, Patancheru, A. P., India, ICRISAT, 1982.

185. Konda, C. R. Parmeswarappa, R., and Rao, T. S. "Pigeon-Pea ICP 8863—A Boon to Fusarium Wilt Epidemic Areas of Karnataka." *International Pigeonpea Newsletter* 36, 1986.

186. Parmeswarappa, R., Rao, T. S., Konda, C. R., Rao, T. M., Raghumurty, M., Emmimath, V. S., and Saguppa, K. K. "Maruthi—A Wilt Resistant Red Gram Variety." *Current Research. Univ. of Agric. Sci., Bangalore* 15 (1986): 69.

187. Prasad, M., and Chaudhary, S. K. "Studies on the Effect of Different Phosphorus Concentrations on the Production of Chlamydospores, Microconidia and Macroconidia in the Culture of *Fusarium udum* Butler." *Proc. Natl. Acad. Sci.* (Sec. B) 36 (1966): 43.

188. Kannaiyan, J., and Nene, Y. L. "Influence of Wilt at Different Growth Stages on Yield Loss in Pigeon-Pea." *Trop. Pest Manag.* 27 (1981): 141.

189. Vasudeva, R. S. "Soil-Borne Plant Diseases and Their Control." *Curr. Sci.* 18 (1949): 114.

190. Reddy, N. P. E., and Chaudhary, K. C. B. "Variation in *Fusarium udum.*" *Indian Phytopathol.* 38 (1985): 172.

191. Zote, K. K., and Dandnaik, B. P. "Screening of Pigeonpea for Multiple Disease Resistance." *Indian Phytopathol.* 40 (1987): 59.

192. Kotasthane, S. R., Om Gupta, and Khare, M. N. "Influence of Fungicidal Seed Treatment and Soil Amendment on the Development of *Fusarium udum* Propagules in Soil and Pigeon-pea Wilt." *Indian Phytopathol.* 40 (1987): 197.

19

STERILITY MOSAIC OF PIGEON PEA

A. M. GHANEKAR, V. K. SHEILA, S. P. S. BENIWAL, M. V. REDDY AND Y. L. NENE

ICRISAT, Hyderabad, A. P. (India)

19-1 INTRODUCTION

Sterility mosaic (SM) is one of the most important diseases of pigeon pea in India. It was first reported from Pusa in the state of Bihar in India more than 50 years ago.[1] Alam[2] gave a detailed description of SM. Capoor[3] established the infectious nature of the disease through graft transmission of the SM pathogen to healthy pigeon pea; he also reported sap transmission of the causal pathogen but this could not be confirmed later by others.[4-5] The disease is not seedborne. The most important contribution of practical significance came from Seth[6] when he showed that the SM pathogen is transmitted under natural conditions by an eriophyid mite, *Aceria cajani* Channa Basavanna. Over the years this finding has been confirmed by other workers.[5,7]

The disease is present in the major pigeon pea producing states of India and, of late, it has become a serious problem in northeastern (especially Bihar and Uttar Pradesh) and southern (especially Tamil Nadu) states of India.[8] The disease appears to be restricted to Asia and is reported from Bangladesh,[9] Burma,[10] Sri Lanka,[9] Thailand,[9] and Nepal.[8,9]

19-2 LOSSES

A susceptible genotype infected at early stages (first 45 days) of crop growth shows near complete sterility and yield loss up to 95%. As the plants become older (>45 days), their susceptibility to the SM pathogen decreases; such plants show partial sterility.[11] In case of early infection, the yield reduction was related to the percentage of infected plants, but in case of late infection, the yield reduction was not correlated to the percentage of infected plants as the plants showed only partial sterility. The disease causes an estimated annual loss of 205,000 tons of grains in India alone.[8]

19-3 SYMPTOMS

The symptoms of SM in a susceptible cultivar are characterized by a bushy and pale green appearance of plants, drastic reduction in leaf size, increased number of secondary and tertiary branches from leaf axils, and complete or partial cessation of reproductive structures[5] (Figure 19-1). Because the infected plants show "sterility" and the leaves show the "mosaic" symptom, the name of the disease is *sterility mosaic*. While screening pigeon-pea germplasm for resistance to SM at ICRISAT, three types of symptoms were commonly seen. The first is severe mosaic on leaflets where the plants do not produce flowers and pods. The second is the ring spot where there is no sterility; this reaction is characterized by green islands surrounded by a chlorotic halo on leaflets (Figure 19-2) and symptoms tend to disappear as the plants mature. It was confirmed that the ring spot reaction was produced by the SM pathogen on certain genotypes.[12] The third is the mild mosaic on leaflets where only a partial sterility is caused.

19-4 BIOLOGY OF THE CAUSAL PATHOGEN

19-4-1 Graft Transmission

Graft transmission of SM was first shown by Capoor[3] before the mite vector was reported. Since the mite vector of SM is extremely small, there is a possibility that it could go unnoticed in the stem pieces to be used as scions. In the authors' laboratory, a tissue implantation method of graft inoculation was successfully done after eliminating the

Figure 19-1   Pigeon pea plants showing severe mosaic symptoms caused by sterility mosaic

Figure 19-2   A pigeon pea plant showing
ring spot symptom caused by sterility
mosaic

mites. In this method, about 1-cm infected stem pieces, used as scions, were pretreated
with 0.3% Metasystox (demeton-S-methyl) and were implanted below the growing point
of healthy pigeon pea plants by making a vertical slit in the stem. Symptoms of SM were
seen after about a month and success of graft transmission was 12%.

19-4-2 Mechanical Sap Transmission

The young leaf tissue from SM infected pigeon pea was ground in cold buffer, in a ratio of
1:10, in a chilled mortar. Using the thick end of a pestle, the crude sap was rubbed onto
leaf surface of different indicator plant species which were earlier dusted with carborun-
dum. Rubbed leaves were then gently rinsed with tap water and the plants were kept in
glasshouse for expression of symptoms. Many buffers, including potassium phosphate
(pH 7.5 to 8.0) and sodium citrate, containing an antioxidant or a tannin inhibitor, were
used in the sap inoculation experiments. In addition to the leaf-rub method of inoculation,
as described above, the air brush method[13] of inoculation was also used. None of the
following indicator hosts, sap inoculated using different buffers, and infected tissue
sources (root, leaf, flower) showed symptoms of SM: *Chenopodium amaranticolor* Coste
and Reyn., *C. quinoa* Willd, *Cucumis sativus* Linn., *Gomphrena globosa* Linn., *Nico-
tiana benthamiana* Domin, *N. clevelandi* A. Gray, *N. glutinosa* Linn., *N. tabacum* var
Xanthi-nc Linn., *N. tabacum* Samsun-NN Linn., *Petunia hybrida* Vilni., *Phaseolus*

vulgaris Linn. cv. Bountiful, *Vicia faba* Linn. cv. Topcrop, Pinto, *Vigna unguiculata* (L,) Walp cv. Early Ramshorn, and C-152. We also failed to transmit the pathogen to pigeon pea.

19-4-3 Seed Transmission

Six thousand seeds of pigeon pea cv. T-21 collected from SM-infected plants were sown in sterilized soil in pots in a screenhouse. The plants were sprayed with 0.025% Metasystox at weekly intervals and were observed for SM symptoms for 5 months. None of the plants showed symptoms of SM. Thus, the disease is not seed transmissible.

19-4-4 Electron Microscopy

Purified preparations. In order to know the nature of the causal pathogen of SM, the infected leaves were partially purified by using different extraction media which included use of buffers containing antioxidants such as 2-mercaptoethanol, polyethylene glycol with sodium chloride, organic solvents such as carbon tetrachloride and chloroform, and sucrose of cesium chloride gradients. The material was subjected to differential centrifugation. Healthy leaf tissue, collected from plants raised in glasshouse, was also processed simultaneously. At different stages during purification, samples were examined in Philips electron microscope Model 200. Purification methods used for other eriophyid mite transmitted viruses such as wheat streak mosaic and rye grass mosaic[14,15] were also followed. Nothing definite could be said about the nature of SM pathogen, although long flexuous rods were often observed in the preparations from diseased plants, but such particles were also sometimes observed in the preparations obtained from apparently healthy plants.

Ultrathin sections. The SM-infected leaves were processed for ultrathin sectioning[16] and examined for the presence of virus-like particles (VLPs) or inclusion bodies. Neither VLPs nor inclusions such as pinwheels or presence of mycoplasma-like organisms (MLOs) were seen.

19-4-5 Effect of Tetracyclines

In case of plant diseases caused by MLOs, treatment of infected plants with tetracyclines usually resulted in remission of symptoms. With a view to check the involvement of MLOs in the SM disease, four tetracycline compounds: Anchromycin, Aureomycin, Terramycin, and Ledermycin, at 500 ppm were used as foliar sprays on SM-infected plants of cv. Sharada. Tetracycline application did not result in any remission of SM symptoms in pigeon pea.

19-5 BIOLOGY OF THE MITE VECTOR

19-5-1 Morphology and Habitat

A. cajani is a wormlike, eriophyid mite, about 200–250 μm in length. It has two pairs of legs, attached one after another, on its anterior end where its mouth parts are also located; on its posterior end are two cirri which act as hold fast (Figure 19-3). They can be seen clearly under a stereomicroscope at a magnification of 40X. The mites have short life cycles of less than 2 weeks that comprise egg, 2 nymphal stages, and adult. Eggs of *A. cajani* (size 30 × 40 μm) could be detected on the growing tips of pigeon pea plants and are milky white, oval, translucent, and slightly smaller than the glands of trichomes. Oldfield et al.[17] furnished detailed information of *A. cajani* with regard to its life cycle and reproduction. Mites feed with puncturing and sucking type mouth parts that consist of slender stylets. Since they do not possess wings and eyes, their dispersal is passive, and in nature it is mainly by wind currents.

    A. cajani is light-shy and is seen to feed on the lower side (abaxial surface) of the leaflets, preferably at the terminals of a pigeon pea plant. When observed under stereo-microscope, mites are seen partially or totally buried in the thick mass of hairs of pigeon pea leaves. *A. cajani* is host-specific to pigeon pea as under natural conditions has so far been observed to survive on pigeon pea, and its wild relatives, *Atylosia scarabaeoides* and *A. cajanifolia*. Presence of a large number of eriophyids on a pigeon pea leaflet goes unnoticed mainly because they do not cause a visible injury to leaves.

19-5-2 Establishment of SM Pathogen-free Mite Colony

Generally mite colonies are found on SM-diseased plants and in the absence of a definite proof for viral nature of SM, a doubt arises whether SM is a case of mite toxaemia. Establishment of mite colonies on healthy susceptible pigeon pea cultivar should elimi-nate the possibility of mite toxaemia being the cause of SM. A procedure followed at ICRISAT to develop SM pathogen-free mite colony is described below.

    The cultivar ICP 8136 that favors mite multiplication, but is resistant to SM, in its primary leaf stage was staple-inoculated with infected leaves carrying mites. It did not

Figure 19-3 Scanning electron micrograph of *Aceria cajani* on pigeon pea leaf

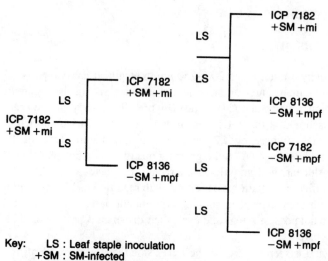

Key: LS : Leaf staple inoculation
        +SM : SM-infected
        −SM : SM-free
        +mi : Infective mites present
       +mpf : Pathogen-free mites present

Figure 19-4 A schematic diagram showing the procedure to raise a sterility mosaic pathogen-free mite colony of *Aceria cajani*

develop SM symptoms up to 30 days after inoculation. Its leaflets carrying mites were staple-inoculated onto a set of plants of ICP 8136 and BDN 1 (ICP 7182) (SM and mite susceptible) in their primary leaf stage and these were watched for symptoms up to 45 days. Both the lines, resistant and susceptible, did not express SM symptoms. A schematic diagram for raising a pathogen-free mite colony is given in Figure 19-4.

It appeared that the new brood of mites that emerged after 30 days on the resistant line ICP 8136 were free of the SM pathogen because the genotype did not allow SM pathogen multiplication and that there is no evidence of transovarial transmission in eriophyid mite vectors.

19-5-3 Pathogen–vector Relationship

The microscopic size and delicate body of eriophyids pose problems in handling, confining, and manipulation for pathogen–vector relationship studies.[18] This is reflected in less number of published papers on these aspects for mite-borne pathogens. So far, such information is only available for wheat streak mosaic virus and its mite vector, *A. tulipae*.[19] The acquisition access, defined as minimum time required by the mite to acquire the pathogen from a suscept, was between 5 to 10 min for SM pathogen and *A. cajani*. The inoculation access, defined as minimum time required by an infective mite to inoculate the pathogen in a suscept host, was found to be 30 min. Transmission of the pathogen by the mites is of persistent type, i. e., pathogen once acquired by the mite is retained for life, provided the vector mite continues to feed on a suscept, such as cv. BDN 1. On the basis of work done at ICRISAT concerning getting a pathogen-free colony of mites on a resistant genotype, it is indicated that the SM pathogen is not transmitted via eggs. Adult and all nymphal stages of mites transmit the SM pathogen.[20]

19-5-4 Multiplication of Mite Vector on SM-resistant and Susceptible Pigeon Pea Lines

These experiments were done by staple-inoculation of SM-infected leaves carrying mites on different resistant and susceptible lines. The average number of mites per square centimeter area on leaflets were counted at 30, 60, and 90 days past inoculation.

In general, resistant genotypes seldom supported continued mite multiplication. In contrast, the susceptible genotypes supported increased mite number, especially in later observations.[21]

19-5-5 Mite Multiplication on *Atylosia spp.*

Atylosia spp. are wild relatives of pigeon pea. Four *Atylosia* spp.; viz., *A. volubilis* (Blanco) Gamble, *A. albicans* Wight and Arnott, *A. lineata* Benth, and *A. sericea* Benth. ex. Bak., showed apparent resistance to the SM pathogen and did not favor mite multiplication. Two *Atylosia* spp., *A. scarabaeoides* (L). Benth. and *A. platycarpa* Benth., did not give consistent results. One species, *A. cajanifolia* Haines, was susceptible to SM pathogen and supported good population of mites.

19-5-6 Are There Biotypes of the Mite Vector or Strains of the SM Pathogen?

Since 1978, through ICAR-ICRISAT Uniform Trial for Pigeon Pea Sterility Mosaic Resistance (IIUTPSMR), several resistant lines identified at ICRISAT Center were tested at 10 different locations in India through the All India Coordinated Pulses Improvement Project (AICPIP). As a result of this study, it was realized that some of the germplasm lines resistant to SM at ICRISAT Center were found susceptible at the other locations. The differential reaction of germplasm lines to SM over locations indicates that there are at least two different biotypes of *A. cajani* or strains of SM pathogen in India. The strains from Dholi, Bangalore, and Vamban fall into one group and those from Badnapur, Hyderabad, Pantnagar, Kanpur, Ludhiana, Faizabad and Varanasi form the second group.

19-6 DISEASE CYCLE IN NATURE

The disease cycle of SM is not fully understood. Since the pathogen is not seedborne, the disease must be getting introduced into the rainy season crop from external sources through the mite vector, *A. cajani*. During summer months (April–May) around Hyderabad, *A. cajani* was fount to survive on *Atylosia scarabaeoides,* a wild relative of pigeon pea commonly seen on field bunds. Though sometimes the *A. scarabaeoides* plants colonized by mites also show mild mosaic mottle symptoms, we have so far failed to transmit the disease from such *A. scarabaeoides* plants to pigeon pea. Under these circumstances, the role of mites found on *A. scarabaeoides* is uncertain. Diseased plants of pigeon pea left over in fields after harvest or growing on field bunds during summer months are definitely a potential source of inoculum for the disease development in the following season. The possibility of such volunteer pigeon peas playing a role in SM disease cycle is greater in northern India where long duration (perennial) pigeon peas are grown and chances for survival of these plants are greater in summer than in peninsular India. In areas where the volunteer pigeon peas are not common, how SM pathogen survives in the absence of pigeon pea during summer months and reappears in the rainy season crop is not known. It was observed that pigeon pea sown late (September) near the kharif plantings develop more disease, indicating the spread of the disease from early infected plants in the early sowings to the late-sown crop.

The observations on disease incidence in pigeon pea fields at ICRISAT indicate that the plants get infected throughout the growing season of the crop, and within an early infected crop there is a secondary spread of the disease in September–October months. The experiments conducted at ICRISAT showed that the disease can spread up 2 km downwind direction from the source of inoculum. The spread against wind direction is very limited (≤ 200 m), indicating that wind assists the spread of the mites.

A large seasonal variation in the incidence of the disease in the farmers' fields is observed in most parts of India. At present there is no information available to explain such variation. In north-eastern parts of India, high incidence of SM is a regular feature and the conditions there appear to be more congenial for survival of the mites and pathogen during the off-season. Also in the areas where vegetation is good and humidity is higher year-round (Pudukkottai in Tamil Nadu, Dharwar in Karnataka and northern-eastern India), the disease incidence is higher than at other places.

19-7 CONTROL

The practical and relatively less expensive method of control is to use disease-resistant varieties. Chemical methods of control, while effective, are expensive.

19-7-1 Host Plant Resistance

Though the causal agent of SM is not yet known, considerable progress has been made on development of inoculation techniques, identification of resistance sources, and development of resistant cultivars for the disease.

Infector hedge inoculation technique.   This field inoculation technique was reported by Nene et al.[22] It consists of advance growing of a hedge of a susceptible cultivar on upwind border of the field which is to be used as screening nursery. Four to eight rows of a susceptible cultivar are sown 4–6 months (December–February) ahead of the normal sowing of the crop (June). When the seedlings are about 10 days old, these are inoculated with SM either by leaf-stapling[7] or spreading the diseased twigs carrying mites on them. The pathogen and mites multiply on these hedge plants and serve as source inoculum for disease spread through wind on to test materials during the cropping season.

Pigeon pea cultivars such as NPWR-15, BDN 1, ICP 8863, susceptible to SM but resistant to fusarium wilt, have been found to be good for the infector hedge. Care should be taken to avoid any insecticide spray or drift on to the infector hedge in addition to providing adequate irrigation during summer months. Once a good hedge is established, it can be effective for 2 to 3 seasons. Frequent pruning of the hedge is done to encourage fresh growth for the colonization of mites.

While planting test materials, rows of a susceptible cultivar (BDN 1 or ICP 8863) are repeated after every 10 rows to serve as indicator rows for the disease spread. The disease spread is seen 3 weeks after the sowing of test materials, first on the materials closer to the hedge, and by the end of the crop season the disease incidence is 100% up to 200 m downwind from the hedge. The susceptible test materials sown closer to the hedge get infected earlier and show severe mosaic coupled with sterility symptoms. The materials further from the hedge sometimes show less severe symptoms and partial sterility due to late infection. But as the reaction of the pigeon pea materials is scored by disease incidence rather than severity, the stage of infection of the plants does not interfere with effective screening.

Spreader row inoculation technique.   This method is also a field inoculation technique wherein several rows of a susceptible cultivar are planted all through the field about 4 months in advance instead of a single hedge. The frequency of spreader to test rows is 1:10. In this method, though the disease spread is more uniform than in the infector hedge method, maintenance of several spreader rows in the field poses land preparation and irrigation problems.

Leaf-stapling technique.   This technique can be used for inoculating plants both in field and in pots. A diseased leaflet is folded on a primary leaf of a healthy seedling in such a way that the undersurface of the diseased leaflet comes in contact with both surfaces of the healthy one, and these are then stapled. The advantage of this method is that it enables inoculation of pigeon pea in its primary leaf stage, and expression of disease symptoms is quick.[7] This technique is very useful to confirm resistance of the lines observed promising under field conditions and for disease inheritance and strain identification studies.

Disease rating scale.   Rating scale quantifies the disease in the field and is of great help in scoring the material when different grades of resistance are noticed. Except one report from Nene et al.,[23] there is no published report of a disease rating scale for SM. Nene et al.[22-23] described a 9-point scale, which has been found very useful for scoring the germplasm material.

Sources of resistance.   Efforts to identify resistant sources were made at New Delhi, Coimbatore, and Pantnagar; however, a systematic effort in this direction was initiated at ICRISAT Center in 1975.[24] By screening all the germplasm at ICRISAT, 326 resistant lines (no visible symptoms) and 97 tolerant lines (ring spot symptoms) were identified. Among the 326 resistant lines, 62 were germplasm accessions while the remaining were selections from the accessions that showed segregation for resistance and susceptibility. Of these, 45 resistant lines have been used by the ICRISAT breeders to incorporate resistance in agronomically superior types. Since 1978 under IIUTPSMR through AICPIP it was possible to retest resistant sources identified at ICRISAT at 10 different locations within the country to confirm their resistance to sterility mosaic. As a result of this exercise, three lines, ICP 786, ICP 10976, and ICP 10977, were found resistant or tolerant across all the locations. These are now being used in India as resistant donor parents. These multilocation trials have also helped in identification of several stable sources of resistance to both SM and Fusarium wilt (Table 19-1).[29] Majority of resistant sources of the SM, identified so far, are from medium and mid-late maturity groups.

Breeding for resistance.   With the availability of good sources of resistance to SM, resistance breeding work is under way at several centers in India. Some of the centers where active work is going on are ICRISAT, Pantnagar, Pudukkotai, Dholi, Badnapur, Rahuri, and Faizabad. Among the old varieties developed, NPWR-15 has some tolerance to SM. The cultivar Bahar is resistant to SM but highly susceptible to fusarium wilt. The recently released ICRISAT early maturing line, ICPL 151, has tolerance to SM. Several lines such as ICPL 146, ICPL 269, ICPL 366, ICPL 8327, DA 11, DA 12, DA 13, AA 15, DA 51, MA 97, Sehore 367, DPPA 84-61-3, and DPPA 84-8-3, Pant A 104, Pant A 8505, Pant A 8508, Bhavanisagar 1, NPRR 1, which are in all-India coordinated trials, have resistance to SM.

Inheritance of resistance   Sharma et al.[25] reported that susceptibility to SM disease was dominant over resistance and tolerance, and that tolerant reaction was dominant over the resistance of certain lines. Two loci and more than two alleles at each locus were the suggested explanations for reactions in F1 and F2 generations in different cross-combinations.

Mechanism of resistance.   Many of the resistant sources identified so far do not favor mite multiplication[21] and hence these appear to be resistant to the vector. The precise reason for less number of mites on resistant lines is not known, but it appears to be due to nonpreference of the mites for the resistant lines. A line of pigeon pea ICP 8136 was, however, noticed resistant to the pathogen but not to the mite vector. Ring spot reaction in some genotypes, such as ICP 2376, appears to be a hypersensitive response of the host to restrict the spread of the pathogen. However, these lines favored mite multiplication on par with those of susceptible lines.

19-7-2 Cultural Practices

Intercropping.   Pigeon pea intercropped with sorghum (4:1) showed more SM than the sole crop. In intercropped pigeon pea, wind velocity was reduced. There could be recycling of the mites already present by the convection air currents, resulting in a higher incidence of the disease.

TABLE 19-1   SOME EXAMPLES OF MULTIPLE-DISEASE RESISTANCE IN PIGEON PEA[29]

Cultivars/lines	Resistance to
ICP 7198, 8024, 8860 to 8862, 9142, 10960, PR 5149, ICPL 83-227	Wilt, sterility mosaic
ICP 11302 to 11304	Wilt, sterility mosaic, Phytophthora blight
ICP 8861, 8862, 10960	Wilt, sterility mosaic, Alternaria blight

Sowing date.   A susceptible cultivar (BDN 1) was sown at monthly intervals all through the year near a hedge of pigeon pea inoculated with SM. The incidence was high in March through October plantings, whereas the incidence was low in December through February plantings. It was concluded that low disease incidence in pigeon pea was due to a low vector population in the infector hedge. Wind velocity and direction, temperature, and rainfall did not affect SM incidence.

19-7-3 Chemicals

Seed dressing.   Seed dressing with higher dose of Furadan (carbofuran) 3 G (25%) was found to protect the plants from SM infection up to 45 days after sowing.

Soil application.   Application of carbofuran (40 kg/ha) and Temik 10G (15 kg/ha) to the soil at the sowing time gave protection to pigeon pea against SM for 75 days past sowing; however, it had no effect against the late SM infections.

Foliar sprays.   Three acaricides, Tedion, Morestan, and Kelthane, at 0.1% concentration, were sprayed on SM-infected plants harboring eriophyid mites. All the three acaricides were highly effective (kill of mites above 90%). It is a common observation that pigeon pea sprayed with acaricides from the early stages do not show much disease.

19-8 INTERACTION OF STERILITY MOSAIC WITH OTHER PIGEON PEA DISEASES

19-8-1 Powdery Mildew

Reddy et al.[26] reported that infection of pigeon pea with SM predisposes them to powdery mildew (*Oidiopsis* sp.) infection.

19-8-2 Fusarium Wilt

In ICRISAT fields, pigeon pea plants affected by SM were observed to show less wilt. Usually, maximum wilt appears at flowering and podding time. SM infection by causing sterility of plants seem to make them less susceptible to wilt.

19-9 LOOKING AHEAD

Though a good deal of information on several aspects of the SM has been generated in the last few years, there are several gaps in our knowledge.

19-9-1 Investigation on the Nature of the Causal Pathogen of SM

New techniques will have to be used to elucidate the cause of SM disease. Search for the association of dsRNA, as demonstrated in other diseases,[27-28] may prove very useful for detection and identification of SM pathogen from infected leaf tissue and infective mites.

Knowledge of the causal pathogen of the SM will enable us to purify the pathogen and develop an antiserum and/or cDNA probes against SM agent. The monoclonal antiserum will be of great help in testing if there are different strains of the SM pathogen, as well as in identifying other hosts of the pathogen.

19-9-2 Disease Cycle

There is a need to understand more thoroughly the life cycle of the SM pathogen. A proposed disease cycle of SM in India is shown in Figure 19-5. It is necessary to ascertain the relative importance of different components of the disease cycle such as *A. scarabaeoides,* long duration pigeon peas, and volunteer pigeon peas in carryover of the pathogen and mites during the off-season, and causing the annual recurrence of the SM disease.

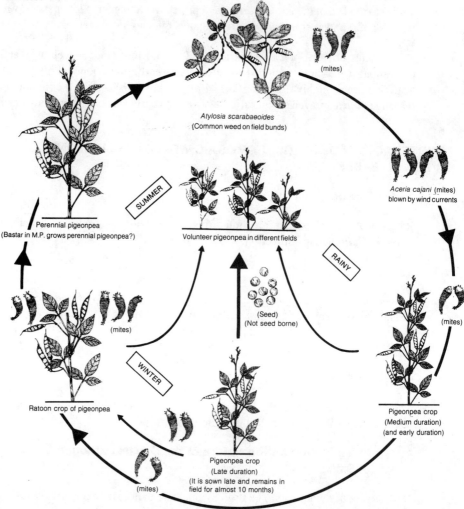

Figure 19-5 Proposed disease cycle for sterility mosaic disease

19-9-3 Mechanism of Resistance

Although it is known that the different resistant sources identified are principally resistant to the mite vector, there is a need to know the precise mechanism of resistance, such as antibiosis, nonpreference, and tolerance operative against the mite vector. Similarly, in the line ICP 8136, resistant to the SM pathogen but susceptible to the mite, the mechanism of resistance to the SM pathogen needs to be investigated.

19-10 REFERENCES

1. Mitra, M. *Report of the Imperial Mycologist, Scientific Reports of the Agricultural Research Institute, Pusa,* 1929–30, 1931, pp. 58–71.

2. Alam, M. "Arhar Sterility." Proceedings 20th Annual Meeting. Indian Science Congress, Poona, *Sect. Agri.* 43 (1933): 15.

3. Capoor, S. P. "Observations on the Sterility Disease of Pigeonpea in Bombay." *Indian J. Agri. Sci.* 22 (1952): 271.

4. Kandaswamy, T. K., and Ramakrishnan, K. "An Epiphytotic of Pigeonpea Sterility Mosaic at Coimbatore." *Madras Agric. J.* 47 (1960): 440.

5. Nene, Y. L. "A Survey of Viral Diseases of Pulse Crops in Uttar Pradesh." *Research Bulletin No. 4,* Pantnagar, Uttar Pradesh, India: Govind Ballabh Pant University of Agriculture and Technology, 1972, 191.

6. Seth, M. L. "Transmission of Pigeonpea Sterility by an Eriophyid Mite." *Indian Phytopathol.* 15 (1962): 225.

7. Nene, Y. L., and Reddy, M. V. "A New Technique to Screen Pigeonpea for Resistance to Sterility Mosaic." *Trop. Grain Legume Bull.* 5 (1976): 23.

8. Kannaiyan, J., Nene, Y. L., Reddy, M. V., Ryan, J. G., and Raju, T. N. "Prevalence of Pigeonpea Diseases and Associated Crop Losses in Asia, Africa and Americas." *Trop. Pest Manag.* 30 (1984): 62.

9. Nene, Y. L., Sheila, V. K., and Sharma, S. B. "A World List of Chickpea (*Cicer arietinum* L.) and Pigeonpea (*Cajanus cajan* (L.) Millsp.) Pathogens." Pulse Pathology, *Progress Report, 32, ICRISAT, Andhra Pradesh-502 324,* India, 1984.

10. Su, U. *Plant Diseases in Burma. Intern. Bull. Plant Prot. Year V.* 141, 1931.

11. Reddy, M. V., and Nene, Y. L. "Estimation of Yield Loss in Pigeonpea Due to Sterility Mosaic," in *Proceedings of the International Workshop on Pigeon Peas*, 15–19 December, 1980, ICRISAT Center, India, Volume 2, Patancheru P.O. Andhra Pradesh 502 324, India, 1981, pp. 305–312.

12. Reddy, M. V., and Nene, Y. L. "Ringspot Symptom: A Genotypic Expression of Pigeonpea Sterility Mosaic." *Trop. Grain Legume Bull.* 15 (1979): 27.

13. Gadh, I. P. S., and Bernier, C. C. "Resistance in Fababean (*Vicia faba*) to Bean Yellow Mosaic Virus." *Plant Dis.* 68 (1984): 109.

14. Brakke, M. K. "Wheat Streak Mosaic Virus." *CMI/AAB Descriptions of Plant Viruses* (1971): 48.

15. Paliwal, Y. C., and Tremaine, J. H. "Multiplication, Purification and Properties of Ryegrass Mosaic Virus." *Phytopathology* 66 (1976): 406.

16. Venable, J. M., and Coggeshall, R. "A Simplified Lead Citrate Stain for Use in Electron Microscopy." *J. Cell Biol.* 25 (1965): 407.

17. Oldfield, G. N., Reddy, M. V., Nene, Y. L., and Reed, W. "Preliminary Studies of the Erio-phyid Vector of Sterility Mosaic." *International Pigeonpea Newsletter* 1 (1981): 25.

18. Paliwal, Y. C. "Mite Transmitted Plant Viruses," in *Vistas in "Plant Pathology*, eds. A. Varma and J. P. Verma, New Delhi: Malhotra Publishing House, 1986), 567–578.

19. Slykhuis, J. T. "*Aceria tulipae* Keifer (Acarina: Eriophyidae) in Relation to the Spread of Wheat Streak Mosaic." *Phytopathology* 45 (1955): 116.

20. Janarthan, R., Navaneethan, G., Subramanian, K. S., and Samuel, G. S. "Some Observations on the Transmission of Sterility Mosaic of Pigeonpea." *Curr. Sci.* 41 (1972): 646.

21. Reddy, M. V., and Nene, Y. L. "Influence of Sterility Mosaic Resistant Pigeonpeas on Multiplication of the Mite Vector." *Indian Phytopathol.* 33 (1980): 61.

22. Nene, Y. L., Reddy, M. V., and Deena, E. "Modified Infector-row Technique to Screen Pigeonpea for Sterility Mosaic Resistance." *International Pigeonpea Newsletter* 1 (1981): 27.

23. Nene, Y. L., Kannaiyan, J., and Reddy, M. V. "Pigeonpea Diseases: Resistance-Screening Techniques." *Information Bulletin No. 9 Patancheru, A.P. 502 324, India:* International Crops Research Institute for the Semi-Arid Tropics 14, 1981.

24. Reddy, M. V. "Diseases of Pigeonpea and Chickpea and Their Management," in *Plant Protection in Field Crops*, eds. M. Veerabhadra Rao and S. Sithanantham (Rajendranagar, Hyderabad, India: Plant Protection Association of India, CPPTI), 1987, 175–184.

25. Sharma, D., Gupta, S. C., Rai, G. S., and Reddy, M. V. "Inheritance of Resistance to Sterility Mosaic Disease in Pigeonpea-I." *Indian J. Genet.* 44 (1984): 84.

26. Reddy, M. V., Kannaiyan, J., and Nene, Y. L. "Increased Susceptibility of Sterility Mosaic Infected Pigeonpeas to Powdery Mildew." *Int. J. Trop. Plant Dis.* 2 (1984): 35.

27. Morris, T. J., and Dodds, J. A. "Isolation and Analysis of Double-Stranded RNA from Virus-Infected Plant and Fungal Tissue." *Phytopathology* 69 (1979): 854.

28. Mirkov, T. E., and Dodds, J. A. "Association of Double-Stranded Ribonucleic Acids with Lettuce Big Vein Disease." *Phytopathology* 75 (1985): 631.

29. Nene, Y. L. "Multiple-disease Resistance in Grain Legumes." *Ann. Rev. Phytopathol.* 26 (1988): 203.

20

CHICK-PEA WILT

BHUSHAN L. JALALI AND HARI CHAND

Department of Plant Pathology
Haryana Agricultural University Hisar-125 004

20-1 INTRODUCTION

Chick-pea (*Cicer arietinum* L.), a major grain legume crop in Asia, Africa, and America with a world cultivated area of 10.21 million hectares and annual world production of 6.15 million tons[1] is attacked by a diverse spectrum of pathogens. Chick-pea wilt was first reported from India by Butler.[2] McKerral[3] considered it to be soilborne, and the causal pathogen belonged to *Fusarium.* Narasimhan[4] observed association of *Fusarium* sp. and *Rhizoctonia* sp. with wilted plants. McRae[5] as well as Prasad and Padwick[6] reported it to be caused by *Fusarium* species. In the studies of Dastur,[7] *Rhizoctonia bataticola* produced wilted plants and he called the disease "Rhizoctonia wilt." Later, the fungus causing chick-pea wilt was named *F. orthoceras* var. *ciceri.*[8]

Association of *Verticillium albo-atrum* has also been observed.[9–10] Erwin[11] reported *F. lateritium* f. sp. *ciceri* to be the cause and questioned the name *F. orthoceras* var. *ciceri.* Following the classification of Snyder and Hansen[12] f. sp. *F. orthoceras* var. *ciceri* was renamed as *F. oxysporum* f. sp. *ciceri*[13] and is now accepted widely.[14] Suryanarayana[15] identified a wilt-like "foot-blight," disease caused by *Phytophthora.* Grewal[16] reported isolation of a virulent race of *Fusarium* that established its pathogenicity. *Operculella padwickii* has also been found to be associated with wilt disease.[17]

20-2 DISTRIBUTION AND ECONOMIC IMPORTANCE

The disease has been reported from several countries (Figure 20-1) including India, Bangladesh, Burma, Ethopia, Mexico, Pakistan, Syria, and Tunisia,[18] Chile, Iran, Sudan, the United States,[19] Peru,[20] USSR,[21] and Malawi.[22] However, chick-pea cultivation is greatly threatened by this disease in India, Iran, Pakistan, Nepal, Burma, Spain, and Tunisia.

All types of chick-peas irrespective of plant type and seed size succumb to the disease. In general, spring-sown crop is more vulnerable to wilt than winter sown.[23] Another dimension is with regard to "early" or "late" wilting; in most of the northern belt of India early wilting occurs around November, and late wilting at flowering stage around February–March. Because of the extreme temperature shift in this region, such a situation prevails. During the prevailing cold spell in December–February months the incidence of wilt remains at its lowest ebb. Once the weather becomes warmer, the disease incidence gains momentum. However, such a phenomenon is not of common occurrence in other parts where the environment remains uniformly warm all through the crop season. With regard to crop losses, no definite data are available. However, rough estimates indicate that losses may hover around 10–15 % each year as a regular feature. In years of severe epidemics, crop losses have gone as high as 60-70%. Nema and Khare[24] observed damage to be up to 61% at seedling stage and 43% at flowering stage. Similarly,

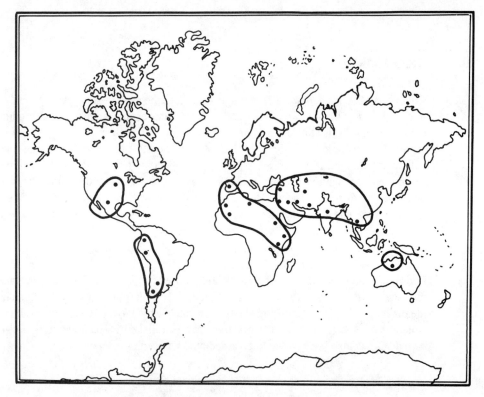

Figure 20-1   Geographical distribution of chick-pea wilt

early wilting reduced the seed number/plant and caused more yield losses than late wilting.[25] The seeds harvested from wilted plants are lighter, wrinkled, and duller than those from healthy plants. The yield losses vary between 10% and 100%, depending on the agroclimatic conditions.[26] Sattar et al.[27] reported an annual loss of 12 million ruppees (approx. US $ 1 million) from Pakistan.

20-3 SYMPTOMS

Generally the disease occurs at two stages of plant growth: (i) seedling stage and (ii) flowering stage or adult stage. The pathogen infects the root system by penetrating the epidermis, cortex, and finding its way into xylem vessels where it colonizes extensively, producing conidia. The pathogen then grows upward in the shoot, and extensive growth of the pathogen in vascular bundles plugs them. The chief symptoms of the disorder are: yellowing and drying of leaves from base upward, drooping of petioles and rachis, improper branching, withering of plants, browning of vascular bundles, and finally wilting of plants.[6,28-29] Chauhan[30] reported the initial symptom of the disease to be acropetal vein clearing of leaves. Nene et al.[31] after making detailed symptomatological studies observed diagnostic symptoms as: when the disease occurs at seedling stage (3-5 weeks after sowing), the seedlings collapse and lie flat on the soil surface, although they retain normal green color. Such diseased seedlings, when uprooted, generaly show uneven shrinking of the stem above and below the collar region. The affected seedlings do not show any external root rot; however, when split open vertically from collar region downward, black discoloration of xylem vessels is visible. The seedlings of highly susceptible cultivars which die within 10 days after emergence may not exhibit black discoloration of internal tissues; nonetheless, internal browning from root tip upward will be observable.

In the case when adult plants (6-8 weeks after sowing) are infected, the diseased plants (Fig. 20-2) exhibit drooping of petioles, rachis, and leaflets. Initially, drooping is observed in the upper part of the plants, but within a short time (1-2 days) it is visible on the entire plant. The lower leaves dry but are not shed at maturity. External root rot is not observed but the central inner portion shows dark brown to black discoloration of xylem below and above collar region. Sometimes only a few branches are affected which result in partial wilting. In the studies of Jalali et al.,[32] the expression of wilt infection appeared to be different in some varieties (H 75-35, ICC 3181, ICC 1153). Initial symptoms started from the lower portion as the leaves turned yellow, and then light brown or straw colored; gradually such symptoms progressed upward. In later stages the leaflets shed and clear rachis could be seen. Typical xylem necrosis in such plants was observed. Murumkar and Chavan[33] noted physiological changes taking place in leaves infected by the fungal pathogen (reduction in chlorophyll and increase in organic acids, polyphenols, and carbohydrates). In a similar study the number of chloroplasts and starch formation in the mesophyll cells decreased following infection by the pathogen.[30]

Figure 20-2 Single chick-pea plant showing characteristic wilt symptoms caused by *Fusarium oxysporum* f. sp. *ciceri* (Courtesy of ICRISAT)

20-4 CAUSAL ORGANISM

Chick-pea wilt is caused by *Fusarium oxysporum* Schlecht. emnd Snyd. & Hans. f. sp. *ciceri* (Padwick) Snyd. & Hans.

Different synthetic and nonsynthetic media have profound influence on cultural and morphological characters of the pathogen. The fungus can tolerate a wide range of pH (optimum being 5.0 to 6.5), and can effectively utilize all nitrate sources.[34]

The fungus is septate, profusely branched; growing on potato-sucrose/dextrose agar at 25°C initially white turning light buff or deep brown later, fluffy or submerged. The growth becomes felted or wrinkled in old cultures. Various types of pigmentations (yellow, brown, crimson) may be observed in culture. On solid medium, microconidia may be usually borne on simple and short conidiophores which arise laterally on the hyphae. They are oval to cylindrical, straight or curved and measure 2.5–3.5 × 5–11 μm. Macroconidia are borne on branched conidiphores, thin walled, 3 to 5 septate, fusoid, pointed at both ends and measure 3.5–4.5 × 25–65 μm. Chlamydospores are formed in old cultures, which are smooth or rough walled, terminal and intercalary, and may be formed singly or in pairs or in chains.[35]

20-4-1 Variation in the Pathogen

Chauhan[36] dealing with 22 isolates of the pathogen, placed them in three groups based on mycelium and colony type; and in five groups on the basis of filterate toxicity and mortality in pot inoculation tests. While studing the reactions of five isolates of the pathogen on ten differentials, Haware and Nene[37] grouped them in three races. In their further study[38] using ten cultivars, the existence of four races among Indian isolates was demonstrated. Similar physiological variations have been observed among seven isolates collected from different parts of India.[32] Colina et al.[39] working with 54 isolates from Spain reported that the race 1 and the new race 5 both produced wilt symptoms in chick-pea plants. Vascular yellowing (50 isolates) was caused by the newly described race 0 which was widely present in Spain. Recently Gupta et al.[35] identified 6 isolates (1-1 to 1-6) based on clearcut differences in cultural characteristics, fungal morphology, and virulence.

20-5 EPIDEMIOLOGY

The fungus may be seedborne and may survive (Figure 20-3) in plant debris in soil.[29,31,40,41,42] Kumar et al.[40] isolated the pathogen from seed samples obtained from five sources. Haware et al.[43] showed the fungus to be in the hilum of the seed in the form of chlamydospore-like structures. The primary infection is through chlamydospores or mycelia. The conidia of the fungus are short lived; however, the chlamydospores can remain viable up to next crop season. The pathogen survives well in roots and stems, even in apparently healthy looking plants growing among diseased chick-pea harboring enough fungus.[42] The fungus, however, did not survive in the roots placed on the soil suface. Plant species other than chick-pea may serve as symptomless carriers of the disease. Haware and Nene[44] found pigeon pea (*Cajanus cajan*), lentil (*Lens culinaris*), and pea (*Pisum sativum* L.) as symptomless carriers of the disease. The pathogen may also parasitize several weeds such as *Cyperus rotendus. Tribulus terrestris, Convolvulus arvensis,* and *Cardiospermum halicacabum.*[31]

The soil type, reaction, moisture, and temperature are known to influence disease development. The greenhouse studies substantiate the fact that disease is more severe in light sandy soils than heavy clay ones.[20,45] Chandra et al.[46] attributed higher disease severity in light sandy soils to low water retention abililty of these soils. Chauhan[47] in his studies noted that the disease intensity increased with lowering pH, being considerably low at pH 9.2. In another such study, he[48] observed that alkaline soils favor incidence of wilt. However, Shaikh[34] reported that the pathogen tolerated a wide range of pH, with optima between 5.0 and 6.5. High soil temperature and deficiency of moisture appear to have a definite bearing on the incidence of the disease.[49-52] Lower levels of soil moisture (10%) kept the plant mortality due to the disease low, though 12% of the plants were damaged, as compared to 83% in soil with moisture at 25% level.[53] Soil temperature relations showed that the disease is optimum at 25°C and is at a lower ebb at 20°C. The

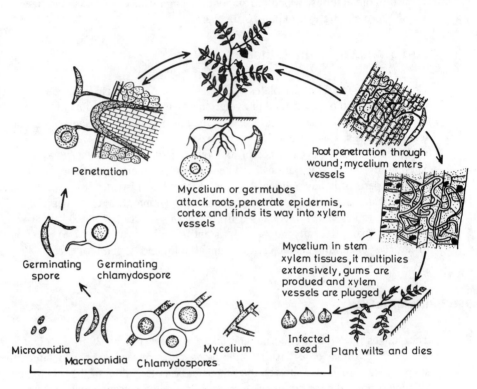

Penetration

Mycelium or germtubes
attack roots, penetrate epidermis,
cortex and finds its way into xylem
vessels

Root penetration through
wound; mycelium enters
vessels

Germinating   Germinating
spore   chlamydospore

Mycelium in stem
xylem tissues, it multiplies
extensively, gums are
produed and xylem
vessels are plugged

Microconidia

Macroconidia   Chlamydospores

Mycelium

Infected
seed   Plant wilts and dies

Infected plant parts remain in soil . Pathogen lives
saprophytically and forms various types of spores and mycelium .
In infected seeds , the pathogen survives in the form of
chlamydospore .

Figure 20-3   Disease cycle of chick-pea wilt

amount of organic matter[54] and humus content[55] of the soil were found inversely related to
wilt incidence.

20-6 DISEASE MANAGEMENT

Management practices directed toward the pathogen for checking the progression of the
disease occurrence could be exclusion and eradication of the pathogen, and to reduce its
inoculum.[56] By the very nature of the pathogen involved, evolving resistant varieties have
so far proved to be the best bet, although other conventional chemical and cultural methods have also yielded good results. Since this crop is grown principally is rainfed areas,
many of the known conventional chemical methods have not found wide adoption.

20-6-1 Varietal Resistance

The observations made at many places showed that cultivar G-24 possesses a fair degree of resistance.[57] Singh et al.[58] identified 11 lines exhibiting resistance. Jalali et al.[59] screened a large number of chick-peas in a multiple disease sick plot. In their tests, the chick-pea line ICCC-10 exhibited multiple resistance to wilt, root rot, and stunt, and DA-1 was wilt and stunt resistant. In a further study, the multiple resistance of these lines was confirmed and some more lines combining resistance to wilt and stunt were identified.[60] The field screening and supplementary tests conducted by Nene and Haware[61] showed wilt resistance in 14 lines. Of the nine *Cicer* spp. tested in pots only *C. judaicum* collections exhibited wilt resistance. Field screening of chick-pea segregating material identified three crosses exhibiting tolerance to the disease.[62] A high level of resistance was exhibited by P. 597, P. 621, P. 3649, P. 4115, P. 4128, P 4245, and P. 4347 in a wilt-sick plot for four years.[63] Of these P. 621 possessed a highly stable resistance, and nature of resistance was considered to be horizontal. Halila et al.[64] reported several kabuli entries (PL-Se-Be 81 series, nos. 48, 78, 86, 87, 103, 116 and 146) which showed no wilt symptoms when grown in field heavily infested with the pathogen. Six lines (ICC series: 11313, 12237, 12239, 12258, 12259, and 12275) remained free from wilt among 34 supposedly resistant germplasm lines from ICRISAT and 3 promising varieties grown in infested soil in pots.[65] Of the 1,390 lines tested in a 40-year-old multiple disease sick plot infested with several fungi including *F. oxysporum* f. sp. *ciceri,* two lines (GG 588 and GG 609) were resistant to wilt as well as to foot rot (*Operculella padwickii*).[66] Similarly, several other studies have reported sources of resistance to the disease from time to time.[67-71] On the basis of multilocation testing for Fusarium wilt and root rots several stable and durable sources of resistance have been identified. Some of these express resistance against other diseases also (Table 20-1).[73]

Screening techniques   Sati and Grewal[72] designed a technique designated as "glass tube technique" which involved placement of seed around the periphery of a hollow glass tube (4 cm in diameter, closed at bottom and filled with soil) buried in the center of a 6- to 8-inch earthen pot filled with soil. The pot as well as the tube were watered to maintain a uniform soil temperature. At the desired age of plants, the glass tube was carefully removed to expose a mat of roots around its periphery and base. The roots thus exposed were then inoculated by filling the gap with sand-maize meal inoculum of the pathogen.

Nene et al.[73] have suggested three techniques, viz. sick plot, pot culture, and water culture. The pot and water culture techniques are usually adopted to supplement field

TABLE 20-1   SOME EXAMPLES OF MULTIPLE-DISEASE RESISTANCE IN CHICK PEA[73]

Cultivars/lines	Resistance to
ICP 12237–12269	Fusarium wilt, dry root rot, black root rot
ICC 1069	Fusarium wilt, Ascochyta blight, Botrytis grey mold
ICC 10466	Fusarium wilt, dry root rot, stunt
ICC 858, 959, 4918, 8933, 9001	Fusarium wilt, Sclerotinia stem rot

screening tests; these are also well suited for inheritance studies. Sick plot method has been widely used and was found efficient.[59,63,64,66] It also provides natural fluctuating environmental conditions and interaction between pathogenic and nonpathogenic microorganisms.

Mechanism of resistance. In most inheritance studies[74–79] the resistance to the disease was found to be governed by a single recessive gene. This gene is inherited independently of the genes for seed coat color and branching habit[74] and successfully transferred from cv. WR-315, CPS-1, and other resistant cultivars to a wide range of genetic backgrounds by hybridization and pedigree and bulk selection.[78] Phillips[80] observed monogenic recessive inheritance in F_2 but this was not confirmed in F_3 and F_4 generations. Lopez[81] reported two pairs of recessive genes conditioning the resistance. Upadhyaya et al.[82] and Smithson et al.[83] observed digenic inheritance of race-1, and observed a ratio of 9:6:1 in F_2 among early wilting, late wilting, and resistance respectively. In a further study, Smithson et al.[84] reported the resistance to race-1 to be trigenic and proposed the genotypic formulae for various wilt resistant cultivars. Singh et al.[108] studied reactions of parents and F_1 and F_2 generations of crosses of chick-pea cvs. K-850 with C-104 and JG-62 and F_3 progenies of K-850 × C-104 to race 1 of *F. oxysporum* f. sp. *ciceri*. They found that K-850 carries a recessive allele for resistance at a locus differnt from and independent of that carried by C-104 and recessive alleles at both loci together confer complete resistance.

Kumar and Jalali[85] noted differences in seed leachates of wilt-susceptible (JG-62) and resistant (CPS-1) genotypes, susceptible cultivar exuding much greater amounts of sugar and amino acids than resistant one. On the other hand, liberation of total phenolics from resistant cultivar was on a much higher gradient as compared to susceptible cultivar. The spore germination of *F. oxysporum* f. sp. *ciceri* in seed leachates of resistant cultivar was drastically reduced.

Root exudates may also determine resistance/susceptibility of the plant to wilt. The conidia of the fungus showed 9% germination in root exudates of resistant cultivar (CPS-1) as compared to 95.4% in presence of exudates from susceptible cultivar (JG-62); the mycelial growth was also affected similarly.[86] In other similar studies Satyaprasad and Rama Rao[87] and Gupta et al.[88] reported that root exudates from susceptible cultivar stimulated the mycelial growth and conidial germination of the wilt fungus whereas those from the resistant cultivar inhibited them. Aspartic acid, asparagine and histidine were detected only in the exudates of the susceptible cultivar, whereas cystine was consistently observed in the exudates of resistant variety.

When the resistant and susceptible plants were compared chemically, the quantities of some minerals (viz. N, P, K, Na, S, and Zn) were greater in resistant varieties than susceptible ones.[88] Ayyar and Iyer[74] suggested that the resistant strains possessed a thick-layer of suberin in the periphery of the root cortex whereas no such layer was formed in susceptible types.

20-6-2 Seed Treatment with Fungicides

Seed dressing with 0.15% Benlate T (30% benomyl + 30% thiram) destroys the seed-borne fungus completely.[43] Benzimidazoles including benomyl act on the pathogen by interfering with nuclear division as they bind with the β-tubulin reuired for the spindle formation.[89] In greenhouse tests, seed treatment with Bavistin or carboxin (0.25%) protected the plants in potted soil infested with pathogen.[90] The fungicides also protected seedlings in the field for 30 days. Bavistin treatment (0.5 g/kg seed) improved seed germination by 16.5%, greatly reduced wilt incidence and increased yield by 23.7%.[91] Seed inoculation with *Rhizobium* followed by seed treatment with Bavistin (o.1%) is more effective in reducing wilt, increasing nodules/plant and yield than Bavistin (0.1%) alone.[92]

Seed treatment with Bavistin + thiram (0.5 + 2.0 g/kg seed) has also been found promising.[32] Rovral (0.2%), Mildothane (0.1%), Brassicol (0.2%), and Dithane M-45 (0.2%) treatments enhanced seed germination and/or seedling vigor; Mildothane exhibiting best responses.[93] Mani and Sethi[94] noted the influences of seed treatment with either Benlate, thiram, carbofuran, or Oftanol on seedling emergence of chick-pea in the presence of *Meloidogyne incognita, F. oxysporum* f. sp. *ciceri* and *F. solani*. Benlate and carbofuran gave the highest seedling emergence (68.82% and 63.58% respectively) as compared to controls (46.2%).

It needs to be emphasized here that the fungicides belonging to benzimidazoles group may become ineffective due to evolution of resistant strains of the pathogen, since these fungitoxicants are site specific with respect to their mode of action.[89] However, dithiocarbamates are known to remain effective for a longer time as they may inhibit about 20 enzyme pathways.[95] Kuzmina[96] observed that antibiotics such as phytobacteriomycin (seed treatment) and trichothecin (dusting) are able to decrease disease incidence. Seed treatments with garlic leaf extract and neem oil are also reported to produce disease free seedlings.[97-98]

20-6-3 Biological Control

Jalali and Thareja[99] observed suppression of the disease in potted plants following soil inoculation with vesicular-arbuscular mycorrhizal fungus (Figure 20-4). The endophyte colonized plant roots extensivley which resulted in reduction in wilt incidence from 80% in nonmycorrhizal to 26.6% in mycorrhizal plants.

In such mycorrihizal plants, phosphate uptake efficiency was also significantly enhanced. Detailed studies carried out on plant growth responses to V-A mycorrhizae have offered positive indications that mycorrhizal colonization has the potential to provide effective protection against fusarial infections, and perhaps other soil pathogens including Rhizoctonias.

20-6-4 Manipulation of Agronomical Practices

Early planted crop usually attract more disease. Several studies have suggested that higher disease control and yields are obtained when the planting is delayed until the last week of October.[46,100-101] The lower disease incidence in late-sown crop was considered to be due to low temperature prevailing during the period of late-sown crop.[102]

Plants spaced at 15–20 cm had much higher disease incidence than those spread at 7.5 cm;[103] this was attributed to the shallower root system in widely spaced plants which were susceptible to wilt when subjected to moisture stress.

Figure 20-4   Biological suppression of chick-pea wilt by V-A mycorrhizal inoculation

Planting of seeds at proper depth (10–12 cm) is helpful in reducing disease incidence,[57] while shallow sown crop seem to attract more disease.[34,46]

Planting the crop with "pora" method[104] using lower seed rates[34] helps to minimize disease, whereas broadcast method of planting increased wilt incidence.[105] Development of wilt is more prominent under moisture stress conditions;[106] one irrigation before flowering decreases disease incidence and increases yield.[104,107] Mixed cropping of chick-pea with wheat and berseem[104] have given measurable disease control.

20-7 CONCLUSIONS

Fusarium wilt continues to be a serious threat to successful cultivation of chick-pea. Now substantial progress has been made to identify with relative ease the distinct diagnostic symptoms associated with "wilt complex". Some preliminary studies have offered indications regarding the existence of physiological races of the pathogen. Techniques of precision need to be utilized for detailed study of the pathogen isolates from divers geographical regions for distribution, and importance of these races should be one of the thrust areas for future research endeavors. Similarly it is important to determine numerical threshold value required for initiating an epidemic under congenial environmental conditions.

The most meaningful strategy in combating this disease lies in disease resistance breeding programs. The need is to tailor varieties with multiple resistance to wilt and root rots. The resistant lines identified in different geographical regions of the world need to be more extensively used for resistance breeding programs.

20-8 REFERENCES

1. F. A. O. *Production Year Book*. Rome, F. A. O. 1983.

2. Butler, E. J. *Fungi and Diseases in Plants*. Calcutta, India: Thacker Spink and Co. 1918, p. 547.

3. McKerral, A. "A Note on *Fusarium* Wilt of Gram in Burma and Measures Taken to Combat It." *Agric. J. India* 28 (1923): 608.

4. Narasimhan, R. "A Preliminary Note on a *Fusarium* Parasitic on Bengal Gram (*Cicer arietinum*)," in *Madras Agric. Dep. Year Book,* 1929, 5.

5. McRae, W. "Report of the Imperial Mycologist." *Sci. Agric. Res. Inst.* Pusa, 1930–31, 78, 1932.

6. Prasad, N., and Padwick, G. W. "The Genus *Fusarium* II. A Species of *Fusarium* as a Cause of Wilt of Gram (*Cicer arietinum* L.)." *Indian J. Agric. Sci.* 9 (1939): 371.

7. Dastur, J. F. "Gram Wilt in the Central Provinces." *Agric. Live-stk.* 5 (1935): 615.

8. Padwick, G. W. "The Genus *Fusarium* III, A Critical Study of the Fungus Causing Wilt of Gram (*Cicer arietinum* L.) and of the Related Species of the Sub-Section Orthoceras with Special Relation to Variability of the Key Characteristics." *Indian J. Agric. Sci.* 10 (1940): 241.

9. Erwin, D. C. "*Verticillium* Wilt of *Cicer arietinum* in Southern California." *Plant Dis. Reptr.* 42 (1958): 1111.

10. Bhatti, M. A., Khan, S. M., and Ajmal, A. H. "Verticillium Wilt of Chickpea." *FAO Plant Prot. Bull.* 31 (1983): 36.

11. Erwin, D. C. "*Fusarium lateritium* f. sp. *ciceri*, Incitant of Fusarium Wilt of *Cicer arietinum.*" *Phytopathology* 48 (1958): 498.

12. Snyder, W. C., and Hansen, H. N. "The Species Concept in *Fusarium.*" *Am. J. Bot.* 27 (1940): 64.

13. Chattopadhyay, S. B., and Sen Gupta, P. K. "Studies on Wilt Diseases of Pulses, I. Variation and Taxonomy of *Fusarium* spp. Associated with Wilt Disease of Pulses." *Indian J. Mycol. Res.* 5 (1967): 45.

14. Booth, C. *The Genus Fusarium.* Kew (Surrey): Commonwealth Mycological Institute, England, 1971, p. 137.

15. Suryanarayana, D. "Foot Blight—A New Disease of Gram." *FAO Pl. Prot. Bull.* 16 (1968): 71.

16. Grewal, J. S. "Screening of Gram Varieties for Resistance to Wilt through Artificial Inoculation of *Fusarium.*" *Proc. IV Annual Workshop on Pulse Crops,* Punjab Agricultural University, Ludhiana, India, 1970, 5.

17. Singh, G., and Bedi, P. S. "Fungi from the Punjab State—II." *PAU J. Res.* 9 (1971): 610.

18. Nene, Y. L., Sheilla, V. K., and Sharma, S. B. "A World List of Chickpea (*Cicer arietinum* L.) and Pigeonpea (*Cajanus cajan*(L.) Millsp.) Pathogens." *ICRISAT Pulse Pathology Progress Report No. 8* (1984): 19.

19. Haware, M. P., Nene, Y. L., and Mathur, S. B. *Seed-Borne Diseases of Chickpea.* Technical Bulletin, Danish Government Institute of Seed Technology for Developing Countries, Copenhagen, No. 1, 1986, 32.

20. Echandi, E. "Wilt of Chickpeas or Garbanzo Beans (*Cicer arietinum*) Incited by *Fusarium oxysporum.*" *Phytopathology* 60 (1970): 1539.

21. Stepanova, M. Yu. "Spread of *Fusarium oxysporum* in Legumes." *Trudy. Ves. Inst. Zashch. Rast.* 29 (1971): 100.

22. Kannaiyan, J. "Diseases of Chickpea in Malawi." *Int. Chickpea Newslett.*" 4 (1981): 16.

23. Hawtin, G. C., and Singh, K. B. "Prospects and Potential of Winter Sowing of Chickpeas in the Mediterranian Region," in *Ascochyta Blight and Winter Sowing of Chickpeas*, eds M. C. Saxena and K. B. Singh. The Hague: Martinus Nijhoff/Dr. W. Junk Publishers, 1984, 16.

24. Nema, K. B., and Khare, M. N. "A Conspectus of Wilt of Bengal Gram in Madhya Pradesh," *Symposium on Wilt Problem and Breeding for Wilt Resistance in Bengal Gram,* Indian Agr. Res. Inst., New Delhi, India, 1973, 4.

25. Haware, M. P., and Nene, Y. L. "Influence of Wilt at Different Stages on the Yield Loss in Chickpea." *Trop. Grain Legume Bull.* 19 (1980): 38.

26. Grewal, J. S., and Pal, M. "Fungal Diseases of Gram and Arhar." *Proc. IV Annual Workshop on Pulse Crops,* P.A.U. Ludhiana, 1970, 168.

27. Sattar, A., Arif, A. G., and Mohy-ud-din, M. "Effect of Soil Temperature and Moisture on the Incidence of Gram Wilt." *Pakistan J. Scientific Res.* 5 (1953): 16.

28. Argikar, G. P. "Gram," in *Pulse Crops of India*, ed. P. Kachroo. (New Delhi: ICAR, 1970), 54.

29. Westerlund, F. V., Campbell, R. N., and Kimble, K. A. "Fungal Root Rots and Wilt of Chickpea in California." *Phytopathology* 64 (1974): 432.

30. Chauhan, S. K. "Observations on Certain Symptoms in Fusarium Wilt of Gram (*Cicer arietinum.* L.)." *Agra Univ. J. Res.* 11 (1962): 285.

31. Nene, Y. L., Kannaiyan, J., Haware, M. P., and Reddy, M. V. "Review of Work Done at ICRISAT on Soil-borne Diseases of Pigeonpea and Chickpea." *Proc. Consultants Group Discussion on the Resistance to Soil Borne Diseases of Legumes*, ICRISAT, Patancheru, 1979, 3, 1980.

32. Jalali, B. L., Sangwan, M. S., and Khirbat, S. K. *Report on Rabi Pulses Pathology,* ICAR Project on Improvement of Pulses, 1980–81, 29.

33. Murumkar, C. V., and Chavan, P. D. "Physiological Changes in Chickpea Leaves Infected by Fusarium Wilt." *Biovigyanam* 11 (1985): 118.

34. Shaikh, M. H. "Studies on Wilt of Gram (*Cicer arietinum* L.) Caused by *Fusarium oxysporum* f. sp. *ciceri* in Marathwada Region." (M.Sc.(Ag.) thesis, Marathwada Krishi Vidyapeeth, Parbhani, India, 1974).

35. Gupta, O., Khare, M. N., and Kotasthane, S. R. "Variability Among Six Isolates of *Fusarium oxysporum* f. sp. *ciceri* Causing Vascular Wilt of Chickpea." *Indian Phytopathol.* 39 (1986): 279.

36. Chauhan, S. K. "Physiological Variations in *Fusarium orthoceras* App. & Wr. var. *ciceri* Padwick Causing Wilt of Gram (*Cicer arietinum* L.)." *Proc. Nat. Acad. Sci.* India Section B, 32 (1962): 78.

37. Haware, M. P., and Nene, Y. L. "Physiologic Races of the Chickpea Wilt Pathogen, Symptomless Carriers of the Chickpea Wilt Fungus." *Int. Chickpea Newslett.* 1 (1979): 7.

38. Haware, M. P., and Nene, Y. L. "Races of *Fusarium oxysporum* f.sp. *ciceri.*" *Plant Dis.* 66 (1982): 809.

39. Colina, J. C., Trapero-Casas, A., and Jimenez-Diaz, R. M. "Races of *Fusarium oxysporum* f.sp. *ciceri* in Andalucia, Southern Spain. *Int. Chickpea Newslett.* 13 (1985): 24.

40. Kumar, A., Jalali, B. L., Panwar, M. S., and Sangwan, M. S. "Fungi Associated with Different Categories of Chickpea Seeds and their Effects on Seed Germination and Seedling Infection." *Int. Chickpea Newslett.* 9 (1983): 19.

41. Sharma, B. L., and Gupta, R. N. "Survival of *Fusarium oxysporum* f.sp. *ciceri* Causing Wilt of Bengal Gram During Summer Months." *Indian Phytopathol.* 36 (1986): 561.

42. Padwick, G. W. "Report of the Imperial Mycologist." *Sci. Rep. Agr. Int.* New Delhi, India, 1939–40, (1941): 94.

43. Haware, M. P., Nene, Y. L., and Rajeshwari, R. "Eradication of *Fusarium oxysporum* f.sp. *ciceri* Transmitted in Chickpea Seed." *Phytopathology* 68 (1978): 1364.

44. Haware, M. P., and Nene, Y. L. "Symptomless Carriers of the Chickpea Wilt *Fusarium.*" *Plant Dis.* 66 (1982): 250.

45. Kotasthane, S. R., Agrawal, P. S., Joshi, L. K., and Singh, L. "Studies on Wilt Complex in Bengal Gram (*Cicer arietinum* L.)." *JNKVV Res. J.* 10 (1979): 257.

46. Chandra, S., Tomer, Y. S., and Malik, B. P. S. "Aspects of Wilt Disease in Gram with Special Reference to Haryana State." *Indian J. Genet. Plant Breed.* 34 (1974): 257.

47. Chauhan, S. K. "Influence of pH in Sand Cultures on Disease Intensity and Crop Loss Correlation in *Fusarium* Wilt of Gram (*Cicer arietinum* L.)." *J. Indian. Bot. Soc.* 41 (1962): 220.

48. Chauhan, S. K. "Influence of Different Soil Temperatures on the Incidence of Fusarium Wilt of Gram (*Cicer arietinum* L.)." *Proc. Indian Acad. Sci.* B 33 (1963): 552.

49. Baker, K. F., and Cook, R. J. *Biological Control of Plant Pathogens.* San Francisco: Freeman and Company, 1974, p. 433.

50. Chauhan, S. K. "Influence of Fusarium Wilt of Gram in Relation to Soil Moisture." *Agra Uni. J. Res.* 12 (1963): 271.

51. Vashistha, B. R. *Botany-Fungi*, 8th ed. (New Delhi: S. Chand & Co. Ltd., 1983).

52. Grewal, J. S., Pal, M., and Kulshrestha, D. D. "Fungi Associated with Gram Wilt." *Indian J. Genet. Plant Breed.* 34 (1974): 242.

53. Sinha, S. "Some Factors of the Soil in Relation to Fusarium Wilt of Bengal gram (*Cicer arietinum* L.)." *Symposium on Wilt Problem and Breeding for Wilt Resistance in Bengal Gram,* Indian Agr. Res. Inst., New Delhi, India, 1973, 9.

54. Chauhan, S. K. "The Interaction of Certain Soil Conditions in Relation to the Occurrence of Fusarium Wilt of Gram." *Indian J. Agric. Sci.* 35 (1965): 52.

55. Chauhan, S. K. "Fusarium Wilt of Gram (*Cicer arietinum* Linn.) in Relation to Organic Matter of Soil." *Vigyana Parishad Anusandhan Patrika* 5 (1962): 73.

56. Sharvelle, E. G. *Plant Disease Control.* Connecticut: AVI Publishing Co. Inc., 1979.

57. Singh, K. B., and Sandhu, T. S. *Cultivation of Gram in Pubjab.* Punja Agricultural University, Ludhiana, 1973, 12.

58. Singh, D. V., Mishra, A. N., and Singh, S. N. "Sources of Resistance to Gram Wilt and Breeding for Wilt Resistance in Bengal Gram in U.P." *Indian J. Genet. Plant Breed.* 34 (1974): 239.

59. Jalali, B. L., Khirbat, S. K., and Sangwan, M. S. "Reaction of Chickpea (*Cicer arietinum*) and Mungbean (*Vigna radiata*) Lines to Different Diseases for Resistant Breeding," in *Proc. Nat. Semi. Disease Resistance in Crop Plants,* eds. N. Shanmugan, R. Jeyarajan, and P. Vidhyasekaran. (Coimbatore: T. N. Agric. Univ., 1980), 131.

60. Khirbat, S. K., Sangwan, M. S., and Jalali, B. L. "Evaluation of Chickpea Genotypes under Multiple Disease Stress Conditions for Resistance Breeding." *Indian Phytopathol.* 37 (1984): 394.

61. Nene, Y. L., and Haware, M. P. "Screening Chickpea for Resistance to Wilt." *Plant Dis.* 64 (1980): 379.

62. Verma, M. M., and Gill, A. S. "Field Screening of Chickpea Segregating Material in a Multiple Disease Sick Plot." *Int. Chickpea Newslett.* 4 (1981): 14.

63. Govil, J. N., and Rana, B. S. "Stability of Host Plant Resistance to Wilt (*Fusarium oxysporum* f.sp. *ciceri*) in Chickpea." *Int. J. Trop. Plant Dis.* 2 (1984): 55.

64. Halila, H. M., Gridley, H. E., and Houdiard, P. "Sources of Resistance to *Fusarium* Wilt in Kabuli Chickpea." *Int. Chickpea Newslett.* 10 (1984): 13.

65. Patel, H. R., Patel, B. K., Thaker, N. A., and Patel, C. C. "Reactions of Few Chickpea Lines to Fusarium Wilt." *Int. Chickpea Newslett.* 13 (1985): 16.

66. Singh, G., Kapoor, S., Gill, A. S., and Singh, K. "Chickpea Varieties Resistant to Fusarium Wilt and Foot Rot." *Indian J. Agric. Sci.* 56 (1986): 344.

67. Singh, D., Lal, S., and Singh, S. N. "Breeding Gram for Resistance to Wilt." *Indian J. Genet. Plant Breed.* 34 (1974): 267.

68. Vidhyasekaran, P., Arjunan, G., Mariappan, V., Ranganathan, K., and Kolandaisamy, S. "Field Tolerance of Some Bengal Gram Types to Wilt and Root Rot." *Indian Phytopathol.* 30 (1978): 537.

69. Kumar, J., Haware, M. P., and Nene, Y. L. "Fusarium Wilt Resistant Lines Developed at ICRISAT." *Int. Chickpea Newslett.* 3 (1980): 5.

70. Haware, M. P., and Nene, Y. L. "Sources of Resistance to Wilt and Root Rots of Chickpea." *Int. Chickpea Newslett.* 3 (1980): 11.

71. Gurha, S. N., and Mishra, D. P. "Donors for Resistance to *Fusarium oxysporum* f.sp. *ciceri* in Chickpea (*Cicer arietinum*)." *Indian J. Mycol. Plant Pathol.* 13 (1985): 229.

72. Sati, K. C., and Grewal, J. S. "A Technique for Seedling Inoculation with Fusaria Associated with Gram (*Cicer arietinum*)." *Indian J. Mycol. Plant Pathol.* 11 (1981): 175.

73. Nene, Y. L. "Multiple-disease Resistance in Grain Legumes." *Ann. Rev. Phytopathol.* 26 (1988): 203.

74. Ayyar, R. V., and Iyer, B. R. "A Preliminary Note on the Mode of Inheritance to Wilt in *Cicer arietinum* L." *Proc. Indian Acad. Sci.* 3 (1936): 438.

75. Pathak, M. M., Singh, K. P., and Lal, B. B. "Inheritance of Resistance to Wilt (*Fusarium oxysporum* f. sp. *ciceri*) in Gram." *Indian J. Farm Sci.* 3 (1975): 10.

76. Tiwari, A. S., Pandey, R. L., Mishra, P. K., and Kotasthane, S. R. "Studies on Wilt Inheritance in Gram (*Cicer arietinum* L.)." *All India Rabi Pulse Workshop*, Bhubneswer, 1978.

77. ICRISAT. *ICRISAT Annual Report*, 1980–81, ICRISAT Patancheru, 1981.

78. Kumar, J., and Haware, M. P. "Inheritance of Resistance of Fusarium Wilt in Chickpea." *Phytopathology* 72 (1982): 1035

79. Sindhu, J. S., Singh, K. P., and Slinkard, A. E. "Inheritance of Resistance to Fusarium Wilt in Chickpeas." *J. Hered.* 74 (1983): 68.

80. Phillips, J. C. "Inheritance of Fusarium Wilt Resistance in Chickpea." *Agron. Abstr. Am. Soc. Agron.* 76, 1983.

81. Lopez, G. H. "Inheritance of the Character Resistance to Wilt (*Fusarium* spp.) in Chickpea (*Cicer arietinum*)." *Agr. Tec. Mex.* 3 (1974): 286.

82. Upadhyaya, H. D., Smithson, J. B., Haware, M. P., and Kumar, P. "Resistance to Wilt in Chickpea. II. Further Evidence for Two Genes for Resistance to Race-1." *Euphytica* 32 (1983): 749.

83. Smithson, J. B., Kumar, J., Haware, M. P., and Singh, H. "Complementation Between Genes for Late Wilting to Race 1 of *Fusarium oxysporum* f.sp.*ciceri* in Chickpea." XV. *Int. Congr.Genet.*, New Delhi, 1983.

84. Smithson, J. B., Kumar, J., and Singh, H. "Inheritance of Resistance to Fusarium Wilt in Chickpea." *Int. Chickpea Newslett.* 9 (1983): 21.

85. Kumar, A., and Jalali, B. L. "Differential Effects of Seed Leachates on Seed-borne Pathogens Associated with Chickpea (*Cicer arietinum*)." *Indian Phytopathol.* 38 (1985): 99.

86. Haware, M. P., and Nene, Y. L. "The Role of Chickpea Root Exudates in Resistance to Fusarium Wilt." *Int. Chickpea Newslett.* 10 (1984): 12.

87. Satyaprasad, K., and Rama Rao, P. "Effect of Chickpea Root Exudates on *Fusarium oxysporum* f.sp. *ciceri.*" *Indian Phytopathol.* 36 (1983): 77.

88. Gupta, O., Kotasthane, S. R., and Khare, M. N. "Studies on Root Exudates of Resistant and Susceptible Cultivars of Chickpea." *Legume Res.* 8 (1985): 42.

89. Davidse, L. C. Biochemical Aspects of Benzimidazoles - Action and Resistance." in *Modern Selective Fungicides—Properties, Applications, Mechanisms of Action,* ed. H. Lyr. (London: Longman Group UK Ltd., 1987), 245.

90. Verma, R. K. "Chemotherapeutic Activity of Five Systemic Fungicides in Gram Seedlings Against Soil-Borne Pathogens." (M.S.(Ag.) thesis, J. N. Agricultural University, Jabalpur, 1976), 78.

91. Shukla, P., Singh, R. R., and Mishra, A. N. "Search for Best Seed Dressing Fungicides to Control Chickpea Wilt." *Pesticides* 15 (1981): 15.

92. Anonymous. "Effect of Seed Treatment with Bavistin and *Rhizobium* on Wilt Incidence, Nodulation and Yield of Gram." *All India Rabi Pulses Workshop*, Project Directorate (Pulses), IARI Regional Station, Kanpur, 1983, 39.

93. Shrisat, A. M., and Kale, U. V. "Effect of Fungicidal Seed Treatment on Germination and Seedling Vigour of Chickpea (*Cicer arietinum* L.)." *Trop. Grain Leg. Bull.* 16 (1979): 29.

94. Mani, A., and Sethi, C. L. "Influence of Seed Treatment on Seedling Emergence of Chickpea in Presence of *Meloidogyne incognita, Fusarium oxysporum* f.sp. *ciceri* and *F.solani.*" *Indian J. Nematol.* 14 (1984): 68.

95. Day, P. R. *Genetics of Host-Parasite Interaction.* (San Francisco: W. H. Freeman & Co., 1974), p. 238.

96. Kuzmina, G. "Antibiotics Against Wilt." *Zashch. Rast. Mosk.* 11 (1966): 31, (in Russian).

97. Singh, U. P., Pathak, K. K., Khare, M. N., and Singh, R. B. "Effect of Leaf Extract of Garlic on *Fusarium oxysporum* f.sp. *ciceri, Sclerotinia sclerotiorum* and on Gram Seeds." *Mycologia* (1979): 556.

98. Singh, U. P., Singh, H. B., and Singh, R. B. "The Fungicidal Effect of Neem (*Azadirachta indica*) Extracts on Some Soil-Borne Pathogens of Gram (*Cicer arietinum*)." *Mycologia* 72 (1980): 1077.

99. Jalali, B. L., and Thareja, M. L. "Suppression of Fusarium Wilt of Chickpea in Vesicular-Arbuscular Mycorrhizal-Inoculated Soils." *Int. Chickpea Newslett.* 4 (1981): 21.

100. Padwick, G. W., and Bhagwagar, P. R. "Wilt of Gram in Relation to Date of Sowing." *Indian J. Agric. Sci.* 13 (1943): 289.

101. Mundkur, B. B. "Report of the Imperial Mycologist." *Sci. Rep. Agr. Res. Inst.*, New Delhi, 57, 1946.

102. Padwick, G. W. "Report of the Imperial Mycologist." *Sci. Rep. Agr. Res. Inst.*, New Delhi, 103, 1940.

103. Bahl, P. N. "Crop Stand and its Bearing on Incidence of Wilt in Bengal Gram." *Indian J. Genet. Plant Breed.* 36 (1976): 351.

104. Saraf, C. S., Ahlawat, I. P. S., and Singh, A. "Agronomic Investigation on Rabi Pulses During 1972–73 at IARI." Report *All India Rabi Pulses Workshop*, Calcutta, 1973.

105. Bedi, K. S., and Pracer, C. S. "Gram Wilt Disease and Its Control in Punjab." *Punjab Farmer* 4 (1952): 296.

106. Kausar, A. G. "Performance of Blight Resistant Gram Selections in Wilt Nursery at Six Locations." *Pak. J. Agric. Sci.* 5 (1968): 264.

107. Sekhon, G. S. "Effect of Time of Irrigation on Yield and Incidence of Wilt Attack on the Gram Crop." *Punjab Farmer* 3 (1952): 122.

108. Singh, H., Kumar, J., Smithson, J. B., and Haware, M. P. "Complementation between Genes for Resistance to Race 1 of *Fusarium oxysporum* f. sp. *ciceri* in Chick Pea." *Plant Pathol.* 36 (1987): 539.

21

ASCOCHYTA BLIGHT OF CHICK-PEA

H. S. CHAUBE AND T. K. MISHRA

*Department of Plant Pathology, G.B. Pant Univ. of Agric. & Technol.
Pantnagar (India)*

21-1 INTRODUCTION

Chick-pea (*Cicer arietinum* L.) is an important grain legume crop of dryland agriculture in Asia, Africa, and Central and South America. The total cultivated area of chick-pea in the world is about 10.1 million ha and annual production is about 6.3 million tons. The average yields per hectare are estimated to be around 623 kg. Of several diseases recorded on chick-pea, Ascochyta blight is considered to be one of the most important.[1] The disease has been reported from more than 25 countries of the world (Figure 21-1).

21-2 LOSSES

Labrousse[2] reported that the disease was very destructive in Morocco in 1929. In prepartitioned India, it was found that annually 25–50% of the crop was destroyed.[3] In Pakistan, the blight appeared in epidemic form in 1978–79 and reduced production by 17%.[4] It appeared again in severe form during 1979–80 and resulted in 48% reduction in total chick-pea production. According to Kovachevski[5] 20–50% of the crop was lost annually in Bulgaria. In the Dnepropetrovsk region of USSR, blight was severe in 1956, sometimes causing 100% loss.[6] In Greece,[7] 10–12% damage was reported during 1957–58. Puerta Romero[8] found that in different provinces of Spain the decrease in yield varied from 25–100%. Georgiou and Papadopoulos[9] found appreciable economic losses being very destructive on local cultivars in Cyprus. Mlaiki and Hamadi[10] reported that in Tuni-

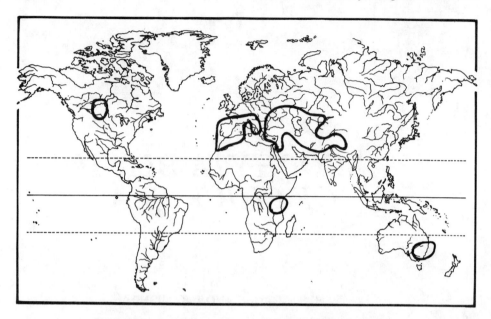

Figure 21-1 Geographical distribution of Ascochyta blight of chick-pea

sia Ascochyta blight may reduce yield by 40%. Epidemics of the disease have also been reported by several other workers.[11-17]

21-3 SYMPTOMS

Several workers[1] have described the symptoms of the disease as it occurs in different countries. The descriptions are remarkably similar. All above-ground parts of the plant are attacked (Figures 21-2 and 21-3). On leaflets, the lesions are round or elongated, bearing irregularly depressed brown spots, and are surrounded by a brownish red margin. On the green pods the lesions are usually circular with dark margins and have pycnidia arranged in concentric circles. When lesions girdle the stem, the portion above the point of attack rapidly dies. If the main stem is girdled at the collar region, the whole plant dies.

21-4 THE PATHOGEN

21-4-1 Asexual Stage

The mycelium of *A. rabiei* is hyaline to brownish and septate. The pycnidia on host are immersed, amphigenous, spherical to subglobose or depressed and generally vary in size from 65–245 μm.[18] The wall is composed of 1–2 layers of elongated pseudoparenchymatous cells. The pycnidium has a prominent ostiole measuring 30–59 μm, pycnidiospores are formed from hyaline, amoulliform phialides from the inner cells of the pycnidium. They are hyaline, oval to oblong, straight or slightly curved, occasionally bicelled, 8.2–

Figure 21-2   Single chick-pea plant showing symptoms caused by *Ascochyta rabiei*

Figure 21-3   Field view of typical Ascochyta blight-affected crop of chick-pea

10.0×4.2 to $4.5 \ \mu m$. Bedi and Aujla[19] reported that on oat meal–agar medium pycnidia developed best at pH 7.6 to 8.6 at 20°C. Besides oat meal–agar, chick-pea seed meal (4–8%) agar has been found to be a good medium for the growth of the fungus and pycnidial production.[20–23] Khalil and Khan[24] developed a new medium, which supports better growth and pycnidial production. Optimum temperature for growth, pycnidial production, and spore germination has been reported to be around 20°C.[19,20,25–27] Temperatures below 10°C and above 30°C are unfavorable to the fungus.[20,26] Kaiser[20] and Nene[1] observed that continuous light resulted in increased sporulation. However, Chauhan and Sinha[26] reported reduced sporulation on infected plants under continuous light. The incubation period varies between 5 and 7 days depending on the temperatures provided.[25–26] It also varies with genotypes inoculated. Spore germination is improved in the presence of N/50 and N/25 malic acid and acidified carbon food.[28]

21-4-2 Sexual Reproduction

The perfect stage of *Ascochyta rabiei* was first reported by Kovachevski[5] as *Mycosphaerella rabiei* Kov. Later, the perfect stage was transferred to genus *Didymella* on the basis of presence of paraphyses with asci which are mostly parallel, not in fascicles, and ascospores being constricted at the septum,[29] and thus, perfect stage of *A. rabiei* was named as *Didymella rabiei* (Kov.) von Arx. Later on, Gorlenki and Bushkova[30] confirmed the presence of the perfect stage in the USSR and Zachos et al.[25] in Greece. However, neither diseased chick-pea debris collected from India, Iran, Turkey, and Pakistan nor the isolates of the fungus cultured on media resulted in the formation of the sexual stage of the fungus.[20] Obviously, a cold winter is a prerequisite for the production of perithecia.

21-4-3 Physiologic Races

Luthra et al.[31] reported six different forms of *A. rabiei*, namely A, B, C, D, E, and F. Forms B, D, E, and F, which were biologically identical, differed from form C morphologically while form A was nonpathogenic. Aujla[32] has reported differences in cultural characters and pathogenic behavior of 11 isolates on different varieties of chick-pea. Later, Bedi and Aujla[33] suggested that the possible existence of physiologic races must be kept in view while testing breeding material for resistance. Kaiser[20] found that isolates of *A. rabiei* from India, Iran, Turkey, and Pakistan varied greatly in growth rate, sporulation, colony appearances, and pathogenicity.

Luthra et al.[34] tested 392 lines and found 3 lines from the United States, namely, Pois chick Nos. 4732, 199, and 281 which showed a high degree of resistance to *A. rabiei* under varying environmental conditions. These lines were named F 8, F 9, and F 10, respectively. Ahmed et al.[35] released a cultivar C-12-34 (progeny of a cross between F 8 x Pb 7) as resistant to blight. It lost its resistance to blight in 1950–51. A new cultivar C 235 was developed which also lost resistance in the epiphytotic year 1968, probably due to the appearance of a new race.[36] The studies carried out at New Delhi,[37–38] India, indicate that 13 morphological forms (out of 268 isolates) could be distinguished as two pathogenic races on the basis of disease reaction produced on three chick-pea cultivars. Resistant reaction was produced by cultivars I 13, EC 26435, and C 235 against race 1 represented by fast-growing and less sporulating isolates. These cultivars were, however, moderately susceptible to susceptible to slow growing and more sporulating isolates representing race 2, except isolate J-101 which incited resistant type of infection on cultivars I 13, and C 235, and was distinguished as a biotype of race 2. Pathogenicity tests with 25 well-known blight resistant lines against race 2 showed that all of them except the highly resistant line 1528-1-1 from Morocco were moderately susceptible to susceptible to race 2. The observations that cultivar C 235, previously thought to be resistant in India, was susceptible to race 2 indicates that race 2 is a newly evolved or introduced race of the pathogen. It is also probable that the highly sporulating isolate of *A. rabiei* reported from Iran[20] may belong to race 2 of the pathogen. Some new blight-resistant lines have been identified in Bulgaria.[39–41]

Simon[42] through pathogenicity tests of isolates of *A. rabiei*, recorded differential responses among isolates and cultivars. In studies conducted in Pakistan, isolate 4 from a farmer's field in Attock district was the most virulent. Two cultivars (CM 72 and ILC 195) showed resistant reaction to this isolate. Isolate 13 from Islamabad, originally isolated from cv. CM 72, showed susceptible reaction on cvs. CM 72, C 44, Pub 1, and C 727. The least virulent isolate was from ILC 200, to which only cvs. Pb 1 and C 727 were susceptible. Cultivars C 727 and Pb 1 were susceptible and cvs ILC 195 and ILC 200 were resistant to all the eight isolates.[43] Porta-Puglia[44] tested 6 isolates of *A. rabiei* from different Italian regions on 18 ICARDA and 3 Italian land races and observed the existence of complex behavior of the pathogen. Bashir et al.[45] studied disease reaction of 46 chick-pea genotypes and one local susceptible check against 6 races of *A. rabiei*.

21-5 PATHOGENESIS

Pandey et al.[46–47] studied the mechanism of fungal penetration, invasion, and subsequent development of *A. rabiei* in leaf and stem tissues. The conidia begin to germinate 12 hr after inoculation. The germ tubes elongate and sometimes form a swelling at the tip. Penetration of stem occurs 24 hr after inoculation directly through the cuticle. For a short distance the hyphae push forward subcuticularly along the junction of epidermal cells before proceeding inward. Near a stoma the hyphae penetrate through the juncture of guard and subsidiary cell, even when the stoma is open. The penetrating hyphae invade the cortical cells resulting in excessive damage to cortex. Till this stage (3 days after inoculation) epidermal cells appear intact, inoculated stem normal with no macroscopic symptoms.

Initial macroscopic symptoms are noticed 4 days after inoculation as yellow specks on the stem surface. Subepidermal mycelium forms dark aggregates and the epidermal cells turn necrotic [Figure 21-4(a)]. Fusion of hyphae is observed at this stage. Entire cortex and part of pith are completely disintegrated by the fifth day [Figure 21-4(b)]. The hyphae in cortical tissue aggregate and become compact. These aggregates later appear as pseudoparenchymatous mycelial mass and eventually differentiate into pycnidia [Figure 21-4(c)]. From the fifth to sixth day after inoculation, pycnidia mature and could be seen as black dots on the surface. Pycnidia protrude out of the epidermis of the stem surface and conidia ooze out in mass through a well-developed ostiole [Figure 21-4(d)].

Pith cells damage 5 days after inoculation coincides with necrosis initiation. By the seventh day most of the nonlignified tissues are destroyed and necrosis is very much

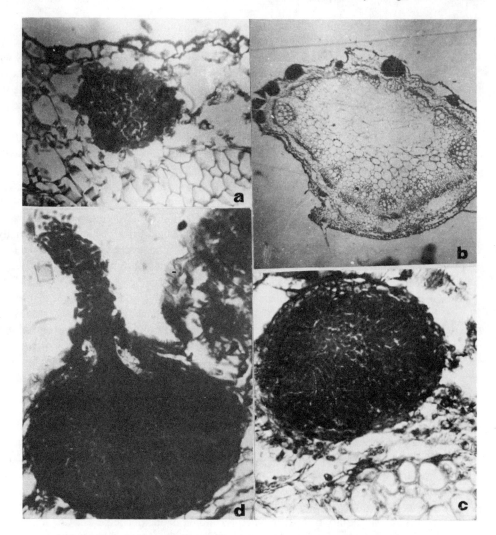

Figure 21-4 (a) T.S. of chick-pea stem 4 days after inoculation showing subepidermal aggregates of fungal mycelium

(b) T.S. of chick-pea stem 5 days after inoculation showing several developing pycnidia and extensive damage of parenchymatous cortical and pith tissues while lignified cells are intact

(c) Maturation of pycnidium 6 days after inoculation

(d) A mature pycnidium with pycnidiospores oozing out of ostiole

evident. There is no effect on lignified tissues, particularly tracheary elements. Nature of damage, excessive breakdown and necrosis, of cellulosic tissues in advance of invading hyphae, indicated involvement of wall degrading enzymes and/or toxins(s).[46,47] Recently Alam et al.[48] identified two toxins (solanapyrones A and B) from *A. rabiei*. Recent studies have also demonstrated pronounced cutinase (Klein-Bolting and Barz, unpublished; cited in Höhl et al.[49]) and other hydrolytic exoenzymes activities (Lorenz, unpublished; cited in

Höhl et al.[49]) in the culture filtrate of *A. rabiei*. More work is needed to demonstrate the role of these toxins/enzymes in pathogenesis.

Recently Höhl et al.[49] performed detailed histopathological study of development of *A. rabiei* in leaf tissues of susceptible and resistant cultivars. Their observation of infection process was more or less similar to one described by Pandey et al.[47] for the stem except that they noticed the formation of appressoria and secretion of mucilaginous exudate, which presumably provided for the fungal cells a tight contact to the host surface, from young germ tubes. Fungal development on the leaf surface (spore germination, germ tube growth, secretion of mucilaginous exudate, and appressoria formation) was essentially identical for both resistant and susceptible plants. Leaves of the susceptible plants were invaded and subepidermally colonized by the fungus and leaf spots and fungal pycnidia could be observed 6-8 days after inoculation. Whereas the resistant cultivars rapidly responded to fungal penetration by rapid cell necrosis and accumulation of phenolic compounds resulting in cessation of fungal colonization.[49]

21-6 EPIDEMIOLOGY

21-6-1 Survival

Crop debris.   Several workers[3,5,20,25,50] have indicated the importance of infected debris as a source of survival (Figure 21-5). Sattar[3] could not determine the absolute importance of infected crop debris in the survival of the fungus. However, Luthra et al. (cited in Nene[1]) considered diseased plant debris as an important source of primary infection. Kaiser[20] found that the fungus survived for over 2 years in naturally infected tissues at 10–35°C and 0–3% relative humidity at the soil surface. However, the fungus lost its viability rapidly at 64–100% relative humidity at soil depth of 10–40 cm. Pandey[22] reported that the fungus survived for over 1 year in infected debris stored at room temperature. In diseased debris stored at different temperatures, the length of survival was for about 2 months at 40°C, while at 20–35°C, the survival was recorded for a period of 6 to 8 months. When infected stem pieces were buried in moist sterilized and natural soils and incubated at 20–40°C, the fungus did not survive beyond 2–3 months. In infected crop debris buried in soil under natural conditions at the depths of 0–20 cm, the length of survival was to the extent of 2 to 3 months only.

Seed.   Luthra and Bedi were the first to demonstrate the seedborne nature of the pathogen[1] (Figure 21-5). They showed that the seed coat and cotyledons of infected seeds contained mycelium. Sattar[3] demonstrated surface contamination of seed with fungus spores and their role in causing infection. He found that 50% of such spores survived on seed for 5 months at 25–30°C, but only 5% of spores survived for 5 months at 35°C. Later other workers[16,20,25,51–55] also confirmed the seedborne nature of the pathogen. Lukashevich[56] showed that the fungus can behave as a saprophyte and spread to noninfected tissues if the harvested material is stored for some time before threshing. Tripathi et al.[57] observed that in infected seeds stored at low temperatures (0–10°C), *A. rabiei* survived for 14–15 months and at 20-30°C, the survival was reduced by 2 to 4 months; at room temperature, the survivability of the fungus declined sharply as an initial seed

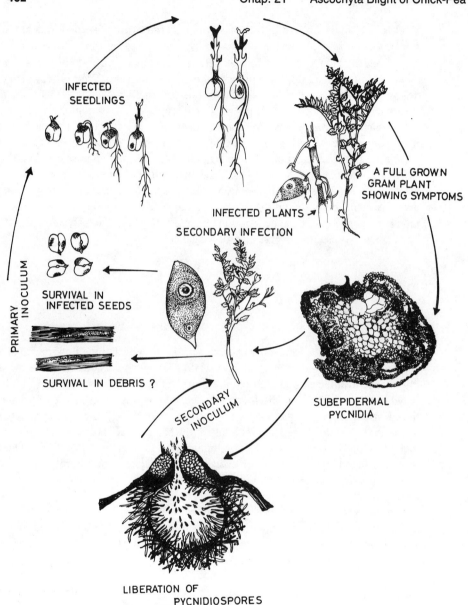

INFECTED
SEEDLINGS

A FULL GROWN
GRAM PLANT
SHOWING SYMPTOMS

INFECTED PLANTS ↗
SECONDARY INFECTION

PRIMARY INOCULUM

SURVIVAL IN
INFECTED SEEDS

SURVIVAL IN DEBRIS ?

SECONDARY INOCULUM

SUBEPIDERMAL
PYCNIDIA

LIBERATION OF
PYCNIDIOSPORES

Figure 21-5   Disease cycle of Ascochyta blight of chick pea

infection of 30–32% dropped down to complete elimination after 12 months.

Site of Infection in seed.   Research by Luthra and Bedi,[1] Halfon-Meiri,[54] Maden et al.,[27] and Vishunavat et al.[58] has provided valuable information on the location of the pathogen in seed. Infection of seed may occur during cool, wet weather while immature or mature seeds are still in the pod, or during the harvesting and threshing operations. Halfon-Meiri[54] observed that 50–80% of the seed from chick-pea pods with *Ascochyta* lesions were infected with *A. rabiei*, but the pathogen could not be detected in seeds from apparently healthy pods collected from diseased plants.

Infected seeds may or may not show signs of infections.[27,54] On seeds, lesions are light–dark brown[23,27,54] and range in size from 1–4 mm diameter. Black pycnidia containing mature spores are observed in several lesions, some of which form concentric zones.[3,27,54] In seeds with lesions, the fungus frequently penetrates the seed coat (Figure 21-6) and could be isolated from cotyledonary tissues.[27,54,58] The pathogen has not been detected in embryo.[27]

21-6-2 Spread

The spread of the disease has been attributed to the pycnidiospores produced at the foci of primary infection, either through crop debris or infected seeds. Sattar[3] had concluded that the aerial infection of seedlings raised from infected seed is systemic and compared it with smut infection of cereals. However, Maden et al.[27] failed to locate the fungus in whole mount preparations from green and healthy looking tissues of the seedlings below or in between the lesions. Chaube and Pandey[59] observed that the fungus starts sporulating on the seeds placed in soil even before the emergence of the radicle and plumule. Plumules which initially form "Crozier" often touch pycnidiospores and in the process of their growth, carry the inoculum to the shoots at the crown region.

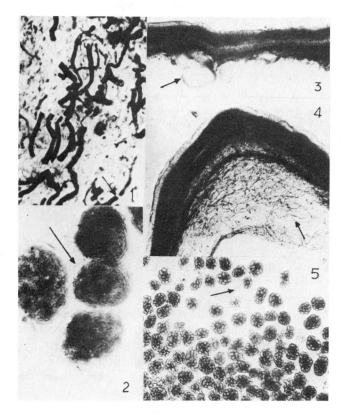

Figure 21-6 Location of *Ascochyta rabiei* in chick pea seeds
(1) Mycelium in seed coat (whole mount) 675 x
(2) Mycelium in cotyledons (whole mount) 675 x
(3) Pycnidia in seed coat (V.S.) 150 x
(4) Mycelium in seed coat (V.S.) 150 x
(5) Mycelium in cotyledons (V.S.) 150 x

It is an accepted fact that the disease spreads rapidly. This indicates the existence of adequate means of transportation of inoculum in nature. Though experimental evidence is lacking, the circumstantial evidence suggests that air currents and to some extent rain splashes play vital roles in the spread of inoculum. In this context the findings of Luthra et al. (cited by Nene[1]) are worth mentioning. According to them, the primary infection foci in a field are limited and isolated, but windy and wet conditions help in the rapid spread of the disease. They further suggested that infected debris, blown off from brittle diseased plants, could be transported by wind for several hundred meters. Temperatures of 20–25°C are best for the buildup of infection.[25,26,60] Chauhan and Sinha[26] in a glasshouse study found 85–98% relative humidity and 20°C to be most favorable. The incubation period under these conditions was 6 days. Khachatryan[53] working in Armenia reported over 60% relative humidity, with 350–400 mm rainfall and an average daily temperature of not less than 15°C, to be congenial for the spread of the disease.

21-6-3 Effect of Host Age and Inoculum Density

Tripathi[23] observed that with the increase in age of plants, there was increase in incubation period and decrease in the incidence and severity of the disease. He advocated that increased toughness of the tissues and age based biochemical changes might be responsible for such decrease in disease severity.

Singh[61] observed that inoculum concentrations have little role to play in the initiation and severity of the disease. However, Tripathi[23] concluded through glasshouse experiments that under congenial conditions even a small amount of inoculum will be enough for initiation and spread of the disease.

21-6-4 Host Range

Most workers[30,60,63] have reported *Cicer* spp. to be the only hosts of *A. rabiei*. However, Kaiser[20] reported that the fungus could infect cowpea (*Vigna sinensis*) and bean (*Phaseolus vulgaris*) when inoculated artificially. Tripathi et al.[64] observed that *A. rabiei* infects only *Cicer arietinum* L.

21-6-5 Disease Cycle

Available evidence suggests that *A. rabiei* neither produces resistant/resting structures nor has any alternate or collateral hosts as it infects only *Cicer* spp. Information on its saprophytic survival is very limited. Lukashevich[56] reported that the fungus can behave as a saprophyte and spread to noninfected tissues if the harvested material is stored for some time before threshing. This does not explain saprophytic survival of the fungus outside the host.

Two major sources of its survival are therefore diseased crop debris left in the field after harvest and/or infested or infected seeds. As already described and discussed, infested and/or infected seeds are undoubtedly the most vital source of primary inoculum. How far diseased debris left in the field play role in perpetuation of the fungus and thereby serving as a source of primary inoculum is uncertain. Available literature reveals that in areas where hot summer and rainy season follows the chick-pea crop season, the perpetuation of the fungus is of very short duration. It appears so because the crop debris left in the field decomposes due to rains and the fungus as such cannot withstand the onslaughts of environment and microbial antagonism. Contrary to it, in areas where climatic conditions between the two chick-pea seasons is dry, the survivability of the fungus will be enough to serve as a source of primary inoculum. In addition, in regions like Eastern Europe and Western Asia, the fungus by producing perithecial stage ensures its survival for the next season. Obviously, cold winter following the chick-pea season ensures survival of *A. rabiei* and in these areas crop debris definitely play role in carry over of the inoculum.

The secondary spread of the pathogen is by pycnidiospores produced at foci or primary infection either through crop debris or infected seed. Air currents and rain splashes help in movement of inoculum from diseased to healthy areas.

21-7 MANAGEMENT

21-7-1 Physical and Cultural Practices

Field sanitation.   *A. rabiei* could survive and multiply on chick-pea refuge,[25,49] thereby providing a potential source of inoculum.[65] Therefore, chick-pea refuge that remains in the field should be destroyed. Deep ploughing will hasten the decomposition of infected straw and remove it as a source of inoculum.[65-66]

Field inspection.   It is essential that seed fields be inspected at periodic intervals till harvest by qualified, trained personnel.[65] The inspection will also be useful in identifying the presence and potential importance of other seed-, soil-, and vector-borne diseases. Field inspections should be coordinated with laboratory tests designed to detect *A. rabiei* and other seedborne pathogens of chick-pea.[65]

Production and use of disease-free seed.   The production and use of healthy seed is essential to prevent introduction of the pathogen into disease-free areas where environmental conditions may favor spread and development of blight. Ascochytosis of chick-pea is dependent on cool, wet weather for its development and spread.[3,20,25] Surface contaminated or internally infected seed is the most important source of spreading and perpetuating the disease.[3,20,23,27] Warm weather impedes disease development and spread.[65] It is preferable to locate seed production fields in arid areas, where little or no rainfall occurs during the flowering and fruiting periods or at harvest. If plants are watered during the growing season, this would be done by furrow irrigation rather than overhead sprinkling.[65]

Effect of solar and dry heat treatment on seedborne inoculum.   Tripathi et al.[67] found that direct exposure of seeds to sunlight on cemented floor for 15 days reduced the recovery of *A. rabiei* by 50% without affecting the germinability. Exposure of seed at 40–50°C reduced the survival of *A. rabiei* by about 40–70%.[68]

Crop rotation.   Since only chick-peas are susceptible to Ascochyta blight, inclusion of cereals in the rotation will bring about sharp reduction in inoculum potential of *A.*

rabiei. Crop rotation may also be effective in reducing the inoculum of other seed- or soilborne chick-pea pathogens.[65]

Effect of date of sowing.   Tripathi et al.[69] conducted field experiments for three successive seasons and concluded that sowing of chick-pea in northern region of U.P. (India) on November 19, resulted in lowest disease severity and highest grain yields.

Effect of row direction and row spacing.   Tripathi[70] concluded through exhaustive trials for 3 years that neither East-West nor North-South row direction had any significant effect on spread and severity of disease. Reddy And Singh[71] reported no effect of inter row spacings on the incidence of Ascochyta blight. However, Tripathi[70] found that a row spacing of 45–60 cm is less favorable for development and spread of the disease.

Effect of intercropping and host nutrition.   Luthra et al.[1] suggested intercropping chick-pea with wheat, barley, mustard (*Brassica campestris*), etc. to reduce disease spread in the crop season. Tripathi[70] reported that sowing of chick-pea in between dwarf and medium height wheat varieties results in comparative low disease incidence and higher grain yield.

Lukashevich[56] suggested the application of potassium fertilizers (45 kg/ha) to reduce disease severity. Tripathi et al.[72] also recommended K_2O, 40-60 Kg/ha.

21-7-2 Host Resistance

Sources of resistance.   More than 50 years ago, three pure lines of *Cicer arietinum* L. were reported to be completely resistant to Ascochyta blight from North Africa.[73] Since then there have been many reports of resistant genetic material from India, Pakistan, Iran, Turkey, Syria, Morocco, North Africa, France, Bulgaria, and the USSR. A comprehensive list of the sources of resistance by Singh et al.[73] is reproduced in Table 21-1 with certain additions/modifications.

Inheritance of resistance.   In 1953, Hafiz and Ashraf[74] reported that the resistance to Ascochyta blight was determined by a single dominant gene in two parents, namely, F8 and F10. Later Satyavir et al.[75] and Singh and Reddy[76] reported that a single dominant gene conditioned the resistance in four parents, ILC 72, ILC 183, ILC 200, and ICC 4935, whereas the resistance in ILC 191 was conferred by a single recessive gene. This was the first report on identification of recessive gene governing resistance to blight. Tewari[77] observed that resistance to Ascochyta blight of chick-pea was monogenic-dominant in the resistant lines EC 26446, PG 82-1, P 919, P 1251-1, and NEC 4551 and monogenic-recessive in resistant line BRG 8.

Mechanism of resistance.   Sattar[3] considered that more malic acid secreted by leaves at flowering/podding time favored infection. However, Hafiz[78] claimed that a resistant cultivar (F8) secreted more malic acid than a susceptible cultivar (Pb 7) and that malic acid was inhibitory to spore germination and germ tube development. He further observed no difference in cuticle thickness between resistant and susceptible types but found higher number of stomata in the resistant type. Very little difference was found in

TABLE 21-1   LIST OF CHICK-PEA LINES REPORTED RESISTANT/TOLERANT

Year	Lines
1931	Two pure lines of *C. arietinum* L. var. *album* and one pure line of *C. arietinum* var. *nigrum*
1938	Pois chiches Nos. 180, 199, 281, 4F 32
1938	Three cultivars from France
1943	Natural hybrid no. 62-18
1949	Lines 99/21, 21, 142
1952	Pois chiches Nos. 4732, 199, 281, renamed as F8, F9 and F10 respectively
1960	Kabauskii, 16; Krasnotkutskii, 195; Krasnogradskii, 1; Ustoichi-vyi, 2; Askokitous-toichiyyi, 1.
1962	C 727
1962	C 235
1964	Bulgarian
1966	Dobrudzauskyi, Plovdiv 19
1967	11/48-7, B.N. 3118, P-36, Broach G.G. pedapuram, Allock 234, 337, 172/3, 436, 84, 241/1.
1970	Gbraztsov chiflik 6 (No. 180), Gbraztsov chiflik 7 (No. 307)
1972	1-13 (12-074-06625)
1972	EC 21629, EC 26414, EC 26435, EC 26446
1974	Selection P 1528-1-1, 1-13 (12-074-06625)
1974	K 1459, K 1469, K 1488, K1502
1976	Code No. 72-012
1976	AUG - 480
1977	Vars. Sovkhozuyi 14, Kubanskivi 199, VIR 32, No. 222, Resursi 216
1978	Four moderately resistant lines
1979	Pch 15, Pch 34, Pch 128
1981	55 lines
1981	CM 68, C 72
1981	21 lines
1982	BRG8, EC 26446, P 1252-1, Pant G82-1
1983	Seven lines
1983	ICC 7389, ICC 8536
1983	25 cultivars
1984	ILC-195 against 8 isolates of the fungus.
1984	ICC-8161, -6269, -6270, -8160, -8189
1985	ICC 4000, ICC 5033, P919, P1528-1-1, P2129, C8, CPI 56566, JM995, E100Y.
1985	Ten *desi* and six *Kabuli* accessions

the acidity of sap collected from resistant and susceptible types. Ahmad et al.[79] reported that resistant types (F8 and F10) were significantly taller, possessed a large number of hairs per unit area of stem and leaf, and had a smaller number of tertiary branches than the susceptible types (Pb 7 and C 7). Kunzru and Sinha[80] reported that an antibiotic principle which diffuses in the inoculation droplets during the course of interaction between *A. rabiei* and pod tissues plays some role in resistance. The antibiotic principle was given a status of phytoalexin and named Cicerin and is composed of two phenolic compounds. Satyavir and Grewal[81–84] found that the resistant cultivar showed higher peroxidase activity, higher L-cystin content, and more phenolic content and higher catalase activity after inoculation. Kaiser[20] and Singh et al.[73] reported that majority of the *desi* lines are black seeded and small with higher proportions of resistance than *Kabuli* lines. Weigand et al.[85] studied the accumulation of phytoalexins and isoflavone glucosides in a resistant and a susceptible cultivar of chick-pea during interaction with *Ascochyta rabiei* and reported that the two cultivars showed no significant differences in the level of isoflavones and isoflavone conjugates. However, the resistant cultivar ILC 3279 rapidly accumulated large amounts of both phytoalexins (medicarpin and maackiain) whereas susceptible cultivar (ILC 1929) produced very small amounts of medicarpin.

21-7-3 Chemical Control

Seed treatment.   In crops such as chick-pea, at the present yield levels, fungicides can only be feasible and economical when used for seed treatment. Seedborne nature of *Ascochyta rabiei* makes fungicidal seed treatment essential and useful.[1,3,27,52-54] Seed treatment with effective fungicides can greatly help in reducing the initial inoculum level and preventing the spread of the disease or races, into areas where they are not present. Seed treatment, especially of moderately resistant cultivars, may be much more effective both by preventing the entry of new races and reducing the pathogen population and thus selection pressure.[86]

Ever since the discovery of seedborne nature of the disease, attempts have been made to eradicate the pathogen by chemical treatments.[1,3,25,52-54,56,86-94] Several workers[89,92] found seed treatment with Calixin M (11% tridemorph + 36% maneb) alone or in combination with benlate to completely eradicate the fungus from the seed. This treatment did not affect the germination and there was no noticeable phytotoxicity. Reddy and Kababeh[93] found seed treatment with thiabendazole was equally effective. Recently, Tripathi et al.[95-96] reported that Calixin M alone or in combination with thiram, Bavistin (carbendazin), or Bavistin + thiram completely eradicated the fungus from seed.

Foliar spray.   Foliar application of fungicides alone or in combination with seed treatment have been recommended for the control of ascochytosis of chick-pea.[97] These fungicides include Boardeaux mixture,[5] Wettable sulphur,[56] zineb,[1] ferbam,[8] maneb,[9] captan,[98] and Daconil.[99-100] Reddy and Singh[101] conducted field trials using three fungicides, namely Bravo 500 (chlorothalonil), Calixin M (11% tridemorph + 35% maneb), and Rubigan (fenarimol) for the control of Ascochyta blight and recommended that spraying with Bravo 500 alone gave almost complete protection. Calixin M alone was only partially effective but gave similar protection and seed yield to Bravo 500 when used alternately with this fungicide, which could be valuable insurance against loss of effectiveness of a single chemical. They also studied the efficacy of one spray of Bravo 500 at different crop stage and found that spraying at early podding reduced pod damage and markedly increased seed yields in ILC-482 (moderately resistant). Gaur and Singh[102] tested 21 chemicals and reported that Delan 75 WP (dithionon), Daconil 2787 W75 (chlorothalonil), Foltaf 80WP (captafol), and Captaf 75 WP (captan) controlled the disease satisfactorily.

Tripathi et al.[96] found that seed treatment with Bavistin + thiram followed by 3 sprays of Bavistin at 10-day intervals resulted in lowest disease severity and highest grain yield followed by seed treatment with Bavistin and Bavistin + Brassicaol with 3 sprays of Bavistin. It would be pertinent to indicate that under highly favorable conditions for disease development, most of these fungicides, even with as many as 12 applications, may not give sufficient protection across the season.

21-7-4 Integrated Management

Based on the information reviewed on various aspects of the disease, the following suggestions are made to minimize the incidence and severity of Ascochyta blight.

1. Since infested and/or infected seed is the most important source of primary inoculum, clean, healthy, and certified seed should be used for raising new crop. The seed must be treated with a suitable fungicide such as Calixin M and/or Bavistin before sowing.

2. In areas where conditions favor the perpetuation of *Ascochyta rabiei* in infected debris, the crop refuge must be destroyed either by burning or deep ploughing.

3. Effective fungicides such as Bravo 500, Bavistin, or zineb should be sprayed at suitable intervals. Even the resistant or tolerant cultivars should be given foliar spray which may help in lengthening the life of cultivars with vertical resistance.

4. Fertilize the crop also with K_2O (40-60 kg/ha) to increase level of tolerance in the host.

5. Sprinkler system of irrigation should be avoided.

21-8 REFERENCES

1. Nene, Y. L. "A Review of Ascochyta Blight of Chickpea (*Cicer arietinum* L.)." *Trop. Pest Manag.* 28 (1982): 61.

2. Labrousse, F. "Anthracnose of the Chickpea (*Cicer arietinum* L.). *Revue de Pathologie vegetable et ol' Entomologie Agricole de France* 27 (1930): 174 (in French).

3. Sattar, A. "On the Occurrence, Perpetuation and Control of Gram Blight Caused by *Ascochyta rabiei* (Pass.) Labr., with Special Reference to Indian Conditions." *Ann. Appl. Biol.* 20 (1933): 612.

4. Malik, B. S., and Tufail, M. "Chickpea Production in Pakistan," *in Ascochyta Blight and Winter Sowing of Chickpea,* eds. M. C. Saxena and K. B. Singh. (ICARDA, Martinus Nijhoff/Dr. W. Junk Publisher, 1984), 229.

5. Kovachevski, I. C. "Parasitic Fungi New for Bulgaria Fourth Contribution." *Travaux de la Societe Bulgare des Science Naturelle* 27 (1935): 13 (in Russian).

6. Nemlienko, F. E., and Lukashevich, A. I. "Agrotechnical Measures Against Ascochytosis of Chickpea." *Plant Protection Moscow* 4 (1957): 31 (in Russian).

7. Demetriades, S. D., Zachos, D. G., Constantinou, P. T., Panagopulos, C. G., and Holevas, C. D. "Brief Reports on the Principal Plant Diseases Observed in Greece During the year 1958." *Annales de I' Institut Phytopathologieque Benaki* 2 (1959): 3 (in French).

8. Puerta Romero, J. "Gram Blight Isolation of the Fungus *Ascochyta rabiei* and Study of the Gram Varieties Possibly Resistant to It." *Boletin de Pathologia Vegetaly-Entomologia Agricola* 27 (1964): 15 (in Spanish).

9. Georgiou, G. P., and Papadopoulos, C. "A Second List of Cyprus Fungi." *Technical Bulletin* 5, 1957.

10. Mlaiki, A., and Hamadi, S. B. "Chickpea Improvement in Tunisia," in *Ascochyta Blight and Winter Sowing of Chickpea*, eds. M. C. Saxena and K. B. Singh. (ICARDA: Martinus Nijhoff/Dr. W. Junk Publisher, 1984), 255.

11. Benlloch, M. "Some Phytopathological Characteristics of the Year 1941." *Boletin de Pathologia vegetal—Y-Entomologia Agricola* 10 1 (1941): 29.

12. Biggs, C.E.G. Annual Report, 1943, Department of Agriculture, Tanganyika Territory, Japan, 8, 1944.

13. Zalpoor, N. "*Mycophaerella rabiei* Kovacevski (*Ascochyta rabiei* Pass.) Labr." *Entomologie et Phytopathologic Applicues* 21 (1963): 10.

14. Kausar, A. G. "Epiphytology of Recent epiphytotics of Gram Blight in Pakistan." *Pakistan J. Agric. Sci.* 2 (1965): 185.

15. Radulescu, E., Capetti, E., Schmidt, E., and Cassian, A. "Contributions to the Study of Anthracnosis of Chickpea." *Lucrari Stuntifice* 14 (1977): 311.

16. Kaiser, W. J. "Occurrence of Three Fungal Diseases of Chickpea in Iran." *F.A.O. Plant Protection Bulletin* 20 (1972): 74.

17. Grewal, J. S. "Ascochyta Blight of Bengal Gram," *in* Prof. R. N. Tondon birthday Celebration Committee, *Advances in Mycology and Plant Pathology,* Indian Phytopathological Society, New Delhi, 1975, 161.

18. Sattar, A. "A Comparative Study of the Fungi Associated with Blight Diseases of Certain Cultivated Leguminous Plants." *Trans. Brit. Mycol. Soc.* 18 (1934): 76.

19. Bedi, P. S., and Aujla, S. S. "Factors Affecting the Mycelial Growth and the Size of the Pycnidia Produced by *Phyllosticta rabiei* (Pass.) Trot. the Incitant of Gram Blight." *Punjab J. Res.* 7 (1970): 606.

20. Kaiser, W. J. "Factors Affecting Growth, Sporulation, Pathogenecity, and Survival of *Ascochyta rabiei.*" *Mycologia* 65 (1973): 444.

21. Reddy, M. V., and Nene, Y. L. "A Case for Induced Mutation in Chickpea for Ascochyta Blight Resistance," in Proc. of the Symp. on the Role of Induced Mutation in Crop Improvement, Osmania University, Hyderabad, India, 1979, 398.

22. Pandey, B. K. "Studies on Chickpea Blight Caused by *A. rabiei* (Pass.) with Special Reference to Survival in Crop Debris." (M.S. Ag. thesis, GBPUA&T, Pantnagar, India, 1984).

23. Tripathi, H. S. "Studies on Ascochyta Blight of Chickpea." (Ph.D. thesis, GBPUA&T, Pantnagar, India, 1985), 142.

24. Khalil, S., and Khan, M. A. "An Improved Agar Growth Medium for *Ascochyta rabiei* (Pass.) Labr." *International Chickpea Newsletter* 14 (1986): 27.

25. Zachos, D. G., Panagopulos, C. G., and Maknis, S. A. "Researches on the Biology, Epidemiology and Control of Anthracnose of Chickpea." *Annales de'l' Institute Phytopathologique Benaki*, 5 (1963): 167 (in French).

26. Chauhan, R. K. S., and Sinha, S. "Effect of Varying Temperature, Humidity, and Light during Incubation in Relation to Disease Development in Blight of Gram (*Cicer arietinum* L.) Caused by *Ascochyta rabiei.*" *Proceedings of the National Science Academy of India* B 37 (1973): 473.

27. Maden, S., Singh, D., Mathur, S. B., and Neergaard, P. "Detection and Location of Seed-Borne Inoculum of *Ascochyta rabiei* and Its Transmission in Chickpea." *Seed Sci. Tech.* 3 (1975): 667.

28. Aslam, M. "A Review of Research Studies on Chickpea Blight Fungus in Pakistan," in *Ascochyta Blight and Winter Sowings of Chickpea,* eds M. C. Saxena and K. B. Singh. (ICARDA: Allepo, Syria, 1984), 255.

29. von Arx, J. A. *The Genera of Fungi Sporulating in Pure Culture.* J. Cramer In Der A.R. Ganter Verlag Kommandit-gesellschaft FL 9490, Vadus, 1974.

30. Gorlenki, M. V., and Bushkova, L. N. "Perfect State of the Causal Agent of Ascochytosis of Chickpea." *Plant Protection* 3 (1958): 60 (in Russian).

31. Luthra, J. C., Sattar, A, and Bedi, K. S. "Variation in *Ascochyta rabiei* (Pass.) Labr., The Causal Fungus of Blight of Gram (*Cicer arietinum* L.)." *Indian J. Agric. Sci.* 9 (1939): 791.

32. Aujla, S. S. "Study on Eleven Isolates of *Phyllosticta rabiei* (Pass.) Trot., The Causal Agent of the Gram Blight in Punjab." *Indian Phytopathol.* 17 (1964): 83.

33. Bedi, P. S., and Aujla, S. S. "Variability in *Phyllosticta rabiei* (Pass.) Trot., the Incitent of Blight diseases of Gram." *Punjab J. Res.* 6 (1969): 6.

34. Luthra, J. C., Sattar, A., and Bedi, K. S. "Determination of Resistance to Blight Disease in Gram Types." *Indian J. Agric. Sci.* 11 (1941): 249.

35. Ahmad, T., Hasanian, S. N., and Sattar, A. "Some Popular Method of Plant Disease Control in Pakistan." *Agriculture Pakistan* 18, 1949.

36. Grewal, J. "Important Fungal Diseases of *Cicer arietinum* L. in India," Seminar on Pulse Production, Karanj, Iran, 35, 1969.

37. Satyavir, S., and Grewal, J. S. "Physiologic Specialization in *Ascochyta rabiei*, the Causal Organism of Gram Blight." *Indian Phytopathol.* 27 (1974): 355.

38. Grewal, J. S. "Evidence of Physiologic Races in *Ascochyta Rabiei* of Chickpea" in *Ascochyta Blight and Winter Sowing of Chickpea*, ed. M. C. Saxena, and K. B. Singh, ICARDA, Martinus Nijhoff/Dr. W. Junk Publisher, 1984, 259.

39. Ganeva, D., and Matsov, B. "Comparative Testing of Introduced and Local Samples of Chickpea." *Rasteniev dni Nanki* 14 (1977): 51.

40. Radkov, P. "Biological and Economic Properties of Some New Varieties of Chickpea." *Rasteniev dni Nanka* 15 (1978): 81.

41. Singh, K. B., Hawtin, G. C., Nene, Y. L., and Reddy, M. V. "Resistance in Chickpea Ascochyta Blight." *Pant Dis.* 65 (1981): 586.

42. Simon, G. "Pathogenicity of Isolates of *Ascochyta rabiei.*" *International Chickpea Newsletter* 7 (1982): 16.

43. Qureshi, S. H., and Alam, S. "Pathogenic Behaviour of *Ascochyta rabiei* Isolates on Different Cultivars of Chickpea in Pakistan." *International Chickpea Newsletter* 11 (1984): 28.

44. Porta–Puglia, A. "Differential Behaviour of Chickpea Lines Towards Some Italian Isolates of *Ascochyta rabiei.*" *International Chickpea Newsletter* 12 (1985): 22.

45. Bashir, M., Alam, S. S., and Qureshi, S. H. "Chickpea Germplasm Evaluation for Resistance to Ascochyta Blight under Artificial Conditions." *International Chickpea Newsletter* 12 (1985): 24.

46. Pandey, B. K., Singh, U. S., and Chaube, H. S. "Development of *Ascochyta rabiei* in the Leaves of a Susceptible Chickpea Cultivar." *Indian Phytopathol.* 38 (1986): 779.

47. Pandey, B. K., Singh, U. S., and Chaube, H. S. "Mode of Infection of Ascochyta Blight of Chickpea Caused by *Ascochyta rabiei.*" *J. Phytopathol.* 119 (1987): 88.

48. Alam, S. S., Bilton, J. N., Slawin, A. M. Z., Williams, D. J., Sheppard, R. N., and Strange, R. N. "Chickpea Blight: Production of the phytotoxins Solanapyrones A and B by *Ascochyta rabiei.*" *Phytochemistry* 28 (1989): 2627.

49. Hohl, B., Pfautsch, M., and Barz, W. "Histology of Disease Development in Susceptible and Resistant Cultivars of Chickpea (*Cicer arietinum* L.) Inoculated with *Ascochyta rabiei.*" *J. Phytopathol.* 129 (1990): 31.

50. Schariff, G. E., Niemann, E., and Ghanea, M. "Chickpea Blight in Iran, *Mycosphaerella rabiei* Kovacevski-*Ascochyta rabiei* (Pass.) Labr." *Entomologie Phytopath.* 25 (1967): 9.

51. Zachos, D. G. "A Case of Parasitism of Chickpea (*Cicer arietinum* L.). Seeds by *Stemphyllium botryosum* Wallr." *Ann. Inst., Phytopath., Benaki,* 6 (1952): 60 (in French).

52. Goblez, M. "Research Work on the Varieties and Areas of Spread of Bacterial and Parasitic Diseases Affecting and Contaminating the Seeds of Cultivated Plants Grown in Certain Provinces of Central Anatolia as Well as the Approximate Degree of Damage Caused by Such diseases." *Ziraat Fukultesi Yayinlavi* Ankara Universitesi 62 (1956): 1 (in Turkish).

53. Khachatryan, M. S. "Seed Transmission of Ascochytosis Infection in Chickpea and the Effectiveness of Treatment." *Sbornik Nauchnykh Trudov Nauchno-Issledo vatel'sk Zemledel Armyarskei* 2 (1961): 147 (in Russian).

54. Halfon-Meiri, A. "Infection of Chickpea Seeds by *Ascochyta rabiei* in Israel." *Plant Dis. Reptr.* 54 (1970): 442.

55. Morrall, R. A. A., and McKenzie, D. L. "A Note on the Inadvertant Introduction to North America of *Ascochyta rabiei*, a Destructive Pathogen of Chickpea." *Plant Dis. Reptr.* 58 (1974): 342.

56. Lukashevich, A. I. "Control Measures Against Ascochytosis of Chickpea. *J. Agric. Sci., Moscow* 5 (1958): 131 (in Russian).

57. Tripathi, H. S., Singh, R. S., and Chaube, H. S. "Survival of *Ascochyta rabiei* in Infected Chickpea Seeds Stored at Different Temperatures." *Indian J. Mycol. Pl. Path.* 17 (1988): 98.

58. Vishunavat, K., Agarwal, V. K., and Singh, R. S. "Location of *Ascochyta rabiei* in Gram Seeds." *Indian Phytopathol.* 38 (1985): 377.

59. Chaube, H. S., and Pandey, B. K. "Transmission of Seed-Borne Inoculum of *Ascochyta rabiei* (Pass) Labr. in Chickpea Seedlings." *Bulletin of Pure and Applied Sciences* 5B (1986): 18.

60. Askerov, I. B. "Ascochytosis of Chickpea." *Zashchita Rastenii of Vrechtelei i Boloznei* 13 (1968): 52 (in Russian).

61. Singh, G. "Screening for Resistance to Aschochyta Blight with Special Reference to the Efficient Screening Technique," in Workshop on *rabi* pulses, All India Coordinated Pulses Improvement Project, Ludhiana, 1982.

62. Bondartzeva-monteverde, V. N., and Vassilievsky, N. I. "A Contribution to the Biology and Morphology of Some Species of *Ascochyta* on Leguminosae." *Acta Institute Botanici Academial Scientiarum* USSR 1938, Ser. 11 (1940): 345 (in Russian).

63. Sprague, R. "Notes on *Phyllosticta rabiei* on Chickpea." *Phytopathol.* 20 (1930): 591.

64. Tripathi, H. S., Singh, R. S., and Chaube, H. S. "Host-Range of *Ascochyta rabiei* (Pass.) Labr., the Causal Agent of Ascochyta Blight in Chickpea." *International Chickpea Newsletter* 16 (1987): 11.

65. Kaiser, W. J. "Control of Ascochyta Blight of Chickpea through Clean Seed," in *Ascochyta Blight and Winter Sowing of Chickpea,* eds M. C. Saxena, and K. B. Singh. ICARDA, Martinus Nijhoff, Dr. W. Junk Publisher, 117, 1984.

66. Chaube, H. S., and Pandey, B. K. "Management of *Ascochyta* Blight of Gram." *Seeds Farms* 11 (1985): 41.

67. Tripathi, H. S., Singh, R. S., and Chaube, H. S. "Effect of Sun Drying on the Recovery of *Ascochyta rabiei* from Infected Chickpea Seeds." *International Chickpea Newsletter* 16 (1987): 13.

68. Tripathi, H. S., Singh, R. S., and Chaube, H. S. "Effect of Dry-Heat Treatment on the Survival of *A. rabiei* (Pass.) Labr. in Infected Chickpea Seeds." *International Chickpea Newsletter* 16 (1987): 13.

69. Tripathi, H. S., Chaube, H. S., and Singh, R. S. "Effect of Date of Sowing on Severity of Ascochyta Blight of Chickpea at Pantnagar, India." *International Chickpea Newsletter* 18 (1988): 23.

70. Tripathi, H. S. Personal communication.

71. Reddy, M. V., and Singh, K. B. "Effect of Inter-Row Spacing on the Reaction of Chickpea Lines to Ascochyta Blight." *International Chickpea Newsletter* 3 (1980): 13.

72. Tripathi, H. S., Singh, R. S., and Chaube, H. S. "Effect of NPK Fertilizers on the Severity of Ascochyta Blight of Chickpea." *Indian Phytopathol.* 40 (1987): 561.

73. Singh, K. B., Nene, Y. L., and Reddy, M. V. "International Screening of Chickpea for Resistance to Ascochyta Blight," in *Ascochyta Blight and Winter Sowing of Chickpea,* eds. M. C. Saxena and K. B. Singh. ICARDA, Martinus Nijhoff/Dr. W. Junk Publisher, 1984, 67.

74. Hafiz, A., and Ashraf, M. "Studies on the Inheritance of Resistance to *Mycosphaerella* Blight in Gram." *Phytopathology* 43 (1953): 580.

75. Satyavir, S., Grewal, J. S., and Gupta, V. P. "Inheritance of Resistance to Ascochyta Blight in Chickpea." *Euphytica* 24 (1975): 209.

76. Singh, K. B., and Reddy, M. V. "Inheritance of Resistance to Ascochyta Blight in Chickpea." *Crop Sci.* 23 (1983): 9–10.

77. Tewari, S. K. "Genetic Architecture of *Ascochyta* Blight Resistance and Certain Quantitative Traits in Chickpea (*Cicer arietinum* L.)." (Ph.D. thesis, GBPUA&T., Pantnagar, 1984), 200.

78. Hafiz, A. "Basis of Resistance in Gram to *Mycosphaerella* Blight." *Phytopathology* 42 (1952): 422.

79. Ahmad, G. C., Hafiz, A., and Ashraf, M. "Association of Morphological Characters with Blight Reaction," in Proc. of the fourth Pakistan Sci. Conf., 1952, 17.

80. Kunzru, R., and Sinha, S., "'Cicerin' a New Phytoalexin Associated with Blight of Gram," in Proceedings of the first International Symposium of Plant Pathology, Indian Phytopath. Soc., New Delhi, 1966, 724.

81. Satyavir, S., and Grewal, J. S. "Peroxidase Activity Associated with Ascochyta Blight of Gram (*Cicer arietinum* L.)." *Phytopathologia Mediterranea* 13 (1974): 174.

82. Satyavir, S., and Grewal, J. S. "Changes in Phenolic Context of Gram Plants Induced by *Ascochyta rabiei* Infection." *Indian Phytopathol.* 27 (1974): 524.

83. Satyavir, S., and Grewal, J. S. "Role of Free Amino Acids in Disease Resistance to Gram Blight." *Indian Phytopathol.* 28 (1975): 286.

84. Satyavir, S., and Grewal, J. S. "Change in Cotalase Activity of Gram Plant Induced by *Ascochyta rabiei* Infection." *Indian Phytopathol.* 28 (1975): 223.

85. Weigand, F., Kostor, J., Weltzein, H. C., and Barz, W. "Accumulation of Phytoalexins and Isoflavone Glucosides in a Resistant and a Susceptible Cultivar of *Cicer arietinum* L. during Infection with *Ascochyta rabiei.*" *J. Phytopathol.* 115 (1986): 214.

86. Zachos, D. G. "Studies on the Disinfection of Chickpea Seeds (*Cicer arietinum* L.) Infected with *Ascochyta rabiei* (Pass.) Labr.." *Annales de l'Institute Phytopathologique Benaki* 5 (1951): 76 (in French).

87. Ibragimov, G. R., Akhmedov, S. A., and Garadogi, S. M. "Use of Phenthiuram and Phenthiuram Molybdate Against Diseases of Fodder Beans and *Cicer arietinum* under Azerbaidzhani Conditions." *Khimiya Sel' Khoz.* 4 (1966): 23.

88. Kaiser, W. J., Okhoval, M., and Mossahebi, G. H. "Effect of Seed Treatment Fungicides on Control of *Ascochyta rabiei* in Chickpea Seed Infected with the Pathogen." *Plant Dis. Reptr.* 57 (1973): 742.

89. Reddy, M. V. "Calixin-M, An Effective Fungicide for Eradication of *Ascochyta rabiei* in Chickpea Seeds." *International Chickpea Newsletter* 3 (1980): 120.

90. Reddy, M. V., Singh, K. B., and Nene, Y. L. "Further Studies of Calixin M in the Control of Seed-Borne Infection of Ascochyta Blight in Chickpea." *International Chickpea Newsletter* 6 (1982): 17.

91. Grewal, J. S. "Control of Important Seed-Borne Pathogens of Chickpea." *Indian J. Genet. Pl. Breeding* 42 (1982): 393.

92. Bhatti, M. A., Malik, B. A., and Hussain, S. A. "Efficacy of Calixin M and Other Fungicides Against Ascochyta Blight in Chickpea." *International Chickpea Newsletter* 8 (1983): 23.

93. Reddy, M. V., and Kababeh, S. "Eradication of *Ascochyta rabiei* from Chickpea Seed with Thiabendazole." *International Chickpea Newsletter* 10 (1984): 17.

94. Aslam, B., M. Akhlaq Hussain, S., Ahmed Malik, B., Tahir, M., and Ashraf Zahid, M. "Effect of Fungicidal Treatment of Diseased Seed against *Ascochyta rabiei* on Germination in Chickpea." *Pakistan J. Agric. Res.* 5 (1984): 23.

95. Tripathi, H. S., Singh, R. S., and Chaube, H. S. "Efficacy of Some Fungicides Against *Ascochyta rabiei* (Pass.) Labr." *International Chickpea Newsletter* 15 (1986): 20.

96. Tripathi, H. S., Singh, R. S., and Chaube, H. S. "Effect of Fungicidal Seed and Foliar Application on Chickpea Ascochyta Blight." *Indian Phytopathol.* 40 (1987): 63.

97. Reting, B., and Tobolsky, J. "A Trial for the Control of Ascochyta in Chickpeas." First Israel Congr. Pl. Pathol. 50, 1967.

98. Satyavir, S., and Grewal, J. S. "Evaluation of Fungicides for the Control of Gram Blight." *Indian Phytopathol.* 27 (1974): 641.

99. Se,Nycirex, M., OLC, UN, and Zavrak, Y. "Preliminary Studies on Chemical Control of Chickpea Anthracnose (*Mycophaerella rabiei* Kov.) in the Black Sea Region, Zirai Mucadale Araes." *Tirma Yilbigi* 11 (1977): 71.

100. Hanounik, S. B. "Influence of Host-Genotype and Chemical Treatment on Severity of Ascochyta Blight in Chickpea." *International Chickpea Newsletter* 2 (1980): 13.

101. Reddy, M. V., and Singh, K. B. "Foliar Application of Bravo 500 for Ascochyta Blight Control." *International Chickpea Newsletter* 8 (1983): 26.

102. Gaur, R. B., and Singh, R. D. "Control of Ascochyta Blight of Chickpea Through Foliar Spray." *International Chickpea Newsletter* 13 (1985): 22.

22

GREY MOLD
OF CHICK-PEA

H. S. CHAUBE, B. K. PANDEY, AND T. K. MISHRA

*Department of Plant Pathology, G. B. Pant University of Agric. & Technol.,
Pantnagar-263145, India*

22-1 INTRODUCTION

The first record of the occurrence of grey mold of chick-pea (*Cicer arietinum* L.) caused by *Botrytis cinerea* Pers. in India was by Shaw and Ajrekar[1] (1915) and later by Butler and Bisby.[2] However, the first record of its widespread occurrence was by Joshi and Singh.[3] They observed complete destruction of a chick-pea variety G-24 during the 1967–68 season. Now the disease is known to occur in almost all major chick-pea growing areas.[4-5] The disease is also known to occur in several countries, notably in Argentina, Australia, Canada, and Colombia.[6-9] Grey mold has been responsible for heavy production losses to chick-pea during the last few years. In January 1979 the disease appeared in an epiphytotic form and destroyed chick-pea crop over an area of 20,000 hectares in 'Tal' area of Bihar (India). In 1980–81 the disease caused 70–100% losses at the central state Farm, Hissar and in parts of Punjab. Geographical distribution of grey mold of chick-pea is shown in Figure 22-1.

22-2 SYMPTOMS

Several workers have described the symptoms of the disease as it occurs in different chick-pea growing regions.[3,8,10,13] The descriptions are remarkably similar.

Under natural conditions, the disease first appears in isolated patches at a time when crop growth has reached its maximum vegetative growth (Figure 22-2), the morning

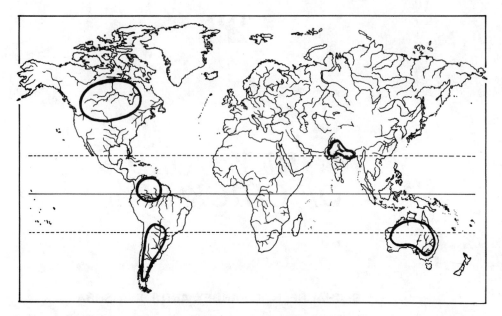

Figure 22-1 Geographical distribution of grey mold of chick-pea.

relative humidity is very high and temperature is low. The fungus attacks leaves, petioles, tender branches, inflorescence, and flowers causing development of grey or dark brown lesions covered with erect, hairy sporophores (Figure 22-2). The branches hang down due to rotting of the tissues. On thick branches lesions are large and stems break at the point where soft rot has occurred. Under conditions of favorable weather persisting for several days, the blighting of foliage spreads rapidly and the entire plant dries up (Figure 22-2).

In the event of severe infection, pods are also attacked. At times, no seeds or only small shriveled seeds are formed in the affected pods. Lesions on the pod are water soaked and irregular. If observed early, greyish-white mycelial growth can be seen (Figure 22-2). Joshi and Singh[3] and later Singh[14] have observed that at a later stage grey mold phase is replaced by a dirty grey or black mass containing dark brown to black sporodochia having minute conidia. They had also observed development of microsclerolial bodies.

22-3 THE PATHOGEN

The fungus *Botrytis cinerea* Pers. ex. Fr. belongs to class Hyphomycetes, order-Moniliales and family-Moniliaceae/Dematiaceae of the subdivision–Deuteromycotina. The asexual (conidial) stage (Figures 22-3 and 22-4) on *Cicer arietinum*[3,12,14] has the following characteristics: mycelium septate, brown, branched, measuring 8–16 μm; conidiophores and conidia produced free; conidiophores light brown, septate, 8–24 μm wide, tips or its branches slightly enlarged and bearing small pointed sterigmata; conidia hyaline, one

Figure 22-2 Symptoms of grey mold of chick-pea
(a) chick-pea plants infected with *Botrytis cinerea* under field condition
(b),(c),(d) Symptoms on foliage and flower of chick-pea
(e) Infected pods
(f) Healthy pods

celled, oval, globose or short cylindrical, borne in clusters, measuring 4–24 × 4–18 μm (average 14 x 8.5 μm) (Figures 22-3, 22-4).[36]

Sporodochial (sclerotial) layer on host is fairly large, (0.5–5.0 mm diameter) consisting of densely interwoven brown septate hyphae and produces round to oval aseptate conidia measuring 4–8 μm. These conidia do not germinate. Sporodochia soon enlarge into hard sclerotia measuring 2–11 x 2–7 (5.6 x 3.8) μm.[3] The perfect stage of *B. cinerea* is *Botrytinia fuckeliana*.[15] However, there is no report of the occurrence of perfect stage of the fungus on chick-pea.

Figure 22-3 Asexual structures of *Botrytis* spp. A young conidium showing the blastic origin of the conidia on the ampullae[36]

Figure 22-4 Conidia of *Botrytis cinerea* showing the sterigmata attachment to the ampullae[36]

22-3-1 Variability

Variability is the very basis of survival, and pathogens in order to cope with the host diversity and fluctuating environmental conditions produce different pathotypes or races. Joshi and Singh[3] and Singh[14] observed formation of sclerotial and/or sporodochial bodies on chick-pea plants infected by *B. cinerea* in *Tarai* region of Nainital, India, but Pandey[12] working at the same location did not observe development of sclerotial and sporodochial bodies either under natural conditions or in vitro studies. He concluded that the isolate of *B. cinerea* existing during 1968–70 does not exist any more. It has possibly been replaced by another isolate which does not produce these bodies or as a result of creation of genetic variability, the fungus has lost its ability to produce these structures. Singh and Bhan[16] based on formation of sclerotia reported the occurrence of different isolates of *B. cinerea* in Punjab (India).

In a detailed study, Rewal[17] isolated *B. cinerea* from different locations in India. The purified isolates were divided into six groups on the basis of variation in growth and sporulation. They were designated as strains B_1 to B_6 and were found to be pathogenic on chick-pea variety JG 62. Strains B_1 and B_2 sporulated profusely while strain B_3 formed few sclerotia in addition to excellent sporulation.[17] Strain B_4 formed white mycelial mat without conidia or sclerotia formation. Strain B_5 did not sporulate but formed sclerotia on prolonged incubation in darkness. Strain B_6 formed sclerotia readily with sparse sporulation. Variation was observed in size and shape of conidia of six strains. Conidia of strains B_1, B_2, B_4, and B_5 were distinctly oval in shape whereas those of strains B_3 and B_6 were globose. Microconidia formation was also observed in spermadochium comprising a compact penicillate cluster of phialides. Strains B_3 and B_6 showed significantly higher growth at 5°C as compared to strains B_1, B_2, B_4, and B_5, which showed less growth and nil to poor sporulation at this temperature.[17] Significantly higher growth of the latter strains was obtained at 25°C. Serological differences were not observed in different strains of *B. cinerea* used in the study.[17] The six strains were classified into five physiologic races on the basis of their reaction on a set of chick-pea differentials.[17] Sporulating strains B_1 and B_2 showed similar disease reaction on all differentials and thus were regarded as one physiologic race. They could be differentiated from strains B_3 and B_4 on variety BG 256 and from strains B_5 and B_6 on variety RSG 3. Strains B_5 and B_6 could be differentiated from each other on variety BGM 413. These results gave concrete evidence regarding the existence of physiologic specialization in *B. cinerea*.[17]

22-4 INFECTION PROCESS

22-4-1 Penetration

To establish a parasitic relationship, the fungus first has to enter the host. Inoculated spores of the fungus on the foliage germinate within 6 to 8 hr.[12] After germination, the germ tubes proliferate saprophytically and mostly form a sort of mycelial mat on the leaf surface (Figure 22-5a). In between proliferation and formation of mycelial mat, the hyphal tips having direct contact with the host surface swell to form appresoria and subsequently the infection hyphae which penetrate directly through the cuticle, form subcuticular as well as subepidermal mycelium (Figure 22-5b-d). In some cases, the hyphae as

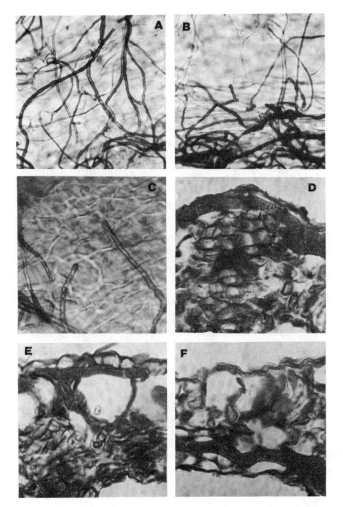

Figure 22-5   (a) Saprophytic growth of the *Botrytis cinerea* on leaf surface
(b) Direct penetration through cuticle
(c) Formation of the appressoria and subcuticular hyphae
(d) Enlarged view of appressoria and subcuticular hyphae
(e),(f) T.S. of chick-pea leaf showing presence of fungal mycelium inside the leaf tissue and extensive damage of mesophyll tissues

such penetrate directly through the host surface. Penetration/entry through stomata has not been observed.[12]

22-4-2 Invasion

After penetration the infection hyphae grow and ramify leaf tissues subcuticularly as well as subepidermally[12] (Figure 22-5d). Mycelial growth takes place in mesophyll cells also. Mycelium after penetration gets thickened and branched[12] (Figure 22-5d-f).

Pathogen causes extensive damage to leaf tissues by destroying epidermal and mesophyll cells, most probably by degrading cell walls, even in advance of invading hypahe. It indicates the involvement of cell wall degrading enzymes like pectinases and cellulases. Only lignified xylem and tracheary elements remain unaffected (Figure 25-5e-f) probably due to inability of the pathogen to degrade lignin.[12]

22-5 SURVIVAL

Survival of the pathogen in the absence of host in case of annual crops like chick-pea is very important for the recurrence of the disease every year. Report of epidemics of Botrytis grey mold in different parts of the world indicates the existence of definite and efficient mechanisms for survival of the fungus between seasons.

22-5-1 Seed

Infected or contaminated seeds can introduce disease into new regions or recontaminate areas with new strains where the disease was already existing. The seedborne nature of *B. cinerea* on chick-pea has been reported by many workers.[11,18-22] Laha and Grewal[11] reported that disease is 8.0% to 18.5% naturally seedborne in few cultivars. Grewal[21] indicated that the fungus was externally and internally seedborne to the extent of 8.2% and 2.5%, respectively.

Laha and Grewal[11] recorded that incidence of *B. cinerea* in seeds decreased more rapidly in seeds stored at normal room conditions than in those stored at cooler place (10°C). Under room condition of storage, the fungus remains viable for more than 7 months, though magnitude of survival was very low (0.2–0.5%). Madhu Meeta et al.[23] reported that the pathogen remains viable in chick-pea seeds up to the next growing season. Singh[13] recorded survival of *B. cinerea* in chick-pea for 8 months. He further recorded that survivability of the fungus was affected by room temperature prevalent during storage.

22-5-2 Survival in Soil

There is only brief information available about survival of pathogen on/in infected/infested plant debris. Being basically a saprophyte, it has been observed to survive saprophytically in soil and/or on infected plant debris and may serve as a source of primary inoculum. It has been reported that *B. cinerea* survived in infected plant debris for 6 months at a soil depth of 6 cm but not at 2 or 4 cm.[23] Singh[13] found that *B. cinerea* did not survive beyond 8 months in chick-pea debris placed at the depth of 10 cm in soil. However, the survival of the fungus was much higher in the crop refuge just placed on the soil surface.

The disease-cycle of Botrytis grey mold of chick-pea is proposed in Figure 22-6.

22-6 MANAGEMENT

22-6-1 Resistance

The cultivation of disease-resistant varieties of chick-pea is probably the cheapest and most effective method of combating disease. Rewal[17] tested a large number of chick-pea varieties/lines for resistance and grey mold under artificial inoculation conditions. Only two lines, ICC 1069 and ICC 5035, were found to be resistant to disease (disease rating 3) on 1–9 scale basis. Chaube et al.,[24] Verma et al.,[25] Rathi et al.,[26] and Singh and Kapoor[27] also reported ICC 1069 to be resistant to grey mold.

Rewal[17] undertook a breeding program to transfer grey mold resistance from black, bold seeded ICC 1069 to brown seeded high-yielding varieties BG 256, BGM 413, BGM 408, and BGM 419 by hybridization followed by single plant selection. Crosses ICC 1069

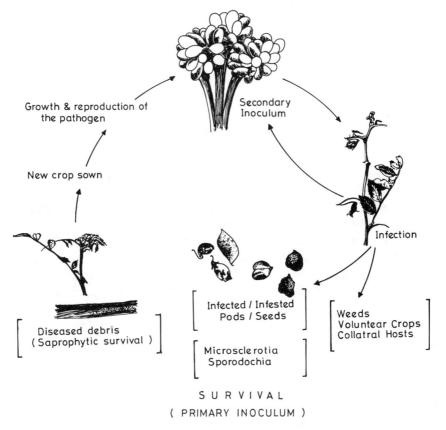

Figure 22-6 Disease cycle of Botrytis grey mold of chick-pea.

x BG 256 and ICC 1069 x BGM 413 showed monogenic dominant resistance in the ratio of 3 R: 1S. However, in crosses, ICC 1069 x BGM 408 and ICC 1069 x BGM 419 a ratio of 13 S: 3 R was obtained indicating the presence of epistatic interaction. This indicated the presence of a type of major gene resistance.[17]

Pandey,[12] in a detailed study on histological alterations in different cultivars of chick-pea in response to infection by *B. cinerea*, observed that under high moisture conditions (glasshouse), disease development was recorded even in germplasms showing resistance under field conditions. However, it was delayed in resistant germplasms.

The only anatomical difference that could be recorded in healthy leaves of the resistant and susceptible genotypes was the compactness of the mesophyll cells.[12] Both pallisade and spongy parenchyma cells in resistant genotype (ICC 10302) were more compact with less intercellular spaces than in moderately resistant (GG-588) and susceptible genotype (H-355). No significant anatomical alterations were observed in any of the genotype up to 48 hr after inoculation. This much time is required for germination, saprophytic growth of the pathogen on leaf surface and penetration.[12]

After 72 hr of inoculation, degradation of mesophyll cells was quite evident in most parts of the leaf of susceptible cultivar which became more pronounced, resulting in necrosis of leaf, after 96 hr. Necrosis was complete after 120 hr of inoculation. In MR and R genotype, breakdown of mesophyll cell was first observed after 96 hr of inoculation. At morphological levels, yellowing was observed after 120 hr of inoculation in MR and R genotypes. Damage of the mesophyll cells became quite pronounced in R genotype after 120 hr of inoculation.[12] Complete degradation of mesophyll tissues of the leaves of moderately resistant and resistant genotypes was observed at or after 144 hr of inoculation.[12]

The pathogen causes extensive damage of the mesophyll cells probably by destroying their cell wall. Nature of damage, extensive breakdown of cellulosic tissues in advance of invading hyphae, indicates involvement of wall degrading enzymes which are reported to be produced by the *B. cinerea*.[28-30] Lignified xylem tracheary elements remained intact. This may be because of the inability of the pathogen to degrade lignin. Pandey[12] further observed that under high humidity *B. cinerea* is able to infect and colonize leaf tissues of the genotypes showing high degree of resistance under field conditions. The only difference was that it was delayed by about 24-48 hr. The results further indicate that leaf surface inhibitor(s), probably phenolic in nature, is the most important factor governing resistance in chick-pea - *B. cinerea* system under field conditions.[12] In the presence of an inhibitor, both spore germination and germ tube growth are delayed for 6-8 hr and this much time should be sufficient to cause desiccation of spores and germ tubes on the leaf surface under natural conditions of the tropics. However, under controlled conditions, where high humidity was provided, there was no desiccation of spores or germ tubes; therefore, there was infection, though delayed by about 24-48 hr. Mohammadi[33] recorded that there was no correlation between resistance to *B. cinerea* and quantity and/or quality of phenolic compounds present in leaves of resistant chick-pea cultivars. However, a positive correlation was observed between total phenolic content of the leaf washing and degree of resistance of genotypes. Nevertheless, involvement of constitutive or induced inhibitory compounds cannot be ruled out, as calluses of the resistant cultivars were only poorly colonized by the pathogen and there was a direct relationship between degree of callus colonization and subsequent sporulation in dual cultures and degree of susceptibility of a genotype.[12]

Kessmann and Barg[34] reported isolation of isoflavon and pterocarpon phytoalexins from cell suspension culture of chick-pea. They are reported to be involved in the resistance of chick-pea against *Ascochyta rabiei*.[35] It would be interesting to find out whether phytoalexins are present in the leaf exudates of the *B. cinerea* inoculated plants and also differential content (constitutive or induced synthesis) of these compounds in calluses of the susceptible and resistant genotypes.

Rewal[17] has made some interesting observations concerning biochemical basis of resistance in chick-pea against *B. cinerea*. He observed that the amount of total soluble sugars and free amino acids was less in healthy plant samples of resistant line ICC 1069 than in disease tolerant BG 257 and susceptible variety BGM 408. Phenol content was higher in resistant line than in the susceptible variety BGM 408. Sugars and phenols decreased after infection. The decrease in phenols was more pronounced in susceptible variety BGM 408. Decline in sugar content was more in resistant line ICC 1069. The free pool of amino acids increased in susceptible variety after infection. Total nitrogen also showed an appreciable decrease in susceptible variety following infection by *B. cinerea*. In healthy plants, amino acid content was higher in susceptible variety than in disease tolerant and resistant line. However, the amount of sulphur containing amino acids, methionine, and cystine was almost double in resistant line as compared to susceptible variety.

22-6-2 Fungicides

B. cinerea attacks all the aerial parts of chick-pea. Young fruit and flowers are particularly susceptible to the disease. Fungicides applied at regular intervals in the expectation of disease can give a considerable measure of control, the so called "insurance approach." Cother[18] observed that seed treatment with thiram or thiram + benomyl reduced incidence of grey mold of chick-pea. Grewal and Laha[5] have also observed eradication of seedborne inoculum by dry seed treatment with Ronilan, Bavistin + thiram or Bavistin alone. They also observed that foliar application of Ronilan, Bavistin or Bavistin + thiram gave complete protection of chick-pea plants against *B. cinerea*. Rewal[17] reported that Bavistan 50 WP and Bavistin + thiram gave very good protection to chick-pea plants even at 14-day spray intervals. Seedborne *B. cinerea* was completely eliminated by seed treatment with Bavistin 50 WP, Bavistin + thiram, Ronilan, and Benlate. He also observed that these seed treatments also protected the seedlings up to 8 weeks after sowing against aerial infection by *B. cinerea*. Pandey[12] observed that seed treatment with Bavistin + thiram (1:4) followed by 3 sprays of Bavistin at 10-day intervals gave excellent control and highest grain yield.

22-7 REFERENCES

1. Shaw, F. J. F., and Ajrekar, S. L. "The Genus *Rhizoctonia* in India." Mem. Dept. Agric. India. *Bot. Ser.* 7 (1915): 117.

2. Butler, E. J., and Bisby, G. R. "The Fungi of India." ICAR New Delhi, Sci. Mono. No. 1, XVIII + 237, 1931.

3. Joshi, M. M., and Singh, R. S. "A Botrytis Grey Mould of Gram." *Indian Phytopathol.* 22 (1969): 125.

4. Singh, G., Kapoor, S., and Singh, K. "Screening Chickpea for Grey Mold Resistance." *Int. Chickpea Newsletter* 7 (1983): 13.

5. Grewal, J. S., and Laha, S. K. "Chemical Control of Botrytis Blight of Chickpea." *Indian Phytopathol.* 36 (1983): 516.

6. Carranza, J. M. "Wilt of Chickpea (*C. arietinum* L.) Caused by *B. cinerea.*" Revista de la Facultad de agronomia, Universidad nacional de la Plata 41 (1965): 135 (in Spanish).

7. Corbin, E. J. "Present Status of Chickpea Research in Australia," in International Workshop on Grain Legumes, Jan. 13–16, 1975, ICRISAT, Hyderabad, India, 87, 1975.

8. Kharbanda, P. D., and Bernier, C. C. "Botrytis Grey Mold of Chickpea in Manitoba." *Pl. Dis. Reptr.* 63 (1979): 662.

9. Nene, Y. L. "A World List of Pigeonpea (*Cajanus cajan* (L.) Millsp.) and Chickpea (*Cicer arietinum* L.) Pathogens." Begumpet, Hyderabad, ICRISAT, 1-11-256, 1978.

10. Haware, M. P., and Nene, Y. L. "Screening of Chickpea for Resistance to Botrytis Grey Mold." *Int. Chickpea Newsletter* 6 (1982): 18.

11. Laha, S. K., and Grewal, J. S. "A New Technique to Induce Abundant Sporulation in *Botrytis cinerea.*" *Indian Phytopathol.* 36 (1983): 409.

12. Pandey, B. K. "Studies on Botrytis Grey Mold of Chickpea (*Cicer arietinum* L.)." (Ph.D. thesis, Plant Pathology, G.B.P.U.A. & T., Pantnagar, 173, 1988).

13. Singh, M. P. "Studies on Survivability of *Botrytis cinerea* Pers. ex. Fr. Causal Agent of Grey Mould of Chickpea." (M. S. thesis (Ag.), Plant Pathology, G.B.P.U.A. & T., Pantnagar, 87 1989).

14. Singh, R. B. "Studies of the Control of Blight and Wilt of Gram." (M. S. thesis (Ag.), Plant Pathology, G.B.P.U.A. & T., Pantnagar, 49, 1970).

15. Groves, J. W., and Loveland, C. A. "The Connection Between *Botrytinia fuckeliana* and *Botrytis cinerea.*" *Mycologia* 45 (1953): 415.

16. Singh, G., and Bhan, L. K. "Chemical Control of Grey Mold of Chickpea." *Int. Chickpea Newsletter* 15 (1986): 18.

17. Rewal, N. "Studies on Variability in *Botrytis cinerea* Pers. Causing Grey Mould of Chickpea and Its Management." (Ph.D. thesis, I.A.R.I., New Delhi, 1987), 104.

18. Cother, E. J. "Isolation of Important Pathogenic Fungi from Seeds of Chickpea." *Seed. Sci. Tech.* 5 (1977): 593.

19. Haware, M. P., Nene, Y. L., and Mathur, S. B. "Seed-borne Diseases of Chickpea." *Tech. Bull.* ICRISAT, India, 32, 1986.

20. Maden, S. "Seed-borne Fungal Diseases of Chickpea in Turkey." *J. Turkis Phytopathol.* 16 (1987): 1.

21. Grewal, J. S. "Diseases of Pulse Crops—An Overview." *Indian Phytopathol.* 41 (1988): 1.

22. Sandhu, R., and Sah, D. N. "Effect of Botrytis Gray Mould on Chickpea Flowers, Pod Formation and Yield in Nepal." *Int. Chickpea Newsletter* 10 (1988): 13.

23. Madhu-Meeta, Bedi, P. S., and Jindal, K. K. "Survival of *Botrytis cinerea* Causal Organism of Grey Mould of Gram in Punjab." *Indian J. Pl. Path.* 16 (1986): 1.

24. Chaube, H. S., Beniwal, S. P. S., Tripathi, H. S., and Nene, Y. L. "Field Screening of Chickpea for Resistance to Botrytis Grey Mold." *Int. Chickpea Newsletter* 2 (1980): 16.

25. Verma, N. N., Singh, G., Sandhu, T. S., Singh, H., Sandhu, S. S., Singh, K., and Bhullar, B. S. "Sources of Resistance to Gram Blight and Gray Mold." *Int. Chickpea Newsletter* 4 (1981): 14.

26. Rathi, Y. P. S., Tripathi, H. S., Chaube, H. S., Beniwal, S. P. S., and Nene, Y. L. "Screening Chickpea for Resistance to Botrytis Gray Mold." *Int. Chickpea Newsletter* 11 (1984): 31.

27. Singh, G., and Kapoor, Shashi. "Screening for Combined Resistance to Botrytis Gray Mold and Ascochyta Blight of Chickpea." *Int. Chickpea Newsletter* 12 (1985): 21.

28. Verhoeff, K., and Warren, J. M. "*In vitro* and *in vivo* Production of Cell Wall Degrading Enzymes by *Botrytis cinerea* from Tomato." *Neth. J. Pl. Path.* 78 (1972): 179.

29. Honcock, J. G., Millar, R. L., and Lorbeer, J. W. "Pectolytic and Cellulolytic Enzymes Produced by *Botrytis allii, B. cinerea* and *B. squamosa in vitro* and *in vivo.*" *Phytopathology* 54 (1964): 928.

30. Honcock, J. G., Millar, R. L., and Lorbeer, J. W. "Role of Pectolytic and Cellulolytic Enzymes in Botrytis Leaf Blight of Onion." *Phytopathology* 54 (1964): 932.

31. O'Brien, T. P., Feder, N., and McCully, N. F. "Polychromatic Staining of Plant Cell Walls by Toludinene Blue O." *Protoplasma* 59 (1964): 368.

32. Jensen, W. A. *Botanical Histochemistry.* San Francisco: W. H. Freeman and Co. 1962.

33. Mohammadi, A. G. "Phenols in Relation to Resistance Against Ascochyta Blight and Botrytis Grey Mould of Chickpea." (M. S. thesis (Ag.), Plant Pathology, G.B.P.U.A. & T., Pantnagar, 94, 1987).

34. Kessmann, H., and Barg, W. "Accumulation of Isoflavones and Pterocarpon Phytoalexins in Cell Suspension Cultures of Different Cultivars of Chickpea (*Cicer arietinum* L.)." *Plant Cell Reptr.* 6 (1987): 55.

35. Kessmann, H., and Barg, W. "Elicitation and Suppression of Phytoalexin and Isoflavone Accumulation in Cotyledons of *Cicer arietinum* L. as Caused by Wounding and Polymeric Compounds from the Fungus *Ascochyta rabiei,*" *J. Phytopathol.* 117 (1986): 321.

36. Jarwis, W. R. *The Biology of Botrytis*, eds. J. R. Coley-Smith, K. Verhoeff, and W. R. Jarvis. London: Academic Press, 1980, p. 318.

INDEX